DIVERSITY OF SELENIUM FUNCTIONS IN HEALTH AND DISEASE

OXIDATIVE STRESS AND DISEASE

Series Editors

LESTER PACKER, PhD
ENRIQUE CADENAS, MD, PhD

UNIVERSITY OF SOUTHERN CALIFORNIA SCHOOL OF PHARMACY
LOS ANGELES, CALIFORNIA

1. Oxidative Stress in Cancer, AIDS, and Neurodegenerative Diseases, *edited by Luc Montagnier, René Olivier, and Catherine Pasquier*
2. Understanding the Process of Aging: The Roles of Mitochondria, Free Radicals, and Antioxidants, *edited by Enrique Cadenas and Lester Packer*
3. Redox Regulation of Cell Signaling and Its Clinical Application, *edited by Lester Packer and Junji Yodoi*
4. Antioxidants in Diabetes Management, *edited by Lester Packer, Peter Rösen, Hans J. Tritschler, George L. King, and Angelo Azzi*
5. Free Radicals in Brain Pathophysiology, *edited by Giuseppe Poli, Enrique Cadenas, and Lester Packer*
6. Nutraceuticals in Health and Disease Prevention, *edited by Klaus Krämer, Peter-Paul Hoppe, and Lester Packer*
7. Environmental Stressors in Health and Disease, *edited by Jürgen Fuchs and Lester Packer*
8. Handbook of Antioxidants: Second Edition, Revised and Expanded, *edited by Enrique Cadenas and Lester Packer*
9. Flavonoids in Health and Disease: Second Edition, Revised and Expanded, *edited by Catherine A. Rice-Evans and Lester Packer*
10. Redox–Genome Interactions in Health and Disease, *edited by Jürgen Fuchs, Maurizio Podda, and Lester Packer*
11. Thiamine: Catalytic Mechanisms in Normal and Disease States, *edited by Frank Jordan and Mulchand S. Patel*
12. Phytochemicals in Health and Disease, *edited by Yongping Bao and Roger Fenwick*
13. Carotenoids in Health and Disease, *edited by Norman I. Krinsky, Susan T. Mayne, and Helmut Sies*
14. Herbal and Traditional Medicine: Molecular Aspects of Health, *edited by Lester Packer, Choon Nam Ong, and Barry Halliwell*
15. Nutrients and Cell Signaling, *edited by Janos Zempleni and Krishnamurti Dakshinamurti*
16. Mitochondria in Health and Disease, *edited by Carolyn D. Berdanier*
17. Nutrigenomics, *edited by Gerald Rimbach, Jürgen Fuchs, and Lester Packer*
18. Oxidative Stress, Inflammation, and Health, *edited by Young-Joon Surh and Lester Packer*

19. Nitric Oxide, Cell Signaling, and Gene Expression, *edited by Santiago Lamas and Enrique Cadenas*

20. Resveratrol in Health and Disease, *edited by Bharat B. Aggarwal and Shishir Shishodia*

21. Oxidative Stress and Age-Related Neurodegeneration, *edited by Yuan Luo and Lester Packer*

22. Molecular Interventions in Lifestyle-Related Diseases, *edited by Midori Hiramatsu, Toshikazu Yoshikawa, and Lester Packer*

23. Oxidative Stress and Inflammatory Mechanisms in Obesity, Diabetes, and the Metabolic Syndrome, *edited by Lester Packer and Helmut Sies*

24. Lipoic Acid: Energy Production, Antioxidant Activity and Health Effects, *edited by Mulchand S. Patel and Lester Packer*

25. Dietary Modulation of Cell Signaling Pathways, *edited by Young-Joon Surh, Zigang Dong, Enrique Cadenas, and Lester Packer*

26. Micronutrients and Brain Health, *edited by Lester Packer, Helmut Sies, Manfred Eggersdorfer, and Enrique Cadenas*

27. Adipose Tissue and Inflammation, *edited by Atif B. Awad and Peter G. Bradford*

28. Herbal Medicine: Biomolecular and Clinical Aspects, Second Edition, *edited by Iris F. F. Benzie and Sissi Wachtel-Galor*

29. Inflammation, Lifestyle and Chronic Diseases: The Silent Link, *edited by Bharat B. Aggarwal, Sunil Krishnan, and Sushovan Guha*

30. Flavonoids and Related Compounds: Bioavailability and Function, *edited by Jeremy P. E. Spencer and Alan Crozier*

31. Mitochondrial Signaling in Health and Disease, *edited by Sten Orrenius, Lester Packer, and Enrique Cadenas*

32. Vitamin D: Oxidative Stress, Immunity, and Aging, *edited by Adrian F. Gombart*

33. Carotenoids and Vitamin A in Translational Medicine, *edited by Olaf Sommerburg, Werner Siems, and Klaus Kraemer*

34. Hormesis in Health and Disease, *edited by Suresh I. S. Rattan and Éric Le Bourg*

35. Liver Metabolism and Fatty Liver Disease, *edited by Oren Tirosh*

36. Nutrition and Epigenetics, *edited by Emily Ho and Frederick Domann*

37. Lipid Oxidation in Health and Disease, *edited by Corinne M. Spickett and Henry Jay Forman*

38. Diversity of Selenium Functions in Health and Disease, *edited by Regina Brigelius-Flohé and Helmut Sies*

DIVERSITY OF SELENIUM FUNCTIONS IN HEALTH AND DISEASE

EDITED BY
REGINA BRIGELIUS-FLOHÉ
HELMUT SIES

CRC Press
Taylor & Francis Group
Boca Raton London New York

CRC Press is an imprint of the
Taylor & Francis Group, an **informa** business

CRC Press
Taylor & Francis Group
6000 Broken Sound Parkway NW, Suite 300
Boca Raton, FL 33487-2742

First issued in paperback 2019

ISBN-13: 978-1-4822-5126-5 (hbk)
ISBN-13: 978-0-367-37750-2 (pbk)

Visit the Taylor & Francis Web site at
http://www.taylorandfrancis.com

and the CRC Press Web site at
http://www.crcpress.com

Contents

Series Preface..xi

Preface.. xiii

Acknowledgments.. xv

Editors..xvii

Contributors xix

Abbreviations... xxiii

SECTION I Introduction

Chapter 1   Who Can Benefit from Selenium?...3

Gerald F. Combs, Jr.

SECTION II Se Compounds as a Source for Selenoprotein Biosynthesis

Chapter 2   Selenium Metabolism.. 19

Yasumitsu Ogra

Chapter 3   The Molecular Regulation of Selenocysteine Incorporation
into Proteins in Eukaryotes 31

Aditi Dubey and Paul R. Copeland

Chapter 4   Selenocysteine tRNA[Ser]Sec: The Central Component
of Selenoprotein Biosynthesis 55

*Bradley A. Carlson, Ryuta Tobe, Petra A. Tsuji, Min-Hyuk Yoo,
Lionel Feigenbaum, Lino Tessarollo, Byeong J. Lee,
Ulrich Schweizer, Vadim N. Gladyshev, and Dolph L. Hatfield*

SECTION III Se Compounds with Specific Functions

Chapter 5   Redox Cycling and the Toxicity of Selenium Compounds:
A Historical View..83

Julian E. Spallholz

Chapter 6   Selenite in Cancer Therapy 109

     Sougat Misra, Marita Wallenberg, Ola Brodin,
     and Mikael Björnstedt

Chapter 7   Forms of Selenium in Cancer Prevention 137

     Karam El-Bayoumy, Raghu Sinha, and John P. Richie, Jr.

SECTION IV   Dual Functions of Selenoproteins in Cancer

Chapter 8   Thioredoxin Reductase 1 173

     Bradley A. Carlson

Chapter 9   Glutathione Peroxidase 2 189

     Anna P. Kipp and Mike F. Müller

Chapter 10 The 15-kDa Selenoprotein (SEP15) 203

     Petra A. Tsuji and Cindy D. Davis

SECTION V   Unexpected Links

Chapter 11 Multifaceted and Intriguing Effects of Selenium
     and Selenoproteins on Glucose Metabolism and Diabetes 217

     Ji-Chang Zhou, Holger Steinbrenner, Margaret P. Rayman,
     and Xin Gen Lei

Chapter 12 Selenoproteins and the Metabolic Syndrome 247

     Lucia A. Seale, Ann Marie Zavacki, and Marla J. Berry

Chapter 13 Thioredoxin Reductase: A Coordinator in Metabolic Activities 273

     Sofi Eriksson and Edward E. Schmidt

Chapter 14 Selenium Mediates a Switch in Macrophage Polarization 293

     K. Sandeep Prabhu, Avinash K. Kudva, and Shakira M. Nelson

SECTION VI   Selenoprotein Polymorphisms and Mutations

Chapter 15 Genetic Polymorphisms in Selenoprotein Genes: Functionality
and Disease Risk 309

*Catherine Méplan, Janaina Lombello Santos Donadio,
and John Hesketh*

Chapter 16 Mutations in *SECISBP2* 343

*Erik Schoenmakers, Carla Moran, Nadia Schoenmakers,
and Krishna Chatterjee*

Index 377

Series Preface

Oxidative stress is an underlying factor in health and disease. In this series of books, the importance of oxidative stress and disease associated with organ systems highlights the scientific evidence and the clinical applications of these concepts. This series is intended for researchers in the basic biomedical sciences and clinicians. The potential of such knowledge for healthy aging and disease prevention warrants further knowledge about how oxidants and antioxidants modulate cell and tissue function. Selenium proteins and enzymes are important in oxidative metabolism and especially in oxidative stress protection. After a concise introductory chapter on "Who Can Benefit from Selenium?," this book presents five different sections that address central aspects on the role of selenium in biology and medicine, thus providing important information on specific functions of selenium compounds and their use as sources for the synthesis of selenoproteins. A section emphasizes the dual functions of selenoproteins in cancer, followed by another section on selenoprotein polymorphism and mutations and disease risk. Highly interesting are the chapters under *Unexpected Links*, which address the role of selenoproteins in diabetes, metabolic syndrome, and macrophage polarization. The flow of the sections and the containing chapters in each section is excellent, and the editors Regina Brigelius-Flohé and Helmut Sies are to be congratulated for producing this excellent, well-organized, and timely book in the expanding field of the role of selenium in health science research.

Lester Packer
Enrique Cadenas
Oxidative Stress and Disease Series Editors

Preface

Low dietary intake of selenium has long been associated with increased risk for developing certain long-term degenerative diseases, such as atherosclerosis, chronic inflammation, cardiovascular diseases, and especially cancer. Whereas the large clinical trial Nutrition Prevention of Cancer (NPC) corroborated this association for lung, colon, and prostate cancer, other studies failed, yielded conflicting results, or even revealed adverse effects of selenium supplementation.

Selenium is an essential trace element. It is present in various distinct compounds, as part of inorganic or organic molecules, or in selenoenzymes. Thus, it is not surprising that elemental selenium does not function by itself but rather in the context of various chemical forms with many different activities. This emerging diversity of functions prompted us to propose the present book on selenium in health and disease. We attempt here to help elucidate why studies undertaken to prevent diseases with "selenium" ended with disappointing outcomes and not rarely with opposite results, that is, disease promotion. This book will show that benefit, failure, or side effects depend on the following:

- The chemical form and dose of selenium
- The selenium status of the individual ingesting "selenium"
- The capacity of the selenium form to serve as a source for selenoprotein biosynthesis
- The function of selenoproteins reacting to a change in the selenium status
- The stage of the disease (mainly cancer) at the time point of intervention
- The genetic background of individuals to be treated

This is the scope of the sections:

Section I. Who benefits from selenium supplementation? The answer clearly is only those with a low selenium status. There is no need for healthy individuals to consume more than the upper limit recommended by various advisory panels.

Section II. Selenium compounds as source for selenoproteins. The various compounds need to be metabolized to selenide for incorporation into selenocysteine, the amino acid present in selenoproteins. Related reviews on selenoprotein biosynthesis in general and the specific tRNA[Ser]Sec provide the basis for the chapters in subsequent sections.

Section III. This section deals with selenium compounds that obviously act against cancer without the need to be incorporated into selenoproteins. Two chapters describe mechanisms of the toxicity of certain selenium compounds and their potential use in cancer therapy. As is not well known, selenide— derived from various selenium compounds—can redox cycle in the presence of oxygen and thiols, thereby producing superoxide, which kills cancer cells.

The third chapter highlights basic mechanisms and structural features of selenium compounds that are critical in chemoprevention.

Section IV. At least three selenoproteins have dual functions in cancer, which means that they can prevent it but also can support cancer development and growth in animal experiments. These are thioredoxin reductase 1 (TrxR1), glutathione peroxidase 2 (GPx2), and selenoprotein 15 (Sel15). Which function predominates depends on the model, the type, and stage of cancer and on the selenium status of the treated animals.

Section V. Unexpected functions of selenoproteins in health and diseases; this relates to the role of TrxR1 as a coordinator in metabolic activities to diabetes, metabolic syndrome, and inflammation. The unexpected results reveal not only side effects but also unanticipated functions of selenium compounds and particular selenoproteins.

Section VI. Polymorphisms, mutations, or deletions in selenoproteins and/or proteins required for selenoprotein biosynthesis can be responsible for different responses to selenium intake, for a change in the functions of the Se-proteome, and can be associated to disease risks. Understanding this interaction has the potential to develop personalized dietary recommendations for selenium.

Thus, this book will focus on current knowledge on those aspects of selenium research that are relevant to its medical use and particularly to chemoprevention of cancer. In this respect, it covers aspects that are different from available general monographs on selenium biochemistry. The narrow scope of this book is to contribute to resolving the present confusion about potential benefits and hazards of selenium in human health. It is hoped that the book will provide a solid scientific basis for optimum use of selenium in preventing or treating human diseases.

Regina Brigelius-Flohé and Helmut Sies

Acknowledgments

We thank the authors of the chapters for their input and enthusiasm. Special thanks go to Dr. Holger Steinbrenner for expert advice. R.B.F and H.S. wish to acknowledge longstanding research support by the Deutsche Forschungsgemeinschaft, Bonn, Germany and H.S. also by the National Foundation for Cancer Research (NFCR), Bethesda, Maryland.

Editors

Regina Brigelius-Flohé earned her PhD in biochemistry in Tübingen and Münster, Germany, in 1978. During a postdoctoral fellowship in Munich and Düsseldorf, together with Helmut Sies, she investigated the cellular thiol-disulfide status in perfused organs under various conditions of oxidative stress. The detection and analysis of mixed disulfides of proteins and glutathione provided a basis for the regulation of enzyme activities by thiol modification, a field that is now expanding and known as redox regulation. In 1984, she changed to the pharmaceutical industry and headed a molecular biology lab and later the department of molecular biology (Grünenthal Ltd., Aachen, Germany). In 1994, she became a professor at the University of Potsdam combined with heading the Department of Biochemistry of Micronutrients at the German Institute of Human Nutrition Potsdam-Rehbruecke (DIfE) where she now is an emeritus professor. Her favorite micronutrients were vitamin E and selenium. Her group was the first to elucidate the metabolism of tocopherols and tocotrienols. Research regarding selenium is related to redox-regulated processes in inflammation and cancer development influenced by selenium and specific selenoproteins, mainly glutathione peroxidase 2.

She was a member of the scientific advisory board of the German Society of Biochemistry and Molecular Biology for 6 years and president of *SFRR-Europe* 2005/2006. She now serves on the editorial boards of several journals in the field of nutrition, micronutrients, free radicals, and redox regulation.

Helmut Sies, MD, PhD(hon), studied medicine at the Universities of Tübingen, Paris and Munich, and did the habilitation for physiological chemistry and physical biochemistry at the University of Munich. He spent sabbaticals at the University of Buenos Aires, Argentina; the University of California at Berkeley, USA; the University of Siena, Italy; and the Heart Research Institute, Sydney, Australia. He was chair of the Department of Biochemistry and Molecular Biology I at Heinrich-Heine-University Düsseldorf, Germany, where he now is an emeritus professor. He also is senior scientist at the Leibniz Research Institute for Environmental Medicine at Düsseldorf and an adjunct professor at the University of Southern California and a professor of biochemistry at King Saud University, Riyadh, Saudi Arabia. He served as president of the Society for Free Radical Research International and of the Oxygen Club of California (OCC). He is a Fellow of the National Foundation for Cancer Research (NFCR), Bethesda, Maryland, USA, and of the Royal College of Physicians (FRCP), London, England. He is a member of the German National Academy of Sciences Leopoldina. His research interests include hydrogen peroxide metabolism, oxidative stress, redox signaling, and micronutrients, notably flavonoids, carotenoids, and selenium.

Contributors

Marla J. Berry
Department of Cell and Molecular
   Biology
John A. Burns School of Medicine
University of Hawaii
Honolulu, Hawaii

Mikael Björnstedt
Karolinska Institutet
Department of Laboratory Medicine
Division of Pathology F42
Karolinska University Hospital
   Huddinge
Stockholm, Sweden

Ola Brodin
Karolinska Institutet
Department of Oncology – Pathology
   (OnkPat, K7)
The Oncology Clinic, Södersjukhuset
Stockholm, Sweden

Bradley A. Carlson
Molecular Biology of Selenium Section
Mouse Cancer Genetics Program
Center for Cancer Research
National Institutes of Health
Bethesda, Maryland

Krishna Chatterjee
University of Cambridge
Metabolic Research Laboratories
Wellcome Trust-MRC Institute of
   Metabolic Science
Addenbrooke's Hospital
Cambridge, United Kingdom

Gerald F. Combs, Jr.
Grand Forks Human Nutrition Research
   Center
USDA-ARS
Grand Forks, North Dakota

Paul R. Copeland
Department of Biochemistry and
   Molecular Biology
Rutgers Robert Wood Johnson Medical
   School
Piscataway, New Jersey

Cindy D. Davis
Office of Dietary Supplements
National Institutes of Health
Bethesda, Maryland

Janaina Lombello Santos Donadio
Departamento de Alimentos e Nutrição
   Experimental
Faculdade de Ciências Farmacêuticas
Universidade de São Paulo
São Paulo, Brazil

and

Institute for Cell and Molecular
   Biosciences
Newcastle University
Newcastle-upon-Tyne, United Kingdom

Aditi Dubey
Department of Biochemistry and
   Molecular Biology
Rutgers Robert Wood Johnson Medical
   School
Piscataway, New Jersey

Karam El-Bayoumy
Department of Biochemistry and
    Molecular Biology
Pennsylvania State University College
    of Medicine
Hershey, Pennsylvania

Sofi Eriksson
Department of Oncology-Pathology
    (ONKPAT, CCK)
Karolinska Institutet
Stockholm, Sweden

Lionel Feigenbaum
Laboratory of Animal Science Program
Science Applications International
    Corporation
Frederick, Maryland

Vadim N. Gladyshev
Division of Genetics
Department of Medicine
Brigham and Women's Hospital
Harvard Medical School
Boston, Massachusetts

Dolph L. Hatfield
Molecular Biology of Selenium Section
Mouse Cancer Genetics Program
Center for Cancer Research
National Institutes of Health
Bethesda, Maryland

John Hesketh
Human Nutrition Research Centre
Institute for Cell and Molecular
    Biosciences
Newcastle University
Newcastle-upon-Tyne, United Kingdom

Anna P. Kipp
Department of Biochemistry of
    Micronutrients
German Institute of Human Nutrition
Potsdam-Rehbruecke (DIfE), Germany

Avinash K. Kudva
Department of Veterinary and
    Biomedical Sciences
Center for Molecular Toxicology and
    Carcinogenesis
and
Center for Molecular Immunology and
    Infectious Disease
The Pennsylvania State University
University Park, Pennsylvania

Byeong J. Lee
School of Biological Sciences
Interdisciplinary Program in
    Bioinformatics
Institute of Molecular Biology and
    Genetics
Seoul National University
Seoul, Korea

Xin Gen Lei
Department of Animal Science
Cornell University
Ithaca, New York

Catherine Méplan
School of Biomedical Sciences
Faculty of Medical Sciences
Human Nutrition Research Centre
Newcastle University
Newcastle-upon-Tyne, United Kingdom

Sougat Misra
Karolinska Institutet
Department of Laboratory Medicine
Division of Pathology F42
Karolinska University Hospital
Huddinge, Stockholm, Sweden

Carla Moran
University of Cambridge
Metabolic Research Laboratories
Wellcome Trust-MRC Institute of
    Metabolic Science
Addenbrooke's Hospital
Cambridge, United Kingdom

Mike F. Müller
Edinburgh Cancer Research Centre
Western General Hospital
Edinburgh, Scotland, United Kingdom

Shakira M. Nelson
Division of Cancer Epidemiology and
  Genetics
National Cancer Institute
National Institutes of Health
Rockville, Maryland

Yasumitsu Ogra
Department of Toxicology and
  Environmental Health
Graduate School of Pharmaceutical
  Sciences
Chiba University
Chiba, Japan

K. Sandeep Prabhu
Department of Veterinary and
  Biomedical Sciences
Center for Molecular Toxicology and
  Carcinogenesis
and
Center for Molecular Immunology and
  Infectious Disease
The Pennsylvania State University
University Park, Pennsylvania

Margaret P. Rayman
School of Biosciences and Medicine
Faculty of Health and Medical Sciences
University of Surrey
Guildford, United Kingdom

John P. Richie, Jr.
Department of Public Health Sciences
Pennsylvania State University College
  of Medicine
Hershey, Pennsylvania

Edward E. Schmidt
Department of Microbiology and
  Immunology
Montana State University
Bozeman, Montana

Erik Schoenmakers
University of Cambridge
Metabolic Research Laboratories
Wellcome Trust-MRC Institute of
  Metabolic Science
Addenbrooke's Hospital
Cambridge, United Kingdom

Nadia Schoenmakers
University of Cambridge
Metabolic Research Laboratories
Wellcome Trust-MRC Institute of
  Metabolic Science
Addenbrooke's Hospital
Cambridge, United Kingdom

Ulrich Schweizer
Institut für Biochemie und
  Molekularbiologie
Rheinische Friedrich-Wilhelms-
  Universität
Bonn, Germany

Lucia A. Seale
Department of Cell and Molecular
  Biology
John A. Burns School of Medicine
University of Hawaii
Honolulu, Hawaii

Raghu Sinha
Department of Biochemistry and
  Molecular Biology
Pennsylvania State University College
  of Medicine
Hershey, Pennsylvania

Julian E. Spallholz
Texas Tech University
Lubbock, Texas

Holger Steinbrenner
Institute of Biochemistry and Molecular
    Biology I
Heinrich-Heine-University
Düsseldorf, Germany

Lino Tessarollo
Neural Development Group
Mouse Cancer Genetics Program
Center for Cancer Research
National Institutes of Health
Bethesda, Maryland

Ryuta Tobe
Molecular Biology of Selenium Section
Mouse Cancer Genetics Program
Center for Cancer Research
National Institutes of Health
Bethesda, Maryland

Petra A. Tsuji
Department of Biological Sciences
Towson University
Towson, Maryland

Marita Wallenberg
Karolinska Institutet
Department of Laboratory Medicine
Division of Pathology F42
Karolinska University Hospital
    Huddinge
Stockholm, Sweden

Min-Hyuk Yoo
Molecular Biology of Selenium Section
Mouse Cancer Genetics Program
Center for Cancer Research
National Institutes of Health
Bethesda, Maryland

Ann Marie Zavacki
Thyroid Section
Division of Endocrinology, Diabetes
    and Hypertension
Brigham and Women's Hospital
Harvard Medical School
Boston, Massachusetts

Ji-Chang Zhou
Molecular Biology Laboratory
Shenzhen Center for Chronic Disease
    Control
Shenzhen, China

Abbreviations

3′UTR	3′ untranslated region
4EBP1	eukaryotic translation initiation factor 4E binding protein 1
8-iso-PGF$_{2a}$	8-iso-prostaglandin-F$_{2a}$
8-OHdG	8-hydroxy-2′-deoxyguanosine
aa	amino acid
ACC	acetyl-CoA carboxylase
ACC1	acetyl-CoA carboxylase 1
ACF	aberrant crypt foci
AE	activator element
AE1	membrane anionic exchanger 1
AIDS	acquired immunodeficiency syndrome
Akt, AKT	protein kinase B
AKT	v-akt murine thymoma viral oncogene
AMPK	5′-adenosine monophosphate-activated protein kinase
AMPKα	AMP-kinase alpha
AOM	azoxymethane
AP-1	activator protein 1
APOE	apolipoprotein E
ASK1	apoptosis signal-regulation kinase
ATBC	alpha-tocopherol, beta-carotene cancer prevention
ATM	ataxia telangiectasia mutated
BAT	brown adipose tissue
Bax	BCL2-associated X protein
BC	breast cancer
Bcl-2	B-cell CLL/lymphoma 2 protein
Bim	BCL2-like 11 protein
bp	base pair
BSC	benzyl selenocyanate
BW	body weight
CDK4	cyclin-dependent kinase 4
CK-MM	muscle-specific creatine kinase
CNS	central nervous system
COX-2	prostaglandin-endoperoxide synthase 2
Cp	chloramphenicol
CRC	colorectal cancer
CRD	cysteine-rich domain
C-terminal	carboxy terminal
Cys	cysteine
D1	type 1 deiodinase
D2	type 2 deiodinase
D3	type 3 deiodinase
DBA	delayed bone age

DHT	dihydrotestosterone
DIO	iodothyronine deiodinase
DMBA	7,12-dimethylbenz[a]anthracene
DNA	deoxyribonucleic acid
DNMT	DNA methyltransferase
Dox	doxycycline
DPMSC	diphenylmethyl selenocyanate
DSE	distal sequence element
ECOG	Eastern Cooperative Oncology Group
eEFSec	eukaryotic elongation factor for Sec
eF4a3	eukaryotic elongation factor
Efsec	selenocysteine tRNA elongation factor gene
e.g.	exempli gratia
eGPx	erythrocyte glutathione peroxidase
eIF2a	eukaryotic translation initiation factor 2A
eIF4a	eukaryotic translation initiation factor 4A
EJC	exon junction complex
EPA	US Environmental Protection Agency
ER	endoplasmic reticulum
ERα	estrogen receptor α
ERβ	estrogen receptor β
Erk	extracellular signal-regulated kinases
fdhF	bacterial formate dehydrogenase gene
FOXO1	forkhead box protein O1
FoxO1a	forkhead box protein class O1a
FPG	fasting plasma glucose
G418	geneticin
GABAergic	related to γ-aminobutyric acid
Gad67	glutamate decarboxylase 1
Gbp-1	guanylate binding protein-1
GDM	gestational diabetes mellitus
GH	growth hormone
GLUT4	glucose transporter 4
GPx	glutathione peroxidase
GPx1	glutathione peroxidase 1
Gpx1	glutathione peroxidase 1 gene
GPx2	glutathione peroxidase 2
GPx3	cytosolic glutathione peroxidase
GPx4	glutathione peroxidase 4
Grx	glutaredoxin
GSH	glutathione
GSIS	glucose-stimulated insulin secretion
GS-Se-SG	selenodiglutathione
GWAS	genome wide association studies
HDAC	histone deacetylase
HDL	high-density lipoprotein

HFD	high fat diet
Hgf	hepatocyte growth factor
HIF-1	hypoxia inducible factor 1
HIF-1a	hypoxia inducible factor 1, alpha subunit
HIV	human immunodeficiency virus
HLA-E	major histocompatibility complex, class I, E
HMG-CoA	3-hydroxy-3-methyl-glutaryl-CoA
HNF-4α or HNF4A	hepatocyte nuclear factor 4 alpha
HOMA	homeostasis model assessment
HOMA-IR	homeostasis model assessment of insulin resistance
HSA	human serum albumin
hTERT	human telomerase reverse transcriptase
ICAM-1	intercellular adhesion molecule 1
IFNγ	interferon gamma
IL-6	interleukin 6
INSR	insulin receptor
IOM	(US) Institute of Medicine
Ipf1	insulin promoter factor 1
IQ	intelligence quotient
IRS	insulin receptor substrate
IV	intravenous
JAK	Janus kinase
JNK	cJun N-terminal kinase
kd	knock-down
Keap1	kelch-like ECH-associated protein 1
LINE-1	long interspersed nucleotide elements
LPS	lipopolysaccharides
m¹A	1-methyladenosine
MAPK	mitogen-activated-protein-kinase
mcm⁵U	5-methoxycarbonylmethyluracil
mcm⁵Um	5-methoxycarbonylmethyluracil-2′-O-methylribose
MEF	mouse embryonic fibroblast
Met	methionine
MMP-2	matrix metallopeptidase 2
MMP-9	matrix metallopeptidase 9
MnSOD	manganese superoxide dismutase
MRP	multi-drug resistance protein
MSRB1/SEPR	methionine-R-sulfoxide reductase 1
MTD	maximal tolerable dose
mTOR	mammalian target of rapamycin
mTORC1	mTOR complex 1
mTORC2	mTOR complex 2
NBT	negative biopsy trial
NES	nuclear export signals
NF-κB	nuclear factor of kappa light polypeptide gene enhancer in B-cells 1

NHANES	(US) National Health and Nutrition Examination Survey
NHANES III	Third National Health and Nutrition Examination Survey
NHL	non-Hodgkin lymphoma
NK	natural killer
NLS	localization signal
NMD	nonsense mediated decay
NNK	4-(methylnitrosamino)-1-(3-pyridyl)-1-butanone
NOAEL	no adverse effect level
NOS	NO synthase
NOX	NADPH oxidase
NPC	Nutritional Prevention of Cancer trial
Nrf2	nuclear factor (erythroid-derived 2)-like 2
NRX	nucleoredoxin
NSCLC	Non-small cell lung cell
nt	nucleotide
N-terminal	amino terminal
Opn1	osteopontin
OR	odds ratio
p21cip1	cyclin-dependent kinase-interacting protein 1
p27kip1	cyclin-dependent kinase inhibitor 1B
p53	tumor protein 53
PACAP	pituitary adenylate cyclase-activating peptide
PARP	poly (ADP-ribose) polymerase
PBISe	S,S′-1,4-phenylenebis(1,2-ethanediyl)bis-isoselenourea
PBIT	S,S′-1,4-phenylenebis(1,2-ethanediyl)bis-isothiourea
PBMC	peripheral blood mononuclear cells
PDX1	pancreatic duodenal homeobox 1
PEPCK	phosphoenolpyruvate carboxykinase
PGC1α	peroxisomal proliferator-activated receptor-γ coactivator 1α
PGJ2	prostaglandin J2
PI3-K	phosphatidylinositol 3-kinase
PI3K	phosphoinositide 3-kinase or phosphatidylinositide 3-kinase
PKC	protein kinase C
PPAR	peroxisome-proliferated activator receptor
PSA	prostate-specific antigen
PSE	proximal sequence element
Pstk	phosphoseryl-tRNA kinase
PTC	premature termination codon
PTEN	phosphatase and tensin homolog
PTP1B	protein-tyrosine phosphatase 1B
PV+	parvalbumin positive
p-XSC	1,4-phenylenebis(methylene) selenocyanate
p-XSeSG	glutathione conjugate of p-XSC
ψU	pseudouridine
RBC	red blood cells

RBD	RNA binding domain
RCTs	randomized, controlled trials
RDBPC	randomized double-blind placebo controlled
REDD1	regulated in development and DNA damage responses
RNA	ribonucleic acid
ROS	reactive oxygen species
rT3	reverse 3, 3′, 5′ triiodothyronine
SAM	S-adenosylmethionine
SARS	severe acute respiratory syndrome
SBP2	SECIS binding protein 2
Scly	selenocysteine lyase
Se	selenium
Sec	selenocysteine
SECIS	selenocysteine insertion sequence
SECISBP2	SECIS binding protein 2
SECISBP2L	SECIS binding protein 2L
SecTRAPs	selenium-compromised thioredoxin reductase-derived apoptotic proteins
SeCys	selenocysteine
SEL15	selenoprotein 15
selA	selenocysteine synthase gene in eubacteria
selB	selenocysteine tRNA specific elongation factor gene in eubacteria
selC	selenocysteine tRNA gene in eubacteria
selD	selenophosphate synthetase gene in eubacteria
SELECT	Selenium and Vitamin E Cancer Prevention Trial
SELH	Selenoprotein H
SELI	Selenoprotein I
SelK or SELK	selenoprotein K
SelM or SELM	selenoprotein M
SelN or SELN	selenoprotein N
SELO	selenoprotein O
SelR	selenoprotein R
SelS or SELS	selenoprotein S
SELSA-1	bis(5-phenylcarbamoyl pentyl) diselenide
SELSA-2	5-phenylcarbamoyl pentyl selenocyanide
SelT or SELT	selenoprotein T
SELV	selenoprotein V
SelW or SELW	selenoprotein W
SeMet	selenomethionine
Sep15, SEP15	15-kDa selenoprotein
SePP, Sepp1, or SEPP1	selenoprotein P
SePP1	selenoprotein P1
SEPP1	selenoprotein P gene
Ser	serine
SerS	seryl-tRNA synthetase

SID	Sec incorporation domain
SMG1	a PI3K-related kinase
SNP	single nucleotide polymorphism
SOD	superoxide dismutase
SOD	superoxide dismutase gene
SPS	selenophosphate synthethase
SPS1	selenophosphate synthetase 1
Sps1	selenophosphate synthetase 1 gene
SPS2	selenophosphate synthetase 2
Sps2	selenophosphate synthetase 2 gene
STAF	Sec tRNA transcription activating factor
STAT3	signal transducer and activator of transcription 3
SWOG	Southwest Oncology Group
T	testosterone
T1DM	type 1 diabetes mellitus
T2	diiodothyronine
T2D	type 2 diabetes
T2DM	type 2 diabetes mellitus
T3	triiodothyronine
T4	thyroxine
tagSNP	tagging SNP
TERT	telomerase reverse transcriptase
TGF-β	tumor growth factor-beta
TGR	thioredoxin glutathione reductase
TH	thyroid hormone
TNF-α	tumor necrosis factor-α
TRIT1	tRNA[Ser]Sec isopentenyl transferase
Trsp	selenocysteine tRNA gene *or* tRNA[Ser]Sec gene
Trsp$^\Delta$	selenocysteine tRNA knockout gene
Trsp$^{c\Delta}$	selenocysteine tRNA conditional knockout gene
Trsp^t	selenocysteine tRNA transgene
Trsp^{tA34}	mutant selenocysteine tRNA transgene at position 34
Trsp^{tG37}	mutant selenocysteine tRNA transgene at position 37
Trx	thioredoxin
TrxR or TRXR	thioredoxin reductase
TrxR1	cytoplasmic thioredoxin reductase 1
TrxR2	mitochondrial thioredoxin reductase 2
TSC	tetraselenocyclophane
TSH	thyroid-stimulating hormone
TUDCA	tauroursodeoxycholic acid
TXNRD	thioredoxin reductase (gene)
UCP-1	uncoupling protein-1
UCP2	uncoupling protein 2
UGGT	UDP-glucose:glycoprotein glucosyltransferase
UK PRECISE	United Kingdom Prevention of Cancer by Intervention of Selenium

UTR	untranslated region
VCAM-1	vascular cell adhesion protein 1
VEGF	vascular endothelial growth factor
WAT	white adipose tissue

Section I

Introduction

1 Who Can Benefit from Selenium?

Gerald F. Combs, Jr.

CONTENTS

1.1 Emergence of Selenium in Nutrition and Health 3
1.2 The Sequence of Questions about Selenium 4
    1.2.1 Is Selenium Essential? 4
    1.2.2 Is Selenium Required by Humans? 4
    1.2.3 How Much Selenium Is Required? 5
    1.2.4 Is Anyone Deficient in Selenium? 5
    1.2.5 What Are the Health Consequences of Low Selenium Status? 6
    1.2.6 Can Selenium Reduce Cancer Risk? 7
    1.2.7 Is Selenium Safe? 8
1.3 The Current Question: Who Can Benefit from Selenium? 9
Acknowledgments 10
References 10

1.1 EMERGENCE OF SELENIUM IN NUTRITION AND HEALTH

The trace element selenium (Se) appeared relatively late on the nutrition–health agenda. It was first reported in the 1930s to be the toxic principle involved in "blind staggers," a neurological condition of horses and cattle grazing on seleniferous plants in the American Northern Plains (Franke, 1934a,b). It was not until the late 1950s that Se was found to prevent disorders of vitamin E deficiency (Schwarz and Foltz, 1957; Schwarz et al., 1957). This made Se the last nutrient to have been recognized as a dietary essential. Accordingly, the roles of Se in biology and health have been researched for less time than any other essential nutrient. Initial investigations of Se in animal nutrition spawned studies of its metabolic mechanisms starting in the 1960s. Then, suggestion that Se may have a role in cancer prevention (Shamberger and Frost, 1969; Shamberger and Willis, 1971) led to studies with a variety of tumor models starting in the late 1960s and, ultimately, to human clinical trials starting in the 1980s. This body of work addressed a sequence of central research questions. The resulting answers served, after the manner of scientific inquiry, to provoke additional questions. Accordingly, questions remain, the major one being who can benefit from increased Se intake? To address that, it is necessary to consider the answers to the progression of scientific questions that have led to it.

1.2   THE SEQUENCE OF QUESTIONS ABOUT SELENIUM

1.2.1   Is Selenium Essential?

In the mid-1950s, Schwarz and colleagues found that the addition of sodium selenite to vitamin E–free diets of rats and chicks prevented vascular and central organ lesions in those models (Patterson et al., 1957; Schwarz and Foltz, 1957; Schwarz et al., 1957). Over the following two decades, those findings were extended to a range of livestock and laboratory species. In each case, selenite and, later, organic Se compounds prevented pathologies caused by deficiencies of the antioxidant vitamin E (Dan and Sondergaard, 1957; Patterson et al., 1957; Schwarz and Foltz, 1957; Schwarz et al., 1957; Nesheim and Scott, 1958; Bonetti and Stripe, 1963; Calvert and Scott, 1963; Ewan and Jenkins, 1967; Scott, 1974). Therefore, Se also became regarded as an antioxidant although the basis of its vitamin E "sparing" effect was not clear.

It was not until the early 1970s that a plausible mechanism was offered for the vitamin E–Se interaction. That came with the discovery by the groups of Rotruck (Rotruck et al., 1973) and Flohé (Flohé et al., 1973) that Se is an essential constituent of a redox-active enzyme, glutathione peroxidase (GPx). Thus, it became apparent that Se and vitamin E function together in cellular antioxidant protection: vitamin E as a chain-breaking antioxidant in membranes, and Se via GPx in reducing reactive oxygen species in the cytosol and mitochondrial matrix space. Dietary deficiencies of either nutrient can, thus, compromise protection against oxidative stress.

Tappel, Wendel, and their colleagues discovered the form of Se at the GPx active site: selenocysteine (SeCys) (Forstrom et al., 1978; Wendel et al., 1978), the Se-analog of the common sulfur amino acid cysteine. Studies of how SeCys is incorporated into GPx resulted in the discovery by Harrison and colleagues (Chambers et al., 1986) that UGA, normally a termination codon, specifies SeCys. Ultimately, SeCys was found to be synthesized on a seryl-tRNA in the presence of a SeCys insertion element (SECIS) and associated transacting factors (Lee et al., 1989); see also Chapter 4.

Subsequent discoveries revealed several more selenoproteins, broadly classified as redox-active enzymes (Reeves and Hoffman, 2009). Each selenoprotein contains SeCys apparently incorporated by the same process. These include at least a dozen enzymes, including four isoforms of GPxs, three of thioredoxin reductases, three of iodothyronine deiodinases, and the Se-transport protein selenoprotein P (SePP). Genomic analyses by Gladyshev and colleagues (Kryukov et al., 2003) have indicated 25 human selenoprotein genes, suggesting that additional SeCys enzymes remain to be characterized.

1.2.2   Is Selenium Required by Humans?

In the 1980s, Chinese researchers reported severe endemic Se deficiency in parts of their country, showing geographic distributions of low-Se soils that corresponded to the prevalence of a cardiomyopathy (Keshan disease) among children and women with very low blood Se levels (<25 ng/ml) (Ge et al., 1983). Subsequent studies demonstrated dramatic reductions in Keshan disease incidence with supplementation using oral sodium selenite (0.5–1 mg Se/child/week) or selenite-fortified table salt (10–15 mcg Se/g salt) (Keshan Disease Research Group, 1979). These reports drew international attention, establishing Se as a nutrient important in human health.

1.2.3 HOW MUCH SELENIUM IS REQUIRED?

The single study by Yang and colleagues (Yang et al., 1986) was the basis of the first RDA for Se. That study found a total of 41 mcg Se/day to support the maximal expression of plasma GPx (GPx3) in a small cohort of Se-deficient Chinese men; that value was adjusted to the 1989 recommendations for 55 and 70 mcg/d for women and men, respectively (National Research Council, 1989). Implicit in that approach was the assumption that the dose-response relationship of the extracellular GPx3 is similar to other metabolically relevant selenoproteins. Advances in research facilitated a more complete approach in revising the RDA in 2000, which considered the amounts of Se necessary to prevent Keshan disease in children and to support maximal expression of GPx3 and SePP (Institute of Medicine, 2000). A more recent study by Burk and colleagues (Xia et al., 2005) indicated that more Se may be needed to maximize plasma SePP than for GPx3 but that RDA-levels of intake are sufficient to support maximal levels of both. Nevertheless, various national advisory panels have produced a wide range of recommended reference intakes (25–125 mcg/d) (Hurst et al., 2013).

1.2.4 IS ANYONE DEFICIENT IN SELENIUM?

Chinese researchers pointed out that Keshan disease was endemic in a mountainous belt extending from the northeast to the south-central parts of that country. They showed that soils in this belt contain very low amounts of Se (<125 ng Se/kg soil) and that foods grown on those soils also contain very low amounts (<4 ng/100 g dry weight) of the element (Ge et al., 1983; Tan et al., 1986). Accordingly, people residing in those locales showed the lowest blood Se levels ever to have been reported: <25 ng/ml.

In the 1960s, studies of deficiency diseases of livestock showed that both New Zealand (Wells, 1976) and Finland (Oksanen, 1965) also had Se-deficient soils. These conditions were reflected in those national food systems, each of which provided only meager amounts of Se. Accordingly, those populations had blood Se levels much lower than observed in other Western countries (Robinson and Thomson, 1988; Kantola et al., 1997). Although neither country had significant malnutrition, each had a history of health problems in livestock that responded to Se supplementation. Each took a different tack in addressing its Se deficiencies: New Zealand approved a variety of Se treatments for livestock (drenches, oral beadlets, diet supplements) and pastures (sprays); Finland approved Se supplements to some livestock feeds and, starting in 1980, addition of Se to agricultural fertilizers (Aspila, 2005).

Comparisons of the Se contents of different national food systems (Combs, 2001; Fairweather-Tait et al., 2011) have suggested that subclinical Se deficiency (insufficient Se to support maximal GPx3 expression, i.e., <50 mcg/day) may affect 10%–50% of residents in most countries for which data are available. Only Canada, Japan, Norway, and the United States showed no evidence of prevalent low Se status.

Low blood Se levels have also been observed in patients with several other diseases. Children with the protein-deficiency diseases, kwashiorkor or marasmus, tend to have low plasma Se levels and may have increased needs for antioxidant nutrients due to the pro-oxidative effects of malnutrition and inflammation. Neonates typically have lower blood Se levels than their mothers, and low plasma Se levels have been associated with increased risks to early wheezing (Devereaux et al., 2007) and to respiratory morbidity among very low birth weight newborns (Darlow et al., 1995).

1.2.5 WHAT ARE THE HEALTH CONSEQUENCES OF LOW SELENIUM STATUS?

Two diseases have been associated with severe endemic Se deficiency in humans: the cardiomyopathy Keshan disease and an osteoarthropathy, Kaschin-Beck disease. Each occurs in rural areas of China and Russia (eastern Siberia) in food systems with exceedingly low Se supplies. While Kaschin-Beck disease remains poorly studied, large-scale intervention trials have demonstrated that Keshan disease can be prevented by Se supplementation (Keshan Disease Research Group, 1979). Still, it is not clear whether Se deficiency is the primary cause of either disease although, in each case, it appears to be, at least, a predisposing factor.

It has been suggested that the etiology of Keshan disease may involve one or more cardiotrophic viruses, several of which have been isolated from hearts of Keshan disease fatalities. One such RNA virus, Coxsackie B4, was found to cause more severe heart damage to Se-deficient mice than it did to Se-adequate mice (Darlow et al., 1995). Further, Beck and colleagues found that a nonpathogenic strain of Coxsackie B3 virus became highly pathogenic after passage through a Se-deficient host (Beck and Levander, 1998; Beck et al., 1998). Increased virulence was associated with discrete changes in the viral genome, corresponding to profiles of myocarditic strains (Beck, 2007). They demonstrated this same phenomenon for another RNA-virus, an influenza strain (Beck et al., 2004). This raises the possibility that severe, endemic Se deficiency may facilitate the development of virulence of other RNA-viruses, such as measles, influenza, hepatitis, severe acute respiratory syndrome (SARS), and acquired immune deficiency syndrome (AIDS)—all global public health problems. Indeed, some cohort studies have shown associations of low Se status and progression of AIDS although interventions with Se in HIV[+] subjects have yielded mixed results (Stone et al., 2010).

Studies in central Africa found the iodine-deficiency diseases, goiter and myxedematous cretinism, to be prevalent among Se-deficient populations (Corvilain et al., 1993). These observations led to the recognition that Se is essential for normal thyroid hormone metabolism, conversion of thyroxine to the active thyroid hormone requiring selenodeiodinases (Köhrle, 2013). Hence, the efficacy of iodine supplementation may be limited in Se-deficient populations.

1.2.6 CAN SELENIUM REDUCE CANCER RISK?

That Se may also be anticarcinogenic was suggested in the late 1960s and early 1970s based on findings of inverse relationships of cancer mortality rates and forage crop Se contents in the United States (Shamberger and Frost, 1969; Shamberger and Willis, 1971; Shamberger et al., 1974; Schrauzer et al., 1977). Subsequent evidence has shown that Se can, indeed, play a role in cancer prevention: Virtually all animal and cell model studies (Ip and Ganther, 1990, 1992; Ip, 1998; Ganther, 1999; Fleming et al., 2001; Lu, 2001; Whanger, 2004; Rayman, 2005) and most clinical intervention trials (Dennert et al., 2011; Fritz et al., 2011; Lee et al., 2011) have demonstrated cancer risk reduction by Se given to nondeficient subjects. This body of research shows that both inorganic and organic Se compounds can be antitumorigenic at doses greater than required to support the maximal expression of the selenoenzymes.

These effects appear to involve actions of selenoenzymes (GPxs and TxRs) as well as Se metabolites (e.g., methylselenol) (Ip and Ganther, 1990, 1992; Ip, 1998; Ganther, 1999; Fleming et al., 2001; Lu, 2001; Diwadkar-Navsariwala and Diamond, 2004; Whanger, 2004; Rayman, 2005; Jackson and Combs, 2008; Zeng, 2009; Jackson and Combs, 2011), the former acting by reducing hydroperoxides to prevent DNA damage and the latter acting in several ways: increasing caspase-mediated apoptosis (Kim et al., 2001; Lu, 2001) inhibiting PI3K/AKT/mTOR and MAPK pathways (Jiang et al., 2004; Unni et al., 2005; Zeng et al., 2010), activating AMPK and increasing expression of tumor-suppressor genes p53 and p21 (Hwang et al., 2006; Wang et al., 2008), upregulating phase II enzymes (Xiao and Parkin, 2006), inactivating PKC to inhibit tumor promotion and cell growth (Cheng, 2009), and attenuating expression of pro-inflammatory factors (Zeng and Botnen, 2007; Vunta et al., 2008) to inhibit IL-6 and the downstream cascade NFκB and JAK/STAT3 signal pathways (Kretz-Remy and Arrigo, 2001; Gasparian et al., 2002; McCarty and Block, 2006; Gazi et al., 2007). Supranutritional intakes of Se can also reduce the spread of malignant cells to distant organs by reducing adhesion of cancer cells to extracellular matrix, inhibiting the urokinase plasminogen activator system and reducing angiogenesis (Yan and Frenkel, 1992; Yan et al., 1997; Zeng et al., 2006; Yan and Demars, 2012). This means that Se may inhibit both primary carcinogenesis and later metastatic cancer spread.

Several clinical trials have been conducted to determine whether Se can reduce cancer risk in humans, yet the clinical significance of Se in cancer prevention remains a subject of current debate. A recent systematic review concluded that the evidence for supplemental Se reducing cancer risk is not convincing (Dennert et al., 2011); however, two other systematic reviews concluded that Se may be effective in cancer prevention for individuals of low to adequate but not high Se status (Fritz et al., 2011; Lee et al., 2011). Each pointed out that the nine randomized controlled trials (RCTs) conducted to date have yielded inconsistent results. This is best demonstrated by the two major ones, the Nutritional Prevention of Cancer Trial (NPC) (Clark et al., 1996) and the Selenium and Vitamin E Cancer Trial (SELECT) (Lippman et al., 2009). The NPC trial found supplemental Se to reduce risks to total carcinomas, total cancer mortality, and cancers of the prostate and colon–rectum by as much as 65% (Clark et al., 1996). In contrast, the much larger SELECT trial (Lippman et al., 2009) found no protective effects of Se against prostate cancer. Although the latter results have been interpreted as countering the former ones, a consideration of the Se status of each of those cohorts shows that, in fact, the SELECT results are fully consistent with those of NPC. Subjects in SELECT, which noted no reduction in prostate cancer risk by Se, had relatively high baseline plasma Se levels averaging (136 ng/ml). In that respect, they were similar to the upper tertile (plasma Se >120 ng/ml) of NPC subjects, which showed no prostate cancer risk reduction by Se. In fact, the NPC trial showed that Se treatment reduced risk *only* for subjects with baseline plasma Se <106 ng/ml (Reid et al., 2002). Thus, the totality of available clinical evidence indicates that Se can reduce cancer risk for individuals of low-to-adequate Se status defined by plasma Se in the range of 70–106 ng/ml.

1.2.7 IS SELENIUM SAFE?

Chronic selenosis was identified in the 1960s among residents of Enshi County, Hubei Province, China, apparently resulting from exceedingly high concentrations of Se in the local food supplies and, in fact, throughout the local environment (Yang et al., 1989a,b). Local soils were found to contain nearly 8 ppm Se; coal contained as much as 84,000 ppm Se (ash was used to amend agricultural soils); locally produced foods contained the highest concentrations of Se ever reported. Drinking water, leaching through seleniferous coal seams, contained >50 ppb Se. In affected villages, residents consumed an estimated 3.2–6.7 mg Se/person/day and showed losses of hair and nails; some also showed skin lesions, hepatomegaly, polyneuritis, and gastrointestinal disturbances. Both the World Health Organization (1996) and the IOM (Institute of Medicine, 2000) set the upper safe limit of Se intake at 400 mcg/day for an adult, slightly lower than the level of 450 mcg/day set by the UK Food Standards Agency (Hurst et al., 2013). These levels are conservative; each is about half the estimate made by Yang and colleagues (Yang et al., 1989b). A review by the US EPA (Poirier, 1994) set the no adverse effect level (NOAEL) for an adult of 853 mcg Se/day; studies by Burk and colleagues found such a level without adverse effects in healthy American adults (Burk et al., 2006).

Questions have also been raised about whether supranutritional Se intakes below such upper limits may increase risk of type 2 diabetes (T2D) (Bleys et al., 2007; Stranges et al., 2007; Laclaustra et al., 2009). Studies in animal models have revealed no negative effects of Se on glycemic control (Sun and Lei, 2013; Sunde and Yan, 2013) or insulin sensitivity (Stapleton, 2000; Mueller and Pallauf, 2006), yet excess T2D was noted among subjects in the upper quintile of plasma Se in US NHANES cohorts (1988–1994, 2003–2004) (Bleys et al., 2007; Laclaustra et al., 2009) and among Se-supplemented subjects in the NPC Trial who achieved plasma Se levels averaging approximately 190 ng/ml (Stranges et al., 2007). These studies suffered from the methodological limitation of using unconfirmed, patient-reported T2D status (in some cases also with history of medication use), which is subject to significant ascertainment bias due to the widespread underdiagnosis and, hence, lack of treatment of T2D and prediabetic conditions. We observed no relationship of fasting glucose and plasma Se in a cohort of subjects in Grand Forks, ND (Combs et al., 2012), and the *only* clinical trial to include unambiguous confirmation of T2D found no effect of supranutritional doses of Se (Rayman et al., 2012); see also Chapter 11.

1.3   THE CURRENT QUESTION: WHO CAN BENEFIT FROM SELENIUM?

In the span of seven decades, Se has moved from being thought of as a toxicant to being considered an essential nutrient with the potential to reduce cancer risk. The elucidation of its roles in nutrition and health has led to fundamental discoveries in biochemistry and molecular biology (the unique metabolism of SeCys and the biosynthesis of selenoproteins; See Chapters 3 and 4), virology (the genomic destabilization of RNA viruses due to oxidative stress), and public health (a role in reducing cancer risk). Unlike the other micronutrients, which were first recognized for the fatal outcomes of their deficiencies, many consequences of Se deprivation are subclinical in nature, requiring other precipitating factors (vitamin E deficiency, viruses, carcinogens) to reveal the effects of suboptimal expression of selenoenzymes and/or insufficient amounts of active Se metabolites. Of course, questions remain.

The most salient current question is "Who can benefit from increased Se intake?" This must be addressed with two groups of people in mind: those who are deficient in the element and those who are not.

First, it is certain that adults with Se intakes less than approximately 50 mcg/day can benefit from increasing those intakes. Clinically, they will respond with increases in circulating levels of GPx3 and SePP to the extent that their baseline plasma Se levels were less than about 70 ng/ml (Xia et al., 2005). In any case, they will show increases in the total Se content of the plasma; those will be greatest if Se is consumed as the dominant food form selenomethionine (SeMet), which is incorporated nonspecifically into proteins and therefore appears in albumin (Combs et al., 2012). Women were found to show greater increases in plasma nonspecific Se than men in response to a regular oral SeMet supplement (Combs et al., 2012). The same study found the plasma Se response to be related to GPx1 T679T genotype (Combs et al., 2012), which have also been associated with differences in cardiovascular risk (Rayman et al., 2012) and, in some studies, cancers (Kato et al., 2008; Kucukgergin et al., 2011; Lubos et al., 2011; Chen et al., 2012; Crawford et al., 2012; Cao et al., 2014; Met et al., 2014); see also Chapter 15.

Whether individuals with apparently adequate Se nutriture may also benefit from increased Se intakes calls for a more qualified consideration. Evidence suggests that supplemental Se offers the potential of reducing cancer risk for adults of apparently adequate Se status but with plasma Se less than approximately 106 ng/ml. Such individuals comprise at least 10% of Americans, most Europeans, and many other adults in other countries (Combs, 2001; Aspila, 2005). When supplemented with SeMet, these individuals will show changes in plasma GPx3 or SePP (Burk et al., 2006; Combs et al., 2012) but will respond with increases in plasma total Se due to increases in nonspecifically bound Se.

This means that large numbers of adults are likely to benefit from increased Se intakes—some to reduce cancer risks, others also to maximize selenoprotein expression. This is not to suggest that any level of supplemental Se may be safe. Indeed, a "U"-shaped dose–risk relationship has been proposed (Waters et al., 2005), and the unresolved question of whether high doses of Se may increase T2D risk warrants attention (Rayman and Stranges, 2013). Resolving the latter issue will demand well-controlled clinical trials with subjects randomized by T2D risk factors (high BMI, elevated fasting glucose) and followed with unequivocal diagnostic indicators of T2D (fasting glucose, HbA1c, oral glucose tolerance). Until then, there would appear to be no justification for any healthy adult, regardless of his/her baseline Se status, to consume more than 300–450 mcg Se/day recommended as upper limits by various national advisory panels (Hurst et al., 2013).

ACKNOWLEDGMENTS

The author gratefully acknowledges the helpful suggestions of the editors, Drs. Regina Brigelius-Flohé and Helmut Sies, and of Drs. David Taussig, Lin Yan, and Huawei Zeng of the Grand Forks Human Nutrition Research Center who critically reviewed this manuscript.

REFERENCES

Aspila, P. 2005. "History of selenium supplemented fertilization in Finland." In Twenty Years of Selenium Fertilization (M. Eurola, ed.) Agrifood Res Rep no. 69:8–13.

Beck, M.A. 2007. "Selenium and vitamin E status: Impact on viral pathology." J Nutr no. 137:1338–40.

Beck, M.A. and O.A. Levander. 1998. "Dietary oxidative stress and the potential of viral infection." Ann Rev Nutr no. 18:93–116.

Beck, M.A., R.S. Esworthy, Y.S. Ho, and F.F. Chu. 1998. "Glutathione peroxidase protects mice from viral-induced myocarditis." FASEB J no. 12:1143–9.

Beck, M.A., J. Handy, and O.A. Levander. 2004. "Host nutritional status: The neglected virulence factor." Trends Microbiol no. 12:417–23.

Bleys, J., A. Navas-Acien, E. Guallar. 2007. "Serum selenium and diabetes in US adults." Diab Care 30:829–34.

Bonetti, E. and R. Stripe. 1963. "Effect of selenium on muscular dystrophy in vitamin E-deficient rats and guinea pigs." Proc Soc Exp Biol Med no. 114:109–17.

Burk, R.F., B.K. Norsworthy, K.E. Hill, A.K. Motley, and D.W. Byrne. 2006. "Effects of chemical form on plasma biomarkers in a high-dose Se supplementation trial." Cancer Epidemiol Biomarkers Prev no. 15(4):804–10.

Calvert, C.C. and M.L. Scott. 1963. "Effect of selenium on the requirement for vitamin E and cystine for the prevention of nutritional muscular dystrophy in the chick." Fed Proc no. 22:318–26.

Cao, M., X. Mu, C. Jiang, G. Yang, H. Chen, and W. Xue. 2014. "Single-nucleotide polymorphisms of GPX1 and MnSOD and susceptibility to bladder cancer: A systematic review and meta-analysis." Tumour Bio no. 35:759–64.

Chambers, I., J. Frampton, P. Goldfarb, N. Affra, W. McBain, and P.R. Harrison. 1986. "The structure of the mouse glutathione peroxidase gene: The selenocysteine in the active site is encoded by the 'termination' codon, TGA." EMBO J no. 5(6):1221–7.

Chen, H., M. Yu, M. Li et al. 2012. "Polymorphic variations in manganese superoxide dismutase (MnSOD), glutathione peroxidase-1 (GPX1), and catalase (CAT) contribute to elevated triglyceride levels in Chinese patients with type 2 diabetes or diabetic cardiovascular disease." Mol Cell Biochem no. 363:85–91.

Cheng, W.H. 2009. "Impact of inorganic nutrients on maintenance of genomic stability." Environ Mol Mutagen no. 50(5):349–60.

Clark, L.C., G.F. Combs, Jr., B.J. Turnbull et al. 1996. "Effects of selenium supplementation for cancer prevention in patients with carcinoma of the skin. A randomized controlled trial." *JAMA* no. 276:1957–63.

Combs, Jr., G.F. 2001. "Selenium in global food systems." *Br J Nutr* no. 85:517–47.

Combs, Jr., G.F., M.I. Jackson, J.C. Watts et al. 2012. "Differential responses to sele-nomethionine supplementation by sex and genotype in healthy adults." *Br J Nutr* no. 107:514–1525.

Corvilain, B., B. Contempre, A.O. Longombe et al. 1993. "Selenium and the thyroid: How the relationship was established." *Am J Clin Nutr* no. 57:244S–8S.

Crawford, A., R.G. Fassett, D.P. Geraghy et al. 2012. "Relationships between single nucleo-tide polymorphisms of antioxidant enzymes and disease." *Gene* no. 501:89–103.

Dan, H., and E. Sondergaard. 1957. "Prophylactic effect of selenium dioxide against degenera-tion (white striation) of muscle in chicks." *Experientia* no. 13:494–502.

Darlow, B.A., T.E. Inder, P.J. Graham et al. 1995. "The relationship of selenium status to respi-ratory outcome in the very low birth weight infant." *Pediatrics* 96:314–9.

Dennert, G., M.B. Zwahlen, M. Vinceti, M.P. Zeegers, M. Horneber. 2011. "Selenium for preventing cancer." *Cochrane Database Syst Rev* no. 11(5):CD005195.

Devereaux, G., G. McNeill, G. Newman et al. 2007. "Early childhood wheezing symptoms in rela-tion to plasma selenium in pregnant mothers and neonates." *Clin Exp Allergy* no. 37:1000–8.

Diwadkar-Navsariwala, V. and A.M. Diamond. 2004. "The link between selenium and chemo-prevention: A case for selenoproteins." *J Nutr* no. 134:2899–902.

Ewan, L.M. and K.J. Jenkins. 1967. "Antidystrophic effect of selenium and other agents on chicks from vitamin E-depleted hens." *J Nutr* no. 93:470–7.

Fairweather-Tait, S.J., Y.P. Bao, M.R. Broadley et al. 2011. "Selenium in human health and disease." *Antioxid Redox Signal* no. 14:1337–83.

Fleming, J., A. Ghose, and P.R. Harrison. 2001. "Molecular mechanisms of cancer prevention by selenium compounds." *Nutr Cancer* no. 40:42–9.

Flohé, L., W.A. Günzler, and H.H. Schock. 1973. "Glutathione peroxidase: A selenoenzyme." *FEBS Lett* no. 32:132–4.

Forstrom, J.W., J.J. Zakowski, and A.L. Tappel. 1978. "Identification of the catalytic site of rat liver GSH-Px as selenocysteine." *Biochemistry* no.17:2639–44.

Franke, K.W. 1934a. "A new toxicant occurring naturally in certain samples of plant food-stuffs. I. Results obtained in preliminary feeding trials." *J Nutr* no. 8:597–608.

Franke, K.W. 1934b. "A new toxicant occurring naturally in certain samples of plant food-stuffs. II. The occurrence of the toxicant in the protein fraction." *J Nutr* no. 8:609–13.

Fritz, H., D. Kennedy, D. Fergusson et al. 2011. "Selenium and lung cancer: A systematic review and meta analysis." *PLoS One* no. 6(11):e26259.

Ganther, H.E. 1999. "Selenium metabolism, selenoproteins and mechanisms of cancer preven-tion: Complexities with thioredoxin reductase." *Carcinogenesis* no. 20:1657–66.

Gasparian, A.V., Y.J. Yao, J. Lu et al. 2002. "Selenium compounds inhibit I kappa B kinase (IKK) and nuclear factor-kappa B (NF-kappa B) in prostate cancer cells." *Mol Cancer Ther* no. 1:1079–87.

Gazi, M.H., A. Gong, K.V. Donkena, and C.Y. Young. 2007. "Sodium selenite inhibits interleukin-6-mediated androgen receptor activation in prostate cancer cells via upregu-lation of c-Jun." *Clin Chim Acta* no. 380:145–50.

Ge, K., A. Xue, J. Bai, and S. Wand. 1983. "Keshan Disease—An endemic cardiomyopathy in China." *Virchow Arch* no. 401:1–11.

Hurst, R., R. Collings, L. Harvey et al. 2013. "EURRECA—Estimating selenium require-ments four deriving dietary reference values." *Crit Rev Food Sci* no. 53:1077–96.

Hwang, J.T., Y.M. Kim, Y.J. Surh et al. 2006. "Selenium regulates cyclooxygenase-2 and extracellular signal-regulated kinase signaling pathways by activating AMP-activated protein kinase in colon cancer cells." *Canc Res* no. 66:10057–63.

Institute of Medicine. Dietary Reference Intakes for Vitamin C, Vitamin E, Selenium and Beta-Carotene and Other Carotenoids, 284–324, Washington: National Academy Press, 2000.

Ip, C. 1998. "Lessons from basic research on selenium and cancer prevention." *J Nutr* no. 128:1845–54.

Ip, C. and H.E. Ganther. 1990. "Activity of methylated forms of selenium in cancer prevention." *Canc Res* no. 50:1206–11.

Ip, C. and H.E. Ganther. 1992. "Biological activities of trimethylselenonium as influenced by arsenite." *J Inorg Biochem* no. 46:215–22.

Jackson, M.I. and G.F. Combs Jr. 2008. "Selenium and anti-carcinogenesis." *Curr Opin Clin Nutr Health Care* no. 11:718–26.

Jackson, M.I. and G.F. Combs Jr. "Selenium as a Cancer Preventive Agent." *In Selenium: Its Molecular Biology and Role in Human Health* (Hatfield, D.L., M.J. Berry, and V.N. Gladyshev, eds.), 313–324, New York: Springer, 2011.

Jiang, C., K.H. Kim, Z. Wang, and J. Lu. 2004. "Methyl selenium-induced vascular endothelial apoptosis is executed by caspases and principally mediated by p38 MAPK pathway." *Nutr Canc* no. 49:174–83.

Kantola, M., E. Mand, A. Viitak et al. 1997. "Selenium contents of serum and human milk from Finland and neighboring countries." *J Trace Elem Exp Med* no. 10:225–32.

Kato, K., M. Oguri, N. Kato et al. 2008. "Assessment of genetic risk factors for thoracic aortic aneurism in hypertensive patients." *Am J Hypertens* no. 21:1023–7.

Keshan Disease Research Group. 1979. "Observations on effect of sodium selenite in prevention of Keshan disease." *Acta Acad Med Sinica* no. 1:75–83.

Kim, T., U. Jung, D.Y. Cho, and A.S. Chung. 2001. "Se-methylselenocysteine induces apoptosis through caspase activation in HL-60 cells." *Carcinogenesis* no. 22:559–65.

Köhrle, J. 2013. "Selenium and the thyroid." *Curr Opin Endocrinol Diabetes Obes* no. 20:441–8.

Kretz-Remy, C. and A.P. Arrigo. 2001. "Selenium: A key element that controls NF-kappa B activation and I kappa B alpha half life." *Biofactors* no. 14:117–25.

Kryukov, G.V., S. Castellano, S.V. Novoselov et al. 2003. "Characterization of mammalian selenoproteins." *Science* no. 300:1439–43.

Kucukgergin, C., M. Gokpinar, O. Sanli, T. Tefik, T. Oktar, and S. Seckin. 2011. "Association between genetic variants in glutathione peroxidase 1 (GPX1) gene, GPx activity and the risk of prostate cancer." *Minerva Urol Nefrol* no. 63:183–90.

Laclaustra, M., A. Navas-Acien, S. Stranges, J.M. Ordovas, and E. Guallar. 2009. "Serum selenium concentrations and diabetes in US adults: National Health and Nutrition Examination Survey (NHANES) 2003–2004." *Environ Health Perspect* no. 117:1409–13.

Lee, B.J., P.J. Worland, J.N. Davis, T.C. Stadtman, and D.L. Hatfield. 1989. "Identification of a selenocysteneyl-tRNA(Ser) in mammalian cells that recognizes the nonsense codon UGA." *J Biol Chem* no. 264:9724–7.

Lee, E.H., S.K. Myung, Y.J. Jeon et al. 2011. "Effects of selenium supplements on cancer prevention: Meta-analysis of randomized controlled trials." *Nutr Canc* no. 63:1185–95.

Lippman, S.M., E.A. Klein, P.J. Goodman et al. 2009. "Effect of selenium and vitamin E on risk of prostate cancer and other cancers: The Selenium and Vitamin E Cancer Prevention Trial (SELECT)." *JAMA* no. 301:39–51.

Lu, J. 2001. "Apoptosis and angiogenesis in cancer prevention by selenium." *Adv Exp Med Biol* no. 492:131–45.

Lubos, E., J. Loscalzo, and D.E. Handy. 2011. "Glutathione peroxidase-1 in health and disease: From molecular mechanisms to therapeutic opportunities." *Antioxid Redox Signal* no. 15:1957–97.

McCarty, M.F. and K.I. Block. 2006. "Preadministration of high-dose salicylates, suppressors of NF-kappaB activation, may increase the chemosensitivity of many cancers: An example of proapoptotic signal modulation therapy." *Integr Cancer Ther* no. 5:252–68.

Met, T., X. Zhang, J. Yang et al. 2014. "The rs1050450 C > T polymorphism of GPX1 is associated with the risk of bladder but not prostate cancer: Evidence from a meta-analysis." *Tumour Biol* no. 35:269–75.

Mueller, A.S. and J. Pallauf. 2006. "Compendium of the antidiabetic effects of supranutritional selenate doses. *In vivo* and *in vitro* investigations with type II diabetic db/db mice." *J Nutr Biochem* no. 17: 548–60.

National Research Council. Recommended Dietary Allowances, 10th Edition, 217–222, Washington: National Academy Press, 1989.

Nesheim, M.C. and M.L. Scott. 1958. "Studies on the nutritive effect of selenium for chicks." *J Nutr* no. 65:601–10.

Oksanen, H.E. 1965. "Studies on nutritional muscular degeneration (NMD) in ruminants." *Acta Vet Scand* no. 40:suppl 2:1–110.

Patterson, E.L., R. Mistrey, and E.L.R. Stokstad. 1957. "Effect of selenium in preventing exudative diathesis in the chick." *Proc Soc Exp Biol Med* no. 95:617–25.

Poirier, K.A. "Summary of the derivation of the reference dose for selenium." In: *Risk Assessment of Essential Elements.* (Mertz, W., C.O. Abernathy, and S.S. Olin, eds.) 157–166, Washington: ILSI Press, 1994.

Rayman, M.P. 2005. "Selenium in cancer prevention: A review of the evidence and mechanism of action." *Proc Nutr Soc* no. 64:527–42.

Rayman, M.P. and S. Stranges. 2013. "Epidemiology of selenium and type 2 diabetes: Can we make sense of it?" *Free Radic Biol Med* no. 65:1557–64.

Rayman, M.P., G. Blundell-Pound, R. Pastor-Barriuso, E. Guallar, H. Steinbrenner, and S. Stranges. 2012. "A randomized trial of selenium supplementation and risk of type-2 diabetes, as assessed by plasma adiponectin." *PLoS One* no. 7:e45269.

Reeves, M.A. and P.R. Hoffman. 2009. "The human selenoproteome: Recent insights into functions and regulation." *Cell Mol Life Sci* no. 66:2457–78.

Reid, M.E., A.J. Duffield-Lillico, L. Garland, B.W. Turnbull, L.C. Clark, and J.R. Marshall. 2002. "Selenium supplementation and lung cancer incidence: An update on the Nutritional Prevention of Cancer Trial." *Cancer Epidem Biomarkers Prev* no. 11(11):1285–91.

Robinson, M.F. and C.D. Thomson. "Effect of a shipment of high-selenium Australian wheat on selenium status of Otago (NZ) residents." In *Trace Elements in Man and Animals, TEMA 6*, 341–342, New York: Springer, 1988.

Rotruck, J.T., A.L. Pope, H.E. Ganther, A.B. Swanson, D.G. Hafeman and W.G. Hoekstra. 1973. "Selenium: Biochemical role as a component of glutathione peroxidase." *Science* no. 179:588–90.

Schrauzer, G.N., D.A. White, and C.J. Schneider. 1977. "Cancer mortality studies—III. Statistical associations with dietary selenium intakes." *Bioinorg Chem* no. 7:23–31.

Schwarz, K. and C.M. Foltz. 1957. "Selenium as an integral part of factor 3 against dietary necrotic liver degeneration." *J Am Chem Soc* no. 79:3292–3.

Schwarz, K., J.G. Bieri, G.M. Briggs, and M.L. Scott. 1957. "Prevention of exudative diathesis in chicks by factor 3 and selenium." *Proc Soc Exp Biol Med* no. 95:621–9.

Scott, M.L. 1974. "Lesions of vitamin E and selenium deficiencies and their pathogenesis." *Folio Vet Lat* no. 4:113–23.

Shamberger, R.J. and D.V. Frost. 1969. "Possible protective effect of selenium against human cancer." *Can Med Assn J* no. 104:82–90.

Shamberger, R.J. and C.E. Willis. 1971. "Selenium distribution and human cancer mortality." *Clin Lab Sci* no. 2:211–9.

Shamberger, R.J., S.A. Tytko and C.E. Willis. 1974. "Antioxidants and cancer. II. Selenium and human cancer mortality in the United States, Canada, and New Zealand." *Trace Subs Environs Health* no. 7:35–43.

Stapleton, S.R. 2000. "Selenium: An insulin-mimetic." *Cell Mol Life Sci* no. 57:1874–9.

Stone, C.A., K. Kawai, R. Kupka, and W.W. Fawzi. 2010. "The role of selenium in HIV infection." *Nutr Rev* no. 68:671–81.

Stranges, S., J.R. Marshall, R. Natarajan et al. 2007. "Effects of long-term selenium supplementation on the incidence of type 2 diabetes: A randomized trial." *Ann Intern Med* no. 147:217–23.

Sun, X. and X.G. Lei. 2013. "Effect of high dietary selenium intake and GPX1 overproduction on mouse susceptibility to gestational diabetes." *FASEB J*, Abst 234.6.

Sunde, R.A. and X. Yan. 2013. "Effects of high-fat, selenium-deficient, and high-selenium diets on diabetes markers in wildtype and glutathione peroxidase 1-null mice." *FASEB J*, Abst 234.5.

Tan, J.A., D.Z. Zheng, S.F. Hou et al. "Selenium ecological chemicogeography and endemic Keshan disease and Kaschin-Beck Disease in China." In *Proceedings of the Third International Symposium on Selenium in Biology and Medicine* (Combs, Jr., G.F., Spallholz, J.E., Levander, O., Oldfield, J.E., eds.), 859–876, Westport: AVI, 1986.

Unni, E., D. Koul, W.K. Yung, and R. Sinha. 2005. "Se-methylselenocysteine inhibits phosphatidylinositol 3-kinase activity of mouse mammary epithelial tumor cells *In vitro*." *Breast Cancer Res* no. 7:R699–707.

Vunta, H., B.J. Belda, R.J. Arner, C. Channa Reddy, J.P. Vanden Heuvel, and S.K. Prabhu. 2008. "Selenium attenuates pro-inflammatory gene expression in macrophages." *Mol Nutr Food Res* no. 52:1316–23.

Wang, Z., H. Hu, G. Li et al. 2008. "Methylseleninic acid inhibits microvascular endothelial G1 cell cycle progression and decreases tumor microvessel density." *Int J Cancer* no. 122:15–24.

Waters, D.J., S. Shen, L.T. Glickman et al. 2005. "Prostate cancer risk and DNA damage: Translational significance of selenium supplementation in a canine model." *Carcinogenesis* no. 26:1256–62.

Wells, N. "Total selenium in top soils." In *New Zealand Soil Bureau Atlas*, Wellington: Government Printer, 1976.

Wendel, A., B. Kerner, and K. Graupe. "The selenium moiety of glutathione peroxidase." In *Functions of Glutathione in Liver and Kidney* (Sies, H. and A. Wendel, eds.) 107–13. Berlin: Springer-Verlag, 1978.

Whanger, P.D. 2004. "Selenium and its relationship to cancer: An update." *Br J Nutr* no. 91:11–28.

World Health Organization. "Selenium." In *Trace Elements in Human Nutrition and Health*, 105–22. Geneva: World Health Organization, 1996.

Xia, Y., K.E. Hill, D.W. Byrne, J. Xu, and R.F. Burk. 2005. "Effectiveness of selenium supplements in a low-selenium area of China." *Am J Clin Nutr* 81:829–34.

Xiao, H. and K.L. Parkin. 2006. "Induction of phase II enzyme activity by various selenium compounds." *Nutr Cancer* no. 55:210–23.

Yan, L. and G.D. Frenkel. 1992. "Inhibition of cell attachment by selenite." *Cancer Res* no. 52:5803–7.

Yan, L. and L.C. Demars. 2012. "Dietary supplementation with methylseleninic acid, but not selenomethionine, reduces spontaneous metastasis of Lewis lung carcinoma in mice." *Int J Cancer* no. 131:1260–6.

Yan, L., J.A. Yee, M.H. McGuire, and G.L. Graef. 1997. "Effect of dietary supplementation of selenite on pulmonary metastasis of melanoma cells in mice." *Nutr Cancer* no. 28:165–9.

Yang, G.Q., J.Z. Zhu, S.J. Liu et al. "Human selenium requirements in China." In *Proc of the Third International Symposium on Selenium in Biology and Medicine* (Combs, G.F., Jr., Spallholz, J.E., Levander, O., Oldfield, J.E., eds.), 589–607, Westport: AVI, 1986.

Yang, G.Q., R. Zhou, S. Yin et al. 1989a. "Studies of safe maximal daily dietary Se-intake in a seleniferous area in China. Part I. Selenium intake and tissue selenium levels of the inhabitants." *J Trace Elem Electrolytes Health Dis* no. 3:77–85.

Yang, G.Q., S. Yin, R. Zhou et al. 1989b. "Studies of safe maximal daily dietary Se-intake in a seleniferous area in China. Part II. Relation between Se-intake and the manifestations of clinical signs and certain biochemical alterations in blood and urine." *J Trace Elem Electrolytes Health Dis* no. 3:123–30.

Zeng, H. 2009. "Selenium as an essential micronutrient: Roles in cell cycle and apoptosis." *Molecules* no. 14:1263–78.

Zeng, H., and J.H. Botnen. 2007. "Selenium is critical for cancer-signaling gene expression but not cell proliferation in human colon Caco-2 cells." *Biofactors* no. 31:155–64.

Zeng, H., M. Briske-Anderson, J.P. Idso, and C.D. Hunt. 2006. "The selenium metabolite methylselenol inhibits the migration and invasion potential of HT1080 tumor cells." *J Nutr* no. 136:1528–32.

Zeng, H., J.H. Botnen, and M. Briske-Anderson. 2010. "Deoxycholic acid and selenium metabolite methylselenol exert common and distinct effects on cell cycle, apoptosis, and MAP kinase pathway in HCT116 human colon cancer cells." *Nutr Cancer* no. 62:85–92.

Section II

Se Compounds as a Source
for Selenoprotein Biosynthesis

2 Selenium Metabolism

Yasumitsu Ogra

CONTENTS

2.1 Introduction 19
2.2 Nutritional Sources of Selenium.. 21
    2.2.1 Animal Metabolites 22
    2.2.2 Plant Metabolites 23
    2.2.3 Other Biological Sources of Selenium... 24
    2.2.4 Elemental Selenium and Inorganic Salts of Selenium...................... 25
2.3 Conclusion 26
Acknowledgments... 26
References.. 26

2.1 INTRODUCTION

Selenium (Se) is an essential trace element in animals. Se is part of the active center of selenoproteins (Suzuki and Ogra, 2002; Whanger, 2002; Böck et al., 2007). Specifically, Se is required in the formation of the selenol group (–SeH), the active center of a selenocysteinyl (SeCys) residue in selenoprotein sequences (Lu and Holmgren, 2009; Reeves and Hoffmann, 2009). In addition, a selenomethionine (SeMet) residue is found in general protein sequences because animals are unable to distinguish methionine (Met) from SeMet in the protein translation process. A protein containing SeMet is called a Se-containing protein to distinguish it from a selenoprotein. Thus, SeCys and SeMet are considered the major nutritional source of Se from animals, and selenometabolites are the minor source. A proposed metabolic pathway of Se in animals is depicted in Figure 2.1. In plants, on the other hand, Se is not essential and exists as a "bystander" mineral. However, its beneficial effects on plant growth have been reported (Freeman et al., 2007). It is known that certain plants, including *Allium* plants belonging to the family *Liliaceae*, such as garlic, onion, wild leek, and shallot, and *Brassica* plants, such as Indian mustard, broccoli, and radish, are able to accumulate Se (Ip et al., 2000; Ravn-Haren et al., 2008; Reid et al., 2008). In those plants, some unique Se-containing amino acids, such as methylselenocysteine (MeSeCys), γ-glutamyl methylselenocysteine (GluMeSeCys), and selenohomolanthionine (SeHLan), are biosynthesized (Montes-Bayón et al., 2002; Ogra et al., 2005, 2007). Hence, those unique selenoamino acids are considered the nutritional source of Se from plants. A proposed metabolic pathway of Se in plants is depicted in Figure 2.2. Microorganisms, and in particular, fungi, are another possible nutritional source of Se. Yeast accumulates Se as SeMet and its

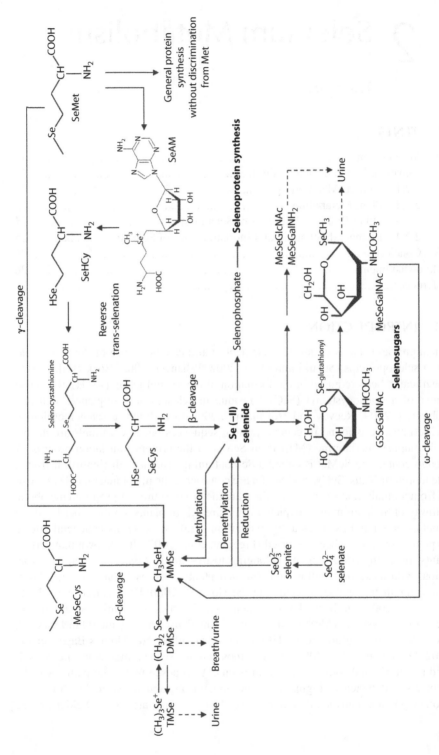

FIGURE 2.1 Metabolic pathway of nutritionally available Se in animals.

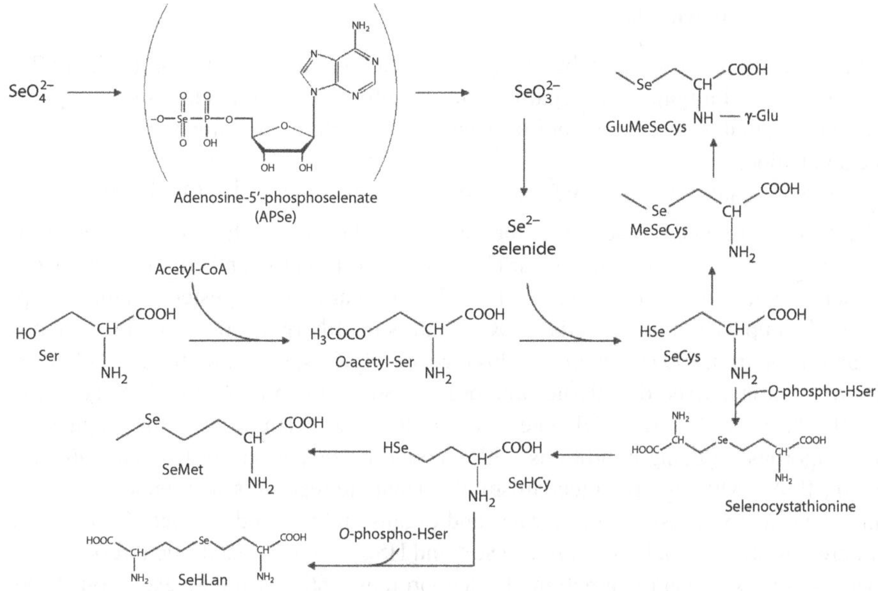

FIGURE 2.2 Metabolic pathway of Se in plants having the capacity to assimilate Se.

derivatives, and thus, it is utilized as a nutritional supplement of Se. Consequently, animals are expected to mainly ingest Se as organoselenium compounds, that is, selenometabolites in the food web. Inorganic Se compounds may also be a Se source from the environment.

In this chapter, I present a review of Se sources for selenoprotein synthesis and their metabolism in animals and plants.

2.2 NUTRITIONAL SOURCES OF SELENIUM

The nutritional availability of Se compounds is defined by the following two criteria. The first criterion is that the Se compounds should be utilizable for selenoprotein synthesis. To evaluate whether a Se compound acts as a Se source or not, the increase of selenoprotein in an animal fed a conventional diet and the recovery of selenoprotein in an animal fed a Se-deficient diet by ingestion of the Se compound are investigated. Then, the amount of selenoprotein and the activity of selenoenzyme are determined. The second criterion is that Se compounds should be excretable in urine as selenosugars. As mentioned later, several selenosugars have been identified as urinary seleno-metabolites (Kobayashi et al., 2002; Ogra et al., 2002; Gammelgaard et al., 2003) with *Se*-methylseleno-*N*-acetyl-galactosamine (MeSeGalNAc) being the most abundant among them (Gammelgaard et al., 2003). Thus, if MeSeGalNAc is detected in urine after the ingestion of a Se compound, that Se compound is considered to be assimilated. At present, speciation analysis with HPLC coupled with inductively coupled argon plasma mass spectrometry is the only technique available for the determination of selenosugars. Se compounds that meet the criteria mentioned above are discussed.

2.2.1 ANIMAL METABOLITES

The currently proposed metabolic pathway of Se in animals is shown in Figure 2.1. Se can exist in inorganic and organic forms in nature. As a human ingests the organic form of Se, that is, selenometabolites from animals, plants, and microorganisms, can be a nutritional source of Se.

Inorganic salts of Se, which consist of mainly selenite $\left(SeO_3^{2-}\right)$ and selenate $\left(SeO_3^{2-}\right)$, are reduced to selenide, the key metabolic intermediate of Se. Selenite is readily reduced by glutathione (GSH) to selenide in cells, and this reduction is a nonenzymatic process (Kobayashi et al., 2001). In animals, ingested selenite is rapidly taken up by red blood cells (RBCs), and is readily reduced by GSH to selenide (Suzuki and Itoh, 1997). Selenide effluxing from RBCs specifically binds to albumin, by which it is delivered to tissues and organs (Shiobara and Suzuki, 1998; Haratake et al., 2008). In contrast, selenate is not reduced to selenide by GSH, suggesting that rigorous reducing conditions with enzyme participation are needed (Shiobara et al., 1999). Although the mechanism of selenate reduction is still unclear, selenate may be directly transferred to tissues and organs and reduced to selenide at the site where it is utilized. Selenide anions (Se^{2-} and HSe^-) are key metabolic intermediates because they stand at the junction of selenoprotein synthesis and Se excretion. There is no evidence that selenide is enzymatically metabolized (oxidized) to selenite and selenate in animals (and also plants); thus, the biotransformation of inorganic Se is a one-way reaction.

Organic selenocompounds in animals are composed of selenoproteins and selenometabolites. SeCys residues are found in meat (Combs, 2001) and milk (Dorea, 2002) proteins (originating from animals), and SeMet residues are present in meat proteins (Cabañero et al., 2005). Those nutritional selenoamino acids are also metabolized to selenide via the specific pathways depicted in Figure 2.1. SeCys is degraded by β-lyase to produce selenide. SeMet is metabolized in a more complicated manner than SeCys (Schrauzer, 2000). In the first metabolic pathway, SeMet liberates methylselenol (CH_3SeH) upon degradation by γ-lyase (Okuno et al., 2005). In the second metabolic pathway, SeMet is transformed into SeCys via the "reverse" trans-selenation pathway (SeMet to SeCys). The transformation involves reverse trans-sulfuration, that is, methionine (Met), S-adenosyl methionine (SAM), S-adenosylhomocysteine, homocysteine, cystathionine, and cysteine (Cys). Indeed, Se-adenosylselenomethionine (SeAM) was actually detected in an animal sample in vitro (Wróbel et al., 2002; Ogra et al., 2009). In the third metabolic pathway, SeMet is incorporated into general proteins without discrimination from Met (McConnell and Hoffman, 1972).

As depicted in Figure 2.1, monomethylselenol (MMSe), dimethylselenide (DMSe), trimethylselenonium ion (TMSe), and selenosugars exist in animals as metabolic intermediates or selenometabolites. As MMSe is reactive, unstable, and volatile, its precursors are used in animal experiments. It was reported that the metabolic intermediate MMSe was used as a nutritional source of Se when its precursor, methylseleninic acid, was administered to animals (Suzuki et al., 2006). All three compounds, that is, methylseleninic acid, dimethylselenoxide, and TMSe, were able to restore glutathione peroxidase activity in Se-depleted animals (Foster et al., 1986b; Magos et al., 1987). This suggests that the complete demethylation to selenide is a normal process in Se metabolism.

When the amount of Se ingested is within the nutritional level, the major urinary metabolite is MeSeGalNAc as identified by HPLC-ICP-MS, ESI-MS-MS, and NMR measurements (Kobayashi et al., 2002). In addition, two minor selenosugars, Se-methylseleno-N-acetyl-glucosamine (MeSeGlcNAc) and Se-methylselenogalactosamine (MeSeGalNH$_2$), were identified by ESI-MS-MS (Bendahl and Gammelgaard, 2004; Gammelgaard and Bendahl, 2004). Se-Glutathionylseleno-N-acetylgalactosamine (GSSeGalNAc) was also identified by ESI-MS-MS in the liver of experimental animals (Kobayashi et al., 2002). This selenosugar is the precursor of the major urinary selenosugar, MeSeGalNAc. The glutathionyl precursors of the two minor selenosugars, MeSeGlcNAc and MeSeGalNH$_2$, have not been identified yet because current mass spectrometers lack sufficient sensitivity for their detection. Those three urinary selenosugars were assimilated and utilized for selenoprotein synthesis (Juresa et al., 2007). It is speculated that such selenosugars as SeMet liberate methylselenol (CH$_3$SeH) by cleavage of the terminal methylselenyl group. Consequently, all selenoproteins and animal selenometabolites are nutritional sources of Se.

2.2.2 PLANT METABOLITES

Although Se is not an essential trace mineral in plants, some Se-accumulating plants are known, and various selenoamino acids have been identified in those plants. In this section, the Se metabolism in Se-accumulating plants is reviewed.

Se existing as selenate is taken up by plants from soil. The first step of Se assimilation in plants is the reduction of selenate to selenite. There is no definitive experimental evidence showing that the reduction of selenate to selenite is catalyzed by the same enzymes as those for sulfate reduction *in vivo*. However, some *in vitro* studies have indicated that the major enzyme in this pathway, ATP sulfurylase, can utilize selenate as its substrate (Murillo and Leustek, 1995). ATP sulfurylase catalyzes the formation of adenosine 5′-phosphosulfate (APS) from ATP and sulfate (Rotte and Leustek, 2000). An *in vitro* study revealed that ATP sulfurylase catalyzed the formation of adenosine 5′-phosphoselenate (APSe) from ATP and selenate (Figure 2.2), and ATP sulfurylase transgenic plants showed increases in the uptake and reduction of and tolerance to selenate (Pilon-Smits et al., 1999; LeDuc et al., 2006). APS is subsequently reduced by APS reductase to produce sulfite (Setya et al., 1996). It was reported that APS reductase transgenic plants also exhibited an increase in selenate reduction (Sors et al., 2005a). This suggests that APS reductase has the capacity to reduce APSe. In addition, there is evidence that APSe can be nonenzymatically reduced to selenide by GSH *in vitro* (Hock Ng and Anderson, 1978; de Souza et al., 1998). Taken together, although it is uncertain whether APSe reduction follows the same pathway as APS reduction, APSe may be nonenzymatically transformed into selenide or selenopersulfide (GSSeH) by GSH to enter the metabolic pathway. As mentioned above, APSe seems to be a key metabolic intermediate in plants. However, there is no direct evidence of its existence in plants.

As Se is a nonessential trace element in plants, the biosynthesis of SeCys tRNA is not necessary. However, SeCys is a key metabolic intermediate for the synthesis of other selenoamino acids. Although selenophosphate acts as a Se donor for the synthesis of SeCys on tRNA in animals (see Chapter 4), selenophosphate is not considered to be a Se donor in plants. Thus, it is supposed that selenide is a Se donor in plants in lieu of selenophosphate. However, there is no solid experimental evidence that selenophosphate is not a Se donor in plants. Serine (Ser) is activated by acetyl-CoA via serine acetyltransferase to form O-acetyl-Ser (Saito, 2004). The activated Ser receives Se in the form of selenide to form SeCys (Sors et al., 2005b).

SeCys can enter the Met biosynthetic pathway to form SeMet via selenocysta-thionine and selenohomocysteine (SeHcy), that is, the "forward" trans-selenation pathway (from SeCys to SeMet) (Läuchli, 1993). Selenocystathionine is formed when SeCys reacts with O-phosphohomoserine, an activated homoserine, in the presence of cystathionine-γ-synthase, and then selenocystathionine is degraded by cystathionine-β-lyase to form SeHcy. It was reported that selenocystathionine also binds to glutamic acid to form γ-glutamylselenocystathionine in monkey pot nut (*Lecythis minor*) (Dernovics et al., 2007). SeHcy is methylated by methyltransferase (Met synthase) to form SeMet with N^5-methyltetrahydrofolate as the methyl group donor (Hesse et al., 2004). As an alternative pathway, SeHcy reacts with another O-phosphohomoserine to form selenohomolanthionine (SeHLan) in Se-enriched Japanese pungent radish (*Raphanus sativus* L. cv. "Yukibijin") (Ogra et al., 2007). Consequently, some Se-accumulating plants accumulated Se as selenocystathionine or SeHLan. An alternative pathway from SeCys involves the formation of MeSeCys by methylation (Neuhierl and Böck, 2002), and MeSeCys is used for the biosynthesis of γ-glutamyl methylselenocysteine (GluMeSeCys) (Ellis et al., 2004).

Plant-specific selenoamino acids, such as MeSeCys and GluMeSeCys, are also nutritional sources of Se. Such methylated selenoamino acids are also metabolized to selenide, namely, MeSeCys and GluMeSeCys liberate methylselenol (CH_3SeH) via β-lyase, and methylselenol is demethylated into selenide to utilize Se for sele-noprotein synthesis in animals (Soda et al., 1984; Tanaka et al., 1985). MeSeCys is more efficiently transformed into methylselenol than SeMet or selenite (Suzuki et al., 2008).

2.2.3 OTHER BIOLOGICAL SOURCES OF SELENIUM

In addition to the nutritional availability of selenometabolites in animals and plants, that of other unique organic selenocompounds is also known (Figure 2.3).

Selenoneine, 2-selenyl-N^α,N^α,N^α-trimethyl-L-histidine, was identified for the first time in the blood of bluefin tuna (Yamashita and Yamashita, 2010) and was also identified in the liver of marine reptiles and fishes (Anan et al., 2011). Marine fishes and reptiles accumulate Se as selenoneine in accordance with the hierarchy of the food chain (Anan et al., 2011). However, the biological role of this unique selenoamino acid is still ambiguous. From the viewpoint of nutritional availability, selenoneine could be the nutritional source of Se because it was reported that the bioavailability of Se in defatted dark muscle of tuna and processed skipjack meat was comparable to that of selenite (Yoshida et al., 1984, 2002). Although selenoneine had not been

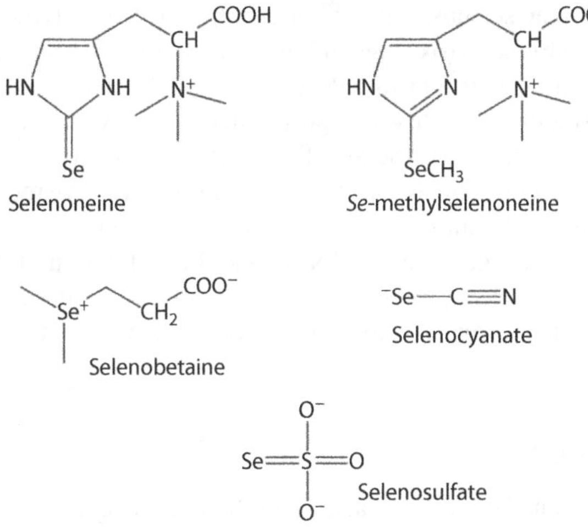

FIGURE 2.3 Unique selenometabolites and inorganic selenocompounds that are a nutritional source of Se.

identified yet at that time, it was shown later that the major Se compound in the dark muscle of tuna and skipjack was selenoneine. This indicates that selenoneine is the major Se source for fish eaters. Recently, selenoneine and its methylated form, that is, *Se*-methylselenoneine, were detected in human blood cell lysate and urine by HPLC-ESI-hybrid linear ion trap-orbital ion trap MS (Klein et al., 2011). Although the biosynthetic pathway of selenoneine in terrestrial animals is still unclear, the authors postulated that selenoneine originated from the diet and was then methylated by methyltransferase to form *Se*-methylselenoneine. As mentioned above, selenoneine could be utilized for selenoprotein synthesis, but excess amounts of selenoneine could be transformed into *Se*-methylselenoneine for detoxification. This postulate seems to be in agreement with the hitherto accepted concept of Se detoxification *in vivo* because reactive Se moieties, such as the -SeH group, are actually masked with a methyl group, for example, SeMet, MeSeCys, and MeSeGalNAc.

Although selenobetaine is not a naturally occurring selenoamino acid, its metabolism has been reported (Foster et al., 1986a; Goeger and Ganther, 1993). Selenobetaine was transformed into DMSe and TMSe *in vivo*. As mentioned above, DMSe and TMSe can be utilized for selenoprotein synthesis. Thus, selenobetaine could be assimilated in selenoproteins and selenometabolites.

2.2.4 Elemental Selenium and Inorganic Salts of Selenium

Selenite and selenate are actually the nutritional sources of inorganic forms of Se. In addition, the bioavailability of other inorganic forms was reported.

Elemental Se nanoparticles can also be utilized for selenoprotein synthesis despite the fact that the detailed mechanism or the metabolic pathway is still unclear (Zhang et al., 2001; Huang et al., 2003).

It was reported that selenosulfate (Figure 2.3) restored activities of selenoenzymes, such as glutathione peroxidase and thioredoxin reductase, in a manner comparable to selenite in Se-deficient rat (Peng et al., 2007). Hence, selenosulfate is a useful nutritional source of Se. It was reported that selenocyanate (Figure 2.3) was excreted in the urine of rat as DMSe and TMSe (Peng et al., 2007). At the time the report was published, selenosugars had not been identified yet. We recently identified selenocyanate in cultured mammalian cell lines as a selenometabolite and detected the major urinary selenosugar, MeSeGalNAc, and the restoration of the amounts of selenoproteins, such as selenoprotein P and glutathione peroxidase, in Se-deficient rats administered selenocyanate (submitted). Thus, selenocyanate is also a nutritional source of Se.

2.3 CONCLUSION

Selenometabolites identified in biota and naturally occurring inorganic selenocompounds are useful nutritional sources of Se even though they are the final urinary metabolites. Although it was not mentioned in this review, plants were also able to metabolize selenometabolites in biota (Ogra et al., 2013). These findings suggest that all selenometabolites and naturally occurring selenocompounds circulate in the environment via biosphere (selenometabolites), geosphere (inorganic selenocompounds), atmosphere (volatile selenocompounds/metabolites), and hydrosphere (selenocompounds/metabolites) although their complete identification and efficacy has yet to be accomplished. The identification of selenometabolites and naturally occurring selenocompounds remains one of the exciting topics in biochemistry and the analytical chemistry of Se.

ACKNOWLEDGMENTS

I wish to thank JSPS KAKENHI Grants (Nos. 23390032, 24659022, and 26293030), MEXT-Supported Program for the Strategic Research Foundation at Private Universities (2013–2017), and Takeda Science Foundation, Japan, for financial support.

REFERENCES

Anan, Y., K. Ishiwata, N. Suzuki, S. Tanabe, and Y. Ogra. 2011. "Speciation and identification of low molecular weight selenium compounds in the liver of sea turtles." *J Anal At Spectrom* no. 26(1):80–5.

Bendahl, L., and B. Gammelgaard. 2004. "Separation and identification of Se-methylselenogalactosamine—A new metabolite in basal human urine—By HPLC-ICP-MS and CE-nano-ESI-(MS)^2." *J Anal At Spectrom* no. 19(8):950–7.

Böck, A., L. Flohé, and J. Köhrle. 2007. "Selenoproteins—Biochemistry and clinical relevance." *Biol Chem* no. 388(10):985–6. doi: 10.1515/BC.2007.148 [doi].

Cabañero, A. I., Y. Madrid, and C. Cámara. 2005. "Enzymatic probe sonication extraction of Se in animal-based food samples: A new perspective on sample preparation for total and Se speciation analysis." *Anal Bioanal Chem* no. 381:373–9.

Combs, G. F., Jr. 2001. "Selenium in global food systems." *Br J Nutr* no. 85(5):517–47. doi: S0007114501000782 [pii].

Dernovics, M., T. García-Barrera, K. Bierła, H. Preud'homme, and R. Łobiński. 2007. "Standardless identification of selenocystathionine and its γ-glutamyl derivatives in monkeypot nuts by 3D liquid chromatography with ICP-MS detection followed by nanoHPLC–Q-TOF-MS/MS." *Analyst* no. 132:439–49.

de Souza, M. P., E. A. Pilon-Smits, C. M. Lytle et al. 1998. "Rate-limiting steps in selenium assimilation and volatilization by indian mustard." *Plant Physiol.* no. 117(4):1487–94.

Dorea, J. G. 2002. "Selenium and breast-feeding." *Br J Nutr* no. 88(5):443–61. doi: 10.1079/BJN2002692.

Ellis, D. R., T. G. Sors, D. G. Brunk et al. 2004. "Production of Se-methylselenocysteine in transgenic plants expressing selenocysteine methyltransferase." *BMC Plant Biol* no. 4:1–11. doi: 10.1186/1471-2229-4-1.

Foster, S. J., R. J. Kraus, and H. E. Ganther. 1986a. "Formation of dimethyl selenide and trimethylselenonium from selenobetaine in the rat." *Arch Biochem Biophys* no. 247(1): 12–9. doi: 0003-9861(86)90527-8.

Foster, S. J., R. J. Kraus, and H. E. Ganther. 1986b. "The metabolism of selenomethionine, Se-methylselenocysteine, their selenonium derivatives, and trimethylselenonium in the rat." *Arch Biochem Biophys* no. 251(1):77–86.

Freeman, J. L., S. D. Lindblom, C. F. Quinn, S. Fakra, M. A. Marcus, and E. A. Pilon-Smits. 2007. "Selenium accumulation protects plants from herbivory by Orthoptera via toxicity and deterrence." *New Phytol* no. 175:490–500.

Gammelgaard, B., and L. Bendahl. 2004. "Selenium speciation in human urine samples by LC- and CE-ICP-MS—Separation and identification of selenosugars." *J Anal At Spectrom* no. 19:135–42.

Gammelgaard, B., K. Grimstrup Madsen, J. Bjerrum, L. Bendahl, O. Jøns, J. Olsen, and U. Sidenius. 2003. "Separation, purification and identification of the major selenium metabolite from human urine by multi-dimensional HPLC-ICP-MS and APCI-MS." *J Anal At Spectrom* no. 18:65–70.

Goeger, D. E., and H. E. Ganther. 1993. "Homocysteine-dependent demethylation of trimethylselenonium ion and selenobetaine with methionine formation." *Arch Biochem Biophys* no. 302(1):222–7. doi: S0003-9861(83)71203-8.

Haratake, M., M. Hongoh, M. Miyauchi, R. Hirakawa, M. Ono, and M. Nakayama. 2008. "Albumin-mediated selenium transfer by a selenotrisulfide relay mechanism." *Inorg Chem* no. 47:6273–80.

Hesse, H., O. Kreft, S. Maimann, M. Zeh, and R. Hoefgen. 2004. "Current understanding of the regulation of methionine biosynthesis in plants." *J Exp Bot* no. 55(404):1799–808. doi: 10.1093/jxb/erh139.

Hock Ng, B., and J. W. Anderson. 1978. "Synthesis of selenocysteine by cysteine synthases from selenium accumulator and non-accumulator plants." *Phytochemistry* no. 17:2069–74.

Huang, B., J. Zhang, J. Hou, and C. Chen. 2003. "Free radical scavenging efficiency of Nano-Se *in vitro*." *Free Radic Biol Med* no. 35:805–13.

Ip, C., M. Birringer, E. Block et al. 2000. "Chemical speciation influences comparative activity of selenium-enriched garlic and yeast in mammary cancer prevention." *J Agric Food Chem* no. 48:2062–70.

Juresa, D., M. Blanusa, K. A. Francesconi, N. Kienzl, and D. Kuehnelt. 2007. "Biological availability of selenosugars in rats." *Chem Biol Interact* no. 168(3):203–10. doi: 10.1016/j.cbi.2007.04.009.

Klein, M., L. Ouerdane, M. Bueno, and F. Pannier. 2011. "Identification in human urine and blood of a novel selenium metabolite, Se-methylselenoneine, a potential biomarker of metabolization in mammals of the naturally occurring selenoneine, by HPLC coupled to electrospray hybrid linear ion trap-orbital ion trap MS." *Metallomics* no. 3(5):513–20. doi: 10.1039/c0mt00060d.

Kobayashi, Y., Y. Ogra, and K. T. Suzuki. 2001. "Speciation and metabolism of selenium injected with 82Se-enriched selenite and selenate in rats." *J Chromatogr B Biomed Sci Appl* no. 760(1):73–81.

Kobayashi, Y., Y. Ogra, K. Ishiwata, H. Takayama, N. Aimi, and K. T. Suzuki. 2002. "Selenosugars are key and urinary metabolites for selenium excretion within the required to low-toxic range." *Proc Natl Acad Sci USA* no. 99(25):15932–6.

Läuchli, A. 1993. "Selenium in plants: Uptake, functions, and environmental toxicity." *Bot Acta* no. 106:455–68.

LeDuc, D. L., M. AbdelSamie, M. Montes-Bayón, C. P. Wu, S. J. Reisinger, and N. Terry. 2006. "Overexpressing both ATP sulfurylase and selenocysteine methyltransferase enhances selenium phytoremediation traits in Indian mustard." *Environ Pollut* no. 144:70–6.

Lu, J., and A. Holmgren. 2009. "Selenoproteins." *J Biol Chem* no. 284:723–7.

Magos, L., S. K. Tandon, M. Webb, and R. Snowden. 1987. "The effects of treatment with selenite before and after the administration of [75Se]selenite on the exhalation of [75Se] dimethylselenide." *Toxicol Lett* no. 36:167–72.

McConnell, K. P., and J. L. Hoffman. 1972. "Methionine-selenomethionine parallels in rat liver polypeptide chain synthesis." *FEBS Lett* no. 24(1):60–2. doi: 0014-5793(72)80826-3 [pii].

Montes-Bayón, M., T. D. Grant, J. Meija, and J. A. Caruso. 2002. "Selenium in plants by mass spectrometric techniques: Developments in bio-analytical methods." *J Anal At Spectrom* no. 17:1015–23.

Murillo, M., and T. Leustek. 1995. "Adenosine-5′-triphosphate-sulfurylase from Arabidopsis thaliana and Escherichia coli are functionally equivalent but structurally and kinetically divergent: Nucleotide sequence of two adenosine-5′-triphosphate-sulfurylase cDNAs from Arabidopsis thaliana and analysis of a recombinant enzyme." *Arch Biochem Biophys* no. 123(1):195–204. doi: 10.1006/abbi.1995.0026.

Neuhierl, B., and A. Böck. 2002. "Selenocysteine methyltransferase." *Methods Enzymol* no. 347:203–7.

Ogra, Y., K. Ishiwata, H. Takayama, N. Aimi, and K. T. Suzuki. 2002. "Identification of a novel selenium metabolite, Se-methyl-N-acetylselenohexosamine, in rat urine by high-performance liquid chromatography—Inductively coupled plasma mass spectrometry and—Electrospray ionization tandem mass spectrometry." *J Chromatogr B Analyt Technol Biomed Life Sci* no. 767(2):301–12.

Ogra, Y., K. Ishiwata, Y. Iwashita, and K. T. Suzuki. 2005. "Simultaneous speciation of selenium and sulfur species in selenized odorless garlic (*Allium sativum* L. Shiro) and shallot (*Allium ascalonicum*) by HPLC–inductively coupled plasma-(octopole reaction system)-mass spectrometry and electrospray ionization-tandem mass spectrometry." *J Chromatogr A* no. 1093:118–25.

Ogra, Y., T. Kitaguchi, K. Ishiwata, N. Suzuki, Y. Iwashita, and K. T. Suzuki. 2007. "Identification of selenohomolanthionine in selenium-enriched Japanese pungent radish." *J Anal At Spectrom* no. 22:1390–6.

Ogra, Y., T. Kitaguchi, K. Ishiwata, N. Suzuki, T. Toida, and K. T. Suzuki. 2009. "Speciation of selenomethionine metabolites in wheat germ extract." *Metallomics* no. 1:78–86.

Ogra, Y., A. Katayama, Y. Ogihara, A. Yawata, and Y. Anan. 2013. "Analysis of animal and plant selenometabolites in roots of a selenium accumulator, Brassica rapa var. peruviridis, by speciation." *Metallomics* no. 5(5):429–36. doi: 10.1039/c2mt20187a.

Okuno, T., S. Motobayashi, H. Ueno, and K. Nakamuro. 2005. "Identification of mouse selenomethionine alpha,gamma-elimination enzyme: Cystathionine gamma-lyase catalyzes its reaction to generate methylselenol." *Biol Trace Elem Res* no. 108:245–57.

Peng, D., J. Zhang, and Q. Liu. 2007. "Effect of sodium selenosulfate on restoring activities of selenium-dependent enzymes and selenium retention compared with sodium selenite *in vitro* and *in vivo*." *Biol Trace Elem Res* no. 117(1–3):77–88. doi: 10.1007/bf02698085.

Pilon-Smits, E. A., S. Hwang, C. Mel Lytle et al. 1999. "Overexpression of ATP sulfurylase in indian mustard leads to increased selenate uptake, reduction, and tolerance." *Plant Physiol* no. 119(1):123–32.

Ravn-Haren, G., B. N. Krath, K. Overvad et al. 2008. "Effect of long-term selenium yeast intervention on activity and gene expression of antioxidant and xenobiotic metabolising enzymes in healthy elderly volunteers from the Danish Prevention of Cancer by Intervention by Selenium (PRECISE) pilot study." *Br J Nutr* no. 99(6):1190–8. doi: 10.1017/S0007114507882948.

Reeves, M. A., and P. R. Hoffmann. 2009. "The human selenoproteome: Recent insights into functions and regulation." *Cell Mol Life Sci* no. 66:2457–78.

Reid, M. E., A. J. Duffield-Lillico, E. Slate et al. 2008. "The nutritional prevention of cancer: 400 mcg per day selenium treatment." *Nutr Cancer* no. 60(2):155–63. doi: 10.1080/01635580701684856.

Rotte, C., and T. Leustek. 2000. "Differential subcellular localization and expression of ATP sulfurylase and 5′-adenylylsulfate reductase during ontogenesis of Arabidopsis leaves indicates that cytosolic and plastid forms of ATP sulfurylase may have specialized functions." *Plant Physiol* no. 124(2):715–24.

Saito, K. 2004. "Sulfur assimilatory metabolism. The long and smelling road." *Plant Physiol* no. 136(1):2443–50. doi: 10.1104/pp.104.046755.

Schrauzer, G. N. 2000. "Selenomethionine: A review of its nutritional significance, metabolism and toxicity." *J Nutr* no. 130:1653–6.

Setya, A., M. Murillo, and T. Leustek. 1996. "Sulfate reduction in higher plants: Molecular evidence for a novel 5′-adenylylsulfate reductase." *Proc Natl Acad Sci USA* no. 93(23):13383–8.

Shiobara, Y., and K. T. Suzuki. 1998. "Binding of selenium (administered as selenite) to albumin after efflux from red blood cells." *J Chromatogr B Biomed Sci Appl* no. 710(1–2):49–56.

Shiobara, Y., Y. Ogra, and K. T. Suzuki. 1999. "Speciation of metabolites of selenate in rats by HPLC-ICP-MS." *Analyst* no. 124(8):1237–41.

Soda, K., N. Esaki, T. Nakamura, N. Karai, P. Chocat, and H. Tanaka. 1984. "Selenocysteine beta-lyase: A novel pyridoxal enzyme." *Prog Clin Biol Res* no. 144A:319–28.

Sors, T. G., D. R. Ellis, G. N. Na et al. 2005a. "Analysis of sulfur and selenium assimilation in Astragalus plants with varying capacities to accumulate selenium." *Plant J* no. 42:785–931

Sors, T. G., D. R. Ellis, and D. E. Salt. 2005b. "Selenium uptake, translocation, assimilation and metabolic fate in plants." *Photosynth Res* no. 86:373–89.

Suzuki, K. T., and M. Itoh. 1997. "Metabolism of selenite labelled with enriched stable isotope in the bloodstream." *J Chromatogr B Biomed Sci Appl* no. 692(1):15–22.

Suzuki, K. T., and Y. Ogra. 2002. "Metabolic pathway for selenium in the body: Speciation by HPLC-ICP MS with enriched Se." *Food Addit Contam* no. 19(10):974–83. doi: 10.1080/02652030210153578.

Suzuki, K. T., Y. Ohta, and N. Suzuki. 2006. "Availability and metabolism of 77Se-methylseleninic acid compared simultaneously with those of three related selenocompounds." *Toxicol Appl Pharmacol* no. 217(1):51–62. doi: 10.1016/j.taap.2006.07.005.

Suzuki, K. T., Y. Tsuji, Y. Ohta, and N. Suzuki. 2008. "Preferential organ distribution of methylselenol source Se-methylselenocysteine relative to methylseleninic acid." *Toxicol Appl Pharmacol* no. 227(1):76–83. doi: 10.1016/j.taap.2007.10.001.

Tanaka, H., N. Esaki, and K. Soda. 1985. "Selenocysteine metabolism in mammals." *Curr Top Cell Regul* no. 27:487–95.

Whanger, P. D. 2002. "Selenocompounds in plants and animals and their biological significance." *J Am Coll Nutr* no. 21:223–32.

Wróbel, K., K. Wróbel, and J. A. Caruso. 2002. "Selenium speciation in low molecular weight fraction of Se-enriched yeast by HPLC-ICP-MS: Detection of selenoadenosylmethionine." *J Anal At Spectrom* no. 17:1048–54.

Yamashita, Y., and M. Yamashita. 2010. "Identification of a novel selenium-containing compound, selenoneine, as the predominant chemical form of organic selenium in the blood of bluefin tuna." *J Biol Chem* no. 285(24):18134–8. doi: 10.1074/jbc.C110.106377.

Yoshida, M., K. Iwami, and K. Yasumoto. 1984. "Determination of nutritional efficiency of selenium contained in processed skipjack meat by comparison with selenite." *J Nutr Sci Vitaminol* no. 30:395–400.

Yoshida, M., M. Abe, K. Fukunaga, and K. Kikuchi. 2002. "Bioavailability of selenium in the defatted dark muscle of tuna." *Food Additives and Contaminants* no. 19(10):990–5.

Zhang, J. S., X. Y. Gao, L. D. Zhang, and Y. P. Bao. 2001. "Biological effects of a nano red elemental selenium." *Biofactors* no. 15:27–38.

3 The Molecular Regulation of Selenocysteine Incorporation into Proteins in Eukaryotes

Aditi Dubey and Paul R. Copeland

CONTENTS

3.1 Introduction 31
3.2 Sec Insertion Process... 33
    3.2.1 Mechanism: A Recoding Event 33
    3.2.2 The Essential Factors.. 34
        3.2.2.1 SECIS Elements.. 34
        3.2.2.2 SECIS Binding Protein 2 36
        3.2.2.3 eEFSec 38
        3.2.2.4 Sec-tRNA$^{[Ser]Sec}$ 39
    3.2.3 Other Factors 39
        3.2.3.1 SECIS-Binding Protein 2L 39
        3.2.3.2 Ribosomal Protein L30.. 40
        3.2.3.3 eIF4a3 41
        3.2.3.4 Sec Redefinition Element.. 42
        3.2.3.5 Other Factors... 42
3.3 Regulation of Selenoprotein Expression.. 43
    3.3.1 Selenium Availability 43
    3.3.2 Nonsense-Mediated Decay 44
    3.3.3 Subcellular Localization.. 46
3.4 Conclusion 47
References... 48

3.1 INTRODUCTION

As an essential micronutrient, selenium is required for a multitude of biological functions. For example, insufficient selenium can lead to male infertility, and several studies have reported on the chemopreventive properties of selenium (Bellinger et al., 2009). Such widespread impact mainly manifests through the action of a limited number of selenoproteins. The requirement for Sec-containing proteins varies across organisms; 21.5% of sequenced bacteria utilize Sec, containing anywhere from one to 31 selenoproteins with *Deltaproteobacteria* and *Clostridia* genera from *Firmicutes* being selenoprotein-rich (Zhang et al., 2006). In addition to bacteria and eukaryotes, selenoproteins are also found in some archaea (Rother et al., 2001). Notably, however, many eubacteria and archaebacteria, as well as all fungi and higher plants, do not utilize Sec and lack the Sec insertion machinery. In addition, although selenoproteins are essential in mice (Bösl et al., 1997), they are dispensable in *Drosophila* (Hirosawa-Takamori et al., 2004). There are 25 reported selenoproteins in humans, and remarkably, most of them are enzymes involved in critical housekeeping functions in the cell.

The high reactivity of Sec in part explains why the expression of selenoproteins is tightly regulated. Housekeeping selenoproteins are expressed constitutively, and others are upregulated in certain conditions, such as cellular stress (Sengupta et al., 2008). Regulation of gene expression occurs at multiple levels in eukaryotes, and these can be broadly grouped by the step at which the regulation occurs: epigenetic, transcriptional, post-transcriptional, and post-translational. Most of the research in selenoprotein biology has focused on the transcriptional, post-transcriptional, and translational control of selenoprotein synthesis. Certain stimuli, such as selenium bioavailability, oxidative stress, and hypoxia, have been shown to alter selenoprotein expression, but their exact mechanism of action remains unclear. Increased ROS production has been implicated as one of the intracellular signals that leads to increased selenoprotein translation during oxidative stress (Touat-Hamici et al., 2014), but how this pathway interacts with the selenoprotein synthesis pathway is still not known. Current research on this topic aims to establish a direct link between these stimuli and the precise step of selenoprotein biosynthesis on which they act.

Apart from containing selenium in the form of Sec, another distinctive feature of selenoproteins is their highly specialized mechanism of translation. Sec is cotranslationally inserted into proteins as a result of a specific recoding event during translation elongation. A canonical UGA stop codon in the coding region gets recoded to Sec in response to a stem-loop structure in the 3′ untranslated region (3′ UTR) of the selenoprotein mRNA called the Sec insertion sequence (SECIS). In most cases, only a single UGA codon must be recoded, but the most notable exception is the mammalian Selenoprotein P (SEPP1), which contains 10 Sec codons. Other factors essential for this process include the SECIS binding protein 2 (SBP2), the specialized elongation factor eEFSec, and its highly specific binding partner, the selenocysteinylated Sec-tRNA$^{[Ser]Sec}$ (Sec-tRNA$^{[Ser]Sec}$). Recently, these were shown to be sufficient for Sec incorporation in a partially reconstituted *in vitro* system (Gupta et al., 2013). Although the basic requirements for the Sec insertion reaction have been established, factors that influence the efficiency and processivity of Sec incorporation remain to be elucidated.

The Sec incorporation machinery itself is unique. For one, eEFSec binds none of the other species of aminoacylated tRNAs, which are bound by eEF1A for canonical translation elongation. As described in Chapter 4, Sec is synthesized on its own tRNA using a group of enzymes rather than utilizing a single synthetase. Understanding the reasons for the evolution of such specialized machinery are actively investigated research avenues. It has, however, become clear that the Sec incorporation machinery itself participates in regulation of selenoprotein expression through features such as composition and spacing of the SECIS element, subcellular localization of SBP2 and eEFSec, 3′ UTR binding proteins, Sec-tRNA[Ser]Sec binding proteins, and ribosome interactions. Given the fact that selenoproteins are essential for normal cell function and viability, it is challenging to study them in *in vivo* or even in cultured environments as slight perturbations in their expression impact several cellular processes. But with careful consideration of experimental conditions and by employing a combination of biochemical, cellular, and genetic approaches, a cohesive picture of selenoprotein expression is beginning to emerge.

3.2 SEC INSERTION PROCESS

3.2.1 MECHANISM: A RECODING EVENT

A recoding event during translation in itself represents a mode of post-transcriptional regulation of gene expression and occurs in several organisms albeit at different frequencies (Baranov et al., 2002). During a recoding event, the ribosome gets programmed to incorporate an amino acid different from that specified by the codon according to the rules of the genetic code (Gesteland and Atkins, 1996). Both sense and stop codons can get recoded to generate more than one protein product for a given mRNA, but certain codons have greater "recodability" than others. In eubacteria and archaebacteria, opal (UGA) and amber (UAG) stop codons are recoded (Ivanova et al., 2014). When UGA and UAG stop codons are present in-frame, their recoding results in the incorporation of unusual amino acids Sec and pyrrolysine (Pyl), respectively. Although Pyl is found only in methanogens (Hao et al., 2002), Sec is prevalent in all three domains of life.

UGA is a versatile codon. In *Mycoplasma*, UGA can be recoded as tryptophan (Trp) due to a single tRNA[ACU] that decodes both UGA and UGG codons (Inamine et al., 1990). It was later found to code for Trp in mitochondria of certain protists (Inagaki et al., 1998) and for Cys in the ciliate *Euplotes* (Meyer et al., 1991). In all these cases, the termination codon was UAA, and the recoding was attributed either to differences in tRNA specificities or to the absence of UGA-recognizing release factors in these organisms (Bertram et al., 2001). In addition, in *E. coli*, UGA can be suppressed at low frequency by tRNA[Trp] in certain codon contexts (Engelberg-Kulka, 1981). The recoding of UGA to Sec is different from these suppression and recoding events and requires the presence of highly specific translational apparatus.

Sec insertion at an in-frame UGA requires several factors. This mechanism described applies mainly to eukaryotes, and although prokaryotic Sec incorporation is similar, there are some key differences (for a complete review of the prokaryotic Sec insertion process, see Böck, 2000). First and foremost, every eukaryotic selenoprotein mRNA contains the SECIS stem-loop structure in the 3′ untranslated region (UTR), which is one of the key features that distinguishes Sec incorporation from nonspecific stop codon read-through. An RNA-binding protein called SECIS binding protein 2 (SBP2) interacts with the SECIS element at its conserved core, and this interaction is required for Sec incorporation to occur (Copeland et al., 2000). Finally, a Sec-specific elongation factor called eEFSec delivers the charged Sec-tRNA[Ser]Sec to the ribosomal A site in response to the above factors. The mechanistic details of eEFSec/Sec-tRNA[Ser]Sec recruitment to the SECIS/SBP2 complex are unknown. As for canonical translation elongation, the delivery of Sec-tRNA[Ser]Sec to the ribosome is presumed to be dependent on GTP hydrolysis, which likely induces structural rearrangements within eEFSec domains to allow for Sec-tRNA[Ser]Sec release.

These components were recently shown to be necessary and sufficient for Sec incorporation at a single UGA in a wheat germ lysate-based *in vitro* translation system (Gupta et al., 2013). Being derived from plants, this system is completely devoid of Sec incorporation machinery, thus providing a blank slate for reconstitution of the Sec incorporation process. Although the efficiency of Sec incorporation was slightly reduced compared to the mammalian counterpart (rabbit reticulocyte lysate [RRL]), incorporation of Sec into a luciferase reporter was robustly dependent on each of the factors and the presence of a SECIS element. Interestingly, however, an attempt to use this system to incorporate the 10 Sec residues into SEPP1 resulted in the successful incorporation of only a single Sec residue, suggesting that additional factors or processes may be required for processive incorporation of multiple Sec residues (Shetty et al., 2014). This is a good example of how this Sec-naïve system can now be used to further investigate not only these basal factors involved in Sec incorporation in greater detail, but also factors influencing efficiency and processivity of the Sec insertion reaction. Although a fully reconstituted *in vitro* system for mammalian Sec incorporation is yet to be developed, this assay provides an important tool for further mechanistic studies in this field.

In eukaryotes, several conditions such as the cellular concentration of tRNA[Ser]Sec, selenium availability, and cellular stress have been shown to increase Sec incorporation at the UGA codon (Berry et al., 1994; Howard et al., 2013). In addition, studies have shown that the ratio of Sec incorporation to termination is affected by the base immediately following the UGA codon (Nasim et al., 2000; Grundner-Culemann et al., 2001). Direct competition between Sec incorporation and termination was shown in an *in vitro* translation system, in which an increasing amount of SBP2 was able to outcompete the eukaryotic release factor 1 (eRF1) and increase Sec incorporation at the Sec codon (Gupta and Copeland, 2007). There may be other unknown factors that contribute to increasing Sec incorporation efficiency and regulating the balance between Sec incorporation and termination. Development of a fully reconstituted *in vitro* system is necessary to take these studies further, and toeprinting can be utilized to directly assess the movement of the ribosome at the UGA codon in response to varying amounts of translation or termination factors and selenium levels.

3.2.2 The Essential Factors

3.2.2.1 SECIS Elements

SECIS elements are the only *cis*-acting factor required for Sec incorporation, and in eukaryotes, they are almost exclusively located in the 3′ UTR (Berry et al., 1993). SECIS elements are not highly conserved at the sequence level but have a consensus stem-loop secondary structure (Walczak et al., 1997; Latrèche et al., 2009). Based on their apical loop structure, two types of SECIS elements have been documented: Form I and Form II. Form I SECIS elements contain a large, "open" apical loop whereas Form II SECIS elements have a smaller apical loop and an internal bulge (Grundner-Culemann et al., 1999). Functional differences between Form I and Form II haven't been reported although Form II occurs more frequently (Chapple et al., 2009). Notably, the SEPP1 3′ UTR contains two SECIS elements: the proximal Form II SECIS and the distal Form I SECIS element (Stoytcheva et al., 2006). Several studies have shown only the proximal Form II SECIS element is required for Sec insertion, while the distal Form I SECIS appears dispensable (Stoytcheva et al., 2006; Fixsen and Howard, 2010; Shetty et al., 2014). Because SePP contains multiple Sec codons, it has been postulated that the two SECIS elements are required for decoding multiple Sec codons, so the SePP 3′ UTR presents a unique opportunity to study the contributions of the two types of SECIS elements to the efficiency and processivity of Sec incorporation.

Despite the lack of high nucleotide sequence conservation, the SECIS element does possess two conserved motifs that are required for Sec incorporation. First is the AUGA:AG SECIS core comprising the center of the stem, and the second is an apical AAR (R = G or A) motif (Berry et al., 1993). The core motif is essential for SBP2 binding although the role of AAR motif is still not known and remains an outstanding question in selenoprotein biology. The SECIS core is formed by a non-Watson-Crick G-A/A-G tandem pairing between the AUGA and the GA on either side (Walczak et al., 1998). This structure falls within the kink-turn (K-turn) class of RNA-binding motifs, found in rRNAs, snoRNAs, and some mRNAs (Klein et al., 2001). K-turn motifs vary in sequence and are involved in protein binding and mediating RNA tertiary structure formation.

Proteins have also been found to contain K-turn binding motifs that serve as RNA-binding domains, called L7Ae binding motifs (Koonin et al., 1994). This domain is present in several ribosomal proteins and in SBP2 (Caban et al., 2007). Interestingly, K-turns have been found to be bent in the presence of magnesium ions and are thought to exist in a dynamic two-state equilibrium in solution (Goody et al., 2004). Additionally, K-turns are flexible and can allow for structural changes in the ribosome (Rázga et al., 2004). Given that SBP2/ribosome interaction is essential for Sec incorporation (Caban et al., 2007) and that SBP2/SECIS interaction is inhibited by magnesium (Copeland and Driscoll, 1999), it is tempting to speculate that the dynamic nature of K-turns regulates the stability of SECIS/SBP2/ribosome interaction, thus potentially regulating the efficiency of Sec incorporation. The current lack of structural data about the Sec incorporation factors also presents a significant hurdle in elucidating the mechanism of Sec incorporation.

SECIS elements are central to regulation of Sec incorporation. SECIS elements bind SBP2 with varying affinities (Donovan et al., 2008; Latrèche et al., 2009) and must be placed more than 51 nucleotides downstream of the Sec-encoding UGA (Martin et al., 1996). Interestingly, the SBP2-binding affinity of a SECIS element does not directly correlate with its recoding potential. Instead, chimeric SECIS elements were used to show that intrinsic features of a SECIS element are more important in determining its efficiency (Latrèche et al., 2009) and its sensitivity to cellular selenium status (Weiss and Sunde, 1998) than to the concentration of eEFSec and SBP2 (Latrèche et al., 2012). Additionally, the role of an additional SECIS binding protein in determining this highly variable recoding efficiency was ruled out by testing SECIS elements in the wheat germ *in vitro* translation system (Gupta et al., 2013), which corroborated the above findings.

3.2.2.2   SECIS Binding Protein 2

SECIS binding protein 2 (SBP2) is another essential component of the eukaryotic Sec incorporation machinery. The importance of SBP2 in mammalian development was underscored by a recent study in which a constitutive deletion of the *SECISBP2* gene in mice caused embryonic lethality (Seeher et al., 2013). *SECISBP2* mutations in humans lead to multisystemic disorders (see Chapter 16). SBP2 was identified in rat testis as a protein that could be specifically cross-linked to GPx4 3′ UTR through the AUGA SECIS core (Lesoon et al., 1997; Copeland et al., 2000) but not to the apical loop. It was shown to stably bind ribosomes through the 28S rRNA and was required for Sec incorporation *in vitro* but not required for general translation (Copeland et al., 2001). Recently, cross-linking was used to map the SBP2 binding site on 28S rRNA to expansion segment 7L (Kossinova et al., 2014). Mammalian SBP2 also contains an N-terminal domain of no known function and a central Sec-incorporation domain (SID) and a C-terminal RNA binding domain (RBD) containing the L7Ae motif (Donovan et al., 2008). For a detailed description of SBP2 domain functions and their possible roles in Sec incorporation, see the review by Donovan and Copeland (2010b) and Chapter 16.

Attempts at structural analysis have shown that 70% of SBP2 is apparently unstructured as indicated by disorder-prediction analysis of its sequence, complemented by nuclear magnetic resonance (NMR) and sedimentation equilibrium analysis (Oliéric et al., 2009). Interestingly, the RBD was reported to be folded and functional in this study, leading to the hypothesis that in order to exert its function SBP2 needs to be in complex with its binding partners, such as SECIS and eEFSec. Such a complex was indeed demonstrated *in vitro* by gel retardation assays in which SBP2 was shown to interact with eEFSec in the presence of the SECIS element (Donovan et al., 2008; Gonzalez-Flores et al., 2012). Therefore, it is likely that a complex of eEFSec/SBP2/SECIS exists in cells, primarily to increase the local concentration of these factors so that instead of stalling at the UGA, the ribosome can proceed with Sec incorporation.

The interactions of SBP2 and the ribosome have generated considerable interest because SBP2 is thought to play a role in programming the ribosome for Sec incorporation. Using selective 2′-hydroxyl acylation analyzed by primer extension (SHAPE) analysis, SBP2 was shown to cause conformational changes in Helix 89 and an expansion segment (ES31) of 80S ribosomes (Caban and Copeland, 2012). Because this region is involved in tRNA accommodation, it is likely that SBP2 plays a key role in eEFSec-mediated delivery of Sec-tRNA[Ser]Sec to the A site of the ribosome. In another report, photoreactive groups were used in cross-linking experiments with minimal SECIS elements to show that SBP2 is bound to ribosomes that are in 48S preinitiation and 80S pretranslocation complexes (Kossinova et al., 2013). Although the latter is not surprising, as it is possible for SBP2 to bind the ribosome before Sec-tRNA[Ser]Sec has been accommodated into the A site, the idea that SBP2 may be bound to the 48S preinitiation complex is novel and would hint at the presence of a subpopulation of ribosomes preprogrammed for Sec incorporation. This is a widely debated topic in the field of translation, and this is also one of the outstanding questions in mechanistic studies of Sec incorporation that require an *in vitro* reconstituted system and crystal structures to move our understanding forward.

Far less is known about the regulation of SBP2 itself. Initial reports suggested that SBP2 is expressed ubiquitously, but not equally, and has at least three transcripts (Copeland et al., 2000; see also Chapter 16). SBP2 is highly expressed in testis and at low levels in other tissues, such as the kidney and heart (Bubenik et al., 2009). Similarly, SBP2 mRNA was detectable in all tissues tested but a smaller 2.5-kb transcript was only enriched in the testis, and this transcript was thought to lack the 3' UTR (Copeland et al., 2000). Because 3' UTRs are regions of considerable post-transcriptional regulation in eukaryotic gene expression (Matoulkova et al., 2012), regulation of SBP2 expression by RNA-binding proteins was investigated. Through cross-linking of conserved regions of SBP2 3' UTR with rabbit reticulocyte lysate, the proximal GU-rich sequence was shown to interact with CUGBP-1 and the distal region was shown to interact with HuR (Bubenik et al., 2009). Although the interactions were demonstrated to be specific, their role in selenoprotein biosynthesis remains to be explored. CUGBP-1 and HuR exist in various ribonucleoprotein complexes in the cell and participate regulating mRNA stability and translation (Abdelmohsen et al., 2008). It will be interesting to see if these proteins regulate SBP2 mRNA during conditions of limiting selenium or oxidative stress.

A later study reported the existence of at least eight splice variants in the 5' region of human *SECISBP2*, which encode 5 isoforms with varying N-terminal sequences (Papp et al., 2008). Most isoforms were cytoplasmic and could be visualized by immunofluorescence. One isoform (mtSBP2) was shown localized to the mitochondria in a small percentage of cells, but as UGA codes for Trp in the mitochondria and there are no SECIS elements (Inagaki et al., 1998), there are no mitochondrial selenoproteins. Therefore, it is difficult to surmise what the biological role for mtSBP2 may be. It is likely that it performs a function that is not related to Sec incorporation. These findings are interesting given that N-terminal domain is dispensable for Sec incorporation, and information specifying the localization of a protein is usually specified by nonconserved sequences (Azad et al., 2001).

In the same study, cellular oxidative stress in the form of UVA irradiation caused an increase in all *SECISBP2* splice variants, but the SBP2 protein product was degraded (Papp et al., 2008). So far, this is the only study to investigate the transcriptional and translational regulation of SBP2 splice variants in response to oxidative stress. There is also some evidence that SBP2 changes its subcellular localization in response to oxidative stress, and this will be discussed below. Although it is interesting that different forms of oxidative stress elicit different responses from SBP2, a comprehensive analysis of all *SECISBP2* variants needs to be performed under different oxidative stress conditions before a model can be developed for SBP2 function and regulation during oxidative stress. Thus, SBP2 seems to be regulated post-transcriptionally, but the impact of this regulation on selenoprotein biosynthesis is yet to be explored.

3.2.2.3 eEFSec

Sec incorporation is a specialized elongation cycle in which a highly specific, dedicated, Sec-specific elongation factor delivers Sec-tRNA[Ser]Sec to the ribosomal A site. eEFSec is a GTP hydrolyzing protein with a prokaryotic counterpart called SelB and is required for the cotranslational incorporation of selenocysteine (Fagegaltier et al., 2000; Tujebajeva et al., 2000). eEFSec specifically binds Sec-tRNA[Ser]Sec but not Ser-tRNA[Ser]Sec and interacts with SBP2 in a SECIS-dependent manner for Sec incorporation to occur (Fagegaltier et al., 2000; Tujebajeva et al., 2000; Zavacki et al., 2003). SBP2/eEFSec/SECIS complex formation has been demonstrated *in vitro*, but *in vivo* it has only been visualized using overexpression in transfected cells (Tujebajeva et al., 2000; Zavacki et al., 2003). Several attempts to detect a stable endogenous complex have been unsuccessful (our unpublished observation and Oliéric et al., 2009), which indicates that the complex may be highly transient and difficult to detect without stabilizing steps such as cross-linking.

Although SelB has been extensively studied, mechanistic details of eEFSec function have lagged behind. eEFSec performs the same general function of tRNA delivery during translation elongation as the canonical elongation factor eEF1A, yet there is only 35% pairwise intraspecies similarity from *Drosophila* to humans. eEFSec also has a unique C-terminal extension that is absent in eEF1A (Leibundgut et al., 2005). This unique C-terminal "Domain IV" of eEFSec was recently shown to be essential for all known functions of eEFSec in the Sec insertion reaction by alanine scanning mutagenesis (Gonzalez-Flores et al., 2012). However, the precise role of the remainder of eEFSec domains, their interaction with each other, and the molecular basis of eEFSec's incredible specificity remain unknown. Moreover, the regulation of eEFSec is a virtually unexplored topic. For detailed analysis of the functions and properties of eEFSec and SelB, see review by Gonzalez-Flores et al. (2013).

EFSec was first identified through sequence homology searches in the EST database against the *M. jannaschii* SelB amino acid sequence (Fagegaltier et al., 2000; Tujebajeva et al., 2000). eEFSec homologs were found in archaea, *Caenorhabditis elegans*, and *Drosophila melanogaster. Drosophila* eEFSec has 40% similarity to eEFSec, and deletion of the eEFSec gene in flies results in loss of selenoprotein expression, but the lifespan and oxidative stress resistance of flies lacking dEFSec are unaffected (Hirosawa-Takamori et al., 2004). This indicates that Sec incorporation is dispensable in *D. melanogaster*, and it is likely because *Drosophila* genes encoding SelH, SelD, and SelK are duplicated to encode nonselenoproteins that contain Cys or Arg residues instead of Sec as shown by in silico analysis (Castellano et al., 2001; Martin-Romero et al., 2001). A later study showed that *D. melanogaster* lacking eEFSec were less resistant to starvation than controls, and this effect appeared to be mediated by SelK (Shchedrina et al., 2011). This suggests that the Sec-containing homolog of SelK is dispensable for viability under normal conditions but is required to resist starvation-induced stress. Further studies in the dEFSec knockout (KO) can be used as a model system to study selenoprotein functions ectopically.

Knockdown of an eEFSec homolog was also used to study the role of selenoproteins in vector-pathogen-host interface in the Gulf-coast tick *Amblyomma maculatum* that were free from or infected with *Rickettsia parkeri*, which causes a mild febrile illness. Lack of eEFSec did not alter the feeding or the ability to reproduce for *Amblyomma*, but significantly reduced the antioxidant capacity of the salivary gland and reduced the transcription of some antioxidant genes in the midgut: SelM, SelN, SelS, SelX, SOD (Adamson et al., 2013). The pathogen burden for *Rickettsia* was twice as high in the salivary glands of eEFSec-lacking *Amblyomma*, indicating that selenoproteins play a role mediating oxidative stress and are important for host-pathogen interactions in the tick.

Thus, these two studies give conclusive proof that expression of selenoproteins *in vivo* is brought about by the action of eEFSec during translation elongation. It can be presumed that although selenoproteins are dispensable for normal development and lifespan of invertebrates, they are required during increased stress, such as during pathogen insult or conditions of starvation. It would also be interesting to see the effects of an eEFSec genetic deletion in the mammalian system. Although we hypothesize that the effect of constitutive eEFSec ablation should be similar to that seen with SBP2 or Sec-tRNA[Ser]Sec KOs, it is likely that we will discover new roles for this protein with conditional KO models. Additionally, eEFSec has been shown to localize the nucleus (de Jesus et al., 2006), suggesting a possible role for the protein in regulating the cytoplasmic versus nuclear pools of Sec-tRNA[Ser]Sec for active channeling between tRNA biogenesis and Sec incorporation machineries (Grosshans et al., 2000). These will be further discussed later in this chapter.

3.2.2.4   Sec-tRNA[Ser]Sec

The final essential factor in Sec incorporation is Sec-tRNA[Ser]Sec, which has its own unique mechanism of synthesis and distinct structural features. These are discussed in detail in Chapter 4.

3.2.3   OTHER FACTORS

3.2.3.1   SECIS-Binding Protein 2L

Selenoprotein expression is regulated by several other factors that are not considered essential for Sec incorporation. A SECIS-binding protein with 46% sequence identity to the C-terminus of human SPB2 was identified in BLAST searches when the SBP2 was first cloned (Copeland et al., 2000). It was named SBP2-like, or SECISBP2L, and was shown to be present as an SBP2 paralogue in vertebrates, likely generated by a whole genome duplication event early in the vertebrate lineage (Donovan and Copeland, 2009). Interestingly, SECISBP2L is present as the only SBP in invertebrate deuterostomes such as sea urchins and sea squirts, suggesting that SECISBP2L might perform the function of SBP2 in these organisms (Donovan and Copeland, 2009). According to the Human Gene Atlas, SECISBP2L is highly expressed in the central nervous system (Su et al., 2004). Discerning the function of this protein is still in its nascent stages.

Given that the Sec-incorporation domain (SID) and the RNA-binding domain (RBD) were the primary regions found to be conserved between SBP2 and SECISBP2L, it was hypothesized that SECISBP2L may direct Sec incorporation in a subset of vertebrates (Donovan and Copeland, 2009). However, the C-terminal portion of human SECISBP2L (CT-SECISBP2L) is bound to the GPx4 SECIS element weakly and was not able to direct Sec incorporation *in vitro* (Copeland et al., 2001; Donovan and Copeland, 2009). Chimeric proteins generated by swapping SBP2 and SECISBP2L SID and RBD showed that although a combination of SECISBP2L SID and SBP2 RBD is inactive for Sec incorporation in a reticulocyte lysate translation (RRL) system, 20% of wild-type CT-SBP2 activity is retained by the protein containing SBP2 SID and SECISBP2L RBD (Copeland et al., 2001; Donovan and Copeland, 2009). Interestingly, although human CT-SECISBP2L did not support Sec incorporation, CT-SECISBP2L from the worm *Capitella teleta* did. Further investigation of SECISBP2L showed that it is capable of binding a range of SECIS elements, albeit with differing and weaker affinities when compared to CT-SBP2 (Donovan and Copeland, 2012). However, this study demonstrated that rather than weak SECIS binding, it was likely the inability of the SID and RBD of SECISBP2L to form stable, SECIS-dependent interactions that contribute to its inactivity. This suggested that although mammalian SBP2 and SECISBP2L may be similar, there is no evidence that SECISBP2L plays a role in Sec incorporation *in vivo*. If this is proven true, it might imply that the mechanism of Sec incorporation may be different in deuterostomes, requiring perhaps additional steps or factors. These findings also suggest that SECISBP2L may not be entirely functionally redundant in organisms that contain SBP2 and may have regulatory functions in selenoprotein expression.

3.2.3.2 Ribosomal Protein L30

The ribosomal protein L30 (RPL30) is a 14.5 kDa large subunit ribosomal protein present in eukaryotic and some archaeal ribosomes (Vilardell et al., 2000). It binds the ribosome at the 28S rRNA through a K-turn motif that belongs to the same family of L7Ae binding motifs as SBP2, albeit with only 20% sequence similarity (Vilardell et al., 2000; Chavatte et al., 2005). Rat RPL30 was shown to bind SECIS elements *in vivo* and *in vitro* and was shown to increase UGA recoding efficiency using luciferase reporter constructs in transiently transfected McArdle 7777 (rat hepatoma) cells (Chavatte et al., 2005). In addition, this study also showed that L30 competes with SBP2 for SECIS binding, and increasing the magnesium concentration displaces SBP2, but not L30, from the SECIS. This suggests that the two proteins may favor different conformations of the SECIS element; SBP2 binds an open conformation whereas L30 may bind both open and closed forms. The fact that SBP2 also binds 28S rRNA might suggest that there may be an active exchange of the two proteins between the SECIS element and the ribosome during Sec incorporation.

In a later study, RNA footprinting was used to show that L30 and SBP2 bind to the SECIS element at both overlapping and unique residues (Bifano et al., 2013). Both SBP2 and L30 require the tandem G/A pairs in the SECIS core, but L30 also binds the 3′ side of the internal loop. Additionally, purified L30 reduced UGA recoding efficiency of a luciferase reporter in an *in vitro* RRL system, and inhibition was rescued by SBP2. This suggests that although these proteins may have distinct mechanisms of SECIS interaction, they are not able to bind simultaneously.

However, because the role of L30 is not understood, it is difficult to model how this competition regulates selenoprotein expression. Cryo-EM structures of wheat germ 80S ribosomes showed L30 to be located at the interface of small and large subunits (Halic et al., 2005). It is possible to conceive of a model in which SBP2 binds 28S rRNA to allow for the assembly of "Sec-competent" ribosomes at the UGA codon, and after the Sec insertion reaction, L30 displaces SBP2 to cause dissociation of the SBP2/SECIS complex from the ribosome to allow for canonical translation elongation to continue. Structural data is sorely needed to verify the above model and ascertain whether SBP2 is also located at the subunit interface of the ribosome by virtue of its association with 28S rRNA.

3.2.3.3 eIF4a3

eIF4a3 is an isoform of eukaryotic initiation factor 4A with RNA-dependent ATPase activity and ATP-dependent RNA helicase activity (Li et al., 1999). It is a member of the DEAD-box protein family and is a component of the exon junction complex (EJC) (Palacios et al., 2004). Additionally, it has been shown to inhibit translation in RRL and is essential for nonsense-mediated mRNA decay (NMD; Palacios et al., 2004). This 48-kDa protein was identified as a SECIS-binding protein in nuclear extracts from McArdle 7777 cells, in which it cross-linked to GPx1 but not to GPx4 SECIS elements (Budiman et al., 2009). From these studies, it appears that eIF4a3 binding may play a key role in determining the hierarchy of selenoprotein expression. An explanation may lie in the vastly different mode of SECIS interaction adopted by eIF4a3. Unlike SBP2 or L30, eIF4a3 binds SECIS elements at two sites, including the internal and apical loop, and the two distinct domains of the protein were both critical for this interaction (Budiman et al., 2011). This study suggested that a conserved uridine in the upper 3′ of the core of Form I SECIS elements plays a role in eIF4a3-SECIS complex stability and accounts for their preferential binding.

The above studies also showed that eIF4a3 prevented SBP2 from binding to the GPx1 SECIS element and inhibited recoding of UGA-containing luciferase reporter constructs containing GPx1 but not GPx4. Higher binding affinity to a particular SECIS element positively correlated with the reduction seen in Sec incorporation (Budiman et al., 2011). This selective SECIS binding suggested that eIF4a3 might play a role in establishing the hierarchy of selenoprotein expression. Indeed, experiments done in McArdle 7777 cells showed that eIF4a3 expression was upregulated in the absence of selenium, which, in turn, reduced GPx1 levels in cells while GPx4 expression was unaffected (Budiman et al., 2009). Therefore, these results indicate that selenoprotein mRNAs are sensitive to NMD when they are bound by eIF4a3, which mediates their repression when selenium is limiting. Given that eIF4a3 is a part of the EJC, which is a key regulator of mRNA stability, it is likely during conditions of limiting selenium that this factor is involved in targeting selenoprotein mRNAs for NMD. A direct link between eIF4a3 binding and NMD of selenoprotein transcripts has yet to be established, but other lines of evidence point to an important role for NMD in regulating selenoprotein expression (see below).

3.2.3.4 Sec Redefinition Element

A Sec codon redefinition element (SRE) is the only other known *cis* acting regulator of selenoprotein expression present in eukaryotes, apart from SECIS elements. SREs are phylogenetically conserved sequences present in the vicinity of the Sec codon, rather than the 3' UTR (Howard et al., 2005). Comparative sequence analysis and RNA secondary structure predictions showed the existence of a hairpin structure located 7 nucleotides 3' of the UGA codon in Selenoprotein N (SelN; Howard et al., 2005). Four missense mutations have been reported in the SRE of the SelN gene (SEPN1) that cause SEPN-1–related myopathies of varying severity (Moghadaszadeh et al., 2001). Out of these, R466Q heterozygous mutation was shown to weaken secondary structure of SelN SRE, resulting in reduced Sec incorporation efficiency and SelN mRNA abundance (Maiti et al., 2008). Although the secondary structure of SRE is not conserved, the distance from the UGA codon is. In another comparative genomics analysis for RNA secondary structures, SRE were predicted to exist in SelN and Selenoprotein T (SelT) as RNA structure involved in genetic recoding (Pedersen et al., 2006).

Although the SRE is not required for Sec incorporation, it increases the read-through efficiency of the UGA codon in SelN by 6% in the absence of a SECIS element and by 12% when both SECIS element and SRE were present (Howard et al., 2005). This was tested using a bicistronic reporter in RRL where the presence of an SRE increased SECIS-dependent read-through of UGA by twofold, and this enhancement was abolished by mutating residues in SRE that eliminate base pairing. That these read-through events were indeed incorporating Sec was later confirmed using ^{75}Se labeling (Howard et al., 2007). Interestingly, this study also showed that the SelN SRE can stimulate read-through at UAG codons in cultured mammalian cells, suggesting that SREs may regulate termination efficiency while promoting Sec incorporation as well. The mechanistic details of this process are unclear, but the presence of SREs in only a subset of selenoprotein mRNA might point to a regulatory role for them in selenoprotein expression. As SREs are reminiscent of bacterial SECIS elements, it is also proposed that they might interact with SBP2 or eEFSec to modulate Sec incorporation (Howard et al., 2007), and because there is not much sequence conservation among SREs, this modulation may be different among selenoproteins.

3.2.3.5 Other Factors

Over the years, several factors have been discovered to play a role in Sec incorporation, and only some of them are discussed above. tRNA[[Ser]Sec]-binding proteins such as SECp43 (Small-Howard et al., 2006), and SECIS-binding proteins such as nucleolin and YB1 (reviewed in Allmang et al., 2009), have been proposed to be involved in Sec incorporation, but their significance and precise role remains unclear.

It is also worthwhile to note that efficient selenoprotein expression may be regulated by eukaryotic translation initiation machinery. Experiments using bicistronic reporters in RRL demonstrated that Sec incorporation does not require a 5' cap or a 3' poly(A) tail, and constructs containing an internal ribosome entry site (IRES) are able to support Sec incorporation (Donovan and Copeland, 2010a). However, the efficiency of Sec insertion was consistently reduced without canonical translation initiation, suggesting a potential regulatory role for an initiation factor during this process.

Finally, the action of aminoglycosides (AMGs) appears to be important in influencing UGA recoding for Sec incorporation. AMGs disrupt translational fidelity in bacterial and mammalian systems by binding to the small ribosomal subunit and promoting stop codon read-through (Manuvakhova et al., 2000; Howard et al., 2004). An AMG called G418 increased the expression of endogenous GPx1 in treated cells by increasing read-through of UGA codons in a SECIS-independent manner, and this effect was more pronounced in the absence of selenium (Handy et al., 2006). In this case, the activity of GPx1 was reduced as instead of incorporating Sec, Arg was inserted at the UGA codon. Thus, AMG-mediated stop codon suppression is able to interfere with Sec insertion *in vivo*.

Also other antibiotics, such as doxycycline and chloramphenicol, are able to increase stop codon read-through and reduce Sec incorporation (Tobe et al., 2013, see also Chapter 4). Doxycycline interacts with the small ribosomal subunit to prevent the attachment of aminoacylated tRNA to the A site, and chloramphenicol binds the large ribosomal subunit and blocks peptidyl transfer. These antibiotics, along with G418, differentially affected Sec insertion and reduced the activities of TrxR1, GPx4, and GPx1 (Tobe et al., 2013). All tested selenoproteins were affected by the three antibiotics. Cys was inserted in GPx4, Arg was inserted in GPx1, and Cys and Trp were inserted in TrxR1 at the Sec codon. Thus, Sec insertion is highly sensitive to alterations in ribosomal conformation, likely because these antibiotics disrupt the ribosomal interactions of the Sec incorporation machinery. It is not clear whether selenium supplementation can be used to mediate these effects as supplementation is likely to increase pools of Sec-tRNA$^{[Ser]Sec}$, but that may not be sufficient to counter the reduced translational fidelity of the ribosome. Taken together, the studies highlight the ribosome as a crucial point of regulation during selenoprotein biosynthesis, and they point to the need for further analysis of selenoprotein expression in patients undergoing antibiotic treatment.

3.3 REGULATION OF SELENOPROTEIN EXPRESSION

3.3.1 Selenium Availability

Selenium participates in its own regulation, and selenium availability is one of the most important factors influencing selenoprotein synthesis. Although there is some evidence that selenium can bind proteins as a cofactor and that it can be nonspecifically incorporated as selenomethionine in place of methionine, the physiological role of these forms of selenium is not known (Behne and Kyriakopoulos, 2001). However, most of the selenium is likely channeled into the Sec incorporation machinery through the Sec-tRNA$^{[Ser]Sec}$ biosynthesis pathway as Sec is the main, active form of selenium present in proteins (reviewed in Hatfield and Gladyshev, 2002; Driscoll and Copeland, 2003). The impact of selenium on transcriptional regulation is largely unexplored and remains an outstanding question. There is some evidence to suggest that the Nrf2/Keap1 pathway may be responsive to selenium levels (Brigelius-Flohé and Kipp, 2013), but most selenium-mediated selenoprotein regulation is exerted post-transcriptionally.

There is a well-established hierarchy of selenoprotein expression in both animal models and cell culture that is dependent on selenium levels, but certain selenoproteins are considered "resistant" to selenium levels and remain relatively unaltered (see recent reviews in Schomburg and Schweizer, 2009; Sunde, 2012). Cellular selenium levels have been shown to directly influence synthesis and aminoacylation of Sec-tRNA[Ser]Sec (see Chapter 4). Polysomal profiling using sucrose density gradients has shown that selenium levels influence UGA recoding efficiency by altering ribosomal loading of a subset of selenoprotein mRNAs (Fletcher et al., 2000; Martin and Berry, 2001). Selenium supplementation increased the ribosomal loading of selenoprotein mRNAs, correlating with increased selenoprotein expression.

More recently, the genome-wide impact of selenium availability on selenoprotein mRNAs was analyzed through ribosomal profiling experiments. In particular, the levels of selenium correlated with the amount of aminoacylated Sec-tRNA[Ser]Sec, and there was ribosomal pausing upstream of UGA-Sec codon on transcripts known to be sensitive to selenium levels (Howard et al., 2013). This evidences provides additional support for a model in which selenium availability feeds into the Sec-tRNA[Ser]Sec biogenesis pathway, resulting in increased local concentration of active Sec-tRNA[Ser]Sec through a direct channeling mechanism between tRNA biogenesis and Sec incorporation machineries. Indeed, it has been suggested that endogenous tRNA is never dissociated from protein translation machinery (Stapulionis and Deutscher, 1995), and similar mechanisms likely operate during Sec incorporation and contribute to its efficiency and processivity *in vivo*. Recent work from our lab has demonstrated that *in vitro* processive Sec incorporation in SePP is indeed increased upon selenium supplementation, but not on supplementation with total aminoacylated tRNA (Shetty et al., 2014). Therefore, there is likely an increase in the local concentration of Sec incorporation factors due to selenium supplementation that results in processive synthesis of SePP. Whether this is the only mode of channeling operating during Sec insertion or if there are cytoskeletal proteins or synthetases that act as mediators of this process remains to be seen.

3.3.2 NONSENSE-MEDIATED DECAY

The effects of selenium availability on mRNA stability are mainly exerted through NMD. NMD is a mRNA surveillance mechanism to eliminate transcripts that contain premature termination codons (PTC), and certain selenoprotein mRNAs are considered "natural targets" for NMD (Maquat, 2004). Failure to incorporate Sec during selenium deficiency results in NMD targeting of GPx1 mRNA, but GPx4 mRNA levels are not affected (Moriarty et al., 1998; Weiss and Sunde, 1997; Weiss Sachdev and Sunde, 2001). In addition, degradation of different selenoprotein mRNAs proceeded with unequal rates during low selenium conditions (Hill et al., 1992), demonstrating that despite the presence of potential PTCs, some selenoprotein mRNAs are resistant to NMD, and others are not. NMD was shown to be dependent upon binding of EJC to the exon–exon junction as intronless GPx1 mRNA synthesized from its cDNA was not susceptible to NMD (Sun et al., 2000). Figure 3.1 shows a model of how the GPx1 mRNA becomes susceptible to NMD during conditions of low selenium.

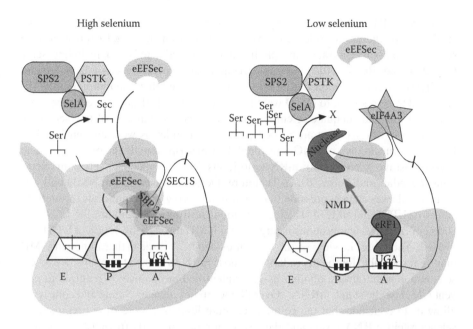

FIGURE 3.1 A model for Sec incorporation on selenium-sensitive mRNAs (e.g., GPx1) during conditions of low and high selenium levels. In conditions of high selenium, there is a higher local concentration of Sec-tRNA[Ser]Sec due to direct channeling from the tRNA biogenesis machinery to the Sec incorporation machinery. During conditions of low selenium, a limited amount of Sec-tRNA[Ser]Sec is present, which triggers mRNA degradation for those mRNAs that are sensitive to NMD.

The Sec codon of *GPx1* is located 105 nucleotides upstream of the only intron in the gene, and moving this intron demonstrated that NMD occurs when the UGA-Sec codon or a UAA codon is located ≥59 bp upstream of the intron but not when located either ≤43 bp upstream or downstream of the intron (Sun et al., 2000). In a broader context, if a PTC is present ~50–55 nucleotides upstream of the last exon–exon junction, the mRNA is targeted for rapid NMD (Maquat, 2004). Based on this criteria, 14 of the 25 selenoprotein mRNAs in humans were predicted to be sensitive to NMD, and 11 were predicted to be resistant (Squires et al., 2007). However, this study used SBP2 knockdowns to assess NMD sensitivity, but a conclusive model did not emerge from the data. A conditional deletion of *SECISBP2* in mouse livers showed that the absence of SBP2 reduces the levels of several selenoprotein mRNAs but that there was no correlation between the loss of an mRNA and its predicted ability to act as an NMD substrate or the type of SECIS element it contained (Seeher et al., 2013). However, because these experiments were not done in the context of altered selenium levels, whether or not SBP2 regulates selenoprotein mRNA stability cannot be determined.

Additionally, NMD factors UPF1 and SMG1 were recently investigated for their role in degradation of selenoprotein mRNAs during selenium deficiency (Seyedali and Berry, 2014). UPF1 is an RNA helicase that forms part of the complex that recognizes a PTC and, when phosphorylated by SMG-1, triggers NMD (Kashima et al., 2006). Increasing selenium levels resulted in an increased expression of a subset of selenoprotein transcripts predicted to be sensitive to NMD, including SelW (Seyedali and Berry, 2014). Immunoprecipitation experiments conducted under varying selenium conditions with UPF1 as a bait showed that selenoprotein mRNAs predicted to be resistant to NMD were associated with UPF1 just as well under conditions of low and adequate selenium. However, under conditions of low selenium, selenoprotein mRNAs that are responsive to selenium levels were also enriched on UPF1, and knockdown of SMG-1 restored the transcript abundance of selenium and NMD-sensitive selenoprotein mRNAs to levels found in adequate selenium conditions (Seyedali and Berry, 2014). Therefore, factors involved in the NMD pathway regulate selenoprotein expression in response to varying selenium levels through the action of UPF1 and SMG-1. During conditions of limiting selenium, the UGA-Sec codon of certain selenoproteins is targeted for NMD through the action of UPF1 and SMG-1. Therefore, in view of the current evidence, it is clear that although NMD may be responsible for the degradation of those mRNAs known to be sensitive to low selenium concentrations, another as-yet unidentified mechanism is required to prevent NMD on the stable mRNAs. Overall, the studies done by Sunde and colleagues allow us to conclude that the cellular selenium level is a more direct regulator of selenoprotein mRNA levels and their susceptibility to NMD (reviewed by Sunde, 2012), but it is more likely a complex phenomenon requiring coordinated action of several factors that regulate selenoprotein expression. Genome-wide studies need to be conducted in the presence and absence of selenium to adequately assess the role of other factors.

3.3.3   SUBCELLULAR LOCALIZATION

In eukaryotes, the subcellular localization of biological molecules such as protein and RNAs provides a context for their physiological function and determines access to binding partners and post-translational modification machinery (Hung and Link, 2011). Subcellular localization is dynamic and frequently changes in response to cues from the cellular environment. There has been some evidence to suggest that selenium levels and oxidative stress play a role in altering the subcellular localization of the Sec incorporation machinery.

The localization of SBP2 has been the most studied out of all components of Sec incorporation machinery. Transfection of tagged versions of SBP2 have demonstrated the protein to be mainly cytoplasmic (Copeland et al., 2000; Papp et al., 2006). The presence of putative nuclear localization signal (NLS) sequences in SBP2 indicated that the protein maybe capable of nuclear-cytoplasmic shuttling (de Jesus et al., 2006). Experiments using the minimum functional fragment of rat SBP2 (aa 399-777) demonstrated that it shuttles between the two compartments and the amounts of nuclear versus cytoplasmic protein varied by the cell line used (de Jesus et al., 2006). Papp et al. later showed that full-length SBP2 is also capable of nuclear-cytoplasmic shuttling, and oxidative stress altered the localization of the protein to predominantly nuclear. In addition, the study also showed the presence of a functional NLS and nuclear export signal (NES) in SBP2 that functions via the CRM1 pathway (Papp et al., 2008). A C-terminal redox-sensitive, cysteine-rich domain (CRD) was identified within SBP2 comprised of residues that could be reversibly oxidized, and it was proposed that oxidative stress promotes the nuclear localization of SBP2 potentially by modification of CRD residues that cause the masking of the NES. Nuclear localization of SBP2 caused a reduction in selenoprotein synthesis, leading to several hypotheses about the role of nuclear localization in regulating selenoprotein expression; see also Chapter 16.

Given that oxidized SBP2 does not bind SECIS elements (Copeland and Driscoll, 1999), it is difficult to conceive of a Sec incorporation complex that assembles in the nucleus. Additionally, oxidative stress has been shown to result in an upregulation of selenoprotein synthesis in cells (Touat-Hamici et al., 2014), so at least some fraction of SBP2 must be present in the cytoplasm to carry out Sec incorporation. Another possibility is that SBP2 has other mRNA targets apart from SECIS elements and participates in transcriptional regulation. In this scenario, the localization of SBP2 to the nucleus might result in increased transcription of selenoprotein mRNAs, which would then be exported to the cytoplasm for their translation involving a separate, cytoplasmic pool of SBP2. Such a model could also explain the reduced abundance of selenoprotein mRNAs seen by Seeher et al. (2013) in *Secisbp2* KO mice. A direct assessment of SBP2-bound transcripts in both nuclear and cellular compartments under varying selenium conditions is needed to address these hypotheses. As of now, the role of nuclear localization of SBP2 in selenoprotein synthesis is an open question.

eEFSec is also a nuclear-cytoplasmic shuttling protein and has been detected in both compartments (de Jesus et al., 2006). SBP2 colocalized with eEFSec in cells, and the extent of its shuttling varied between cell lines, much like SBP2 (de Jesus et al., 2006). This study identified a functional NLS in the C-terminus of eEFSec and a functional NES in the more conserved N-terminal region of the protein. Mutations in the NLS sequence abrogated nuclear localization of eEFSec, and deleting the NES-containing N-terminal domain resulted in predominantly nuclear localization (de Jesus et al., 2006). The C-terminus of eEFSec was previously shown to be sufficient for Sec incorporation in transfected cells (Zavacki et al., 2003), but the direct impact of changes in eEF-Sec localization on selenoprotein expression has not been shown. Other studies have shown the nuclear presence of factors involved in generating mature Sec-tRNA[Ser]Sec (Small-Howard et al., 2006). It is possible that eEFSec in the nucleus acts to channel the mature tRNA from the tRNA biogenesis machinery to Sec incorporation machinery by interacting with Sec-tRNA[Ser]Sec in the nucleus. If indeed this model is true, then in conditions of selenium depletion, a clear reduction in eEFSec/Sec-tRNA[Ser]Sec interaction should be observed. Much work needs to be done to reconcile the nuclear roles of Sec incorporation factors and their relevance to selenoprotein synthesis.

3.4 CONCLUSION

Selenocysteine became the dogma-breaking expansion of genetic code when it was first discovered. From traditional biochemical studies to genetic models and genome-wide analyses, our knowledge of the 21st amino acid has increased considerably over the years. The varying roles of selenoproteins, the specialized mechanism of Sec incorporation and its regulation by several factors, all point to the crucial importance of selenium and selenoproteins in normal human physiology and health. The intricately complex process required for Sec incorporation introduces a multitude of potential regulatory points into the production of selenoproteins. Considering that both the absence and overabundance of selenoproteins have deleterious effects on human health, it is perhaps not surprising that the system is highly regulated. Having spent the last two decades deciphering the basic mechanism of Sec incorporation, we are at the beginning of a new era in the study of selenoprotein biosynthesis—one that promises to decipher the multitude of regulatory mechanisms that control selenium utilization, many of which are likely affected by disease and some of which may be manipulable with as-yet undiscovered ways to modify the system.

REFERENCES

Abdelmohsen, K., Kuwano, Y., Kim, H.H., and Gorospe, M. 2008. "Posttranscriptional gene regulation by RNA-binding proteins during oxidative stress: Implications for cellular senescence." *Biol Chem* no. 389 (3):243–55.

Adamson, S.W., Browning, R.E., Budachetri, K., Ribeiro, J.M., and Karim, S. 2013. "Knockdown of selenocysteine-specific elongation factor in Amblyomma maculatum alters the pathogen burden of Rickettsia parkeri with epigenetic control by the Sin3 histone deacetylase corepressor complex." *PLoS One* no. 8 (11):e82012.

Allmang, C., Wurth, L., and Krol, A. 2009. "The selenium to selenoprotein pathway in eukaryotes: More molecular partners than anticipated." *Biochim Biophys Acta* no. 1790 (11):1415–23.

Azad, A.K., Stanford, D.R., Sarkar, S., and Hopper, A.K. 2001. "Role of nuclear pools of aminoacyl-tRNA synthetases in tRNA nuclear export." *Mol Biol Cell* no. 12 (5):1381–92.

Baranov, P.V., Gesteland, R.F., and Atkins, J.F. 2002. "Recoding: Translational bifurcations in gene expression." *Gene* no. 286 (2):187–201.

Behne, D., and Kyriakopoulos, A. 2001. "Mammalian selenium-containing proteins." *Annu Rev Nutr* no. 21:453–73.

Bellinger, F.P., Raman, A.V., Reeves, M.A., and Berry, M.J. 2009. "Regulation and function of selenoproteins in human disease." *Biochem J* no. 422 (1):11–22.

Berry, M.J., Banu, L., Harney, J.W., and Larsen, P.R. 1993. "Functional characterization of the eukaryotic SECIS elements which direct selenocysteine insertion at UGA codons." *EMBO J* no. 12 (8):3315–22.

Berry, M.J., Harney, J.W., Ohama, T., and Hatfield, D.L. 1994. "Selenocysteine insertion or termination: Factors affecting UGA codon fate and complementary anticodon:codon mutations." *Nucleic Acids Res* no. 22 (18):3753–9.

Bertram, G., Innes, S., Minella, O., Richardson, J., and Stansfield, I. 2001. "Endless possibilities: Translation termination and stop codon recognition." *Microbiology* no. 147 (2):255–69.

Bifano, A.L., Atassi, T., Ferrara, T., and Driscoll, D.M. 2013. "Identification of nucleotides and amino acids that mediate the interaction between ribosomal protein L30 and the SECIS element." *BMC Mol Biol* no. 14 (1):12.

Böck, A. 2000. "Biosynthesis of selenoproteins—An overview." *Biofactors* no. 11 (1–2):77–8.

Bösl, M.R., Takaku, K., Oshima, M., Nishimura, S., and Taketo, M.M. 1997. "Early embryonic lethality caused by targeted disruption of the mouse selenocysteine tRNA gene (Trsp)." *Proc Natl Acad Sci U S A* no. 94 (11):5531–4.

Brigelius-Flohé, R., and Kipp, A.P. 2013. "Selenium in the redox regulation of the Nrf2 and the Wnt pathway." *Methods Enzymol* no. 527:65–86.

Bubenik, J.L., Ladd, A.N., Gerber, C.A., Budiman, M.E., and Driscoll, D.M. 2009. "Known turnover and translation regulatory RNA-binding proteins interact with the 3′ UTR of SECIS-binding protein 2." *RNA Biol* no. 6 (1):73–83.

Budiman, M.E., Bubenik, J.L., Miniard, A.C. et al. 2009. "Eukaryotic initiation factor 4a3 is a selenium-regulated RNA-binding protein that selectively inhibits selenocysteine incorporation." *Mol Cell* no. 35 (4):479–89.

Budiman, M.E., Bubenik, J.L., and Driscoll, D.M. 2011. "Identification of a signature motif for the eIF4a3-SECIS interaction." *Nucleic Acids Res* no. 39 (17):7730–9.

Caban, K., and Copeland, P.R. 2012. "Selenocysteine insertion sequence (SECIS)-binding protein 2 alters conformational dynamics of residues involved in tRNA accommodation in 80 S ribosomes." *J Biol Chem* no. 287 (13):10664–73.

Caban, K., Kinzy, S.A., and Copeland, P.R. 2007. "The L7Ae RNA binding motif is a multifunctional domain required for the ribosome-dependent Sec incorporation activity of Sec insertion sequence binding protein 2." *Mol Cell Biol* no. 27 (18):6350–60.

Castellano, S., Morozova, N., Morey, M. et al. 2001. "In silico identification of novel selenoproteins in the Drosophila melanogaster genome." *EMBO Reports* no. 2 (8):697–702.

Chapple, C.E., Guigó, R., and Krol, A. 2009. "SECISaln, a web-based tool for the creation of structure-based alignments of eukaryotic SECIS elements." *Bioinformatics* no. 25 (5): 674–5.

Chavatte, L., Brown, B.A., and Driscoll, D.M. 2005. "Ribosomal protein L30 is a component of the UGA-selenocysteine recoding machinery in eukaryotes." *Nat Struct Mol Biol* no. 12 (5):408–16.

Copeland, P.R., and Driscoll, D.M. 1999. "Purification, redox sensitivity, and RNA binding properties of SECIS-binding protein 2, a protein involved in selenoprotein biosynthesis." *J Biol Chem* no. 274 (36):25447–54.

Copeland, P.R., Fletcher, J.E., Carlson, B.A., Hatfield, D.L., and Driscoll, D.M. 2000. "A novel RNA binding protein, SBP2, is required for the translation of mammalian selenoprotein mRNAs." *EMBO J* no. 19 (2):306–14.

Copeland, P.R., Stepanik, V.A., and Driscoll, D.M. 2001. "Insight into mammalian selenocysteine insertion: Domain structure and ribosome binding properties of Sec insertion sequence binding protein 2." *Mol Cell Biol* no. 21 (5):1491–8.

de Jesus, L.A., Hoffmann, P.R., Michaud, T. et al. 2006. "Nuclear assembly of UGA decoding complexes on selenoprotein mRNAs: A mechanism for eluding nonsense-mediated decay?" *Mol Cell Biol* no. 26 (5):1795–805.

Donovan, J., and Copeland, P.R. 2009. "Evolutionary history of selenocysteine incorporation from the perspective of SECIS binding proteins." *BMC Evol Biol* no. 9 (1):229. doi: 10.1/86/1471-2148-9-229.

Donovan, J., and Copeland, P.R. 2010a. "The efficiency of selenocysteine incorporation is regulated by translation initiation factors." *J Mol Biol* no. 400 (4):659–64.

Donovan, J., and Copeland, P.R. 2010b. "Threading the needle: Getting selenocysteine into proteins." *Antioxid Redox Signal* no. 12 (7):881–92.

Donovan, J., and Copeland, P.R. 2012. "Selenocysteine insertion sequence binding protein 2L is implicated as a novel post-transcriptional regulator of selenoprotein expression." *PLoS One* no. 7 (4):e35581.

Donovan, J., Caban, K., Ranaweera, R., Gonzales-Flores, J.N., and Copeland, P.R. 2008. "A novel protein domain induces high affinity selenocysteine insertion sequence binding and elongation factor recruitment." *J Biol Chem* no. 283 (50):35129–39.

Driscoll, D.M., and Copeland, P.R. 2003. "Mechanism and regulation of selenoprotein synthesis." *Annu Rev Nutr* no. 23:17–40.

Engelberg-Kulka, H. 1981. "UGA suppression by normal tRNA Trp in Escherichia coli: Codon context effects." *Nucleic Acids Res* no. 9 (4):983–91.

Fagegaltier, D., Hubert, N., Yamada, K., Mizutani, T., Carbon, P., and Krol, A. 2000. "Characterization of mSelB, a novel mammalian elongation factor for selenoprotein translation." *EMBO J* no. 19 (17):4796–805.

Fixsen, S.M., and Howard, M.T. 2010. "Processive selenocysteine incorporation during synthesis of eukaryotic selenoproteins." *J Mol Biol* no. 399 (3):385–96.

Fletcher, J.E., Copeland, P.R., and Driscoll, D.M. 2000. "Polysome distribution of phospholipid hydroperoxide glutathione peroxidase mRNA: Evidence for a block in elongation at the UGA/selenocysteine codon." *RNA* no. 6 (11):1573–84.

Gesteland, R.F., and Atkins, J.F. 1996. "Recoding: Dynamic reprogramming of translation." *Annu Rev Biochem* no. 65:741–68.

Gonzalez-Flores, J.N., Gupta, N., Demong, L.W., and Copeland, P.R. 2012. "The selenocysteine-specific elongation factor contains a novel and multi-functional domain." *J Biol Chem* no. 287 (46):38936–45.

Gonzalez-Flores, J.N., Shetty, S.P., Dubey, A., and Copeland, P.R. 2013. "The molecular biology of selenocysteine." *BioMolecular Concepts* no. 4 (4):349–65.

Goody, T.A., Melcher, S.E., Norman, D.G., and Lilley, D.M. 2004. "The kink-turn motif in RNA is dimorphic, and metal ion-dependent." *RNA* no. 10 (2):254–64.

Grosshans, H., Hurt, E., and Simos, G. 2000. "An aminoacylation-dependent nuclear tRNA export pathway in yeast." *Genes Dev* no. 14 (7):830–40.

Grundner-Culemann, E., Martin, G.W., Harney, J.W., and Berry, M.J. 1999. "Two distinct SECIS structures capable of directing selenocysteine incorporation in eukaryotes." *RNA* no. 5 (5):625–35.

Grundner-Culemann, E., Martin, G.W., Tujebajeva, R., Harney, J.W., and Berry, M.J. 2001. "Interplay between termination and translation machinery in eukaryotic selenoprotein synthesis." *J Mol Biol* no. 310 (4):699–707.

Gupta, M., and Copeland, P.R. 2007. "Functional analysis of the interplay between translation termination, selenocysteine codon context, and selenocysteine insertion sequence-binding protein 2." *J Biol Chem* no. 282 (51):36797–807.

Gupta, N., Demong, L.W., Banda, S., and Copeland, P.R. 2013. "Reconstitution of selenocysteine incorporation reveals intrinsic regulation by SECIS elements." *J Mol Biol* no. 425 (14):2415–22.

Halic, M., Becker, T., Frank, J., Spahn, C.M., and Beckmann, R. 2005. "Localization and dynamic behavior of ribosomal protein L30e." *Nat Struct Mol Biol* no. 12 (5):467–8.

Handy, D.E., Hang, G., Scolaro, J. et al. 2006. "Aminoglycosides decrease glutathione peroxidase-1 activity by interfering with selenocysteine incorporation." *J Biol Chem* no. 281 (6):3382–8.

Hao, B., Gong, W., Ferguson, T.K., James, C.M., Krzycki, J.A., and Chan, M.K. 2002. "A new UAG-encoded residue in the structure of a methanogen methyltransferase." *Science* no. 296 (5572):1462–6.

Hatfield, D.L., and Gladyshev, V.N. 2002. "How selenium has altered our understanding of the genetic code." *Mol Cell Biol* no. 22 (11):3565–76.

Hill, K.E., Lyons, P.R., and Burk, R.F. 1992. "Differential regulation of rat liver selenoprotein mRNAs in selenium deficiency." *Biochem Biophys Res Commun* no. 185 (1):260–3.

Hirosawa-Takamori, M., Chung, H.R., and Jäckle, H. 2004. "Conserved selenoprotein synthesis is not critical for oxidative stress defence and the lifespan of Drosophila." *EMBO Rep* no. 5 (3):317–22.

Howard, M.T., Anderson, C.B., Fass, U. et al. 2004. "Readthrough of dystrophin stop codon mutations induced by aminoglycosides." *Ann Neurol* no. 55 (3):422–6.

Howard, M.T., Aggarwal, G., Anderson, C.B., Khatri, S., Flanigan, K.M., and Atkins, J.F. 2005. "Recoding elements located adjacent to a subset of eukaryal selenocysteine-specifying UGA codons." *EMBO J* no. 24 (8):1596–607.

Howard, M.T., Moyle, M.W., Aggarwal, G., Carlson, B.A., and Anderson, C.B. 2007. "A recoding element that stimulates decoding of UGA codons by Sec tRNA[Ser]Sec." *RNA* no. 13 (6):912–20.

Howard, M.T., Carlson, B.A., Anderson, C.B., and Hatfield, D.L. 2013. "Translational redefinition of UGA codons is regulated by selenium availability." *J Biol Chem* no. 288 (27):19401–13.

Hung, M.-C., and Link, W. 2011. "Protein localization in disease and therapy." *J Cell Sci* no. 124 (20):3381–92.

Inagaki, Y., Ehara, M., Watanabe, K.I., Hayashi-Ishimaru, Y., and Ohama, T. 1998. "Directionally evolving genetic code: The UGA codon from stop to tryptophan in mitochondria." *J Mol Evol* no. 47 (4):378–84.

Inamine, J.M., Ho, K.C., Loechel, S., and Hu, P.C. 1990. "Evidence that UGA is read as a tryptophan codon rather than as a stop codon by Mycoplasma pneumoniae, Mycoplasma genitalium, and Mycoplasma gallisepticum." *J Bacteriol* no. 172 (1):504–6.

Ivanova, N.N., Schwientek, P., Tripp, H.J. et al. 2014. "Stop codon reassignments in the wild." *Science* no. 344 (6186):909–13.

Kashima, I., Yamashita, A., Izumi, N. et al. 2006. "Binding of a novel SMG-1-Upf1-eRF1-eRF3 complex (SURF) to the exon junction complex triggers Upf1 phosphorylation and nonsense-mediated mRNA decay." *Genes Dev* no. 20 (3):355–67.

Klein, D.J., Schmeing, T.M., Moore, P.B., and Steitz, T.A. 2001. "The kink-turn: A new RNA secondary structure motif." *EMBO J* no. 20 (15):4214–21.

Koonin, E.V., Bork, P., and Sander, C. 1994. "A novel RNA-binding motif in omnipotent suppressors of translation termination, ribosomal proteins and a ribosome modification enzyme?" *Nucleic Acids Res* no. 22 (11):2166–7.

Kossinova, O., Malygin, A., Krol, A., and Karpova, G. 2013. "A novel insight into the mechanism of mammalian selenoprotein synthesis." *RNA* no. 19 (8):1147–58.

Kossinova, O., Malygin, A., Krol, A., and Karpova, G. 2014. "The SBP2 protein central to selenoprotein synthesis contacts the human ribosome at expansion segment 7L of the 28S rRNA." *RNA* no. 20 (7):1046–56.

Latrèche, L., Jean-Jean, O., Driscoll, D.M., and Chavatte, L. 2009. "Novel structural determinants in human SECIS elements modulate the translational recoding of UGA as selenocysteine." *Nucleic Acids Res* no. 37 (17):5868–80.

Latrèche, L., Duhieu, S., Touat-Hamici, Z., Jean-Jean, O., and Chavatte, L. 2012. "The differential expression of glutathione peroxidase 1 and 4 depends on the nature of the SECIS element." *RNA Biol* no. 9 (5):681–90.

Leibundgut, M., Frick, C., Thanbichler, M., Böck, A., and Ban, N. 2005. "Selenocysteine tRNA-specific elongation factor SelB is a structural chimaera of elongation and initiation factors." *EMBO J* no. 24 (1):11–22.

Lesoon, A., Mehta, A., Singh, R., Chisolm, G.M., and Driscoll, D.M. 1997. "An RNA-binding protein recognizes a mammalian selenocysteine insertion sequence element required for cotranslational incorporation of selenocysteine." *Mol Cell Biol* no. 17 (4):1977–85.

Li, Q., Imataka, H., Morino, S. et al. 1999. "Eukaryotic translation initiation factor 4AIII (eIF4AIII) is functionally distinct from eIF4AI and eIF4AII." *Mol Cell Biol* no. 19 (11):7336–46.

Maiti, B., Arbogast, S., Allamand, V. et al. 2008. "A mutation in the SEPN1 selenocysteine redefinition element (SRE) reduces selenocysteine incorporation and leads to SEPN1-related myopathy." *Hum Mutat* no. 30:411–416.

Manuvakhova, M., Keeling, K., and Bedwell, D.M. 2000. "Aminoglycoside antibiotics mediate context-dependent suppression of termination codons in a mammalian translation system." *RNA* no. 6 (7):1044–55.

Maquat, L.E. 2004. "Nonsense-mediated mRNA decay: Splicing, translation and mRNP dynamics." *Nat Rev Mol Cell Biol* no. 5 (2):89–99.

Martin, G.W., and Berry, M.J. 2001. "Selenocysteine codons decrease polysome association on endogenous selenoprotein mRNAs." *Genes Cells* no. 6 (2):121–9.

Martin, G.W., Harney, J.W., and Berry, M.J. 1996. "Selenocysteine incorporation in eukaryotes: Insights into mechanism and efficiency from sequence, structure, and spacing proximity studies of the type 1 deiodinase SECIS element." *RNA* no. 2 (2):171–82.

Martin-Romero, F.J., Kryukov, G.V., Lobanov, A.V. et al. 2001. "Selenium metabolism in Drosophila: Selenoproteins, selenoprotein mRNA expression, fertility, and mortality." *J Biol Chem* no. 276 (32):29798–29804.

Matoulkova, E., Michalova, E., Vojtesek, B., and Hrstka, R. 2012. "The role of the 3′ untranslated region in post-transcriptional regulation of protein expression in mammalian cells." *RNA Biol* no. 9 (5):563–76.

Meyer, F., Schmidt, H.J., Plümper, E. et al. 1991. "UGA is translated as cysteine in pheromone 3 of Euplotes octocarinatus." *Proc Natl Acad Sci U S A* no. 88 (9):3758–61.

Moghadaszadeh, B., Petit, N., Jaillard, C. et al. 2001. "Mutations in SEPN1 cause congenital muscular dystrophy with spinal rigidity and restrictive respiratory syndrome." *Nat Genet* no. 29 (1):17–8.

Moriarty, P.M., Reddy, C.C., and Maquat, L.E. 1998. "Selenium deficiency reduces the abundance of mRNA for Se-dependent glutathione peroxidase 1 by a UGA-dependent mechanism likely to be nonsense codon-mediated decay of cytoplasmic mRNA." *Mol Cell Biol* no. 18 (5):2932–9.

Nasim, M.T., Jaenecke, S., Belduz, A., Kollmus, H., Flohé, L., and McCarthy, J.E. 2000. "Eukaryotic selenocysteine incorporation follows a nonprocessive mechanism that competes with translational termination." *J Biol Chem* no. 275 (20):14846–52.

Oliéric, V., Wolff, P., Takeuchi, A. et al. 2009. "SECIS-binding protein 2, a key player in selenoprotein synthesis, is an intrinsically disordered protein." *Biochimie* no. 91 (8):1003–9.

Palacios, I.M., Gatfield, D., St Johnston, D., and Izaurralde, E. 2004. "An eIF4AIII-containing complex required for mRNA localization and nonsense-mediated mRNA decay." *Nature* no. 427 (6976):753–7.

Papp, L.V., Lu, J., Striebel, F., Kennedy, D., Holmgren, A., and Khanna, K.K. 2006. "The redox state of SECIS binding protein 2 controls its localization and selenocysteine incorporation function." *Mol Cell Biol* no. 26 (13):4895–910.

Papp, L.V., Wang, J., Kennedy, D. et al. 2008. "Functional characterization of alternatively spliced human SECISBP2 transcript variants." *Nucleic Acids Res* no. 36 (22):7192–206.

Pedersen, J.S., Bejerano, G., Siepel, A. et al. 2006. "Identification and classification of conserved RNA secondary structures in the human genome." *PLoS Comput Biol* no. 2 (4) e33.

Rázga, F., Spackova, N., Réblova, K., Koca, J., Leontis, N.B., and Sponer, J. 2004. "Ribosomal RNA kink-turn motif—A flexible molecular hinge." *J Biomol Struct Dyn* no. 22 (2):183–94.

Rother, M., Resch, A., Wilting, R., and Böck, A. 2001. "Selenoprotein synthesis in archaea." *Biofactors* no. 14 (1–4):75–83.

Schomburg, L., and Schweizer, U. 2009. "Hierarchical regulation of selenoprotein expression and sex-specific effects of selenium." *Biochem Biophys Acta* no. 1790 (11):1453–1462.

Seeher, S., Atassi, T., Mahdi, Y. et al. 2013. "Secisbp2 is essential for embryonic development and enhances selenoprotein expression." *Antioxid Redox Signal* no. 21 (6):835–49.

Sengupta, A., Carlson, B.A., Hoffmann, V.J., Gladyshev, V.N., and Hatfield, D.L. 2008. "Loss of housekeeping selenoprotein expression in mouse liver modulates lipoprotein metabolism." *Biochem Biophys Res Commun* no. 365 (3):446–52.

Seyedali, A., and Berry, M.J. 2014. "Nonsense-mediated decay factors are involved in the regulation of selenoprotein mRNA levels during selenium deficiency." *RNA* no. 20 (8):1248–56.

Shchedrina, V.A., Kabil, H., Vorbruggen, G. et al. 2011. "Analyses of fruit flies that do not express selenoproteins or express the mouse selenoprotein, methionine sulfoxide reductase B1, reveal a role of selenoproteins in stress resistance." *J Biol Chem* no. 286 (34):29449–61.

Shetty, S.P., Shah, R., and Copeland, P.R. 2014. "Regulation of selenocysteine incorporation into the selenium transport protein, Selenoprotein P." *J Biol Chem* no. 289 (36):25317–26.

Small-Howard, A., Morozova, N., Stoytcheva, Z. et al. 2006. "Supramolecular complexes mediate selenocysteine incorporation *in vivo*." *Mol Cell Biol* no. 26 (6):2337–46.

Squires, J.E., Stoytchev, I., Forry, E.P., and Berry, M.J. 2007. "SBP2 binding affinity is a major determinant in differential selenoprotein mRNA translation and sensitivity to nonsense-mediated decay." *Mol Cell Biol* no. 27 (22):7848–55.

Stapulionis, R., and Deutscher, M.P. 1995. "A channeled tRNA cycle during mammalian protein synthesis." *Proc Natl Acad Sci U S A* no. 92 (16):7158–61.

Stoytcheva, Z., Tujebajeva, R.M., Harney, J.W., and Berry, M.J. 2006. "Efficient incorporation of multiple selenocysteines involves an inefficient decoding step serving as a potential translational checkpoint and ribosome bottleneck." *Mol Cell Biol* no. 26 (24):9177–84.

Su, A.I., Wiltshire, T., Batalov, S. et al. 2004. "A gene atlas of the mouse and human protein-encoding transcriptomes." *Proc Natl Acad Sci U S A* no. 101 (16):6062–7.

Sun, X., Moriarty, P.M., and Maquat, L.E. 2000. "Nonsense-mediated decay of glutathione peroxidase 1 mRNA in the cytoplasm depends on intron position." *EMBO J* no. 19 (17):4734–44.

Sunde, R.A. 2012. "Selenoproteins: Hierarchy, requirements, and biomarkers." In *Selenium: Its molecular biology and role in human health*. (D.L. Hatfield, M.J. Berry, V.N. Galdyshev, eds.) 137–152. New York: Springer Science+Business Media, LLC.

Tobe, R., Naranjo-Suarez, S., Everley, R.A. et al. 2013. "High error rates in selenocysteine insertion in mammalian cells treated with the antibiotic doxycycline, chloramphenicol, or geneticin." *J Biol Chem* no. 288 (21):14709–15.

Touat-Hamici, Z., Legrain, Y., Bulteau, A.L., and Chavatte, L. 2014. "Selective up-regulation of human selenoproteins in response to oxidative stress." *J Biol Chem* no. 289 (21):14750–61.

Tujebajeva, R.M., Copeland, P.R., Xu, X.M. et al. 2000. "Decoding apparatus for eukaryotic selenocysteine insertion." *EMBO Rep* no. 1 (2):158–63.

Vilardell, J., Yu, S.J., and Warner, J.R. 2000. "Multiple functions of an evolutionarily conserved RNA binding domain." *Mol Cell* no. 5 (4):761–6.

Walczak, R., Hubert, N., Carbon, P., and Krol, A. 1997. "Solution structure of SECIS, the mRNA element required for eukaryotic selenocysteine insertion—Interaction studies with the SECIS-binding protein SBP." *Biomed Environ Sci* no. 10 (2–3):177–81.

Walczak, R., Carbon, P., and Krol, A. 1998. "An essential non-Watson-Crick base pair motif in 3′ UTR to mediate selenoprotein translation." *RNA* no. 4 (1):74–84.

Weiss, S.L., and Sunde, R.A. 1997. "Selenium regulation of classical glutathione peroxidase expression requires the 3′ untranslated region in Chinese hamster ovary cells." *J Nutr* no. 127 (7):1304–10.

Weiss, S.L., and Sunde, R.A. 1998. "Cis-acting elements are required for selenium regulation of glutathione peroxidase-1 mRNA levels." *RNA* no. 4 (7):816–27.

Weiss Sachdev, S., and Sunde, R.A. 2001. "Selenium regulation of transcript abundance and translational efficiency of glutathione peroxidase-1 and -4 in rat liver." *Biochem J* no. 357 (Pt 3):851–8.

Zavacki, A.M., Mansell, J.B., Chung, M., Klimovitsky, B., Harney, J.W., and Berry, M.J. 2003. "Coupled tRNA(Sec)-dependent assembly of the selenocysteine decoding apparatus." *Mol Cell* no. 11 (3):773–81.

Zhang, Y., Romero, H., Salinas, G., and Gladyshev, V.N. 2006. "Dynamic evolution of selenocysteine utilization in bacteria: A balance between selenoprotein loss and evolution of selenocysteine from redox active cysteine residues." *Genome Biol* no. 7 (10):R94.

4 Selenocysteine tRNA[Ser]Sec

The Central Component of Selenoprotein Biosynthesis

Bradley A. Carlson, Ryuta Tobe, Petra A. Tsuji,
Min-Hyuk Yoo, Lionel Feigenbaum,
Lino Tessarollo, Byeong J. Lee, Ulrich Schweizer,
Vadim N. Gladyshev, and Dolph L. Hatfield

CONTENTS

4.1 Introduction 56
4.2 Sec tRNA[Ser]Sec 57
    4.2.1 Early History and Modified Bases... 57
    4.2.2 Secondary Structure and Modified Base Synthesis................... 57
    4.2.3 The Sec tRNA[Ser]Sec Gene (*Trsp*) and Its Transcription............ 59
4.3 UGA Is a Codon for Sec 60
4.4 Sec Biosynthesis 61
4.5 Selenophosphate Synthetase 1, A SelD Paralog 63
4.6 Sec Insertion into Protein 64
4.7 Mouse Models Involving the tRNA[Ser]Sec Gene (*Trsp*) and Transgene
    (*Trsp*^t) Elucidate Functions of Selenoproteins in Cancer, Health,
    and Development 65
    4.7.1 *Trsp*^t, *Trsp*^{tG37}, and *Trsp*^{tA34} Mice 66
    4.7.2 *Trsp*^Δ and *Trsp*^{cΔ} Mice 68
    4.7.3 *Trsp*^Δ and *Trsp*^{cΔ} Mice Complemented with Transgenes, *Trsp*^t,
    Trsp^{tG37}, or *Trsp*^{tA34} 68
4.8 Concluding Remarks 71
Acknowledgments.. 72
References.. 72

4.1 INTRODUCTION

Selenium was first shown to be an essential element in the diet of mammals in 1958 when Schwarz and Foltz (1958) reported that this element prevented liver necrosis in rats. Prior to this historic experiment, selenium had been described as a toxin and even a carcinogen (Franke, 1934; Combs and Combs, 1986). Shortly after Schwarz and Foltz's discovery, the importance of this element in supplementing the diets of livestock became quickly realized, wherein selenium deficiency was associated with a variety of disorders (Oldfield, 2006). Supplementation of the diets of domestic animals with selenium throughout the world was estimated to save the livestock industry in the hundreds of millions of dollars (Combs and Combs, 1986).

In the ensuing 40 years following Schwarz and Foltz's discovery of selenium as an essential micronutrient in the diet of rats, this fascinating element was implicated in providing many health benefits in humans and other mammals in addition to those found in livestock. For example, evidence suggested that selenium acted as a cancer chemopreventive agent, had roles in preventing heart disease and other cardiovascular and muscle disorders, delayed the onset of AIDS in HIV-positive patients, slowed the aging process, inhibited viral expression, and had roles in mammalian development, immune function, and male reproduction (Papp et al., 2010; Fairweather-Tait et al., 2011; Hatfield et al., 2012; Rayman, 2012). During the period from the initial study of demonstrating selenium as a vital element in the diet of mammals to the early 1980s, two human diseases were linked with low selenium status, Keshan disease and Kashin-Beck disease (Combs and Combs, 1986). Keshan disease is a cardiomyopathy reported in children in rural areas of China, and Kashin-Beck disease is a chronic osteochondropathy reported in southwestern and northeastern China. Interestingly, Keshan disease has been virtually eradicated by selenium intervention (Loscalzo, 2014); see also Chapter 1.

The selenium field has been expanding at a rapid pace over the past 15–20 years with special emphasis on understanding the molecular biology and metabolic roles of this element (Papp et al., 2010; Fairweather-Tait et al., 2011; Hatfield et al., 2012; Rayman, 2012). Of particular emphasis has been elucidation of the mechanisms by which selenium makes its way into protein and its roles in health, development, and cancer as a chemopreventive agent (Hatfield et al., 2012), and more recently, as a cancer chemopromoting agent (see Chapters 8–10 in this book). Although both small molecular weight selenocompounds and selenoproteins have been implicated in providing the health benefits of selenium, selenoproteins appear to be the more important contributing components (Schweizer and Schomburg, 2005; Brigelius-Flohé, 2008; Papp et al., 2010; Fairweather-Tait et al., 2011; Hatfield et al., 2012; Rayman, 2012). Thus, the pathway involved in selenium's journey from selenite into these selenium-containing proteins is an alluring one and has occupied the time and efforts of many laboratories in the selenium field. One of the major components that plays a central role in carrying selenite on its back through this pathway and into selenoproteins is selenocysteine (Sec) tRNA.

Sec tRNA is designated tRNA[Ser]Sec because it is first aminoacylated with serine (Ser) by seryl-tRNA synthetase (SerS), and historically, tRNAs were named according to the amino acid attached to them by their corresponding aminoacyl-tRNA synthetase. However, unlike any other known tRNA in eukaryotes, Sec is biosynthesized on its tRNA with the Ser moiety serving as the backbone of Sec, and the resulting selenium-containing amino acid is then inserted into protein. Therefore, Sec tRNA is appropriately named tRNA[Ser]Sec. These features, the many novel characteristics of the tRNA, its role as the central component of selenoprotein synthesis, and the tRNA[Ser]Sec gene (designated *Trsp*) and transgene (designated *Trsp^t*) modifications that have been used to generate a variety of mouse models are the subjects of this chapter.

4.2 SEC tRNA[Ser]Sec

4.2.1 EARLY HISTORY AND MODIFIED BASES

Sec tRNA[Ser]Sec was first reported as a minor seryl-tRNA from rat and rooster livers that formed phosphoseryl-tRNA (Maenpaa and Bernfield, 1970) and a minor seryl-tRNA from bovine, rabbit, and chicken livers that decoded specifically the protein synthesis stop codon, UGA (Hatfield and Portugal, 1970). The fact that this seryl-tRNA decoded UGA, which was also referred to as a nonsense or opal codon, tRNA[Ser]Sec was initially thought to be the first nonsense or opal suppressor tRNA found in higher eukaryotes (Hatfield and Portugal, 1970). The primary sequence of the UGA-decoding Ser tRNA was found to be 90 nucleotides in length and contained four modified bases (Diamond et al., 1981, 1993). Thus, the unique features of this tRNA were that tRNA[Ser]Sec was exceptionally long and highly undermodified compared to all other tRNAs, which are about 75 nucleotides in length and contain as many as 15–17 modified nucleosides.

The phosphoseryl-tRNA (Maenpaa and Bernfield, 1970) and UGA-decoding seryl-tRNA (Hatfield and Portugal, 1970) were subsequently reported to be the same tRNA and found to exist in two isoforms (Hatfield et al., 1982). The primary sequences of the two isoforms were determined, and the four modified bases were identified as 5-methoxycarbonylmethyluracil (mcm^{5}U), N^6-isopentenyladenosine (i^{6}A), pseudouridine (ψU), and 1-methyladenosine (m^{1}A) at positions 34, 37, 55, and 58, respectively (Diamond et al., 1993). The isoforms differed by a single 2′-*O*-hydroxylmethyl group at nucleoside 34, designated Um34, and thus, the nucleoside at position 34 was 5-methoxycarbonylmethyluracil-2′-*O*-methylribose (mcm^{5}Um).

4.2.2 SECONDARY STRUCTURE AND MODIFIED BASE SYNTHESIS

The secondary structure of tRNA[Ser]Sec, which also has several novel features compared to other tRNAs, is shown in a cloverleaf model in Figure 4.1. Unlike canonical tRNAs, tRNA[Ser]Sec has nine base pairs in the acceptor stem and four in the TψC stem and thus resides in a 9/4 cloverleaf form compared to a 7/5 cloverleaf form found in all other tRNAs (Böck et al., 1991; Sturchler et al., 1993). The D-stem may have up to six base pairs compared to three to four in other tRNAs but, interestingly, lacks the dihydrouracil base found in the D-loop in canonical tRNAs. The atypically long extra arm in tRNA[Ser]Sec accounts for most of the bases that make it a much longer tRNA. However, the 13 bases found in the acceptor/TψC stems of all tRNAs[Ser]Sec compared to 12 in other tRNAs also account for its longer length.

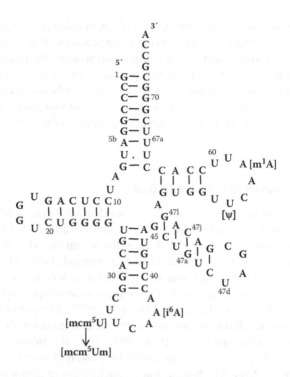

FIGURE 4.1 Primary sequence of bovine liver Sec tRNA[Ser]Sec shown in a cloverleaf model. There are 90 bases in tRNA[Ser]Sec, and bases are numbered as shown in the figure (see also Sturchler et al., 1993). The acceptor stem constitutes the paired 5′ and 3′ terminal bases; the D stem and loop constitute the six paired and four unpaired bases of the left portion of the tRNA; the anticodon stem and loop, the six paired and seven unpaired bases of the lower portion of the tRNA; the variable stem and loop, the five paired and four unpaired bases; and the TψC stem and loop, the four paired and seven unpaired bases of the right portion of the tRNA.

The addition of Um34 to tRNA[Ser]Sec is a highly specialized event in generating the final modification of tRNA[Ser]Sec maturation, resulting in the formation of mcm5Um (Hatfield et al., 2006). The presence of Um34 in tRNA[Ser]Sec dramatically alters tertiary structure (Diamond et al., 1993). Its synthesis is stringently dependent on the correct primary and tertiary structure (Kim et al., 2000). Each modified base at positions 34, 37, 55, and 58 are first synthesized prior to Um34 synthesis, and alteration in tertiary structure also prevented the attachment of this methyl group. Furthermore, tRNA[Ser]Sec Um34 synthesis is dependent on selenium status, wherein the mcm5Um isoform is highly expressed under conditions of selenium sufficiency and poorly expressed under conditions of selenium deficiency (Hatfield et al., 1991; Diamond et al., 1993; Chittum et al., 1997). Conversely, the fraction of the mcm5U isoform is enriched under conditions of selenium deficiency and is reduced under conditions of selenium sufficiency.

The synthesis of the modified bases in tRNA[Ser]Sec has been reconstituted in *Xenopus* oocytes (Choi et al., 1994; Sturchler et al., 1994). Ψ and m1A are synthesized in the nucleus, and the partially modified tRNA[Ser]Sec is transported to the cytoplasm where i[6]A and mcm[5]U are added. The synthesis of i[6]A at position 37 on tRNA[Ser]Sec was recently shown to occur in mammalian cells by demonstrating that the enzyme, tRNA[Ser]Sec isopentenyl transferase 1 (TRIT1), transfers dimethylallyl pyrophosphate to tRNA[Ser]Sec (Fradejas et al., 2013). The biosynthesis of mcm[5]U on tRNA has been elegantly established in *Arabidopsis* (Leihne et al., 2011), and the role of the corresponding mammalian tRNA methyltransferase, ALKBH8, in the synthesis of tRNA[Ser]Sec mcm[5]U has been reported (Songe-Moller et al., 2010). The only remaining modification on tRNA[Ser]Sec is the addition of Um34, and whether its synthesis occurs in the cytoplasm or nucleus is not known.

4.2.3   THE SEC TRNA[Ser]Sec GENE (*TRSP*) AND ITS TRANSCRIPTION

Trsp occurs in single copy in all organisms in which it has been found except in zebrafish in which it occurs in two copies (Xu et al., 1999). *Trsp* and/or tRNA[Ser]Sec have/has been sequenced from chickens, frogs, fruit flies, nematodes (Hatfield et al., 2006), *Chlamydomonas* (Rao et al., 2003), *Dictyostelium*, *Tetrahymena* (Shrimali et al., 2005), *Taxoplasma gondii*, several species of *Plasmodium*, and a variety of other eukaryotes (Carlson et al., 2006). *Trsp* was mapped to chromosome 19 in humans and localized to bands q13.2–q13.3 and ordered with respect to other genes in that region (McBride et al., 1987). The mouse tRNA[Ser]Sec gene was mapped to chromosome 7 (Ohama et al., 1994), and its chromosomal position is similar to that in humans providing another example of human-mouse synteny.

The regulatory elements governing the transcription of *Trsp* are, for the most part, unlike those of any other known tRNA. The major elements governing the expression of canonical tRNAs occur within the intragenic promoter A and B boxes. Although *Trsp* also contains an intragenic A box and B box, the A box has two additional bases, a T and a C at positions 36 and 37, respectively (Hatfield et al., 1983). Furthermore, the efficiency of transcription of *Trsp* has been shown to depend very heavily on three upstream regulatory elements and less on the intragenic promoter elements (see below), and the role of the two internal promoters in *Trsp* transcription has not been resolved (Carbon and Krol, 1991) and is still controversial (Hatfield et al., 1999).

Trsp has a *TATA*-like sequence that begins at or near position −20 and encompasses about 15 upstream bases in all tRNA[Ser]Sec genes sequenced, and this region has been shown to have a prominent role in transcription efficiency both *in vivo* (*Xenopus* oocytes) (Lee et al., 1989a; Carbon and Krol, 1991) and *in vitro* (HeLa cell extracts) (Lee et al., 1989b). There are two other *Trsp* upstream elements, a proximal sequence element (*PSE*) that begins near position −46 and extends to approximately −66 (Park et al., 1995; Hatfield et al., 1999), and a distal sequence element (*DSE*) that begins around position −195 and extends to approximately −210 (Myslinski et al., 1992). The *DSE* consists of an *activator element* (*AE*) containing an SPH motif and an octamer sequence, which is essential for optimal transcription in *Xenopus* oocytes (Myslinski et al., 1992) but is not functional in *Xenopus* oocyte extracts (Park et al., 1996). Disrupting the function of the *DSE* in mice by inserting a large fragment between the *PSE* and *DSE* caused embryonic lethality due to the loss of *Trsp* transcription and subsequent reduced generation of selenoprotein transcripts (Kelly et al.,

2005). The Sec tRNA transcription activating factor (STAF), which binds to the activator element (*AE*), has been characterized in frogs (Schuster et al., 1995) and mice (see Adachi et al., 2000, and references therein) and shown to have roles in numerous genes transcribed by RNA Pol II and III (see Schaub et al., 1999; Barski et al., 2004, and references therein). Deletion of the *AE* region, making a transgenic mouse carrying the deleted Δ*AE* that in turn was used to breed with a mouse lacking one copy of *Trsp* (designated *Trsp*Δ) for two generations yielded a *Trsp*Δ mouse that was dependent on the Δ*AE* transgenic mice for expression of its tRNA[Ser]Sec (Carlson et al., 2009a). Interestingly, this mouse expressed tRNA[Ser]Sec in varying amounts in different organs and tissues, wherein some organs expressed more total tRNA[Ser]Sec and other organs and tissues less. The level of the mcm⁵Um isoform was always less in all tissues and organs examined in the Δ*AE* transgenic mouse than in the corresponding control, wild-type tissues, and organs (Carlson et al., 2009a). The mechanism of how the mutant *AE* affected the tRNA[Ser]Sec population in this manner was not resolved.

Trsp is an RNA Pol III transcribed gene in mammals and *Xenopus* (Hatfield et al., 1999), but more recently, the tRNA[Ser]Sec gene in *Trypanosoma brucei* was shown to be transcribed by Pol II (Aeby et al., 2010). Interestingly, *Trsp* is transcribed unlike any other known tRNA gene in that transcription begins at the first nucleotide within the coding sequence, and all other tRNAs are transcribed with a leader sequence that must be processed (Lee et al., 1987). Sec tRNA[Ser]Sec, like other tRNAs, however, has a 3'-trailer sequence that is processed to yield the completed transcript. The termination sequence for transcription of *Trsp* is located about 25 nucleotides downstream of the coding sequence in mammals and is a region consisting of mostly repeating Ts, approximately 7–10 nucleotides in length (Hatfield et al., 1983; Lee et al., 1987). The fact that transcription begins within the coding region of *Trsp* results in the transcript having a triphosphate attached to its 5'-terminal G. The triphosphate remains intact on tRNA[Ser]Sec through processing of the 3'-terminal sequence and is transported to the cytoplasm (Lee et al., 1987). Whether it remains intact in the biosynthesis of Sec on the tRNA and in the insertion of Sec into protein and/or is required for some other specific function or functions are not known.

4.3   UGA IS A CODON FOR SEC

The genetic code was deciphered in 1966, and 61 of the 64 possible code words were assigned one of the 20 canonical amino acids, and three other codons, UAA, UAG, and UGA, terminated protein synthesis (Nirenberg et al., 1966). Only one code word, AUG, was reported to have a dual function of initiating protein synthesis as it corresponded to the methionine (Met) residue at the N-terminal position of proteins and also coded for Met at internal positions of proteins. Although numerous variations were subsequently found in how some codons were used to decode specific amino acids by certain organisms, including UGA being used to decode tryptophan or cysteine (Cys) (see Watanabe and Yokobori, 2011, and references therein), there did not appear to be additional room in the genetic code for expansion to embrace another amino acid. Such proteinogenic amino acids as phosphoserine, hydroxyproline, etc., were known to be modified post-translationally. Thus, it was indeed surprising when TGA was detected in genes for mammalian glutathione peroxidase 1

(*Gpx1*) (Chambers et al., 1986) and bacterial formate dehydrogenase (*fdhF*) (Zinoni et al., 1986) that coincided with Sec in the corresponding proteins because UGA was known to dictate termination of protein synthesis in mammals and *Escherichia coli*. These latter two studies provided the first indication that the genetic code could encompass a 21st amino acid.

Böck and collaborators subsequently mutated *TGA* in *fdhF* to a number of other codons that, in turn, dictated the synonymous amino acid being inserted, which was not modified to Sec in the completed protein (Zinoni et al., 1987), providing further evidence that Sec was the 21st amino acid (Söll, 1988). However, when the specific UGA decoding phosphoseryl-tRNA in eukaryotes (Hatfield et al., 1982) and the tRNA encoded in *selC* that was essential for synthesizing selenoproteins in *E. coli* (Leinfelder et al., 1988) were found to synthesize Sec on the corresponding tRNAs (Lee et al., 1989b; Leinfelder et al., 1989), Sec was unequivocally demonstrated as the 21st amino acid in the genetic code.

4.4   SEC BIOSYNTHESIS

The initial step in the biosynthesis of Sec on its tRNA in all organisms that synthesize selenoproteins begins with the attachment of serine by SerS (Figure 4.2). In bacteria, there is a single protein, designated SelA, that converts the Ser moiety on seryl-tRNA[Ser]Sec to selenocysteyl-tRNA[Ser]Sec through an intermediate that is likely dehydroalanyl-tRNA[Ser]Sec (Yoshizawa and Böck, 2009) (Figure 4.2a). Upon converting serine to dehydroalanine, SelA accepts the activated selenium donor, selenophosphate, resulting in Sec attached to tRNA[Ser]Sec. Activated selenium is synthesized by SelD, selenophosphate synthetase, from selenide and ATP (Glass et al., 1993).

In eukaryotes and archaea, there is an additional step in the biosynthesis of Sec that does not occur in bacteria. Phosphoseryl-tRNA[Ser]Sec is an intermediate between the aminoacylation of tRNA[Ser]Sec with Ser by seryl-tRNA synthetase and the synthesis of the acceptor of selenophosphate, which is, as in bacteria, likely dehydroalanine (Figure 4.2b). The kinase that converts seryl-tRNA[Ser]Sec to phosphoseryl-tRNA[Ser]Sec (Maenpaa and Bernfield, 1970) had been tenuous to isolate and characterize for more than 30 years until the phosphoseryl-tRNA kinase gene (*Pstk*) was detected by bioinformatics and cloned and the protein, PSTK, expressed and characterized (Carlson et al., 2004b). As noted above, phosphoseryl-tRNA[Ser]Sec is converted to the selenophosphate acceptor by the SelA homologue in archaea and eukaryotes, designated Sec synthase (SecS) (Xu et al., 2007a). Upon addition of activated selenium, selenocysteyl-tRNA[Ser]Sec is formed completing Sec biosynthesis (Yuan et al., 2006; Xu et al., 2007a) (Figure 4.2b). Selenophosphate is synthesized by selenophosphate synthetase, SPS2 (Xu et al., 2007b), which is a selenoprotein in higher eukaryotes (Guimaraes et al., 1996). The fact that SPS2 is a selenoprotein suggests that it might have an autoregulatory role in selenoprotein synthesis (Guimaraes et al., 1996).

Cys was more recently reported to be synthesized de novo in mammals using the Sec biosynthetic machinery (Xu et al., 2010). This amino acid was known to arise intracellularly in mammals from Met via homocysteine and Ser or to be absorbed from the environment and thus was regarded as a semiessential amino acid. However,

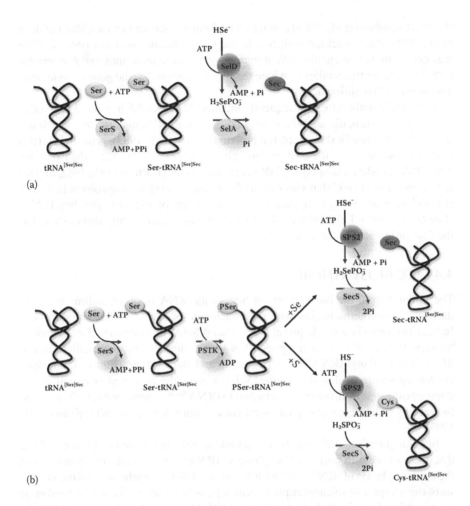

FIGURE 4.2 Biosynthesis of Sec on its tRNA, tRNA[Ser]Sec, in (a) bacteria and in (b) archaea and eukaryotes. Thiophosphate can replace selenophosphate in Sec biosynthesis resulting in a *de novo* biosynthetic pathway of Cys in eukaryotes as shown in the lower panel of (b). The details of the biosynthesis of Sec and Cys are given in the text.

Holmgren and coworkers reported that Cys occurred in place of Sec in thioredoxin reductase 1 (TrxR1) in liver of rats that were maintained on a selenium-deficient diet (Lu et al., 2009). The mechanism of how Cys occurred in TrxR1 was not resolved. We subsequently found that by supplementing the medium of mammalian cells with thiophosphate (SPO_3), we could target the insertion of Cys in TrxR1 (Xu et al., 2010). Cys was synthesized on tRNA[Ser]Sec by SecS that occurred when SPO_3 replaced selenophosphate as the active donor to the intermediate, dehydroalanine, to yield Cys-tRNA[Ser]Sec (Figure 4.2b). Cys-tRNA[Ser]Sec then decoded UGA in TrxR1 mRNA resulting in Cys replacement of Sec. Interestingly, such replacement also occurred naturally *in vivo* in mouse liver TrxR1 of mice maintained on adequate levels of selenium in their diet, but the level of replacement was only about 10% of the total amount of Sec (Xu et al., 2010). When the mice were placed on a selenium-deficient diet, the level of Cys/Sec replacement was approximately 50%. This study revealed a novel use of the Sec insertion machinery resulting in de novo synthesis of Cys and its insertion at UGA codons in the housekeeping selenoproteins, TrxR1 and TrxR2. In addition, the study suggested new biological functions of SPO_3 and sulfide in mammals.

Misreading of UGA codons in selenoproteins and replacement of Sec with other amino acids such as arginine, tryptophan, and Cys was also found to occur following exposure of mammalian cells to the antibiotics doxycycline (Dox), chloramphenicol (Cp), and geneticin (G418) (Handy et al., 2006; Tobe et al., 2013). The degree of misreading of Sec by other amino acids was examined in three selenoproteins, TrxR1, glutathione peroxidase 1 (GPx1), and glutathione peroxidase 4 (GPx4), which are translated differentially by the two Sec tRNA[Ser]Sec isoforms. TrxR1 is synthesized by the mcm⁵U isoform, GPx1 by the mcm⁵Um isoform, and GPx4 by both isoforms (Carlson et al., 2005a,b). Interestingly, these three proteins were differentially affected by varying errors in Sec insertion at UGA in the presence of Dox, Cp, or G418, wherein GPx1 was the least affected by translation errors, TrxR1 most affected, and GPx4 was in between the other two selenoproteins (Tobe et al., 2013). In addition, TrxR1 manifested a high degree of truncation. These results were consistent with the differential use of two Sec tRNA[Ser]Sec isoforms and their roles in supporting the accuracy of Sec insertion and illustrated the vulnerability of misreading of UGA in cells exposed to antibiotics (Tobe et al., 2013).

4.5 SELENOPHOSPHATE SYNTHETASE 1, A SELD PARALOG

It should also be noted that another homologue of SelD, selenophosphate synthetase 1 (SPS1), was identified in higher eukaryotes and was originally proposed to function as selenophosphate synthetase (Kim and Stadtman, 1995; Low et al., 1995). However, subsequent studies demonstrated that selenoprotein synthesis was not restored in an *E. coli selD* mutant when complemented by *Sps1* but was restored when complemented by a *Sps2*(Sec) → *Sps2*(Cys) mutant (Kim et al., 1997, 1999). Supplementing the culture medium of the *E. coli selD* mutant encoding *Sps1* with L-Sec resulted in growth (Tamura et al., 2004). These studies suggested that SPS2 is essential for synthesis of selenophosphate and that SPS1 may have a role in recycling Sec via a selenium salvage pathway but not in the synthesis of selenophosphate. Subsequently, SPS2 was unequivocally shown to synthesize selenophosphate whereas SPS1 did not synthesize the active selenium donor (Xu et al., 2007a,b).

The cellular function of SPS1 has not been resolved. It is essential in the development of *Drosophila melanogaster* (Alsina et al., 1999), but indirect evidence, wherein selenoprotein synthesis is removed in this organism by knocking out the Sec elongation factor, *Efsec*, resulted in no phenotype (Hirosawa-Takamori et al., 2004), suggesting that selenoprotein synthesis and SPS2 are not essential. In addition, several insects, such as *Bombyx mori, Tribolium castaneum*, and several species of fruit flies, but not *D. melanogaster*, do not encode the selenoprotein synthesizing machinery but do carry orthologs of *Sps1* in their genomes (Lobanov et al., 2008). These studies suggest that SPS1 is an essential protein and that it does not have a function in selenoprotein synthesis.

Further examination of SPS1 in SL2 *Drosophila* cells has shed light on the possible role of this protein in insects. The targeted removal of *Sps1* in SL2 cells led to growth inhibition, increased intracellular glutamine levels that induced megamitochondrial formation and decreased expression of genes involved in pyridoxal phosphate (PLP) metabolism (Shim et al., 2009; Lee et al., 2011). Specific inhibition of PLP biosynthesis also slowed SL2 cell growth and caused megamitochondrial formation, suggesting that SPS1 regulates vitamin B6 biosynthesis in *Drosophila*. Interestingly, SPS1 has five splice variants in humans that appear to have different subcellular locations and expression patterns (Kim et al., 2010).

4.6   SEC INSERTION INTO PROTEIN

After Sec is synthesized on tRNA[Ser]Sec, this amino acid is now ready to be incorporated into protein in response to UGA Sec codons. However, because UGA is also used as a genetic code word for termination of protein synthesis in all organisms that make selenoproteins, including mammals and *E. coli*, UGA codons with different functions must be distinguished to either insert Sec or to stop protein synthesis. One feature of all eukaryotic selenoprotein mRNAs is that they have a specific cis-containing stem-loop structure, designated the Sec Insertion Sequence (SECIS) element (Berry et al., 1991). There are two classes of eukaryotic SECIS elements, designated Type I and II (Krol, 2002). In bacteria, a single type of SECIS element exists within the coding region of selenoprotein mRNAs that resides immediately downstream of the UGA codon (Seeher et al., 2012). The specific elongation factor, specified SelB in bacteria, interacts with Sec-tRNA[Ser]Sec, the resulting complex attaches to the SECIS element, which then proceeds to combine with the ribosome for synthesis of the selenoprotein and insertion of Sec dictated by the UGA Sec codon.

In archaea and eukaryotes, the recoding of UGA for Sec insertion involves additional factors and is far more multifarious. EFsec complexes with Sec-tRNA[Ser]Sec, which, in turn, binds to the SECIS binding protein 2 (SBP2) attached to the SECIS element, and this complex then combines with the ribosome for synthesizing the selenoprotein and insertion of Sec in response to UGA-Sec (Chavatte et al., 2005). There are additional factors that play a role in Sec insertion, which include ribosomal protein L30 (Chavatte et al., 2005), eukaryotic initiation factor (eIF4a3) (Budiman et al., 2009), and nucleolin (Miniard et al., 2010). Of these three additional factors, the role of ribosomal protein L30 is the most well characterized as it is bound to the ribosome and apparently functions as a portion of the basic machinery governing Sec insertion. On the other hand, eIF4a3 and nucleolin likely have regulatory roles mitigating selenoprotein synthesis. A possible mechanism of how Sec may be inserted into protein involving Sec-tRNA[Ser]Sec and each of these factors is presented in Figure 4.3 (see also Chapter 3).

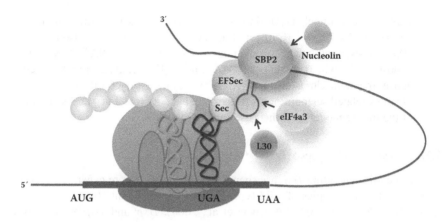

FIGURE 4.3   Incorporation of Sec into protein. The incorporation of Sec into protein involves a multifarious complex as described in the text and in Chapter 3.

4.7   MOUSE MODELS INVOLVING THE tRNA[SER]SEC GENE (*Trsp*) AND TRANSGENE (*Trsp*[t]) ELUCIDATE FUNCTIONS OF SELENOPROTEINS IN CANCER, HEALTH, AND DEVELOPMENT

Selenoproteins are the only known class of proteins whose biosynthesis is dependent on a single tRNA, tRNA[Ser]Sec. The synthesis of selenoproteins can, therefore, be modulated by altering the expression of tRNA[Ser]Sec or even abolished completely by targeting the removal of *Trsp*. Such approaches to perturbing the synthesis of seleno-proteins would provide a means of elucidating the roles of this unique protein class and determining their significance in health and development. Furthermore, a major debate existed in the selenium field at the turn of the 21st century about whether small-molecular-weight selenium-containing compounds or selenium-containing proteins played the more important role in the many benefits attributed to this element (Schweizer and Schomburg, 2005; Brigelius-Flohé, 2008; Papp et al., 2010; Fairweather-Tait et al., 2011; Hatfield et al., 2012; Rayman, 2012).

We took advantage of the fact that selenoprotein synthesis is dependent on tRNA[Ser]Sec and developed a number of mouse models to examine the role of this protein class in health and development (Hatfield et al., 2006). In addition, these mouse models would elucidate the issue of which selenium components, small-molecular-weight selenocompounds or selenoproteins, had the more significant role in health and development.

The mouse models involved the following:

1. Transgenes that encoded either wild-type *Trsp* (specified *Trsp*[t]) or mutant forms of *Trsp* (A37 → G37 [Moustafa et al., 2001, 2003] and T34 → A34 [Carlson et al., 2007], specified *Trsp*[tG37] and *Trsp*[tA34], respectively)
2. Conditional knockout of *Trsp* (specified *Trsp*[cΔ]) that targeted specific organs and tissues (Kumaraswamy et al., 2003)

3. $Trsp^{c\Delta}$ or total knockout of $Trsp$ (specified $Trsp^{\Delta}$) that was complemented by $Trsp^{t}$ or by $Trsp^{tG37}$ (Carlson et al., 2005a,b) or $Trsp^{tA34}$ (Carlson et al., 2007). N^6-isopentyladenosine cannot be synthesized on tRNA[Ser]Sec with G at position 37, which prevents Um34 synthesis (Kim et al., 2000). As noted above, the mcm⁵Um isoform of tRNA[Ser]Sec is necessary for the expression of stress-related selenoproteins (Carlson et al., 2005a,b). Further discussion of the mouse models and their uses are given below.

4.7.1 $T_{RSP}{}^{T}$, $T_{RSP}{}^{TG37}$, AND $T_{RSP}{}^{TA34}$ MICE

Transgenic mice encoding $Trsp$ transgenes (specified $Trsp^{t}$) were generated carrying from a few to many copies of wild-type, $Trsp^{t}$ or either mutant, $Trsp^{tG37}$ (Moustafa et al., 2001, 2003) or $Trsp^{tA34}$ (Carlson et al., 2007). $Trsp^{t}$ and $Trsp^{tG37}$ mice were prepared in the initial studies, and their effects on selenoprotein synthesis examined (Moustafa et al., 2001, 2003). The resulting transgenic mice were the first examples of such mice engineered to encode functional transgenes. Overexpression of $Trsp^{t}$ had little or no effect on the expression of selenoprotein levels in various tissues and organs examined, suggesting that the amounts of tRNA[Ser]Sec are not limiting for selenoprotein biosynthesis. However, the amounts of several selenoproteins were decreased in a protein and organ-specific fashion in $Trsp^{tG37}$ mice, wherein GPx1 was most affected and TrxR1 and TrxR2 were the least affected in liver and testis, respectively. Because the mRNA levels of most selenoproteins were unaffected in $Trsp^{tG37}$ mice, the data clearly showed that the defect in selenoprotein synthesis occurred at the level of translation (Moustafa et al., 2001, 2003). In subsequent studies involving $Trsp^{tG37}$ mice, it was clearly demonstrated that stress-related selenoproteins (e.g., GPx1, GPx3, SelT, and SelW) were synthesized by the isoform containing Um34, mcm⁵Um, and that housekeeping selenoproteins (e.g., TrxR1 and TrxR2) were synthesized by the non-Um34 isoform, mcm⁵U, and some selenoproteins (e.g., GPx4 and SelP) appear to be synthesized by both isoforms (Carlson et al., 2005a,b). Furthermore, the available evidence suggests the Um34 tRNA[Ser]Sec methylase has a strict requirement for selenocysteyl-tRNA[Ser]Sec without any alterations in its primary or tertiary structure (Kim et al., 2000, 2011).

Mice carrying the $Trsp^{tA34}$ transgene were also generated, but interestingly, no more than 12 copies of the transgene were tolerated by the transgenic mouse (Carlson et al., 2007) whereas mice carrying the $Trsp^{tG37}$ transgene could encode many more copies without an apparent adverse effect (Moustafa et al., 2001, 2003). The reason for the reduced number of transgenes in $Trsp^{tA34}$ mice was most likely due to the fact that A in the wobble position of the anticodon is converted to I (inosine), which decodes genetic code words terminating in U, C, and A, and thus, $tRNA_{ICA}^{[Ser]Sec}$ would read the Cys codons, UGU and UGA, and the Sec codon, UGA. The higher number of transgenes would, of course, result in higher amounts of the $tRNA_{ICA}^{[Ser]Sec}$ product that would more effectively compete with Cys-tRNA, likely inserting serine in place of Cys. The possible insertion of Ser in place of Cys in selenoproteins and in other non-selenium-containing proteins is further discussed below. Mice carrying $Trsp^{tG37}$ or $Trsp^{tA34}$ were used for elucidating the loss of stress-related selenoproteins in health and developmental issues. The various uses of these mouse models are summarized in Table 4.1.

TABLE 4.1
Roles of Stress-Related Selenoproteins in Health and Development Elucidated by Blocking Their Synthesis

Model Description	Major Findings[a]
Trsp^t *Trsp^tG37*	First transgenic mouse encoding tRNA transgenes. Little or no variation in selenoprotein expression in *Trsp^t* mice; thus, tRNA[Ser]Sec levels not limiting in selenoprotein synthesis. Stress-related selenoprotein expression decreased in a protein- and tissue-specific manner in *Trsp^tG37* mice, providing an important tool for elucidating the role of this protein subclass in health and development (Moustafa et al., 2001).
Trsp^tG37	Enhanced skeletal muscle adaptation after exercise enhanced growth that was blocked by inhibition of the target of rapamycin (mTOR) pathway. Muscles of transgenic mice exhibited increased site-specific phosphorylation on both Akt and p70 ribosomal S6 kinase before ablation (Hornberger et al., 2003).
Trsp^tG37 Colon targeted by azoxymethane exposure	Enhanced azoxymethane-induced aberrant crypt formation (a preneoplastic lesion for colon cancer). First study suggesting that stress-related selenoproteins reduce colon cancer incidence (Irons et al., 2006).
Trsp^tG37/Tag	[C3(1)/Tag] is a prostate cancer driver gene. Accelerated development of lesions associated with prostate cancer progression. First study suggesting stress-related selenoproteins have a role in preventing prostate cancer (Diwadkar-Navsariwala et al., 2006).
Trsp^tA34	Stress-related selenoprotein expression decreased in a protein- and tissue-specific manner. A total of only 14 copies of *Trsp^tA34* tolerated likely due to mistranslation (see text and Carlson et al., 2007).
Trsp^tG37 Lung targeted by exposure to influenza virus	Minor changes observed in the immune system of *Trsp^tG37* mice and lung pathology similar in these and control mice suggesting that stress-related selenoproteins (e.g., GPx1) have a limited role in protecting mice from influenza viral infection (Sheridan et al., 2007).
Trsp^tG37 *Trsp^tA34*	First study to show that housekeeping selenoproteins are essential in lipoprotein biosynthesis and metabolism. *Trsp^tG37* and *Trsp^tA34* mice used individually and compared to *Trsp* liver knockout mice showing that selenoproteins essential in lipoprotein synthesis and metabolism (Sengupta et al., 2008a).
Trsp^tG37	Higher incidence of micronuclei formation in erythrocytes following exposure to X-rays. Data indicate a role of stress-related selenoproteins in protecting DNA from damage (Baliga et al., 2008).
Trsp^tG37 Varying dietary Se levels Liver targeted by diethylnitrosamine (DEN) exposure	Mice maintained on Se-deficient, adequate, and supplemented diets. Increased incidence of liver tumors in *Trsp^tG37* mice on Se-adequate diets, wherein Se-deficient and -supplemented levels protected against tumor formation. Se-deficient *Trsp^tG37* mice manifested neurological phenotype (Kasaikina et al., 2013).

(Continued)

TABLE 4.1 (CONTINUED)
Roles of Stress-Related Selenoproteins in Health and Development Elucidated by Blocking Their Synthesis

Model Description	Major Findings[a]
$Trsp^{tG37}$ $Trsp^{tG37}/TGF\alpha$ Varying dietary Se levels or TPSC[2]	Mice maintained on Se-deficient, -adequate, and -supplemented diets or diets with TPSC.[b] Widespread pyogranuloma formation, severe neurological phenotype associated with early morbidity and mortality in $Trsp^{tG37}$ and bitransgenic mice on Se-deficient or TPSC diets. Liver tumors significantly enhanced in $TGF\alpha$ (liver cancer driver gene) mice irrespective of selenium or selenoprotein status (Moustafa et al., 2013).
$Trsp^{tG37}$	Reduced synthesis of stress-related selenoproteins caused by overexpression of $Trsp^{tG37}$ mutant tRNA promoted glucose intolerance and led to a diabetes-like phenotype (Labunskyy et al., 2011).
$Trsp^{tG37}$ Varying dietary Se levels	Mice maintained on Se-deficient, -adequate, and -supplemented diets. Translational mechanisms controlling selenoprotein synthesis in liver examined by ribosome profiling. Under Se-deficiency conditions, housekeeping selenoproteins preferentially synthesized. Stress-related selenoproteins require mcm^{5}Um isoform for translation and increased ribosome density occurs at or upstream of the UGA codon. Overall determining factor of selenoprotein translation proficiency is the effectiveness of Sec insertion and not the initiation of translation (Howard et al., 2013).

[a] The studies shown in this column describe the major findings observed in altered mice, and the findings described are relative to the corresponding control mice in the study.

[b] TPSC, triphenylselenonium chloride, a nonmetabolized selenium compound.

4.7.2 $T_{RSP}{}^{\Delta}$ AND $T_{RSP}{}^{c\Delta}$ MICE

Knockout of $Trsp$ ($Trsp^{\Delta}$) is embryonic lethal (Bosl et al., 1997) and, therefore, to elucidate the developmental role of selenoproteins and their functions in specific tissues and organs, mice carrying the conditional knockout of this gene ($Trsp^{c\Delta}$) were generated using $loxP$-Cre technology (Kumaraswamy et al., 2003). In the initial study, the removal of $Trsp$ was targeted in mouse mammary epithelium during lactation when the mammary gland consists of about 90% epithelium (Kumaraswamy et al., 2003). Selenoprotein expression was reduced substantially, but not completely, most likely due to contamination from other cell types that included fat, myoepithelium, and fibroblasts. This early study clearly demonstrated that the selenoprotein population could be targeted for removal by deleting $Trsp$ and provided the foundation for examining the role of selenoproteins in health and development in a variety of tissues and organs as shown in Table 4.2.

4.7.3 $T_{RSP}{}^{\Delta}$ AND $T_{RSP}{}^{c\Delta}$ MICE COMPLEMENTED
WITH TRANSGENES, $T_{RSP}{}^{T}$, $T_{RSP}{}^{TG37}$, OR $T_{RSP}{}^{TA34}$

Both $Trsp^{\Delta}$ and $Trsp^{c\Delta}$ mice have been complemented with $Trsp$ transgenes. Interestingly, $Trsp^{\Delta}$ mice can be rescued with $Trsp^{tG37}$, and the resulting mice appear

TABLE 4.2
Roles of Selenoproteins in Health and Development Elucidated by Knocking Out the Gene for tRNA[Ser]Sec

Targeted Organ or Tissue	Cre Promoter	Major Findings[a]
Mammary gland	MMTV-Cre; Wap-Cre	First conditional Trsp[Δ] mouse providing an important tool for elucidating the role of selenoproteins in health and development (Kumaraswamy et al., 2003). MMTV-Cre mice treated with DMBA[b] had significantly more tumors, suggesting that selenoproteins protect against carcinogen-induced mammary cancer (Hudson et al., 2012).
Liver	Alb-Cre	Death between 1 and 3 months of age due to severe hepatocellular degeneration and necrosis showing that selenoproteins have a role in proper liver function (Carlson et al., 2004a). SelP (and GPx3) reduced in serum and kidney supporting a Se transport role for liver-derived SelP (Schweizer et al., 2005). Loss of Trsp in liver was compensated for by an enhanced expression of phase II response genes (Sengupta et al., 2008b). Mice with selenoprotein loss in liver used as a control to monitor Se pools in kidney due to reduction of GPx3 imported from liver (Malinouski et al., 2012). In hepatocytes, Secisbp2 gene inactivation is less detrimental than Trsp inactivation (Seeher et al., 2014a).
Endothelial cells	TieTek2-Cre	Embryonic lethal. 14.5 dpc embryos were smaller, more fragile, poorly developed vascular system, underdeveloped limbs, tails, and heads. Selenoproteins have a role in endothelial cell development and function (Shrimali et al., 2007).
Heart and skeletal muscle	MCK-Cre	Death from acute myocardial failure on day 12 after birth. Selenoproteins play a role in preventing heart disease (Shrimali et al., 2007).
Kidney	NPHS2-Cre	No increase in oxidative stress or nephropathy in podocyte selenoprotein-deficient mice (Blauwkamp et al., 2008).
T cells	LCK-Cre	Reduction of mature T cells and a defect in T cell–dependent antibody responses. Antioxidant hyperproduction and suppression of T cell proliferation in response to T cell receptor stimulation. Selenoproteins have a role in immune function (Shrimali et al., 2008).
Macrophage	LysM-Cre	Elevated oxidative stress and transcriptional induction of cytoprotective antioxidant and detoxification enzyme genes. Accumulation of ROS levels and impaired invasiveness. Altered expression of several extracellular matrix and fibrosis-associated genes. Selenoproteins have a role in immune function (Carlson et al., 2009b). Selenoproteins are essential for the balance of pro- and

(Continued)

TABLE 4.2 (CONTINUED)
Roles of Selenoproteins in Health and Development Elucidated by Knocking Out the Gene for tRNA[Ser]Sec

Targeted Organ or Tissue	Cre Promoter	Major Findings[a]
		anti-inflammatory oxylipids during inflammation (Mattmiller et al., 2014). Selenoproteins protect mice from DSS-induced colitis by alleviating inflammation (Kaushal et al., 2014). When infected with the parasite *N. brasiliensis*, Se-supplemented knockout mice showed a complete abrogation in M2 marker expression with a significant increase in intestinal worms and fecal eggs. Selenoproteins play a role in the epigenetic modulation of proinflammatory genes (Narayan et al., 2014).
Osteochondroprogenitor	*Col2a1-Cre*	Postnatal growth retardation, chondrodyplasia, chondronecrosis, and delayed skeletal ossification characteristic of Kashin-Beck disease. First mouse model for Kashin-Beck disease (Downey et al., 2009).
Neurons	*Tα1-Cre; CamK-Cre*	Enhanced neuronal excitation followed by neurodegeneration of hippocampus. Cerebellar hypoplasia associated with degeneration of Purkinje and granule cells. Cerebellar interneurons essentially absent. Selenoproteins have a role in neuronal function (Wirth et al., 2010). Striatal interneuron density was reduced in mice with impaired selenoprotein expression (Seeher et al., 2014b).
Skin	*K14-Cre*	Runt phenotype, premature death, alopecia with flaky and fragile skin, epidermal hyperplasia with disturbed hair cycle and an early regression of hair follicles. Selenoproteins have a role in skin and hair follicle development (Sengupta et al., 2010).
Thyroid	*Pax8-Cre* and *Tg-Cre^ER*	Mice showed increased oxidative stress in thyroid. Gross morphology remained intact for at least 6 months. Thyroid hormone levels remained normal in knockout mice; thyrotropin levels moderately elevated (Chiu-Ugalde et al., 2012).
Prostate	*PB-Cre4*	Mice develop PIN-like lesions and microinvasive carcinoma by 24 weeks that was associated with loss of basement membrane and increased cell cycle and apoptotic activity (Luchman et al., 2014).

Source: Nelson, S. M., J. L. James, N. Kaushal, B. A. Carlson, A. Gunderson, J. Urban Jr, and K. S. Prabhu. Selenoprotein enhanced expression of M2 macrophages increases clearance of *Nippostrongylus brasiliensis*. Submitted.

[a] The studies shown in this column describe the major findings observed in altered mice and the findings described are relative to the corresponding control mice in the study.

[b] DMBA, 7,12-dimethylbenz[a]anthracene.

phenotypically normal although they lack stress-related selenoprotein expression (Carlson et al., 2005a,b). In addition, they expressed other selenoproteins, such as GPx4 in reduced levels. $Trsp^t/Trsp^{tG37}$ males were found to have low fertility and generate sperm with distorted morphology that accounted for their reduced fertility. $Trsp^{\Delta}$ mice were rescued with $Trsp^t$, and as expected, the selenoprotein expression in these mice were virtually identical to that found in wild-type, control mice (Carlson et al., 2005a,b). $Trsp^{\Delta}$ mice apparently cannot be rescued with $Trsp^{tA34}$, most likely because the resulting tRNA[Ser]Sec generated from this tRNA contains I in the wobble position of the anticodon that translates UGU and UGC codons in addition to UGA (see Carlson et al., 2005b, and above).

$Trsp^{c\Delta}$ mice that were targeted for loss of selenoproteins in liver and complemented with either $Trsp^t$, $Trsp^{tG37}$, or $Trsp^{tA34}$ were also generated (Carlson et al., 2005b). As expected, selenoprotein expression was fully restored in hepatocyte $Trsp^{c\Delta}$ mice by introducing $Trsp^t$ and was partially restored by introducing $Trsp^{tG37}$ or $Trsp^{tA34}$. Because both $Trsp^{tG37}$ and $Trsp^{tA34}$ transgenes produced tRNAs[Ser]Sec lacking Um34, this observation provided further evidence that the 2'-O-modification on tRNA[Ser]Sec is required for the synthesis of stress-related selenoproteins. Only low copy numbers of the $Trsp^{tA34}$ transgene were tolerated without $Trsp$ whereas both low and high copy numbers of $Trsp^{tG37}$ restored housekeeping selenoprotein expression whether $Trsp$ was or was not present (Carlson et al., 2005b). Interestingly, metabolic labeling of liver $Trsp^{c\Delta}$ mice carrying $Trsp^{tA34}$ with ^{75}Se showed that only selenoproteins incorporated the label. This observation suggested that tRNA$_{ICA}^{[Ser]Sec}$ did not randomly insert Sec in place of Cys in protein but only in specific selenoproteins. However, what was not resolved in this study was whether seryl-tRNA[Ser]Sec, which is known to suppress UGA stop codons (Diamond et al., 1993; Chittum et al., 1998), can decode UGU and UGC randomly in proteins.

4.8 CONCLUDING REMARKS

Elucidation of the function of selenium at the molecular level and, in particular, in selenoproteins has moved at a rapid pace in the last several years. Many aspects of the role of tRNA[Ser]Sec in selenoprotein synthesis have been revealed, but there is still much to be done. For example, much of the transcription and maturation of tRNA[Ser]Sec is known as reviewed herein, but there are still numerous parameters that need to be further investigated. The B box appears to have a regulatory role in tRNA[Ser]Sec transcription, and the A box, with two extra bases, does not (Carbon and Krol, 1991). However, the roles of these two internal promoter regions, which are highly significant in all other known tRNAs, require further investigation in tRNA[Ser]Sec transcription. The role of STAF in the expression of tRNA[Ser]Sec and how the loss of this protein enhanced tRNA$_{mcmU}^{[Ser]Sec}$ expression in select tissues and organs but reduced levels in others and reduced synthesis of tRNA$_{mcm^5Um}^{[Ser]Sec}$ in all tissues and organs examined needs further study (Carlson et al., 2009a). The tRNA[Ser]Sec Um34 methylase has not been identified nor has the role of selenium in the expression of this methylase. Whether selenium may be involved elsewhere in insertion of Sec from tRNA$_{mcm^5Um}^{[Ser]Sec}$ and not in Um34 methylation per se or in preventing the enhanced degradation of tRNA$_{mcm^5Um}^{[Ser]Sec}$ during selenium deficiency has not been resolved. Although the removal of all selenoproteins in specific tissues and organs has been widely studied with the $Trsp^{c\Delta}$ mouse model in elucidating the role of selenoproteins in health and development (Table 4.2), there are many other tissues and organs that can be targeted to illuminate the role of selenoproteins in development or in challenging the resulting mouse environmentally to explicate the role of selenoproteins in health. One use of mouse models involving tRNA[Ser]Sec, which could be further investigated to elucidate the role of stress-related selenoproteins in many aspects of health, including cancer, is to target a specific tissue with $Trsp^{c\Delta}$, wherein the mouse carries a low copy number of $Trsp^{tG37}$ transgenes and then is challenged environmentally.

ACKNOWLEDGMENTS

This work was supported by the Intramural Research Program of the National Institutes of Health, NCI, Center for Cancer Research to D.L.H., NIH grants CA080946, GM061603 and GM065204 to V.N.G., and Towson University's Jess and Mildred Fisher College of Science and Mathematics to P.A.T., who is a Jess and Mildred Fisher Endowed Chair of Biological Sciences.

REFERENCES

Adachi, K., M. Katsuyama, S. Song, and T. Oka. 2000. "Genomic organization, chromosomal mapping and promoter analysis of the mouse selenocysteine tRNA gene transcription-activating factor (mStaf) gene." *Biochem J* no. 346 Pt 1:45–51.

Aeby, E., E. Ullu, H. Yepiskoposyan et al. 2010. "tRNASec is transcribed by RNA polymerase II in Trypanosoma brucei but not in humans." *Nucleic Acids Res* no. 38 (17):5833–43. doi: 10.1093/nar/gkq345.

Alsina, B., M. Corominas, M. J. Berry, J. Baguna, and F. Serras. 1999. "Disruption of selenoprotein biosynthesis affects cell proliferation in the imaginal discs and brain of Drosophila melanogaster." *J Cell Sci* no. 112 (Pt 17):2875–84.

Baliga, M. S., V. Diwadkar-Navsariwala, T. Koh et al. 2008. "Selenoprotein deficiency enhances radiation-induced micronuclei formation." *Mol Nutr Food Res* no. 52 (11):1300–4. doi: 10.1002/mnfr.200800020.

Barski, O. A., V. Z. Papusha, G. R. Kunkel, and K. H. Gabbay. 2004. "Regulation of aldehyde reductase expression by STAF and CHOP." *Genomics* no. 83 (1):119–29.

Berry, M. J., L. Banu, Y. Y. Chen et al. 1991. "Recognition of UGA as a selenocysteine codon in type I deiodinase requires sequences in the 3′ untranslated region." *Nature* no. 353 (6341):273–6. doi: 10.1038/353273a0.

Blauwkamp, M. N., J. Yu, M. A. Schin et al. 2008. "Podocyte specific knock out of selenoproteins does not enhance nephropathy in streptozotocin diabetic C57BL/6 mice." *BMC Nephrol* no. 9:7. doi: 10.1186/1471-2369-9-7.

Böck, A., K. Forchhammer, J. Heider, and C. Baron. 1991. "Selenoprotein synthesis: An expansion of the genetic code." *Trends Biochem Sci* no. 16 (12):463–7.

Bosl, M. R., K. Takaku, M. Oshima, S. Nishimura, and M. M. Taketo. 1997. "Early embryonic lethality caused by targeted disruption of the mouse selenocysteine tRNA gene (Trsp)." *Proc Natl Acad Sci U S A* no. 94 (11):5531–4.

Brigelius-Flohé, R. 2008. "Selenium compounds and selenoproteins in cancer." *Chem Biodivers* no. 5 (3):389–95. doi: 10.1002/cbdv.200890039.

Budiman, M. E., J. L. Bubenik, A. C. Miniard et al. 2009. "Eukaryotic initiation factor 4a3 is a selenium-regulated RNA-binding protein that selectively inhibits selenocysteine incorporation." *Mol Cell* no. 35 (4):479–89. doi: 10.1016/j.molcel.2009.06.026.

Carbon, P., and A. Krol. 1991. "Transcription of the Xenopus laevis selenocysteine tRNA(Ser) Sec gene: A system that combines an internal B box and upstream elements also found in U6 snRNA genes." *EMBO J* no. 10 (3):599–606.

Carlson, B. A., S. V. Novoselov, E. Kumaraswamy et al. 2004a. "Specific excision of the selenocysteine tRNA[Ser]Sec (Trsp) gene in mouse liver demonstrates an essential role of selenoproteins in liver function." *J Biol Chem* no. 279 (9):8011–7. doi: 10.1074/jbc .M310470200.

Carlson, B. A., X. M. Xu, G. V. Kryukov et al. 2004b. "Identification and characterization of phosphoseryl-tRNA[Ser]Sec kinase." *Proc Natl Acad Sci U S A* no. 101 (35):12848–53. doi: 10.1073/pnas.0402636101.

Carlson, B. A., X. M. Xu, V. N. Gladyshev, and D. L. Hatfield. 2005a. "Selective rescue of selenoprotein expression in mice lacking a highly specialized methyl group in selenocysteine tRNA." *J Biol Chem* no. 280 (7):5542–8. doi: 10.1074/jbc.M411725200.

Carlson, B. A., X.-M. Xu, V. N. Gladyshev, and D. L. Hatfield. 2005b. "Um34 in selenocysteine tRNA is required for the expression of stress-related selenoproteins in mammals." In *Topics in Current Genetics*, 431–38. Berlin-Heidelberg: Springer-Verlag.

Carlson, B. A., X. M. Xu, R. Shrimali et al. 2006. "Mammalian and other eukaryotic tRNAs." In *Selenium: Its molecular biology and role in human health.* (D. L. Hatfield, M. J. Berry, and V. N. Gladyshev, eds.) 29–37. New York: Springer Science + Business Media, LLC.

Carlson, B. A., M. E. Moustafa, A. Sengupta et al. 2007. "Selective restoration of the selenoprotein population in a mouse hepatocyte selenoproteinless background with different mutant selenocysteine tRNAs lacking Um34." *J Biol Chem* no. 282 (45):32591–602. doi: 10.1074/jbc.M707036200.

Carlson, B. A., U. Schweizer, C. Perella et al. 2009a. "The selenocysteine tRNA STAF-binding region is essential for adequate selenocysteine tRNA status, selenoprotein expression and early age survival of mice." *Biochem J* no. 418 (1):61–71. doi: 10.1042 /BJ20081304.

Carlson, B. A., M. H. Yoo, Y. Sano et al. 2009b. "Selenoproteins regulate macrophage invasiveness and extracellular matrix-related gene expression." *BMC Immunol* no. 10:57. doi: 10.1186/1471-2172-10-57.

Chambers, I., J. Frampton, P. Goldfarb et al. 1986. "The structure of the mouse glutathione peroxidase gene: The selenocysteine in the active site is encoded by the 'termination' codon, TGA." *EMBO J* no. 5 (6):1221–7.

Chavatte, L., B. A. Brown, and D. M. Driscoll. 2005. "Ribosomal protein L30 is a component of the UGA-selenocysteine recoding machinery in eukaryotes." *Nat Struct Mol Biol* no. 12 (5):408–16. doi: 10.1038/nsmb922.

Chittum, H. S., K. E. Hill, B. A. Carlson et al. 1997. "Replenishment of selenium deficient rats with selenium results in redistribution of the selenocysteine tRNA population in a tissue specific manner." *Biochim Biophys Acta* no. 1359 (1):25–34.

Chittum, H. S., W. S. Lane, B. A. Carlson et al. 1998. "Rabbit beta-globin is extended beyond its UGA stop codon by multiple suppressions and translational reading gaps." *Biochemistry* no. 37 (31):10866–70. doi: 10.1021/bi981042r.

Chiu-Ugalde, J., E. K. Wirth, M. O. Klein et al. 2012. "Thyroid function is maintained despite increased oxidative stress in mice lacking selenoprotein biosynthesis in thyroid epithelial cells." *Antioxid Redox Signal* no. 17 (6):902–13. doi: 10.1089/ars.2011.4055.

Choi, I. S., A. M. Diamond, P. F. Crain et al. 1994. "Reconstitution of the biosynthetic pathway of selenocysteine tRNAs in Xenopus oocytes." *Biochemistry* no. 33 (2):601–5.

Combs, G. F. and S. B. Combs, eds. 1986. *The Role of Selenium in Nutrition.* New York: Academic Press.

Diamond, A., B. Dudock, and D. Hatfield. 1981. "Structure and properties of a bovine liver UGA suppressor serine tRNA with a tryptophan anticodon." *Cell* no. 25 (2):497–506.

Diamond, A. M., I. S. Choi, P. F. Crain et al. 1993. "Dietary selenium affects methylation of the wobble nucleoside in the anticodon of selenocysteine tRNA([Ser]Sec)." *J Biol Chem* no. 268 (19):14215–23.

Diwadkar-Navsariwala, V., G. S. Prins, S. M. Swanson et al. 2006. "Selenoprotein deficiency accelerates prostate carcinogenesis in a transgenic model." *Proc Natl Acad Sci U S A* no. 103 (21):8179–84. doi: 10.1073/pnas.0508218103.

Downey, C. M., C. R. Horton, B. A. Carlson et al. 2009. "Osteo-chondroprogenitor-specific deletion of the selenocysteine tRNA gene, Trsp, leads to chondronecrosis and abnormal skeletal development: A putative model for Kashin-Beck disease." *PLoS Genet* no. 5 (8):e1000616. doi: 10.1371/journal.pgen.1000616.

Fairweather-Tait, S. J., Y. Bao, M. R. Broadley et al. 2011. "Selenium in human health and disease." *Antioxid Redox Signal* no. 14 (7):1337–83. doi: 10.1089/ars.2010.3275.

Fradejas, N., B. A. Carlson, E. Rijntjes et al. 2013. "Mammalian Trit1 is a tRNA([Ser]Sec)-isopentenyl transferase required for full selenoprotein expression." *Biochem J* no. 450 (2):427–32. doi: 10.1042/BJ20121713.

Franke, K. W. 1934. "A new toxicant occurring naturally in certain samples of plant foodstuffs I. Results obtained in preliminary feeding trials." *J Nutr* no. 8:597–608.

Glass, R. S., W. P. Singh, W. Jung et al. 1993. "Monoselenophosphate: Synthesis, characterization, and identity with the prokaryotic biological selenium donor, compound SePX." *Biochemistry* no. 32 (47):12555–9.

Guimaraes, M. J., D. Peterson, A. Vicari et al. 1996. "Identification of a novel selD homolog from eukaryotes, bacteria, and archaea: Is there an autoregulatory mechanism in seleno-cysteine metabolism?" *Proc Natl Acad Sci U S A* no. 93 (26):15086–91.

Handy, D. E., G. Hang, J. Scolaro et al. 2006. "Aminoglycosides decrease glutathione peroxi-dase-1 activity by interfering with selenocysteine incorporation." *J Biol Chem* no. 281 (6):3382–8. doi: 10.1074/jbc.M511295200.

Hatfield, D., and F. H. Portugal. 1970. "Seryl-tRNA in mammalian tissues: Chromatographic differences in brain and liver and a specific response to the codon, UGA." *Proc Natl Acad Sci U S A* no. 67 (3):1200–6.

Hatfield, D., A. Diamond, and B. Dudock. 1982. "Opal suppressor serine tRNAs from bovine liver form phosphoseryl-tRNA." *Proc Natl Acad Sci U S A* no. 79 (20):6215–9.

Hatfield, D. L., B. S. Dudock, and F. C. Eden. 1983. "Characterization and nucleotide sequence of a chicken gene encoding an opal suppressor tRNA and its flanking DNA segments." *Proc Natl Acad Sci U S A* no. 80 (16):4940–4.

Hatfield, D., B. J. Lee, L. Hampton, and A. M. Diamond. 1991. "Selenium induces changes in the selenocysteine tRNA[Ser]Sec population in mammalian cells." *Nucleic Acids Res* no. 19 (4):939–43.

Hatfield, D., V. Gladyshev, J. M. Park et al. 1999. "Biosynthesis of selenocysteine and its incorporation into protein as the 21st amino acid." In *Comprehensive Natural Products Chemistry*, 353–80. Oxford: Elsevier Sc. Ltd.

Hatfield, D. L., B. A. Carlson, X. M. Xu, H. Mix, and V. N. Gladyshev. 2006. "Selenocysteine incorporation machinery and the role of selenoproteins in development and health." *Prog Nucleic Acid Res Mol Biol* no. 81:97–142. doi: 10.1016/S0079-6603(06)81003-2.

Hatfield, D. L., M. J. Berry and V. N. Gladyshev, eds. 2012. *Selenium: Its Molecular Biology and Role in Human Health.* New York: Springer Science + Business Media, LLC.

Hirosawa-Takamori, M., H. R. Chung, and H. Jackle. 2004. "Conserved selenoprotein synthe-sis is not critical for oxidative stress defence and the lifespan of Drosophila." *EMBO Rep* no. 5 (3):317–22. doi: 10.1038/sj.embor.7400097.

Hornberger, T. A., T. J. McLoughlin, J. K. Leszczynski et al. 2003. "Selenoprotein-deficient transgenic mice exhibit enhanced exercise-induced muscle growth." *J Nutr* no. 133 (10):3091–7.

Howard, M. T., B. A. Carlson, C. B. Anderson, and D. L. Hatfield. 2013. "Translational redefinition of UGA codons is regulated by selenium availability." *J Biol Chem* no. 288 (27):19401–13. doi: 10.1074/jbc.M113.481051.

Hudson, T. S., B. A. Carlson, M. J. Hoeneroff et al. 2012. "Selenoproteins reduce susceptibility to DMBA-induced mammary carcinogenesis." *Carcinogenesis* no. 33 (6):1225–30. doi: 10.1093/carcin/bgs129.

Irons, R., B. A. Carlson, D. L. Hatfield, and C. D. Davis. 2006. "Both selenoproteins and low molecular weight selenocompounds reduce colon cancer risk in mice with genetically impaired selenoprotein expression." *J Nutr* no. 136 (5):1311–7.

Kasaikina, M. V., A. A. Turanov, A. Avanesov et al. 2013. "Contrasting roles of dietary selenium and selenoproteins in chemically induced hepatocarcinogenesis." *Carcinogenesis* no. 34 (5):1089–95. doi: 10.1093/carcin/bgt011.

Kaushal, N., A. K. Kudva, A. D. Patterson et al. 2014. "Crucial role of macrophage selenoproteins in experimental colitis." *J Immunol* no. 193 (7):3683–92. doi: 10.4049/jimmunol.1400347.

Kelly, V. P., T. Suzuki, O. Nakajima et al. 2005. "The distal sequence element of the selenocysteine tRNA gene is a tissue-dependent enhancer essential for mouse embryogenesis." *Mol Cell Biol* no. 25 (9):3658–69. doi: 10.1128/MCB.25.9.3658-3669.2005.

Kim, I. Y., and T. C. Stadtman. 1995. "Selenophosphate synthetase: Detection in extracts of rat tissues by immunoblot assay and partial purification of the enzyme from the archaean Methanococcus vannielii." *Proc Natl Acad Sci U S A* no. 92 (17):7710–3.

Kim, I. Y., M. J. Guimaraes, A. Zlotnik, J. F. Bazan, and T. C. Stadtman. 1997. "Fetal mouse selenophosphate synthetase 2 (SPS2): Characterization of the cysteine mutant form overproduced in a baculovirus-insect cell system." *Proc Natl Acad Sci U S A* no. 94 (2):418–21.

Kim, J. Y., K. H. Lee, M. S. Shim et al. 2010. "Human selenophosphate synthetase 1 has five splice variants with unique interactions, subcellular localizations and expression patterns." *Biochem Biophys Res Commun* no. 397 (1):53–8. doi: 10.1016/j.bbrc.2010.05.055.

Kim, J. Y., B. A. Carlson, X. M. Xu et al. 2011. "Inhibition of selenocysteine tRNA[Ser]Sec aminoacylation provides evidence that aminoacylation is required for regulatory methylation of this tRNA." *Biochem Biophys Res Commun* no. 409 (4):814–9. doi: 10.1016/j.bbrc.2011.05.096.

Kim, L. K., T. Mtsufuji, S. Matsufuji et al. 2000. "Methylation of the ribosyl moiety at position 34 of selenocysteine tRNA[Ser]Sec is governed by both primary and tertiary structure." *RNA*. no. 6 (9):1306–15.

Kim, T. S., M. H. Yu, Y. W. Chung et al. 1999. "Fetal mouse selenophosphate synthetase 2 (SPS2): Biological activities of mutant forms in Escherichia coli." *Mol Cells* no. 9 (4):422–8.

Krol, A. 2002. "Evolutionarily different RNA motifs and RNA-protein complexes to achieve selenoprotein synthesis." *Biochimie* no. 84 (8):765–74.

Kumaraswamy, E., B. A. Carlson, F. Morgan et al. 2003. "Selective removal of the selenocysteine tRNA [Ser]Sec gene (Trsp) in mouse mammary epithelium." *Mol Cell Biol* no. 23 (5):1477–88.

Labunskyy, V. M., B. C. Lee, D. E. Handy et al. 2011. "Both maximal expression of selenoproteins and selenoprotein deficiency can promote development of type 2 diabetes-like phenotype in mice." *Antioxid Redox Signal* no. 14 (12):2327–36. doi: 10.1089/ars.2010.3526.

Lee, B. J., P. de la Pena, J. A. Tobian, M. Zasloff, and D. Hatfield. 1987. "Unique pathway of expression of an opal suppressor phosphoserine tRNA." *Proc Natl Acad Sci U S A* no. 84 (18):6384–8.

Lee, B. J., S. G. Kang, and D. Hatfield. 1989a. "Transcription of Xenopus selenocysteine tRNA Ser (formerly designated opal suppressor phosphoserine tRNA) gene is directed by multiple 5'-extragenic regulatory elements." *J Biol Chem* no. 264 (16):9696–702.

Lee, B. J., P. J. Worland, J. N. Davis, T. C. Stadtman, and D. L. Hatfield. 1989b. "Identification of a selenocysteyl-tRNA(Ser) in mammalian cells that recognizes the nonsense codon, UGA." *J Biol Chem* no. 264 (17):9724–7.

Lee, K. H., M. S. Shim, J. Y. Kim et al. 2011. "Drosophila selenophosphate synthetase 1 regulates vitamin B6 metabolism: Prediction and confirmation." *BMC Genomics* no. 12:426. doi: 10.1186/1471-2164-12-426.

Leihne, V., F. Kirpekar, C. B. Vagbo et al. 2011. "Roles of Trm9- and ALKBH8-like proteins in the formation of modified wobble uridines in Arabidopsis tRNA." *Nucleic Acids Res* no. 39 (17):7688–701. doi: 10.1093/nar/gkr406.

Leinfelder, W., K. Forchhammer, F. Zinoni et al. 1988. "Escherichia coli genes whose products are involved in selenium metabolism." *J Bacteriol* no. 170 (2):540–6.

Leinfelder, W., T. C. Stadtman, and A. Böck. 1989. "Occurrence *in vivo* of selenocysteyl-tRNA(SERUCA) in Escherichia coli. Effect of sel mutations." *J Biol Chem* no. 264 (17):9720–3.

Lobanov, A. V., D. L. Hatfield, and V. N. Gladyshev. 2008. "Selenoproteinless animals: Selenophosphate synthetase SPS1 functions in a pathway unrelated to selenocysteine biosynthesis." *Protein Sci* no. 17 (1):176–82. doi: 10.1110/ps.073261508.

Loscalzo, J. 2014. "Keshan disease, selenium deficiency, and the selenoproteome." *N Engl J Med* no. 370 (18):1756–60. doi: 10.1056/NEJMcibr1402199.

Low, S. C., J. W. Harney, and M. J. Berry. 1995. "Cloning and functional characterization of human selenophosphate synthetase, an essential component of selenoprotein synthesis." *J Biol Chem* no. 270 (37):21659–64.

Lu, J., L. Zhong, M. E. Lonn et al. 2009. "Penultimate selenocysteine residue replaced by cysteine in thioredoxin reductase from selenium-deficient rat liver." *FASEB J* no. 23 (8):2394–402. doi: 10.1096/fj.08-127662.

Luchman, H. A., M. L. Villemaire, T. A. Bismar, B. A. Carlson, and F. R. Jirik. 2014. "Prostate epithelium-specific deletion of the selenocysteine tRNA gene Trsp leads to early onset intraepithelial neoplasia." *Am J Pathol* no. 184 (3):871–7. doi: 10.1016/j.ajpath.2013.11.025.

Maenpaa, P. H., and M. R. Bernfield. 1970. "A specific hepatic transfer RNA for phosphoserine." *Proc Natl Acad Sci U S A* no. 67 (2):688–95.

Malinouski, M., S. Kehr, L. Finney et al. 2012. "High-resolution imaging of selenium in kidneys: A localized selenium pool associated with glutathione peroxidase 3." *Antioxid Redox Signal* no. 16 (3):185–92. doi: 10.1089/ars.2011.3997.

Mattmiller, S. A., B. A. Carlson, J. C. Gandy, and L. M. Sordillo. 2014. "Reduced macrophage selenoprotein expression alters oxidized lipid metabolite biosynthesis from arachidonic and linoleic acid." *J Nutr Biochem* no. 25 (6):647–54. doi: 10.1016/j.jnutbio.2014.02.005.

McBride, O. W., M. Rajagopalan, and D. Hatfield. 1987. "Opal suppressor phosphoserine tRNA gene and pseudogene are located on human chromosomes 19 and 22, respectively." *J Biol Chem* no. 262 (23):11163–6.

Miniard, A. C., L. M. Middleton, M. E. Budiman, C. A. Gerber, and D. M. Driscoll. 2010. "Nucleolin binds to a subset of selenoprotein mRNAs and regulates their expression." *Nucleic Acids Res* no. 38 (14):4807–20. doi: 10.1093/nar/gkq247.

Moustafa, M. E., B. A. Carlson, M. A. El-Saadani et al. 2001. "Selective inhibition of selenocysteine tRNA maturation and selenoprotein synthesis in transgenic mice expressing isopentenyladenosine-deficient selenocysteine tRNA." *Mol Cell Biol* no. 21 (11):3840–52. doi: 10.1128/MCB.21.11.3840-3852.2001.

Moustafa, M. E., E. Kumaraswamy, N. Zhong et al. 2003. "Models for assessing the role of selenoproteins in health." *J Nutr* no. 133 (7 Suppl):2494S–6S.

Moustafa, M. E., B. A. Carlson, M. R. Anver et al. 2013. "Selenium and selenoprotein deficiencies induce widespread pyogranuloma formation in mice, while high levels of dietary
selenium decrease liver tumor size driven by TGFalpha." *PLoS One* no. 8 (2):e57389.
doi: 10.1371/journal.pone.0057389.

Myslinski, E., A. Krol, and P. Carbon. 1992. "Optimal tRNA((Ser)Sec) gene activity requires
an upstream SPH motif." *Nucleic Acids Res* no. 20 (2):203–9.

Narayan, V., K. C. Ravindra, C. Liao, N. Kaushal, B. A. Carlson, and K. S. Prabhu. 2014.
"Epigenetic regulation of inflammatory gene expression by selenium." *J Nutr Biochem*
(epub ahead of print). doi: 10.1016/j.jnutbio.2014.09.009.

Nirenberg, M., T. Caskey, R. Marshall et al. 1966. "The RNA code and protein synthesis."
Cold Spring Harb Symp Quant Biol no. 31:11–24.

Ohama, T., I. S. Choi, D. L. Hatfield, and K. R. Johnson. 1994. "Mouse selenocysteine
tRNA([Ser]Sec) gene (Trsp) and its localization on chromosome 7." *Genomics* no. 19
(3):595–6. doi: 10.1006/geno.1994.1116.

Oldfield, J. E. Selenium: A historical perspective. 2006. In *Selenium: Its Molecular Biology
and Role in Human Health (2nd Edition)*, 1–6. New York: Springer Science + Business
Media, LLC.

Papp, L. V., A. Holmgren, and K. K. Khanna. 2010. "Selenium and selenoproteins in health
and disease." *Antioxid Redox Signal* no. 12 (7):793–5. doi: 10.1089/ars.2009.2973.

Park, J. M., I. S. Choi, S. G. Kang et al. 1995. "Upstream promoter elements are sufficient for
selenocysteine tRNA[Ser]Sec gene transcription and to determine the transcription start
point." *Gene* no. 162 (1):13–9.

Park, J. M., E. S. Yang, D. L. Hatfield, and B. J. Lee. 1996. "Analysis of the selenocysteine tRNA[Ser]Sec gene transcription *in vitro* using Xenopus oocyte extracts." *Biochem
Biophys Res Commun* no. 226 (1):231–6. doi: 10.1006/bbrc.1996.1338.

Rao, M., B. A. Carlson, S. V. Novoselov et al. 2003. "Chlamydomonas reinhardtii selenocysteine tRNA[Ser]Sec." *RNA* no. 9 (8):923–30.

Rayman, M. P. 2012. "Selenium and human health." *Lancet* no. 379 (9822):1256–68. doi:
10.1016/S0140-6736(11)61452-9.

Schaub, M., A. Krol, and P. Carbon. 1999. "Flexible zinc finger requirement for binding of
the transcriptional activator staf to U6 small nuclear RNA and tRNA(Sec) promoters."
J Biol Chem no. 274 (34):24241–9.

Schuster, C., E. Myslinski, A. Krol, and P. Carbon. 1995. "Staf, a novel zinc finger protein that
activates the RNA polymerase III promoter of the selenocysteine tRNA gene." *EMBO
J* no. 14 (15):3777–87.

Schwarz, K., and C. M. Foltz. 1958. "Factor 3 activity of selenium compounds." *J Biol Chem*
no. 233 (1):245–51.

Schweizer, U., and L. Schomburg. 2005. "New insights into the physiological actions of
selenoproteins from genetically modified mice." *IUBMB Life* no. 57 (11):737–44. doi:
10.1080/15216540500364255.

Schweizer, U., F. Streckfuss, P. Pelt et al. 2005. "Hepatically derived selenoprotein P is a key
factor for kidney but not for brain selenium supply." *Biochem J* no. 386 (Pt 2):221–6.
doi: 10.1042/BJ20041973.

Seeher, S., Y. Mahdi, and U. Schweizer. 2012. "Post-transcriptional control of selenoprotein
biosynthesis." *Curr Protein Pept Sci* no. 13 (4):337–46.

Seeher, S., T. Atassi, Y. Mahdi et al. 2014a. "Secisbp2 is essential for embryonic development
and enhances selenoprotein expression." *Antioxid Redox Signal* no. 21(6):835–49. doi:
10.1089/ars.2013.5358.

Seeher, S., B. A. Carlson, A. C. Miniard et al. 2014b. "Impaired selenoprotein expression in
brain triggers striatal neuronal loss leading to coordination defects in mice." *Biochem J*
no. 462(1):67–75. doi: 10.1042/BJ20140423.

Sengupta, A., B. A. Carlson, V. J. Hoffmann, V. N. Gladyshev, and D. L. Hatfield. 2008a. "Loss of housekeeping selenoprotein expression in mouse liver modulates lipoprotein metabolism." *Biochem Biophys Res Commun* no. 365 (3):446–52. doi: 10.1016/j .bbrc.2007.10.189.

Sengupta, A., B. A. Carlson, J. A. Weaver et al. 2008b. "A functional link between housekeeping selenoproteins and phase II enzymes." *Biochem J* no. 413 (1):151–61. doi: 10.1042 /BJ20080277.

Sengupta, A., U. F. Lichti, B. A. Carlson et al. 2010. "Selenoproteins are essential for proper keratinocyte function and skin development." *PLoS One* no. 5 (8):e12249. doi: 10.1371 /journal.pone.0012249.

Sheridan, P. A., N. Zhong, B. A. Carlson et al. 2007. "Decreased selenoprotein expression alters the immune response during influenza virus infection in mice." *J Nutr* no. 137 (6):1466–71.

Shim, M. S., J. Y. Kim, H. K. Jung et al. 2009. "Elevation of glutamine level by selenophosphate synthetase 1 knockdown induces megamitochondrial formation in Drosophila cells." *J Biol Chem.* no. 284 (47):32881–94. doi: 10.1074/jbc.M109.026492.

Shrimali, R. K., A. V. Lobanov, X. M. Xu et al. 2005. "Selenocysteine tRNA identification in the model organisms Dictyostelium discoideum and Tetrahymena thermophila." *Biochem Biophys Res Commun* no. 329 (1):147–51. doi: 10.1016/j.bbrc.2005.01.120.

Shrimali, R. K., J. A. Weaver, G. F. Miller et al. 2007. "Selenoprotein expression is essential in endothelial cell development and cardiac muscle function." *Neuromuscul Disord* no. 17 (2):135–42. doi: 10.1016/j.nmd.2006.10.006.

Shrimali, R. K., R. D. Irons, B. A. Carlson et al. 2008. "Selenoproteins mediate T cell immunity through an antioxidant mechanism." *J Biol Chem* no. 283 (29):20181–5. doi: 10.1074/jbc.M802559200.

Söll, D. 1988. "Genetic code: Enter a new amino acid." *Nature* no. 331 (6158):662–3. doi: 10.1038/331662a0.

Songe-Moller, L., E. van den Born, V. Leihne et al. 2010. "Mammalian ALKBH8 possesses tRNA methyltransferase activity required for the biogenesis of multiple wobble uridine modifications implicated in translational decoding." *Mol Cell Biol* no. 30 (7):1814–27. doi: 10.1128/MCB.01602-09.

Sturchler, C., E. Westhof, P. Carbon, and A. Krol. 1993. "Unique secondary and tertiary structural features of the eucaryotic selenocysteine tRNA(Sec)." *Nucleic Acids Res* no. 21 (5):1073–9.

Sturchler, C., A. Lescure, G. Keith, P. Carbon, and A. Krol. 1994. "Base modification pattern at the wobble position of Xenopus selenocysteine tRNA(Sec)." *Nucleic Acids Res* no. 22 (8):1354–8.

Tamura, T., S. Yamamoto, M. Takahata et al. 2004. "Selenophosphate synthetase genes from lung adenocarcinoma cells: Sps1 for recycling L-selenocysteine and Sps2 for selenite assimilation." *Proc Natl Acad Sci U S A* no. 101 (46):16162–7. doi: 10.1073 /pnas.0406313101.

Tobe, R., S. Naranjo-Suarez, R. A. Everley et al. 2013. "High error rates in selenocysteine insertion in mammalian cells treated with the antibiotic doxycycline, chloramphenicol, or geneticin." *J Biol Chem* no. 288 (21):14709–15. doi: 10.1074/jbc.M112.446666.

Watanabe, K., and S. Yokobori. 2011. "tRNA modification and genetic code variations in animal mitochondria." *J Nucleic Acids* no. 2011:623095. doi: 10.4061/2011/623095.

Wirth, E. K., M. Conrad, J. Winterer et al. 2010. "Neuronal selenoprotein expression is required for interneuron development and prevents seizures and neurodegeneration." *FASEB J* no. 24 (3):844–52. doi: 10.1096/fj.09-143974.

Xu, X. M., X. Zhou, B. A. Carlson et al. 1999. "The zebrafish genome contains two distinct selenocysteine tRNA[Ser]sec genes." *FEBS Lett* no. 454 (1–2):16–20.

Xu, X. M., B. A. Carlson, H. Mix et al. 2007a. "Biosynthesis of selenocysteine on its tRNA in eukaryotes." *PLoS Biol* no. 5 (1):e4. doi: 10.1371/journal.pbio.0050004.

Xu, X. M., B. A. Carlson, R. Irons et al. 2007b. "Selenophosphate synthetase 2 is essential for selenoprotein biosynthesis." *Biochem J* no. 404 (1):115–20. doi: 10.1042/BJ20070165.

Xu, X. M., A. A. Turanov, B. A. Carlson et al. 2010. "Targeted insertion of cysteine by decoding UGA codons with mammalian selenocysteine machinery." *Proc Natl Acad Sci U S A* no. 107 (50):21430–4. doi: 10.1073/pnas.1009947107.

Yoshizawa, S., and A. Böck. 2009. "The many levels of control on bacterial selenoprotein synthesis." *Biochim Biophys Acta* no. 1790 (11):1404–14. doi: 10.1016/j.bbagen.2009.03.010.

Yuan, J., S. Palioura, J. C. Salazar et al. 2006. "RNA-dependent conversion of phosphoserine forms selenocysteine in eukaryotes and archaea." *Proc Natl Acad Sci U S A* no. 103 (50):18923–7. doi: 10.1073/pnas.0609703104.

Zinoni, F., A. Birkmann, T. C. Stadtman, and A. Böck. 1986. "Nucleotide sequence and expression of the selenocysteine-containing polypeptide of formate dehydrogenase (formate-hydrogen-lyase-linked) from Escherichia coli." *Proc Natl Acad Sci U S A* no. 83 (13):4650–4.

Zinoni, F., A. Birkmann, W. Leinfelder, and A. Böck. 1987. "Cotranslational insertion of selenocysteine into formate dehydrogenase from Escherichia coli directed by a UGA codon." *Proc Natl Acad Sci U S A* no. 84 (10):3156–60.

Section III

Se Compounds with Specific Functions

5 Redox Cycling and the Toxicity of Selenium Compounds
A Historical View

Julian E. Spallholz

CONTENTS

5.1 Introduction 83
    5.1.1 Discovery of Selenium: Chemistry and Toxicity 83
5.2 An Approximately 750-Year-Old Unanswered Question:
    Why Is Selenium Toxic? 86
5.3 Why Is Selenium Toxic? "Superoxide Generation" 88
    5.3.1 Superoxide Generation Determination 89
    5.3.2 Redox Cycling Selenium, The "Magic Bullet" for Cancer Treatment 95
    5.3.3 Are Selenides Really "Magic Bullets"? 98
5.4 Other and More Recent Selenium Drug Developments 100
5.5 Conclusions and Future Perspectives 103
Acknowledgments 103
References 104

Follow the laws of nature and you'll never go wrong.

Forrest C. Shaklee

5.1 INTRODUCTION

5.1.1 DISCOVERY OF SELENIUM: CHEMISTRY AND TOXICITY

In 1816, Jöns Jakob Berzelius (Figure 5.1) and a group of five other individuals took over at auction an old distillery beside Gripsholm Castle near Mariefried that housed Sweden's first chemical factories. It was in these factories that acetic and sulfuric acids were made in lead-lined chambers and produced to make white lead paint. Berzelius introduced the scientific analyses of products into the production methods of the factory. Sulfur was burned to produce sulfur dioxide, which, when combined with nitrogen dioxide and water, formed sulfuric acid. Alternatively, in place of elemental sulfur,

83

(a)   (b)

FIGURE 5.1 Jöns Jakob Berzelius (1779–1848) and memorial headstone, Stockholm. (a) From Berzelius Exposition, Stockholm, 2011, with permission by Mallory Boylan. (b) Memorial headstone, 2011, by Julian Spallholz.

iron pyrites could be burned to produce the sulfur dioxide. The lead-lined chambers contained the sulfuric acid and dissipated the heat from the synthetic exothermic reaction of sulfuric acid synthesis. In 1817, Berzelius isolated from the lead slimes what he thought was an arsenic compound from the iron pyrite process of making sulfur dioxide, and that process was discontinued. A red precipitate was discovered in the sulfur, which he first thought to be tellurium, but the starting materials for the synthesis of the sulfuric acid contained no detectable tellurium. The red element smelled of horseradish upon being burned. Berzelius must have soon realized that he had discovered a new element, and he named the element selenium (Se) for the Goddess of the moon, Selene. The new element was so named as its chemical properties were found to be similar to that of sulfur and that of tellurium named for the Earth, and selenium fit nicely just above tellurium and below sulfur in the Periodic Table. Element 34 was thus discovered, and the results of Berzelius's discovery were published in 1817 (Berzelius, 1817).

In the 1930s, selenium was discovered to be accumulated and concentrated from soils by some plants (Franke, 1934). Within the Great Plains states of the United States, plants, principally of the genus *Astragalus* (Beath and Lehnert, 1917), *Stanleya*, *Xylorrhiza*, and *Oonopsis*, when growing on seleniferous shales, concentrated the more toxic inorganic selenium (Rosenfeld and Beath, 1946) into the less toxic organic selenium compounds (Shrift and Virupaksha, 1965). Such concentrations of selenium by the "selenium accumulator plants" could exceed 1 g Se/kg of dry matter, and selenium became known to be toxic among grazing livestock, principally cattle and sheep, but also horses (Figure 5.2). Manifestation of selenium toxicity affected the nervous system, growth of hair, and other keratotic tissues. The sloughing off of horses' hooves observed in cases of selenosis in the Great Plains states was apparently also noted in horses by Marco Polo while traveling the Silk Road in the thirteenth century as attributed by his fellow prisoner and biographer, Rustichello da Pisa (In French, 1298; translated by William Marsden, 1818, https://archive.org/stream/travelsofmarcopo92polo#page/60/mode/2up, and again in the translation by Thomas Wright, London, 1892, from the 1818 translation by Marsden).

FIGURE 5.2 *Astragalus bisulcatus* (a) and *Stanleya pinnata* (b); selenium accumulator plants. (Courtesy of the US Geological Survey.)

Until 1957, selenium had only been known for its toxicity. Quite by chance, Klaus Schwarz, a German physician specializing in liver diseases was doing research on a dietary induced condition, liver necrosis in rats (Schwarz, 1954). He had discovered that small amounts of selenium, along with or separately from vitamin E and sulfur amino acids, could prevent the liver necrosis. Prior to this discovery, selenium had been dubbed by Dr. Schwarz "Factor-3" as it was the third of the three factors having been isolated from brewer's yeast that prevented the liver necrosis (Schwarz and Foltz, 1957, 1958).

At the University of Wisconsin in 1973, graduate student John Rotruck, working under the supervision of Dr. William Hoekstra, discovered the first mammalian enzyme, glutathione peroxidase (GPx) isolated from rat erythrocytes to contain selenium (Rotruck et al., 1973). GPx had been originally isolated from bovine erythrocytes in 1959 by Gordon Mills (Mills, 1959), and crystalline GPx was rapidly confirmed to contain 4 gram-atoms of selenium/molecule in the custody of Leopold Flohé (Flohé et al., 1973). Dr. Flohé commented on his surprise to find "that awful poison selenium in his beloved enzyme" at a 1976 meeting of the Selenium-Tellurium Development Association on the campus of Notre Dame University.

Prevention of dietary liver necrosis in rats and, thereafter, exudative diathesis in chickens with dietary selenium supplements changed forever the biological selenium landscape. Like other nutritionally important elements before only known for their toxicity (Stohs and Bagchi, 1995), selenium seemed, too, to be nutritionally important. The later discovery, in 1973, that selenium was not just a cofactor of GPx, but that it resided at the catalytic site of the enzyme as an analog of cysteine, namely selenocysteine, again changed everything (Forstrom et al., 1978; Wendel et al., 1978; Zakowski et al., 1978). More and more enzymes and proteins were being found to contain selenium as selenocysteine, and the static understanding of the genetic DNA/RNA code was being changed to accommodate selenocysteine—what was to become the 21st amino acid (Böck et al., 1991a,b). Ten years after GPx was determined to contain selenium, Thomas Jukes coined the term to selenium "an essential poison" (Jukes, 1983). As Paracelsus (Philippus Aureolus Theophrastus Bombastus von Hohenheim, 1493–1541) had stated so many years ago, "the dose makes the poison," a statement that certainly applied to selenium. I would add to Paracelsus' insight on toxicity that the chemical forms of selenium were also to become important in understanding its more universal toxicity.

5.2   AN APPROXIMATELY 750-YEAR-OLD UNANSWERED QUESTION: WHY IS SELENIUM TOXIC?

The question above, about 750 years after Marco Polo and certainly unknown to him, was to become an overriding interest following my 1965 graduation from Colorado State University with a BS in biological sciences. In the winter quarter of 1966, I took general biochemistry taught by Professor John L. Martin. Together with the bioenergetics from the fall semester, all of my rather dead undergraduate biology major came to life. Having enjoyed biochemistry and Dr. Martin as an instructor, I approached him about accepting me as a graduate student. Dr. Martin's major research interest at the time was the metabolism of selenium in plants and how toxic levels of selenium were metabolized in both plants and animals. This was a full six to seven years before the discovery of selenium in GPx in 1973. Extracts of selenium accumulator plants and HCl protein hydrolyzates of chick tissues were chromatographed on a Beckman-120 Amino Acid Analyzer with Se-75 from selenite being followed by an attached liquid scintillation counter (Martin et al., 1971). I learned to make buffers and ninhydrin and pack 4-foot Beckman columns with ion exchange resins. From selenium accumulator plants, we typically found selenocystathionine and Se-methylselenocysteine. From animal protein hydrolyzates, selenocystine commonly eluted closely with cystine. Such experiments were made possible in part by obtaining selenium accumulator plant samples from the University of Wyoming while visiting with Professor J. W. Hamilton (Hamilton, 1975).

My very first animal experiment with selenium was with baby chicks initially fed 10 and then 20 ppm Se as selenite added to a commercial chick feed. Feeding 10 ppm Se initially and then increasing the feeding to 20 ppm Se after 3 weeks, in comparison to control chicks, which looked like mature chickens, the yellow chicks did not grow very much, and mature feathers would occasionally grow out from the chick's yellow down. I had formulated the diets and was surprised by how little selenium it took to stunt the growth of the chicks (Martin and Spallholz, 1968, 1970). That was 1967, and from that time forward, I would ask Dr. Martin's peers, "Why is selenium toxic?" I certainly asked and, of course, discussed it with Dr. Martin. I certainly asked Howard Ganther as he had his own ideas about why selenium was toxic: selenotrisulfide formation (Ganther, 1968; Ganther and Corcoran, 1969). I am positive I asked Al Tappel, and we likely discussed selenite catalysis (Tsen and Tappel, 1958). I certainly asked Kenneth McConnell (McConnell and Roth, 1968) as he, like Dr. Martin, was showing the formation of selenocystine, selenomethionine, and an unknown selenium compound in animal tissue hydrolyzates by chromatography using Se-75 (Martin and Gerlach, 1969).

Thus, beginning in 1960 and into the 1970s, because selenium was much like sulfur, selenium was thought to possibly be toxic by replacement of the sulfur in sulfur compounds. Events over the next decade, the 1970s, would leave retrospective "experimental published fossils" of the cause of selenium toxicity. The first was the confirmation of the existence of free radicals in biological systems by Irwin Fridovich and Joe M. McCord, recognizing that hemocuprein was, in fact, an enzyme, superoxide dismutase (McCord and Fridovich, 1969a,b, 1988). The free radical idea of aging had been to that point in time but the theory of Dr. Denham Harman (1956). Until the discovery of superoxide and superoxide dismutase by McCord and Fridovich, the situation remained that no one really knew for sure or had published why selenium was toxic. But, in the form of selenite, selenium was known to be reduced to elemental selenium (Tsen and Tappel, 1958) or to be methylated to be less toxic (Ganther, 1966).

In 1980, the late John Milner clearly showed that selenium compounds administered to mice inoculated with Ehrlich ascites tumor cells (EATC) responded differently to selenium treatment and without harm to the cancer host animal. Injections of different inorganic and organic selenium compounds limited EATC growth and survival of mice in both a dose- and time-dependent manner. Data published in *Science* and presented by Dr. Milner at the Second International Symposium on Selenium Biology and Medicine in May 1980 in Lubbock, Texas, demonstrated differences in the effects of selenium compounds administered peritoneally with EATCs on the survival of mice (Greeder and Milner, 1980) and EATCs (Poirier and Milner, 1979).

In the summer of 1980, I was given an invitation and opportunity to do research of my liking at the Forrest E. Shaklee Corporation Research Center in Hayward, California. The Research Center had a Hitachi SEM, and I had wondered what was happening to all of John Milner's EATCs in those mice treated with all those selenium compounds. The mouse data was clearly on my mind at the time (Spallholz et al., 1982). Dr. Milner's paper in *Science* (Greeder and Milner, 1980) and his presentation in Lubbock showed he had treated EATCs in mice with selenium dioxide, selenite, selenate, selenocystine, and selenomethionine, each with a different survival outcome for the mice. So I decided with the Shaklee research invitation to mimic and repeat Dr. Milner's experiments with those selenium compounds he had used against EATCs. I decided to use rat erythrocytes as test cells as I had some previous experience with selenium compounds and erythrocyte hemolysis.

With selenium compounds in hand and a SEM technician assigned to help me that summer of 1980, I drove to Hayward to the Forrest Shaklee Research Center. The rats from which I collected erythrocytes were housed on the University of California Berkeley campus. It was there where I would collect the rat blood, place it on ice, and returned it to Hayward. There, erythrocytes were incubated with the selenium compounds and periodically sampled upon treatment up to 3 hours. Cells were fixed with buffered formalin on glass coverslips, dried to critical point with CO_2, sputtered with gold, and photographed using a Hitachi SEM and attached B&W Polaroid camera. Seeing in the electron micrographs damage to some erythrocyte membranes, I expanded the visual SEM effort to include bacteria and yeast cells. The results, all visually preserved in SEM photographs, made it clear to me that I had confirmed the experimental EATC mice data of John Milner almost perfectly in an *in vitro* system. In the extremes, selenite and selenocystine were visually cytotoxic to erythrocytes, *E. coli*, *B. subtilis*, and Fleishman yeast cells in a time- and dose-dependent manner. Selenomethionine as reported by John Milner appeared to be nontoxic to EATCs with all mice dying, and selenomethionine at extended times were not damaging any of my treated erythrocytes as assessed by SEM (Hu et al., 1982; Hu and Spallholz, 1983). Why, I asked myself, is selenite and selenocystine visually toxic to all cells and selenomethionine visually not toxic under my experimental conditions? The summer of 1980 had come and gone, and still I did not have, and nobody else I knew and had discussed the matter with had a good answer to my lingering question, "Why is selenium toxic?" Moreover, at the end of the summer of 1980, the question was now larger: "Why are selenium compounds so differently toxic?"

The answer to selenite toxicity for me began to be answered in early January 1982. In December 1981, returning from four weeks in China, I stopped in Honolulu to visit with Dr. Larry Piette. He was chairman of the biochemistry/biophysics department at the University of Hawaii and had been my major professor (1968–1971). It was at the University of Hawaii that I was introduced to stable free radicals and nitroxides and learned to do organic chemistry and electron spin resonance spectroscopy (Spallholz and Piette, 1972). I had brought with me several of the SEM Polaroid prints of selenium-generated cell damage taken at the Shaklee Research Center. Showing him the photographs, I asked his opinion of what he thought might have caused the cell damage. "I would guess it's due to free radicals", he said to me. So the idea was instilled in me that the cell damage I had observed from all the various cells exposed to selenite was possibly of "free radical" origin. It also appeared probable to me that selenomethionine did not produce "free radicals" because I had not observed any erythrocyte or other cells damage from it. At the 1982 FASEB meeting later that year in New Orleans, I presented my visual SEM findings of cell damage by selenium compounds (Hu et al., 1982). I returned from New Orleans to perform selenium and Se-yeast (from Nutrition 21, La Jolla, CA) toxicity studies in animals (Spallholz and Raftery, 1983). In addition, I found that alfalfa and other seed germination could also be used to quantify the differences in the toxicity of selenium compounds. Two years later, the 1988 Third International Selenium in Biology and Medicine Meeting in Tübingen, Germany, changed and forever altered my understanding of the question why is selenium toxic?

5.3   WHY IS SELENIUM TOXIC? "SUPEROXIDE GENERATION"

Tübingen is a beautiful German town situated on the Neckar River upstream from Heidelberg not far from the home of my great-grandfather in Stuttgart. The International Selenium meeting was hosted by Dr. Albrecht Wendel (Wendel, 1989). This particular selenium meeting became, for me, transformational as I strolled through the poster session. I came upon the poster by Dr. Yushi Seko showing the generation of a free radical, superoxide, by selenite (Equation 5.1) (Seko et al., 1989; Kitahara et al., 1993; Seko and Imura, 1997).

$$4GSH \quad GSSG \qquad GSH \quad GSSG \qquad GSH \quad GSSG \qquad O_2 \quad O_2^{\cdot -}$$
$$SeO_3^{2-} \longrightarrow GSSeSG \longrightarrow GSSeH \longrightarrow H_2Se \longrightarrow Se^0 \quad (5.1)$$

At that moment, all of my experimental experiences with selenite toxicity, initially with plants and chicks, the toxicity differences in selenium compounds affecting plant seed germination, Se-yeast fed rats, the 1980 summer cell work at the Shaklee Research Center, my January 1982 conversation with Dr. Larry Piette at Waikiki, other animal toxicity experiments with mice and rats of my own, and from the literature all gelled, and it was clear to me within moments why selenium compounds might be toxic. Dr. Seiko's poster just papered a plausible answer to my long sought question, why is selenium toxic? It was the generation of the superoxide "free radical" as originally suggested to me by Dr. Piette. The generation of superoxide as the cause of selenium toxicity was to be questioned later by others, but for me, it was clear; selenite at least was toxic as it generates superoxide. I also immediately recognized the likelihood that other selenium compounds that were experimentally toxic were all also toxic because they too likely generated superoxide. I knew from the literature that selenite reacted with GSH (Ganther, 1971), and as the major thiol of cells, it also declined in cells when treated with selenite (Anundi et al., 1982; Frenkel and Falvey, 1988). The Seko equation was very similar, yet not identical, to the equation of the enhanced catalytic sulfide reduction of methylene blue (MB) by selenium as reported in 1947 by Feigl and West (1947), the selenite catalysis of Tsen and Tappel (1958) and to Gerhard Schrauzer's 1974 methylene blue assay for distinguishing between inorganic and organic selenium compounds, particularly selenite and selenomethionine, in dietary supplemental yeast (Rhead and Schrauzer, 1974). I had also seen the elemental selenium many times in the Seko equation; that orange-red precipitate formed within minutes upon mixing selenite (but not selenate!) with GSH or ascorbic acid. I had also seen it in the cell walls of *E. coli* cultured in the presence of selenite, and I had seen it in the roots of germinating plant seeds. I did not see in the equation the plausibility of the precise stoichiometry presented in the reaction by Dr. Seko, nor did I see selenomethionine generating superoxide by replacing selenite in the equation. Selenomethionine was just never very toxic in any animal experiments—not in cell or tissue cultures, not to EATCs in Milner's experiments, and not to erythrocyte membranes. Dr. Seko's poster did provide me with a general comprehensive overview, a working hypothesis of selenium toxicity (Seko, 1986). For me, selenite as presented by Dr. Seko's poster looked much like glutathione peroxidase and its catalytic selenocysteine active site. Both consumed GSH, and although H_2Se was shown to produce superoxide, selenium enzymes produced either water or organic alcohols, neither being toxic. In biological terms, this catalytic difference seemed to me significant. Although GPx detoxified the dismutation product of superoxide, H_2O_2, producing water, other selenols obviously kept producing superoxide. In retrospect, this fundamental difference in catalysis explained the curiosity of selenium deficiency, for example, oxidative membrane damage, mimicking selenium intoxication. In chemical terms, selenium's enzymatic activity and selenium toxicity, in my view, were *now* both attributes of simply selenide's (RSe-) catalytic activity. I had only to return to the laboratory to experimentally determine if my newly formed hypothesis was correct. Upon my return from Germany, I was able to place an order for a luminometer to test my hypothesis. On September 16, 1992, I gave a seminar titled "Why Is Selenium Toxic?"

5.3.1 SUPEROXIDE GENERATION DETERMINATION

While waiting on the delivery of the luminometer, my student J. J. Chen, JJ as she was known, and I along with Mallory Boylan turned to the 1974 methylene blue reduction assay of Rhead and Schrauzer (1974). This assay catalyzes the reduction of MB by selenite and other ions in the presence of thiols. Reduction is monitored spectro-photometrically at 670 nm. Using the MB assay, we confirmed the reduction of MB by selenite as well as selenocystine, both compounds having been observed in the summer of 1980 by SEM to damage rat erythrocytes. It was possible to also show that the addition of superoxide dismutase to the assay could inhibit the catalytic MB reduction by both selenite and selenocystine (Dickson and Tappel, 1969; McCord and Fridovich, 1969b). With some excitement, we concluded that the 1947 Feigl and West selenium-sulfide and the 1974 Rhead and Schrauzer MB reduction selenium assays were mediated by superoxide as presented by Dr. Seko. On one occasion, JJ came to me and said the MB assay was not working. Indeed it was not working looking at the spectrophotometer, and I inquired if the assay solution had been freshly made. She so acknowledged, and I pointed to the full carboy of distilled water across the laboratory. I said there is no oxygen in the water, and after bubbling oxygen through the assay solution from a small O_2 gas tank, the assay worked fine. This incident became an unplanned confirmation that the reduction of MB was being mediated by superoxide.

As the MB assay requires the presence of both thiols and oxygen for reduction, the necessity of measuring the rate of reduction of MB in the absence of selenite or another selenium compound is necessary and involves a number of considerations as earlier reported by Bernal et al. (1990). We had also discovered that a high or low blank, rapid or slow reduction of MB by GSH alone was dependent upon the amount of GSH in the assay cocktail. This suggested that the source of the elections for MB reduction, as in the Seko poster at Tübingen, was GSH. This also reinforced my existing opinion that H_2Se was not likely the intermediate generator of superoxide by the *in vitro* reaction of selenite with GSH as shown by Dr. Seko's poster because no H_2Se would have been formed by the reaction of selenocystine with GSH *in vitro*. And yet selenocystine, too, reduced MB, which could be quenched by addition of superoxide dismutase to the cocktail. In reconsideration of Dr. Seko's poster equa-tion, the Feigl and West paper (1947), and the formation of selenotrisulfides from thiols, as first described by Howard Ganther, I again drew from such considerations that selenides, GSSeH from the reduction of GSSeSG and RSeH from the reduction of selenocystine by GSH, were the catalytic selenium species generating superoxide. GSH was being repeatedly oxidized by a catalytic selenium species, now referred to as "redox cycling," and in consequence, the overall reaction is not a stoichiometric event. Selenium was just a catalyst as noted by Feigl and West in 1947, by Tsen and Tappel in 1958, and again in 1987 by Tappel as used by Dr. Schrauzer in his com-mercial chemical assay kit to distinguish inorganic selenite from organic selenium, selenomethionine (Hu and Tappel, 1987). First author Maio-Lin Hu had been my graduate and postdoctoral student at Texas Tech University. Now a postdoctoral stu-dent with Professor Tappel, he likely understood this selenium chemistry very well. Their 1987 paper cites an "unidentified" small selenium molecule as the catalytic species: "The higher (catalytic) activity was not attributed to Se-containing proteins, but to an unknown small molecular-weight factor."

Upon arrival of the luminometer at the university, lucigenin chemiluminescence (CL) replaced the Hitachi UV/VIS spectrophotometer that monitored MB selenium assays. J. J. Chen, Mallory Boylan, and myself all conspired to confirm all the same experiments done for measuring superoxide using MB reduction by selenite and selenocystine by now using the lucigenin CL assay for superoxide detection. We then commenced a broader screening program just to determine which selenium compounds would generate superoxide and which selenium compounds did not generate superoxide in the CL assay. Selenite but not selenate generated superoxide. Selenocystine and other aliphatic and aromatic diselenides generated superoxide. One of the best organic generators of superoxide was selenocystamine from Sigma, and it became our "standard" generator to confirm the CL assay cocktail was working. Selenomethionine, which did not reduce MB, did not generate any CL activity by luminometry, confirming the use of the commercial MB assay as developed by Dr. Schrauzer to distinguish selenomethionine containing Se-yeast from Se-yeast doped with selenite.

Following Klaus Schwarz's death from a stroke in 1978, for whom I had been his last postdoctoral student (1974–1978), I rescued and had at hand a number of the many mono- and diselenides originally synthesized in Dr. Schwarz's laboratory by Dr. Keshar Pathak at the Long Beach, California, VA Hospital or by Dr. Arne Fredga of Uppsala, Sweden (Schwarz, 1969; Schwarz and Fredga, 1974, 1975; Schwarz and Pathak, 1975). The Se compounds I had retained were all synthesized in Dr. Schwarz's laboratory and had been tested for toxicity in mice in Germany by Dr. Leopold Flohé (personal account). Dr. Schwarz's goal was to find a biological source of selenium for selenium deficiency diseases then known, which was less toxic than selenite or selenate, the latter being chronically toxic *in vivo* (Schwarz and Fredga, 1974). My goal was to assess each of the Schwarz library of compounds at hand for their capability to generate superoxide in our CL assay. To briefly summarize, all the diselenides I had at hand at the time, or even later, proved to generate superoxide, some better than others, and all monoselenides that I had from the Long Beach laboratory of Dr. Schwarz did not generate superoxide. This was in keeping with our earlier observations that selenomethionine and Se-methylselenocysteine, that is, monoselenides, did not reduce MB or generate superoxide in our CL assay. Many of the compounds of the Schwarz library had a terrible odor but none worse than dimethyldiselenide purchased from Sigma (Sigma Chemical Co., St. Louis, MO, USA). Even this very foul-smelling diselenide generated superoxide in our CL assay (Spallholz et al., 2001).

Following the generalized discovery that all diselenides might generate superoxide in the presence of GSH and other thiols and that all monoselenides might not generate superoxide, I sought to review some of the animal toxicity data as reported by Schwarz for his di- and monoselenide compounds. To my surprise, diselenides, as reported by Schwarz, were almost uniformly more toxic *in vivo* than the corresponding monoselenides (Schwarz, 1969). The correlation was strikingly beautiful. Selenium compounds that generated superoxide were toxic, and if they did not generate superoxide, they were not toxic per se [sic]. Accordingly, the toxicity of the nutritional selenium compounds, such as selenomethionine and Se-methylselenocysteine, had to arise from their metabolism, which, in my view, had to generate some form of selenol, RSeH. This we know now to be true. Methylselenol (CH_3SeH) (Ip et al., 2000, 2002), really methylselenolate at physiological pH, (CH_3Se-) is generated from dimethyldiselenide upon reduction and from selenomethionine and Se-methylselenocysteine by gamma- and beta-lyases. These enzymes, when present in tissues, very likely account for the *in vivo* toxicity of selenomethionine and Se-methylselenocysteine (Spallholz et al., 2004).

By 1992, it was possible, as presented below, to rank, with some certainty, the relative toxicity of naturally occurring and manmade selenium compounds based on literature data, SEM micrographs of cells, MB reduction, CL, and animal and seed germination toxicity studies conducted at Texas Tech and elsewhere as follows:

Selenodiglutathione > selenite = selenite = selenium dioxide > selenocystine and other diselenides (RSeSeR) ≫ monoselenides (RSeR) ≫ methylselenides; selenomethionine = Se-methylselenocysteine ≫ trimethylselenonium ion ≫ elemental selenium

This overall concept of selenium toxicity was presented at the Fifth International Selenium Symposium in Nashville, Tennessee (Spallholz and Whittam, 1992), and the initial list of CL and non-CL selenium compounds was presented at the 1996 Selenium–Tellurium Development–sponsored meeting in Scottsdale, Arizona. The selenium compounds below in (A) are a partial list of those selenium compounds having been shown to generate superoxide by chemiluminescence, by methylene blue, or cytochrome c reduction and those selenium compounds in (B) that do not generate superoxide as assessed by chemiluminescence, by methylene blue, or cyto-chrome c reduction:

(A) Selenite, selenium dioxide, selenodiglutathione, selenocystine, selenocysta-mine, diselenodiproprionic acid, diphenyldiselenide, diphenylmethyldiselenide, dibenzydiselenide, 1,4-phenyl-bis-(methyleneselenocyante, 6-propylselenourea, dimethyldiselenide, methylseleninic acid, acetoacetoxyethyl methacrylate (AAEMA)

(B) Elemental selenium, selenate, dimethylselenoxide, selenopyridine, seleno-methionine, se-methylselenocysteine, trimethylselenonium ion, KSeCN

The top placement of selenodiglutathione as being more toxic than selenite (excluding possibly H_2Se as HSe^-) was initially based upon the comparative sele-nium toxicity *in vitro* from the literature and later corroborated by the comparative CL data of Goswami et al. (2012). A comparison of the CL generated by selenodi-glutathione and selenate revealed that selenodiglutathione more rapidly generates superoxide and generates more superoxide per unit of time than does selenite. This CL difference resided in the starting time delay of measurable superoxide generation, which is accounted for by the time lag in the formation of selenodi-glutathione from selenite and GSH. With selenate, the reaction does not occur at all *in vitro*. To be toxic, selenate has to be first reduced to selenite *in vivo* to react with GSH or other thiols. From selenodiglutathione, additional GSH reduces the selenotrisulfide, forming a catalytic selenopersulfide (GSSe-) as originally sug-gested by Feigl and West in 1947 and as discussed in detail by Tappel in 1958 and later synthesized by Ganther (Ganther, 1968; Ganther and Corcoran, 1969). Selenodiglutathione generates more total CL than selenite followed by the organic selenides, methylseleninic acid, and diselenodiproprionic acid. Lastly, selenomethionine generates no measurable CL.

Superoxide generation measured by MB reduction or CL by selenides correlates rather nicely with selenium toxicity and visual cell or tissue damage *in vitro* as well as toxicity *in vivo*. What had emerged from this collective student–faculty effort in the early 1990s was that organic selenolates (RSe⁻) could nonstoichiometrically redox cycle generating superoxide.

Sometime in the 1992–1993 time frame, I received information from Howard Ganther drawing my attention to a new publication by Jean Chaudière showing that diselenides could generate superoxide (Chaudière et al., 1992). The authors showed that diselenides could generate superoxide and were suggesting that this selenium redox chemistry could account for selenium's toxicity *in vivo*. The Chaudière selenium redox chemistry is presented in Figure 5.3 showing the reduction of selenocystamine forming two equivalent selenides transferring an electron to oxygen, forming superoxide, and a second, reportedly faster electron reduction to superoxide forming H_2O_2 and water.

The paper by Chaudière et al. (1992) explains in eloquent detail the nearly universal CL activity we had observed with almost all diselenides. Moreover, it made clear that the formation of superoxide depended on the concentration and catalytic activity of the selenides, which, in turn, depends on the electrostatic effect on selenium and, thus, is affected by the substituting R moiety in Figure 5.3 (Ip et al., 1995). The authors also reported that RSe⁻ could be carboxymethylated with iodoacetate as had been done and presented in J. J. Chen's 1997 dissertation and later published for reduced selenocystamine, diselenodiproprionic acid, as well as for GSSeH generated from selenite (Chen et al., 2007). Figure 5.3, like no others, explains the nearly universal CL activity we had previously observed for diselenides. The Chaudière et al. data also confirmed that the cytotoxicity of selenium compounds corresponded to our observed catalytic CL activity, that is, the amount of measurable superoxide generated.

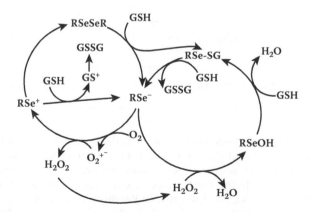

FIGURE 5.3   Selenide redox cycling and superoxide generating by selenocystamine. (From Chaudière, J. et al., *Arch Biochem Biophys* no. 296 (1):328–36, 1992.)

The concept of redox cycling–based selenium toxicity was later expanded to iso-thiocyanates (R-CNS) and isoselenocyanates (R-CNSe) by Gimbor et al. (2010). Once again, the catalytic activity of the compounds was affected by the Se-substituting R (Figure 5.4), with the selenium analogs of the isothiocyanates always being cata-lytically more active. This is in keeping with Clement Ip's and Howard Ganther's comparison of sulfur and selenium cyanates in cancer prevention (Ip et al., 1995), and the differences between the isothiocyanate- and isoselenocyanate-generated CL correlated with the differences in carcinostatic activity *in vitro* and *in vivo* (Crampsie et al., 2012).

Intending to write a review "On the Nature of Selenium Toxicity and Carcinostatic Activity" in the early 1990s, I combed the literature for any associations of super-oxide to the sulfur and selenium redox chemistry as I understood it. A paper I had completely missed revealed that the redox superoxide chemistry we were postulating for selenium had analogously been reported years before. It resides in the 1974 paper by H. P. Misra, "Generation of Superoxide Free Radical During the Auto-oxidation of Thiols." This paper, on its own merits alone, convinced me at that time, that sel-enides were just catalysts accelerating the generation of superoxide and other ROS from thiols. When the draft of the paper was in review in 1993 for FRBM, I ran into Dr. Anthony (Tony) Diplock in a restaurant in Bethesda, Maryland, prior to attending a selenium workshop. Over lunch, he told me he had received my paper for review and thought it very good. Yet he told me I was "suggesting" that selenium might be toxic because it produced superoxide to which I, of course, agreed. He said to me,

Isothiocyanate	Reactivity (GSH × 10^{-2})/sec	cLogP*	Isoselenocyanate	Reactivity (GSH × 10^{-2})/sec	cLogP*
BITC 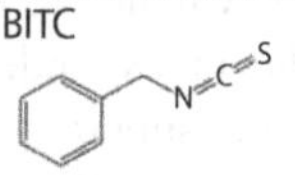	8.05 ± 0.63	3.0	ISC-1 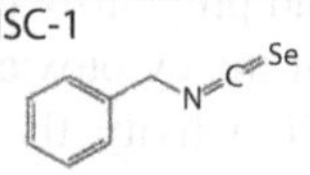	16.19 ± 4.82	2.1
PEITC 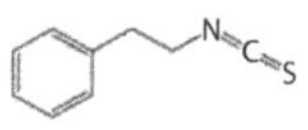	2.23 ± 0.11	3.4	ISC-2 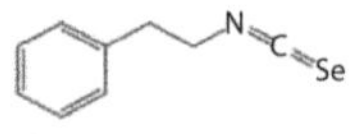	7.40 ± 0.58	2.5
PBITC 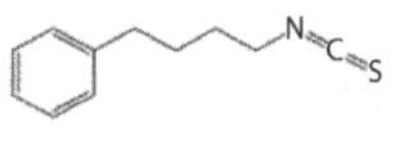	1.23 ± 0.09	4.2	ISC-4 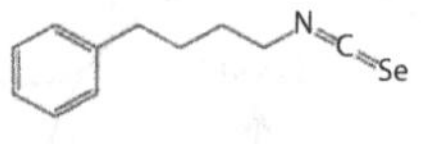	5.40 ± 0.34	3.3
PHITC 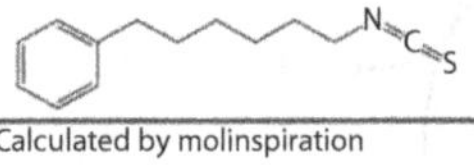	0.96 ± 0.05	5.2	ISC-6 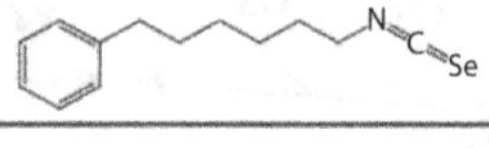	4.35 ± 0.07	4.3

*Calculated by molinspiration

FIGURE 5.4 Comparison of oxidation of GSH by isothiocyanates and isoselenocyanates as a function of aromatic carbon chain length. Note catalytic reactivity of the cyanates by R sub-stitution and higher catalytic activity of isoselenocyanates over isothiocyanates. Differences range from about two- to fivefold increases for isoselenocyanates. (Modified from Gimbor, M. et al., "Isothiocyanates vs. isoselenocyanates: Reactivity towards glutathione and cyste-inyl residues in cell and cell-free systems." In: *Proceedings of the 101st Annual Meeting of the American Association for Cancer Research*; 2010 Apr 17–21; Washington, DC. Philadelphia (PA): AACR; Cancer Res 2010;70(8 Suppl):Abstract nr 3785, 2010.)

"Then, why not just come out and say it." I ended up doing just what Dr. Diplock had suggested. In fact, selenium compounds, when they are toxic, are toxic because their cell or tissue concentration is sufficient to generate superoxide manifesting toxicity as reviewed (Spallholz, 1994). This 1994 review remains an often-cited contribution to understanding selenium toxicity today, which also contains some of the original 1980 SEM Polaroid cell micrographs from the Forrest C. Shaklee Research Center, including some visual evidence that GPx might be capable of generating superoxide in the absence of substrate, H_2O_2, an observation not pursued.

5.3.2 Redox Cycling Selenium, the "Magic Bullet" for Cancer Treatment

The earliest use of selenium to treat a patient's cancer known to me is contained in a letter, dated May 6, 1912 (Box 5.1). The letter by Dr. A. Brennan references a publication by Drs. Lanciere and Thiroloix of Bordeaux in *Le Province Medical*, April 13, 1912 (Paris). The letter reports the successful use of administered selenium to treat a tongue epithelioma although the form of selenium used is not made clear in the letter. The injections caused a transient temperature rise but, otherwise, no obvious signs of patient poisoning. The letter further mentions a statement by a Professor Netter on a remission of rectal cancer with voluminous ganglions in a man 61 years old. It ends with the optimistic outlook "that in a short time cancer will be, if not cured, at least put entirely under control, and its spread stopped, so that it will be considered almost of no consequence." More than 100 years later, the use of selenite for cancer treatment is still under clinical evaluation (see Björnstedt, 2013, and Chapter 6).

The 1912 letter also reveals that the idea to treat cancer with the "known poison" selenium was evidently not uncommon in the early 1900s. A remark on the right side of the letter says, "The German doctor who did the first experimental work with selenium in connection with cancer was Wassermann, not Ehrlich." It was August Paul von Wassermann (1866–1925), who around 1910, systematically investigated the rather successful efficacy of "Eosin-Selen" on mice tumors. Nevertheless, the name of Paul Ehrlich (1854–1915) is intimately linked to Wassermann's and later attempts to fight cancer with Ehrlich's concept to direct a poison to its target in order to achieve selectivity. In Wassermann's Eosin-Selen, eosin was to be the moiety targeting the killing selenium to the cells as eosin would be absorbed by the cancer cells. Ehrlich's idea of targeting drugs to enhance their selectivity is presented in his December 11, 1908, Nobel Prize lecture and later became more widely known as the "magic bullet" concept by the movie with Edward G. Robinson starring as Paul Ehrlich (Figure 5.5).

By 1986, or perhaps earlier, the idea occurred to me that the "magic bullet" concept of Ehrlich's might be adaptable by complexing selenium with a targeting drug for treating cancer. At the 1986 FASEB meeting in St. Louis, Missouri, I discussed with protein chemist Dr. Gary Matsueda of Bristol-Myers-Squibb how one might attach selenite to a protein. He sat down with me and helped map out on paper a possible way to attach selenite to a protein. The chemicals were purchased, but I never found the time to attempt the attachment. Five years later, the idea reemerged while visiting Merrill Lynch to check on Dow Jones Averages. Upon exiting the Merrill Lynch office, I overheard other clients speak about investing in a corporation that

BOX 5.1   TEXT OF THE CANCER TREATMENT LETTER FROM 1912

John Crevus Library, 110 N. Wabash Ave. Chicago, May 6, 1912, Mr. S.D. Gruble (Griffith?) Dear Sir, In your letter of March 30—you asked me to let you know of anything in this cancer cure progress that I noted: I see in a high class French Journal *Le Province Medicale* of 13 April 1912 (Paris) Drs. Lanciere and Thiroloix, of Bordeaux, state that they have treated a patient 39 years old who had a voluminous tongue cancer (Epithelioma). The treatment was as under: On 19th Dec. last an injection of 4 centimeters of Colloidal sletrice (selenite?) Selenium prepared in a special way- the injection was repeated every 8 days, the dose being slightly increased, up to the 16 Feb; the injection was made intravenously and is followed by shivering & a rise in temperature which last about 7 hours, but otherwise in the interval between injections there is no kind of disturbance. After the first injection the immobility of the tongue and the pain were lessoned. On 25th Jan the cancerous excrescence on the side of the right far increased in size and after some days began to fluctuate without any pain- it was punctured and a large amount of liquid drawn off.—Punctures continued to be made to the 16 Feb. when it was entered that the ganglion had almost completely disappeared and was presented under the skin only by a little earlier than a cherry stone. The above account was submitted by Drs. Lancier and Mirolorix to the Medical Society of the Hospital of Paris. At the meeting of this body on 11 March 1912. Professor Netter stated the case was a man 61 years old who had a rectal cancer with voluminous ganglions. Weekly injections in the muscles of the region of colloidal selenium similar to that used by the above doctors were made. After injection the ganglions were affected and the painful stools eased as well as general condition improved. It was too soon to report the effect on the tumor—If you desire a full translation of this article I can make it; but I do not think any more than the above is necessary. I think that in a short time cancer will be, if not cured, at least put entirely under control, and its spread stopped, so that if will be considered almost of no consequence—Yours very truly, Dr. A. Brennan (Photocopy originating from Klaus Schwarz.)

had a diagnostic prostate cancer antibody. The word "antibody" caught my attention as I had wanted to attach selenite to an antibody since or before 1986. I apologetically interrupted their conversation to confirm the organization's name, the Cytogen Corporation. With no Internet at the time, I looked up the Cytogen Corporation in Dun and Bradstreet, wrote down the name of the VP for research and the company's address in Princeton, New Jersey. My brief inquiry indicated that they were attaching Tc 99 to a prostate-specific monoclonal antibody, 7E11-C5. Knowing that selenocystine generated superoxide, I called the VP for research at Cytogen and explained the redox chemistry of selenium and asked if they would like to change their diagnostic antibody to a potential treatment antibody? I was asked to send a proposal, which I did, and was soon contacted by the company and on an airplane. I was invited to Princeton and the company's headquarters having been chauffeured to Princeton south on the New Jersey Turnpike following a flight to Newark, New Jersey. After dinner and the discussions that followed the next day, the plan was for me to send Cytogen selenocystine to attach to their antibody, and they would return the antibody to me to determine if the complex generated superoxide. The four blinded samples upon arrival weeks later were all checked for CL by luminometry without success. As the volumes of the samples were so limited, I resorted back to the MB assay for

FIGURE 5.5 Paul Ehrlich and his "magic bullet" movie story starring Edward G. Robinson. (Photo courtesy of Paul Ehrlich from Wikipedia website. EGR photo from Google Images.)

detection of superoxide, knowing that if the selenium concentration was too low for CL detection, time was on my side using the MB assay. The MB assay was set up in 3-ml test tubes on an early afternoon in 1991, and the entire remaining volume of the samples sent by the Cytogen Corporation were Pasteur pipetted into MB. When 5 p.m. rolled around, I could see no differences in the color of fully oxidized MB. Disappointed, I went home, leaving the samples and control tubes on my desk. Upon entering my office the next morning, all the tubes were still dark blue except one, which was perfectly colorless. Delighted to see one completely clear colorless tube, I called the Cytogen Corporation, and they confirmed that my only clear tube contained the antibody to which they had attached selenocystine. Paul Ehrlich again came to mind, and briefly I thought once again that redox selenium on the Cytogen antibody might really be the "magic" in the envisioned therapeutic "magic bullet."

5.3.3 Are Selenides Really "Magic Bullets"?

Returning to Princeton again, we discussed an arrangement with Texas Tech University's Office of Research Services for the Cytogen Corporation to patent the invention. The somewhat complicated arrangement was never fulfilled. Cytogen, within weeks, advised me that they were not in a financial position to proceed and were withdrawing from filing a patent application. I received a release of the IP from the Cytogen Corporation and later also from Texas Tech University. I was now in a position to file a US Patent application based upon the Cytogen Corporation's conjugation of selenocystine to their 7E11-C5 prostate antibody. On May 3, 1994, I mailed a US Patent application, which was granted in 1998 (Spallholz and Reid, 1998).

By 1994, anticipated collaborative work with the Cytogen Corporation had reverted to their supplying the monoclonal antibody, B72.3, against the pancarcinoma antigen TAG-72. To the TAG-72 monoclonal antibody, we were now able to label the carbohydrates of the heavy chains with selenocystine as Cytogen had originally done. My doctoral student Lugen Chen would go on to label and show the effectiveness of another selenocystine Se-labeled monoclonal antibody B72.3. This antibody was tested against colon cancer cells *in vitro* and *in vivo* in a nude mouse xenograph model. Success *in vivo* and possibly full arrest of LS174T colon tumors in nude mice was limited by a lack of sufficient antibody. This experimentation culminated prematurely in Lugen Chen's 1997 dissertation (Chen, 1997). TEMs of combined Se and nano-gold-labeled 7E11-C5 antibody were also followed by Lugen as part of his dissertation into DU-145 prostate cancer cells, in which mitochondria were observed in the TEM micrographs to be destroyed from the inside. The mitochondria were seemingly "blown apart" intracellularly, and this reminded me of selenite-induced "mitochondrial swelling" first reported by Orville Levander (Levander et al., 1973). By 1995, a US Patent had been filed, and my 1994 review in FRBM had been accepted. Prior to the Patent application, Lugen Chen had also covalently attached selenium to a polyclonal antibody against human erythrocytes and showed that the selenium-labeled antibody but not the native antibody would lyse erythrocytes (Chen and Spallholz, 1995). This was determined by hemolysis as had been originally done earlier for various selenium compounds *in vitro* (Yan and Spallholz, 1991; Yan et al., 1991). Viable cell counts and real-time visual microscopy on videotape recorded changes over time in cell morphology by the Se-labeled erythrocyte membrane antibody. Membrane destruction with the Se-mabs, as seen in the summer 1980 with selenite and selenocystine, seemed like "magic."

In 1994, I met Dr. Ted Reid, professor in the TTU Health Sciences Center's Ophthalmology Department at the University Center. Ted and I had met socially but had never talked about science. He told me he was trying to attach heparin to intraocular lenses (IOLs) to reduce the incidence of secondary cataracts, cell growth on the IOLs. I told him of my experience in attachment of redox cycling selenium to proteins via the carbohydrates on monoclonal antibodies. He asked me whether selenium might be attached to cellulose, and I thought that might be possible and acknowledged in the affirmative. Within 48 hours of leaving lunch, selenocystamine had been covalently attached to a cellulose wick sponge. The white wick sponge, now an orange-yellow sponge following a rinse, was found to be catalytic in generating superoxide. The sponge could be removed from the cocktail of the luminometer, rinsed, and, when reinserted into a fresh cocktail, it would continue generating superoxide unabated. This was of particular interest because it implied that the selenide anion did not have to be fully in solution to generate superoxide.

This observation suggested many redox selenium surface applications for medical devices, whereby selenium might be cytotoxic to bacterial colonization as selenite was toxic in solution to bacteria and cancer cells *in vitro*.

Covalent attachment of selenium, principally selenocystamine, directly to IOLs proved to be unsuccessful. Ted and I therefore discussed inclusion of selenium into the IOLs matrix by addition of selenium into the polymethylmethacrylate (PMMA) polymer prior to lens formation. Having a neighbor who owned a plastic molding plant, I made arrangements to discuss the inclusion of selenium into a molded polymer. While touring the manufacturing facility, I noticed a number of plastic-handled screwdrivers being soaked in a bath of steaming US Army Green pigment. Asking about the process, the screwdriver handles had been cast as clear plastic handles, and then various colored pigments could be infused by heating the handles in the dyes in boiling water over a few days. My reaction to this process was immediate; we might infuse selenium compounds into plastics, and they might possibly be catalytic on the surface. On a Saturday morning in 1995, I enclosed selenocystamine and an IOL in water in a closed scintillation vial and left it overnight in a heating oven. Checking it on Sunday, the IOL had a light pale yellow color, and it redox cycled, generating superoxide in our CL assay. Disappointment set in, however, when the CL activity was lost over time from the IOL by apparent leaching of the water-soluble selenocystamine from the lens surface. Having worked with some of the water-insoluble diselenides that generated CL, like benzyldiselenide, I seized upon the idea of a hydrophobic attachment of diselenides to plastics while reflecting back upon the green US Army screwdriver handles. Over several weeks of trial and error, I found that the nonpolar diselenides dissolved in preheated ethanol in scintillation vials would incorporate the selenium compounds into IOLs, into silicon, and also into other plastics on the order of seconds to minutes. The selenium attachment depended on the solvent, selenium concentration, time, and temperature. Such hydrophobic attachment of water-insoluble diselenides resulted in almost no loss of catalytic activity from the polymer or silicon surface. While my colleage Steven Mathews went on to prevent biofilms on contact lenses in rabbits' eyes

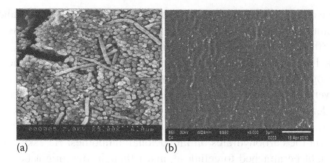

(a)   (b)

FIGURE 5.6 Dental sealants containing redox cycling selenium used to occlude tooth fissure decay. (a) SEM of a mixed patient oral bacterial biofilm from a dental sealant without selenium. (b) Mixed patient oral bacterial biofilm from selenium containing dental sealant. Magnifications identical by SEM; 15 April 2010. (Courtesy of Element 34 Technologies, Inc.; Lubbock, Texas.)

(Mathews et al., 2006), my preference was the hydrophobic route of selenium attachment. The process was controllable, fast, and a near permanent attachment methodology to insoluble polymers without any intervening required attachment chemistry. Whichever attachment was successful, it could be shown by SEM microscopy and the quantization of adherent bacteria by bacterial plaque formation that redox cycling selenium attached to water-insoluble devices could prevent bacterial biofilm formation in a concentration-dependent manner (Tran et al., 2009). This is shown for control (no absorbed selenium) and selenium containing dental occlusive sealants in (Figure 5.6) removed after days from a dental patient's mouth. Generally, only a fraction of 1% of redox cycling selenium by weight is required to be included in a sealant or on the surface of a medical device to prevent bacterial biofilm formation.

This redox selenium chemistry, approved by the US FDA for dental use in 2008, now commercially available, holds great potential to prevent bacterial biofilm formation and colonization. Its application has not yet been widely or successfully adapted although it seems to be an alternative to using silver as an antibacterial agent because selenium is an essential nutrient and silver is toxic. Silver also reacts directly with selenium and is nutritionally nonessential as well as costly. The website below provides some history of the development of selenium-coated dental devices, ligatures, and the incorporation of redox cycling selenium into dental sealants http://e34tech.com/about-us/.

5.4 OTHER AND MORE RECENT SELENIUM DRUG DEVELOPMENTS

At the Selenium in Biology and Medicine Conference held in 2010 in Kyoto, Japan, I reviewed the status of adapting redox-cycling selenium to pharmaceuticals (Spallholz and Sharma, 2010). Beyond the applications described above, other and some more recent applications are included below.

In 1996–1998, attachment of redox-cycling selenium was made to a steroid, dehydrotestosterone, and shown to be cytotoxic to prostate cancer cells as well as an attachment of redox selenium to an anti-HIV peptide directed to the GP-120 viral coat protein.

The peptide, which has been selected from the literature, was effective against HIV replication in cell culture as performed in California. Using a bacterial phage peptide library, Phat Tran, a graduate student at the Texas Tech Health Sciences Center, developed peptides from the library, attached redox-cycling selenium covalently, and showed that the selenium-containing peptides could arrest bacterial cell growth (Tran, 2008). In 2008, redox-cycling selenium was attached to both folic acid and biotin-forming redox active selenofolate and selenobiotin vitamers. These selenium compounds were shown to both redox cycle in our CL assay generating superoxide and were both time- and dose-dependently cytotoxic to MDA-MD-231 and MCF-7 cancer cells *in vitro* (Tsai, 2008).

More recently, as noted, Dr. Arun Sharma and his associates at Penn State University at Hershey, Pennsylvania, have demonstrated that isoselenocyanates also redox cycle with GSH *in vitro* and are cytotoxic to cancer cells in a dose- and time-dependent manner. Dr. Sharma in 2012 has also reported the attachment of redox-cycling selenium to Temozolomide (TMZ) (Figure 5.7), a widely used drug in the treatment of gliomas and various melanomas (Cheng et al., 2012).

Temozolomide, which normally alkylates DNA, was more potent in a mouse xenograft model in inhibiting a mouse glioma and melanoma when it contained redox-active selenium than Temozolomide alone. We have recently confirmed this observation *in vitro* in MCF-7 cells (unpublished data). These authors concluded and we concur that incorporation of Se into TMZ rendered greater potency to this chemotherapeutic agent.

By late 2009, a redox-active selenium modified Bolton-Hunter reagent had been synthesized by Eburon Organics, USA, Inc., and used to effectively attach redox selenium to proteins (Figure 5.8).

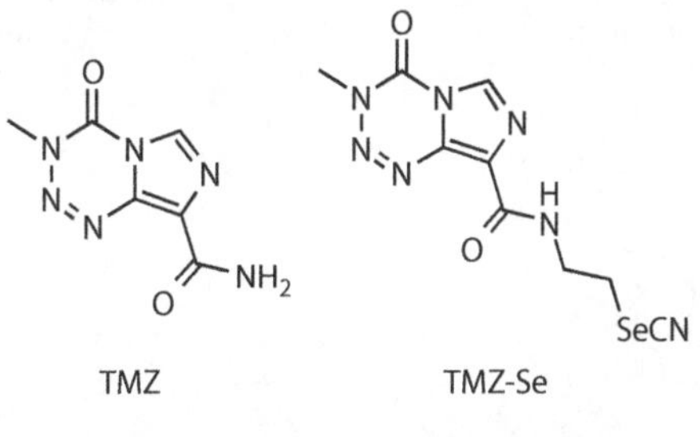

TMZ   TMZ-Se

FIGURE 5.7 Modification of DNA alkylating Temozolomide with redox cycling Selenium. Temozolomide (TMZ) and Seleno-Temozolomide (TMZ-Se). The addition of redox cycling selenium to Temozolomide improves drug therapeutic antitumor potency. TMZ-Se redox cycles in CL assay. (From Cheng, Y. et al., *PLoS One* no. 7 (4):e35104. doi: 10.1371/journal .pone.0035104, 2012.)

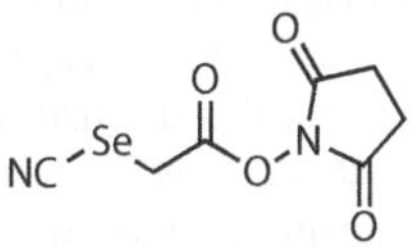

FIGURE 5.8 Selenium modified Bolton-Hunter Reagent. The modified Bolton-Hunter Reagent attaches redox-cycling selenium to terminal amines and Lysine residues of proteins under alkaline conditions, pH about 8.4.

This chemistry attaches selenium as a selenocyanate ($R-CH_2SeCN$) to terminal amines and epsilon-amino lysine residues of proteins as initially used in the 1970s to attach iodine for radioimmunoassays. This selenium chemistry has been investigated by Goswami et al. after using the Se-modified Bolton-Hunter reagent to attach selenium to transferrins and human serum albumin (Goswami and Spallholz, 2013). The transferrin receptor is overexpressed in some leukemias, and the selenium was targeted to both K562 and THP-1 leukemia cells. The selenium-labeled transferrin as well as human serum albumin arrested both cell lines in a time- and dose-dependent manner as selenofolate and selenobiotin had done years before against MD-MDA-281 and MCF-7 cancer cells (Tsai, 2008). Leukemia cell viability appeared to be lost by selenium-labeled proteins in comparison to native proteins as assayed for induced apoptosis.

Most recently, the selenium-modified Bolton-Hunter reagent has been used to attach selenocyanates to Trastuzumab (Herceptin) and Avastin (Genentech, Inc.). Bapat et al. (2013) modified the 2012–2013 transferrin research protocols and now have underway experiments to compare Trastuzumab (Herceptin) with and without conjugated selenium to see whether the efficacy of Trastuzumab treatment might be improved against Her2/nu breast cancer cells *in vitro*. The attachment of redox-cycling selenium by the modified Bolton-Hunter regent renders all proteins and monoclonal antibodies following exhaustive dialysis colored, as shown for Trastuzumab (Herceptin) in Figure 5.9. The amount of selenium attached to Trastuzumab in the photograph is about 27 g-atoms Se/molecule. Continuing *in vitro* studies indicate

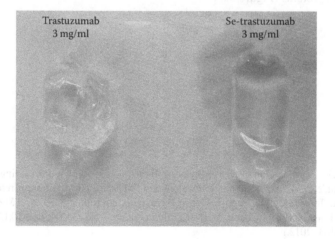

FIGURE 5.9 Trastuzumab (Herceptin) and Trastuzumab labeled with redox cycling selenium (Se-Trastuzumab) using Se-modified Bolton-Hunter Reagent shown in Figure 5.8. Trastuzumab before and after selenium covalent attachment and exhaustive dialysis. Protein concentration is 3 mg/ml and selenium concentration is about 27 μg Se/mg protein; by ICP-MS. (From Bapat, P. et al., "Cytotoxic effects of seleno-trastuzumab on trastuzumab resistant JIMT-1 breast cancer cell line." Meeting of the Free Radicals in Biology and Medicine Meeting, November, San Antonio, TX, 2013; Bapat, P. et al., "Cytotoxic effects of seleno-trastuzumab, trastuzumab (herceptin) and T-DM-1 (kadcyla) on trastuzumab resistant JIMT-1 breast cancer cells." Free Radic Biol Med Meeting Abstract, November, Seattle, WA, 2014.)

that such cytotoxic improvements in monoclonal antibodies are materially possible and significant, including apoptosis induced in otherwise Herceptin resistant JIMT-1 cells (Bapat et al., 2014).

5.5 CONCLUSIONS AND FUTURE PERSPECTIVES

As the 21st amino acid (Böck et al., 1991b), selenocysteine is incorporated into a number of antioxidant enzymes and other proteins. Unlike cysteine, selenocysteine is mostly ionized as the selenide anion (RSe^-) at physiological pH. In the glutathione peroxidases (GPx 1–4), selenocysteine is the catalyst transferring two electrons to H_2O_2 to make water using the cosubstrate GSH. Nonenzymatic organic selenides, GSSe-, other selenides, RSe-, and isoselenocyanates, depending on the R moiety, also catalyze oxidation of GSH and other thiols, but instead of producing water, they generate superoxide via one-electron reductions with ambient oxygen. This chemistry fully accounts for selenium's toxicity. Both the activity of selenium in enzymes and selenium toxicity depend on the catalytic properties of a selenide anion with only the products of catalysis, water or superoxide, being different. But what a difference one electron makes.

Selenium is not unlike many other minerals and nonmetals that are toxic because they also catalytically induce oxidative stress (Stohs and Bagchi, 1995). However, selenium *is* different. Unlike transition and heavy metals, it can be covalently incorporated as an organic catalyst in new and existing drugs and proteins. It is the versatility of organic selenium chemistry in replacing halides, its redox chemistry, and its similarity in replacing sulfur that offers the possibilities for prevention of biofilms and the conversion of existing drugs and natural products to more effective cytotoxic drugs. This potential is not limited to just small molecules, but proteins, such as monoclonal antibodies, that can now be equipped with redox-cycling selenium, perhaps fulfilling some of Paul Ehrlich's vision of the "magic bullet." The formation of selenium "drugs" against viruses, bacteria, or cancer is no longer "magic"; it is now completely rational to proceed with this research based upon the collective redox-cycling selenium literature. The "magic" still on the horizon and yet to be fully obtained is the optimized targeting. As recently reviewed by Wallenberg (Wallenberg et al., 2014) the way forward should be clear to those that have an interest in the applied clinical pharmacology of redox selenium. It needs to be explored by those that have the resources and vision to possibly better treat viral, antibiotic-resistant bacterial infections, malaria perhaps, and cancers, and to those conditions for which there are no therapies at all, such as Ebola.

ACKNOWLEDGMENTS

Acknowledgment with appreciation goes to the past students who have worked with and taught me new discoveries. They are listed as the senior author on abstracts, manuscripts, theses, and dissertations. Many friends and colleagues with whom I have spent time, communicated, and/or collaborated with have now passed on; I miss them all. To name some would seem unjust. A special thanks goes to Dr. Arun Sharma of Penn State, Hershey. He has worked with me, and he is pushing the selenium drug boundary as much as anyone. Dr. Mikael Björnstedt is a selenophile par excellence. His family complains that when we have been all together all we talk about is selenium, selenium, selenium. Many know of this disease. To Leopold Flohé, thank you as a colleague since we first met at the VA Hospital with Klaus Schwarz. I think even you now believe in this selenium redox chemistry. Much of the research presented are ideas originating from Dr. Mallory Boylan's, my wife's, insight into clinical medicine and selenium lessons. If you have been left out of the acknowledgments, it is not because I do not appreciate all that you may have done, it is that there are just so many of you. The list is long, and life is short.

REFERENCES

Anundi, I., J. Hogberg, and A. Stahl. 1982. "Involvement of glutathione reductase in selenite metabolism and toxicity, studied in isolated rat hepatocytes." *Arch Toxicol* no. 50 (2):113–23.

Bapat, P., D. Goswami, A. Shastri et al. 2013. "Cytotoxic effects of seleno-trastuzumab on trastuzumab resistant JIMT-1 breast cancer cell line." Meeting of the Free Radicals in Biology and Medicine Meeting, November, San Antonio, TX.

Bapat, P., L. M. Boylan, E. Cobos et al. 2014. "Cytotoxic effects of seleno-trastuzumab, trastuzumab (herceptin) and T-DM-1 (kadcyla) on trastuzumab resistant JIMT-1 breast cancer cells." Free Radic Biol Med Meeting Abstract, November, Seattle, WA.

Beath, O. A., and E. H. Lehnert. 1917. "The poisonous properties of the two-grooved milk vetch (Astragalus bisulcatus)." *Wyoming Agr Expt Stat, Bulletin* no. 112.

Bernal, J. L., M. J. Del Nozal, L. Deban et al. 1990. "Modification of the methylene blue method for spectrophotometric selenium determination." *Talanta* no. 37 (9):931–6.

Berzelius, J. J. 1817. "Sur deux metaux nouveaux (Litium et Selenium)." *Schweigger J* no. 21:1818–23.

Björnstedt, M. 2013. "Pharmacokinetics and toxicity of sodium selenite in a clinical phase I study: Examining its value as a treatment against cancer." 10th International Symposium on Selenium in Biology and Medicine, Berlin, Germany.

Böck, A., K. Forchhammer, J. Heider, and C. Baron. 1991a. "Selenoprotein synthesis: An expansion of the genetic code." *Trends Biochem Sci* no. 16 (12):463–7.

Böck, A., K. Forchhammer, J. Heider et al. 1991b. "Selenocysteine: The 21st amino acid." *Mol Microbiol* no. 5 (3):515–20.

Chaudière, J., O. Courtin, and J. Leclaire. 1992. "Glutathione oxidase activity of selenocystamine: A mechanistic study." *Arch Biochem Biophys* no. 296 (1):328–36.

Chen, J. J., L. M. Boylan, C. K. Wu, and J. E. Spallholz. 2007. "Oxidation of glutathione and superoxide generation by inorganic and organic selenium compounds." *Biofactors* no. 31 (1):55–66.

Chen, L. 1997. "Cytotoxic activity of a selenium labeled antibody-selenium conjugate against tumor cells." Dissertation, Texas Tech University, Lubbock, TX.

Chen, L., and J. E. Spallholz. 1995. "Cytolysis of human erythrocytes by a covalent antibody-selenium immunoconjugate." *Free Radic Biol Med* no. 19 (6):713–24.

Cheng, Y., U. H. Sk, Y. Zhang et al. 2012. "Rational incorporation of selenium into temozolomide elicits superior antitumor activity associated with both apoptotic and autophagic cell death." *PLoS One* no. 7 (4):e35104. doi: 10.1371/journal.pone.0035104.

Crampsie, M. A., M. K. Pandey, D. Desai, J. Spallholz, S. Amin, and A. K. Sharma. 2012. "Phenylalkyl isoselenocyanates vs phenylalkyl isothiocyanates: Thiol reactivity and its implications." *Chem Biol Interact* no. 200 (1):28–37. doi: 10.1016/j.cbi.2012.08.022.

Dickson, R. C., and A. L. Tappel. 1969. "Reduction of selenocystine by cysteine or glutathione." *Arch Biochem Biophys* no. 130 (1):547–50.

Feigl, F., and P. W. West. 1947. "Test for selenium based on catalytic effect." *Biochemistry* no. 19 (5):351–3.

Flohé, L., W. A. Günzler, and H. H. Schock. 1973. "Glutathione peroxidase: A selenoenzyme." *FEBS Lett* no. 32 (1):132–4.

Forstrom, J. W., J. J. Zakowski, and A. L. Tappel. 1978. "Identification of the catalytic site of rat liver glutathione peroxidase as selenocysteine." *Biochemistry* no. 17 (13):2639–44.

Franke, K. W. 1934. "A new toxicant occurring naturally in certain samples of plant foodstuffs." *J Nutr* no. 8:597–608.

Frenkel, G. D., and D. Falvey. 1988. "Evidence for the involvement of sulfhydryl compounds in the inhibition of cellular DNA synthesis by selenite." *Mol Pharmacol* no. 34 (4):573–7.

Ganther, H. E. 1966. "Enzymic synthesis of dimethylselenide from sodium selenite in mouse liver extracts." *Biochemistry* no. 5 (3):1089–98.

Ganther, H. E. 1968. "Selenotrisulfides. Formation by the reaction of thiols with selenious acid." *Biochemistry* no. 7 (8):2898–905.

Ganther, H. E. 1971. "Reduction of the selenotrisulfide derivative of glutathione to a persulfide analog by glutathione reductase." *Biochemistry* no. 10 (22):4089–98.

Ganther, H. E., and C. Corcoran. 1969. "Selenotrisulfides. II. Cross-linking of reduced pancreatic ribonuclease with selenium." *Biochemistry* no. 8 (6):2557–63.

Gimbor, M., D. M. Desai, J. E. Spallholz, S. Amin, and A. K. Sharma. 2010. "Isothiocyanates vs. isoselenocyanates: Reactivity towards glutathione and cysteinyl residues in cell and cell-free systems." In: *Proceedings of the 101st Annual Meeting of the American Association for Cancer Research*; 2010 Apr 17–21; Washington, DC. Philadelphia (PA): AACR; Cancer Res 2010;70(8 Suppl):Abstract nr 3785.

Goswami, D., and J. E. Spallholz. 2013. "Seleno-transferrin as a pseudo-monoclonal antibody for treatment of leukemia." Soc Free Radic Biol Med Meeting, November, San Antonio, TX.

Goswami, D., L. M. Boylan, and J. E. Spallholz. 2012. "Superoxide generation by selenium compounds and the putative observance of a free radical chain reaction." Soc Free Radic Biol Med Meeting, November, San Diego, CA.

Greeder, G. A., and J. A. Milner. 1980. "Factors influencing the inhibitory effect of selenium on mice inoculated with Ehrlich ascites tumor cells." *Science* no. 209 (4458):825–7.

Hamilton, J. W. 1975. "Chemical examination of seleniferous cabbage Brassica oleracea capitata." *J Agric Food Chem* no. 23 (6):1150–2.

Harman, D. 1956. "Aging: A theory based on free radical and radiation chemistry." *J Gerontol* no. 11 (3):298–300.

Hu, M. L., and J. E. Spallholz. 1983. "*In vitro* hemolysis of rat erythrocytes by selenium compounds." *Biochem Pharmacol* no. 32 (6):957–61.

Hu, M. L., and A. L. Tappel. 1987. "Selenium as a sulfhydryl redox catalyst and survey of potential selenium-dependent enzymes." *J Inorg Biochem* no. 30 (3):239–48.

Hu, M. L., J. E. Spallholz, M. Grisham, and J. Everse. 1982. "Cytolytic activity of selenium compounds and glutathione peroxidase towards rat erythrocytes." *Fed Proc* no. 41:787.

Ip, C., S. Vadhanavikit, and H. Ganther. 1995. "Cancer chemoprevention by aliphatic selenocyanates: Effect of chain length on inhibition of mammary tumors and DMBA adducts." *Carcinogenesis* no. 16 (1):35–8.

Ip, C., H. J. Thompson, Z. Zhu, and H. E. Ganther. 2000. "*In vitro* and *in vivo* studies of methylseleninic acid: Evidence that a monomethylated selenium metabolite is critical for cancer chemoprevention." *Cancer Res* no. 60 (11):2882–6.

Ip, C., Y. Dong, and H. E. Ganther. 2002. "New concepts in selenium chemoprevention." *Cancer Metastasis Rev* no. 21 (3–4):281–9.

Jukes, T. H. 1983. "Selenium, an 'essential poison.'" *J Appl Biochem* no. 5 (4–5):233–4.

Kitahara, J., Y. Seko, and N. Imura. 1993. "Possible involvement of active oxygen species in selenite toxicity in isolated rat hepatocytes." *Arch Toxicol* no. 67 (7):497–501.

Levander, O. A., V. C. Morris, and D. J. Higgs. 1973. "Acceleration of thiol-induced swelling of rat liver mitochondria by selenium." *Biochemistry* no. 12 (23):4586–90.

Martin, J. L., and J. E. Spallholz. 1968. "The effects of selenium toxicity on the free amino acids of chick liver." VII. International Congress of Nutrition, Prague, Czechoslovakia.

Martin, J. L., and M. L. Gerlach. 1969. "Separate elution by ion-exchange chromatography of some biologically important selenoamino acids." *Anal Biochem* no. 29 (2):257–64.

Martin, J. L., and J. E. Spallholz. 1970. "The effects of selenium toxicity on the metabolism of glutathione in chick liver." *Fed Proc* no. 29:499.

Martin, J. L., A. Shrift, and M. L. Gerlach. 1971. "Use of 75-Se-selenite for the study of selenium metabolism in Astragalus." *Biochemistry* no. 10:945–52.

Mathews, S. M., J. E. Spallholz, M. J. Grimson, R. R. Dubielzig, T. Gray, and T. W. Reid. 2006. "Prevention of bacterial colonization of contact lenses with covalently attached selenium and effects on the rabbit cornea." *Cornea* no. 25 (7):806–14. doi: 10.1097/01 .ico.0000224636.57062.90.

McConnell, K. P., and D. M. Roth. 1968. "Incorporation of selenium-75-labeled rat plasma proteins into rat liver ribosomes." *Arch Biochem Biophys* no. 125 (1):29–34.

McCord, J. M., and I. Fridovich. 1969a. "Superoxide dismutase. An enzymic function for erythrocuprein (hemocuprein)." *J Biol Chem* no. 244 (22):6049–55.

McCord, J. M., and I. Fridovich. 1969b. "The utility of superoxide dismutase in studying free radical reactions. I. Radicals generated by the interaction of sulfite, dimethyl sulfoxide, and oxygen." *J Biol Chem* no. 244 (22):6056–63.

McCord, J. M., and I. Fridovich. 1988. "Superoxide dismutase: The first twenty years (1968–1988)." *Free Radic Biol Med* no. 5 (5–6):363–9.

Mills, G. C. 1959. "The purification and properties of glutathione peroxidase of erythrocytes." *J Biol Chem* no. 234 (3):502–6.

Misra, H. P. 1974. "Generation of superoxide free radical during the autoxidation of thiols." *J Biol Chem* no. 249 (7):2151–5.

Poirier, K. A., and J. A. Milner. 1979. "The effect of various seleno-compounds on ehrlich ascites tumor cells." *Biol Trace Elem Res* no. 1 (1):25–34. doi: 10.1007/BF02783840.

Rhead, W. J., and G. N. Schrauzer. 1974. "The selenium catalyzed reduction of methylene blue by thiols." *Bioinorg Chem* no. 3 (3):225–42.

Rosenfeld, I., and O. A. Beath. 1946. "Pathology of selenium poisoning." *Wyoming Agr Expt Sta, Bulletin* no. 275:1–27.

Rotruck, J. T., A. L. Pope, H. E. Ganther, A. B. Swanson, D. G. Hafeman, and W. G. Hoekstra. 1973. "Selenium: Biochemical role as a component of glutathione peroxidase." *Science* no. 179 (4073):588–90.

Schwarz, K. 1954. "Factors protecting against dietary necrotic liver degeneration." *Ann N Y Acad Sci* no. 57:878–88.

Schwarz, K. 1969. "Biological potency of organic selenium compounds. I. Aliphatic mono-seleno- and diseleno-dicarboxylic acids." *J Biol Chem* no. 244 (8):2103–10.

Schwarz, K., and C. M. Foltz. 1957. "Selenium as an integral part of factor 3 against dietary necrotic liver degeneration." *Am Chem Soc J* no. 79:3292–3.

Schwarz, K., and C. M. Foltz. 1958. "Factor 3 activity of selenium compounds." *J Biol Chem* no. 233 (1):245–51.

Schwarz, K., and A. Fredga. 1974. "Biological potency of organic selenium compounds. V. Diselenides of alcohols and amines, and some selenium-containing ketones." *Bioinorg Chem* no. 3 (2):153–9.

Schwarz, K., and A. Fredga. 1975. "Biological potency of organic selenium compounds: VI. Aliphatic seleninic acids and carboxyseleninic acids." *Bioinorg Chem* no. 4 (3):235–43.

Schwarz, K., and K. D. Pathak. 1975. "The biological essentiality of selenium and the development of biologically active organoselenium compounds of minimum toxicity." *Chemica Scripta* no. 8A:85–95.

Seko, Y. 1986. "Hemolysis of mouse erythrocytes by methylmercury and selenite *in vitro*. iv, decrease in sulfhydral concentration and lipid peroxidation by selenite." *Tiekyo Med J.* 9:287–97 (in Japanese with English Abstract).

Seko, Y., and N. Imura. 1997. "Active oxygen generation as a possible mechanism of selenium toxicity." *Biomed Environ Sci* no. 10 (2–3):333–9.

Seko, Y., Y. Saito, J. Kitahara, and N. Imura. 1989. "Active oxygen generation by the reaction of selenite with reduced glutathione *in vitro*." In *Proceedings of the Fourth International Symposium on Selenium in Biology and Medicine*, edited by A. Wendel. Berlin, Heidelberg, New York: Springer-Verlag.

Shrift, A., and T. K. Virupaksha. 1965. "Seleno-amino acids in selenium-accumulating plants." *Biochim Biophys Acta* no. 100:65–75.

Spallholz, J. E. 1994. "On the nature of selenium toxicity and carcinostatic activity." *Free Radic Biol Med* no. 17 (1):45–64.

Spallholz, J. E., and L. H. Piette. 1972. "Interaction of spin-labeled analogues of 1, 10-phenanthroline and iodoacetamides with horse liver alcohol dehydrogenase." *Arch Biochem Biophys* no. 148 (2):596–606.

Spallholz, J. E., and A. Raftery. 1983. "Comparative toxicity of sodium selenite and selenium yeast in the fisher 344 rat." Federation of American Societies for Experimental Biology; Summer Workshop on Trace Elements. Saxton River, Vermont, July 3–8.

Spallholz, J. E., and J. H. Whittam. 1992. "Selenium toxicity interpreted from biological, catalytic, chemiluminescent and scanning electron microscopic data." 5th International Symposium on *Selenium in Biology and Medicine*. Nashville, TN. p.111, abstract 55.

Spallholz, J. E., and T. W. Reid. 1998. "Method for the preparation of free radical pharmaceuticals, diagnostic and devices using selenium conjugates." United States Patent Number 5,783,454. Filed May 13, 1994.

Spallholz, J. E., and Sharma, K. 2010. "Prevention of bacterial biofilm formation and experimental drug development using selenium redox chemistry." 9th International Meeting of *Selenium in Biology and Medicine*. Kyoto, Japan.

Spallholz, J. E., R. Fritas, and J. Whittam. 1982. "Cytolytic activity of selenium compounds and glutathione peroxidase assessed by scanning electron microscopy." Fed Proc 41:529.

Spallholz, J. E., B. J. Shriver, and T. W. Reid. 2001. "Dimethyldiselenide and methylseleninic acid generate superoxide in an *in vitro* chemiluminescence assay in the presence of glutathione: Implications for the anticarcinogenic activity of L-selenomethionine and L-Se-methylselenocysteine." *Nutr Cancer* no. 40 (1):34–41. doi: 10.1207/S15327914 NC401_8.

Spallholz, J. E., V. P. Palace, and T. W. Reid. 2004. "Methioninase and selenomethionine but not Se-methylselenocysteine generate methylselenol and superoxide in an *in vitro* chemiluminescent assay: Implications for the nutritional carcinostatic activity of selenoamino acids." *Biochem Pharmacol* no. 67 (3):547–54. doi: 10.1016/j.bcp.2003 .09.004.

Stohs, S. J., and D. Bagchi. 1995. "Oxidative mechanisms in the toxicity of metal ions." *Free Radic Biol Med* no. 18 (2):321–36.

Tran, P. 2008. "New selenium antimicrobials and material coating against bacteria and bacterial biofilms." Doctoral Dissertation, Texas Tech University, Lubbock, TX.

Tran, P. L., A. A. Hammond, T. Mosley et al. 2009. "Organoselenium coating on cellulose inhibits the formation of biofilms by Pseudomonas aeruginosa and Staphylococcus aureus." *Appl Environ Microbiol* no. 75 (11):3586–92. doi: 10.1128/AEM.02683-08.

Tsai, P. 2008. "Cytotoxicity of selenocyanobiotin against MDA-MB-231 breast cancer cells." Masters Thesis, Texas Tech University, Lubbock, TX.

Tsen, C. C., and A. L. Tappel. 1958. "Catalytic oxidation of glutathione and other sulfhydryl compounds by selenite." *J Biol Chem* no. 233 (5):1230–2.

Wallenberg, M., S. Misra, and M. Bjornstedt. 2014. "Selenium cytotoxicity in cancer." *Basic Clin Pharmacol Toxicol* no. 114 (5):377–86. doi: 10.1111/bcpt.12207.

Wendel, A., ed. 1989. *Selenium in Biology and Medicine.* Berlin, Heidelberg, New York: Springer-Verlag.

Wendel, A., B. Kerner, and K. Graupe. 1978. "The selenium moiety of glutathione peroxidase." In Conference der Gesellschaft für Biologische Chemie. *Functions of Glutathione in Liver and Kidney,* edited by H. Sies and A. Wendel, 107–13. Berlin, Heidelberg, New York: Springer-Verlag.

Yan, L., and J. E. Spallholz. 1991. "Free radical generation by selenium compounds." *FASEB J* no. 5:A581.

Yan, L., J. A. Yee, L. M. Boylan, and J. E. Spallholz. 1991. "Effect of selenium compounds and thiols on human mammary tumor cells." *Biol Trace Elem Res* no. 30 (2):145–62.

Zakowski, J. J., J. W. Forstrom, R. A. Condell, and A. L. Tappel. 1978. "Attachment of seleno-cysteine in the catalytic site of glutathione peroxidase." *Biochem Biophys Res Commun* no. 84 (1):248–53.

6 Selenite in Cancer Therapy

Sougat Misra, Marita Wallenberg,
Ola Brodin, and Mikael Björnstedt

CONTENTS

6.1 Introduction 110
6.2 Selenite at the Crossroad between Antioxidant and Pro-Oxidant
Functions 111
6.3 Selenite: A Potent Chemotherapeutic Agent 112
    6.3.1 *In Vitro* Studies 112
    6.3.2 *In Vivo* Studies with Animal Models 113
6.4 The Molecular Basis of Selenite Cytotoxicity in Cancer 115
    6.4.1 Selenite Transport in Cancer Cells 115
    6.4.2 Generation of Reactive Oxygen Species 117
    6.4.3 Modulation of Important Cancer-Associated Signaling Pathways
        by Selenite 118
        6.4.3.1 Translational Machinery 118
        6.4.3.2 Cell Cycle 119
        6.4.3.3 p53 119
        6.4.3.4 HIF-1α 120
        6.4.3.5 NF-κB 120
        6.4.3.6 AP-1 121
        6.4.3.7 Epigenetic Signature 121
    6.4.4 Possible Mechanisms and Evidence for Potentiation
        of Cytotoxic Effects of Selenite with Chemotherapeutic Drugs 123
6.5 Human Studies 124
    6.5.1 Rationale for Use of Selenium in Treatment of Cancer Patients 124
    6.5.2 Selenium in Combination with Surgery, Radiotherapy,
        and Chemotherapy 125
    6.5.3 Use of Selenium as a Chemotherapeutic Drug in Humans 125
    6.5.4 Phase I Study to Assess Safety and to Determine the Maximal
        Tolerable Dose 129
6.6 Concluding Remarks 129
Acknowledgments 130
References 130

6.1 INTRODUCTION

Almost a century ago, the highly unique properties of selenium to inhibit the growth of neoplastic cells and the great potential of selenium in the treatment of neoplasia were described (Watson-Williams, 1919). Nearly 40 years later, seleno-cystine was shown to be efficient in the treatment of leukemia (Weisberger and Suhrland, 1956). In spite of these positive indications, the number of human trials using selenium compounds in the treatment of malignancies is very limited. The reasons for this fact are difficult to understand, but one possible explanation is the simplicity of the compounds and of the concept of using a single element and simple naturally occurring molecules of this element. At the end of the 1940s, modern oncology developed rapidly, and a number of cytostatic substances emerged from chemical weapons from World War II, including nitrogen–mustard gas derivatives. It is possible that the focus on these substances caused the concept of selenium as a cytostatic drug to be forgotten. However, numerous *in vitro* and animal studies have provided further and solid proof of the cytotoxic potential of redox-active selenium compounds during the past decades. There are two distinct pathways of selenium focused cancer research, that is, prevention and treatment, between which the dominating path has been prevention. These two lines of research are often mixed, leading to confusion, which is not beneficial for the field. Selenium in the prevention of cancer follows completely different mechanisms compared to selenium as a therapeutic agent (Steinbrenner et al., 2013). In the former role, the antioxidant properties are crucial as are the connection to selenoproteins. The field of selenoproteins has dominated selenium research for decades. However, selenoproteins that are the base for the physiological, antioxidant, and preventive effects of selenium are likely of minor importance in the antiproliferative and cytotoxic effects of selenium, that is, in the role of selenium as a cytostatic drug. Selenite is the most widely studied selenium compound. Due to the high reactivity and first passage effect in the liver, the compound requires intravenous administration to achieve a fast increase in redox-active selenium species in the blood. There are numerous reports on the antiproliferative effects of selenium *in vitro*, and in these laboratory investigations, mostly selenite has been applied, and recently a new class of methylated selenium compounds is used in various tumor models and cell lines with striking results. However, in the few human cancer trials that have been published, redox-active selenium compounds or precursors of redox-active selenium compounds, that is, selenite and methylated species, for which there are an abundance of positive data, have not been used. Instead, the relatively inert SeMet has been applied. Furthermore, vitamin E has, for some reason, been combined in high amounts of selenium, demonstrating the common misinterpretation of antioxidant to pro-oxidant effects. This chapter will summarize the current knowledge of the therapeutic potential of selenite. A major aim is to clarify mechanisms and the differences in preventive and chemo-therapeutic effects and to emphasize the importance of using redox-active selenium compounds in cancer therapeutics.

6.2   SELENITE AT THE CROSSROAD BETWEEN ANTIOXIDANT AND PRO-OXIDANT FUNCTIONS

How can an essential trace element, present naturally in 25 selenoproteins, serve as a cytostatic agent? These dual and contradictory effects are explained by the complex chemistry of selenium and the fact that selenium can form reactive selenolates that might either catalyze the reduction of hydroperoxides to alcohols or participate in redox cycles with the resulting massive formation of reactive oxygen species (ROS), depending on the concentration, redox environment of the cell, and access to oxygen (Kumar et al., 1992; Björnstedt et al., 1995b). The preventive effects are readily understood and relatively easily explained, and the connection to selenoproteins is obvious. Not only do selenoproteins exert antioxidant effects, but also low molecular weight compounds, for example, selenite and selenocysteine (Björnstedt et al., 1995a; Kumar et al., 2011). One important feature is the strict and narrow concentration dependence. Selenite, like all redox-active selenium compounds, could be considered as a pro-drug because the metabolites are the active forms. One central metabolite is selenide (see also Chapter 2). Selenide is formed from selenite by reduction and is the critical metabolite as a precursor for the synthesis of selenium-containing redox-active enzymes. However, in the presence of oxygen and thiols, selenide will redox cycle, causing a nonstoichiometric production of ROS (Kumar et al., 1992) (Figure 6.1).

The nature of the ROS formed is debated because several reports suggest superoxide is formed, but others report that the major ROS in fact is hydroxyl radicals (Park et al., 2012). Hydroxyl radicals may cause DNA strand breaks and thus damage rapidly dividing cells (i.e., cancer cells).

After reduction of selenite to selenide, this metabolite will either undergo methylation to form the primary anticancer selenium species methylselenol or is consumed for the synthesis of selenoproteins or create ROS by redox cycles. ROS formation is

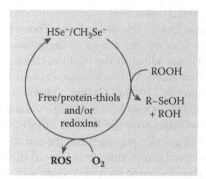

FIGURE 6.1   Antioxidant and pro-oxidant mechanisms of selenide and selenols (HSe^-/$CHSe^-$).

considered one primary antiproliferative mechanism that could be applied in chemo-therapy, but there are likely multiple mechanisms.

6.3   SELENITE: A POTENT CHEMOTHERAPEUTIC AGENT

For no other metalloids is the spectrum of its biological effects as enigmatic as is the case for selenium. During the past 100 years, the view on selenium has changed from an antitumor drug (Watson-Williams, 1919) to a (mis)interpreted carcinogen (Nelson et al., 1943) and later as an essential trace element (Schwarz and Foltz, 1957) with chemopreventive effects. Subsequently, many studies have identified the distinctive anticarcinogenic properties of certain selenium com-pounds with varying efficacy. It is until recently when the prospect of therapeutic application of selenium in cancer has been explored. Of all the studied com-pounds, selenite is one of the most promising candidates with distinct antitumor effects. In this section, findings from several preclinical studies with selenite are presented, focusing on the toxicity and its efficacy as an antitumor agent both *in vitro* and *in vivo*. We have also explored if pharmacological doses represent a therapeutic window.

6.3.1   *IN VITRO* STUDIES

One of the early important findings from *in vitro* experiments is the prominent tumor specificity of selenite to tumor cells. Different selenium compounds, includ-ing selenite (up to 4.34 μM), were used at increasing concentrations to treat four dif-ferent canine mammary tumor cells lines and one non-neoplastic primary cell (Fico et al., 1986). Potent growth inhibition and toxicity of selenite in tumor cells, except one cell line, was reported, and no appreciable toxicity and growth inhibition was recorded in non-neoplastic primary cells at identical exposure concentrations and time. It was found that high cell density increased the toxicity of selenite, a prop-erty consistently observed by us (unpublished data). The implications for this find-ing suggest that cell-secreted extracellular factor(s) (related to the number of cells) may have a potentiating effect on selenite toxicity. Corroborating this concept, the enhancement of selenite toxicity was shown in the promyelocytic cell line HL60 when glutathione (GSH) was added to the medium (Batist et al., 1986). Recently, we validated the role of extracellular thiols (cysteine) as the key modulatory factor of selenite toxicity (Olm et al., 2009a). Caffrey and Frenkel (1992) demonstrated that the difference in selenite toxicity between drug-resistant and nonresistant ovar-ian tumor cells was associated with the level of intracellular GSH. Although an exceptionally high level of selenite was used and the extracellular thiol content was not attributed in the observed effect, this study showed the importance of intracel-lular thiol status in mediating selenite toxicity. Presumably, the apparent differences in selenite toxicity *in vitro* depend on several factors influencing the intracellular and extracellular redox environment. A successful chemotherapeutic application of selenite warrants a substantial therapeutic margin. To represent this, a summary of selenite toxicity in different cancer cell lines and isolated normal cells is presented in Table 6.1.

TABLE 6.1
Toxicity of Selenite in Different Cell Lines of Multiple Origins

Name	Tissue of Origin/Type	IC$_{50}$/Time	Reference
		Normal Cells	
MRC-5	Embryonic lung	>8.1 µM/72 h[a]	Watrach et al. (1984)
PrEC	Normal human prostate	>500 µM/24 h; >500 µM/48 h; 27 µM/72 h	Menter et al. (2000)
PrSt		85 µM/24 h; 20 µM/48 h; 38 µM/72 h	
PrSM		58 µM/24 h; 20 µM/48 h; 22 µM/72 h	
N/A	Normal human prostate	>250 µM/48 h[a]	Husbeck et al. (2005)
N/A	Normal human prostate	>100 µM/48 h[a]	Husbeck et al. (2006)
Astrocytes	Fetal cerebrum	7.5 µM/24 h	Kim et al. (2007)
		Cancer Cells	
NB4	Acute promyelocytic leukemia	~20 µM/24 h	Li et al. (2003)
U87MG, T98G, A172, U343, U251	Glioma	1–3.5 µM/24 h[a]	Kim et al. (2007)
H157	Non-small cell lung cancer	~5 µM/20 h	Selenius et al. (2008)
U1810		7.5 µM/20 h	
H611		50 µM/20 h	
U1906L	Small cell lung cancer	~5 µM/20 h	
U1906E		~5 µM/20 h	
H157	Non-small cell lung cancer	4.5 µM/24 h	Wallenberg et al. (2010)
A549	Lung adenocarcinoma	5.0 µM/24 h	Weekley et al. (2011)
HeLa	Cervical cancer	5.0 µM/24 h	Wallenberg et al. (2014)

[a] Denotes IC$_{50}$ data are not available and an approximation has been made from the presented toxicity data.

6.3.2 *IN VIVO* STUDIES WITH ANIMAL MODELS

Early *in vivo* animal studies have demonstrated the capability of selenite to inhibit the growth of spontaneous, transplanted, and chemically induced tumors. Clayton and Baumann first reported the evidence for anticarcinogenic effects of selenite in an azo dye–induced liver carcinogenesis rat model (1949). Rats were given selenium as sodium selenite at a concentration of 5 ppm during 4 weeks between two 4-week periods of feeding with azo dye. The authors reported decreased development of liver tumors by 50% in animals fed with a selenite-spiked diet in two independent experiments. We have also demonstrated that selenite treatment (1 ppm and 5 ppm) decreased tumor development in the promotion and progression phases in a chemically induced hepatocarcinogenesis model (Björkhem-Bergman et al., 2005).

In cerebellar deficient folia (CDF) mice, leukemic cells (L1210) were transplanted into the peritoneal cavity and subsequently treated with selenite ranging from 0 to 50 µg selenite/day for 6 days. Mice receiving 40 µg selenite/day i.p. had a prolonged lifespan, which was comparable to those receiving 50 µg selenite/day (Milner and Hsu, 1981). No effect of liver weight or red and white blood cells was reported in mice that received below 40 µg selenite/day. In another transplanted tumor model, Ehrlich ascites tumor cell (EAT)–bearing HAP(ICR) BR mice were treated with 5 µg of selenium as sodium selenite i.p. on days 0, 1, 3, and 5 (Poirier and Milner, 1983). The authors reported no incidence of EAT cells in the peritoneum after 5 weeks in these mice. Long-term follow up in these mice showed development of tumor in 90% of the mice following selenite treatment, but these mice survived significantly longer compared to the untreated control group. One of the important findings was that the site of the tumor and route of delivery were important factors in the protective effects of selenite against the development of tumors. When mice were inoculated with EAT cells subcutaneously, the protective effect against tumor incidence by selenite was diminished. The authors suggested that the antitumor effects of selenium might have been improved with prolonged treatment. Watrach et al. (1984) studied effects of subcutaneous selenite injection on the growth of tumors originating from transplanted MCF-7 and MDA-MB-231 cells in nude BALB/c mice. The authors reported more than 80% reduction in tumor growth when the mice received 0.8 or 1.5 µg selenite/g body weight three times a week for five weeks. There were neither any changes of liver or kidney weight nor the histological features in treated animals.

The dose of selenite is crucial for its anticancer effect. Using a transgenic mouse model for liver cancer (expressing c-Myc and TGFα) that develops hepatic tumors between 6 and 8 months of age, both high-dose feeding of selenite (2.25 ppm) and selenium deficiency decreased the frequencies of carcinoma development and incidence of tumor/mice (Novoselov et al., 2005). An opposite result was found in mice receiving 0.1 and 0.4 ppm. Importantly, mice fed with 0.4 ppm Se had the highest rate of tumor proliferation (mitotic index), and mice given 2.25 ppm had the highest rate of apoptosis and inflammatory infiltrates in the liver tumor. These observations provide indications of the multiple functionalities of selenite in cancer. Selenite in lower to moderate doses may function as a nutrient/growth factor to tumors, and in higher doses, selenite can promote and activate the immune system against tumor development and induce cell death.

In selenoprotein-deficient mice with a mutant selenocysteine tRNA gene (Trsp^{A37G} mice, see Chapter 4), the effect of selenium deprivation and a high dose of dietary selenite on diethylnitrosamine-induced hepatocarcinogenesis was studied (Kasaikina et al., 2013). Similar to the earlier study (Novoselov et al., 2005), both selenium deficiency and high dietary supplementation of selenite protected against any tumor development. Corroborating these findings, it was found that a high level of dietary selenite decreased liver tumor size driven by TGF-β (Moustafa et al., 2013). In these studies, the authors described the effect of selenium supplementation and cancer development as a complex interaction between the levels of selenoproteins and selenium compounds.

It is increasingly clear that selenite at high doses exerts antitumor effects in all animal studies mentioned above. These studies highlight that the route of administration, genetic background, and selenoprotein levels are important determinants of the observed antitumor effects. In the context of selenoproteins with antioxidant functionalities, one may consider their critical cytoprotective roles during the early stages of carcinogenesis by providing protection against DNA damage. As a high dose of selenite does not affect the selenoprotein expression to a large extent (Novoselov et al., 2005), it is tempting to conclude that a high dose of selenite certainly exerts notable antitumor effects at physiological selenoprotein optima.

6.4 THE MOLECULAR BASIS OF SELENITE CYTOTOXICITY IN CANCER

In the previous sections, we have presented a brief overview of experimental evidence both *in vitro* and *in vivo* reflecting the efficacy of selenite as a potent cancer chemotherapeutic agent. The therapeutic usage of selenite in cancer warrants in-depth understanding of its pharmacokinetics and pharmacodynamics. The following sections describe updated information on the above aspects.

6.4.1 SELENITE TRANSPORT IN CANCER CELLS

Selenoprotein P is the major selenium transport protein in human plasma. It represents about 65% of total plasma selenium (Åkesson et al., 1994). However, it is essential to understand the physiological transport of selenite and its derivatives at therapeutic doses of selenite when used as an anticancer agent. Oral administration results in reduction of selenite by intestinal microbiota (Krittaphol et al., 2011), the composition of which can substantially be altered by selenite itself (Kasaikina et al., 2011) as previously shown in animal models. In conjunction, the first pass metabolism of selenite by the liver results in chemically diverse metabolites, all of which are not well characterized. Therefore, intravenous (IV) administration of high doses of selenite appears to be a better therapeutic delivery strategy without perturbing the intestinal microbiota that play critical roles in gut immunity (Round and Mazmanian, 2009). An important consideration concerning IV administration of selenite consists of understanding the predominant chemical forms of selenium in the blood. This is due to inherent and spontaneous reactivity of selenite with free –SH groups, including protein thiols and ascorbic acid (ranging from 30 to 150 μM) in the plasma. In general, the plasma-free thiol pool is much more oxidized in comparison to the intracellular milieu, and the reduced form ranges from 12 to 20 μM (reviewed in Turell et al., 2013). But, free cysteine (cysteine 34) residue in human serum albumin (HSA) constitutes the major protein thiol source, accounting for about 0.6 mM free thiol equivalent. About 75% of these cysteine residues of HSA remain reduced under normal conditions. It is therefore envisaged that these free thiol moieties may play major roles in reducing IV administered selenite. The uptake of selenite by red blood cells (RBCs) is also important for consideration. Apart from a high number of RBCs in blood, there is abundant expression of membrane anionic exchanger 1 (band 3 protein, AE1), implicated in selenite transport (Galanter et al., 1993). High influx of selenite is expected in RBCs, given that a single erythrocyte contains about

~10^6 AE1 protein (Kopito and Lodish, 1985) and approximate influx of 7500 selenite ions/transporter/second (Galanter et al., 1993). Selenite is subsequently reduced in RBCs and effluxed into the plasma by a proposed relay mechanism, in which free cysteine residues of AE1 are implicated to play important roles (Haratake et al., 2009). Together, these findings suggest that AE1 in RBCs plays an important role for the influx of selenite.

Reduction of selenite to elemental selenium requires six molar equivalents of free –SH group per mole of selenite (Cui et al., 2008). It is evident from the above discussion that blood components have enough reducing capacity to reduce selenite even when administered at pharmacological concentrations. Therefore, it is important to obtain a comprehensive understanding of selenium speciation in plasma to determine the available selenium species for uptake at a distant site following IV administered selenite. So far, there is no published information on the speciation of selenium in human blood following a high dose of selenite administration. A recent study on speciation analysis in blood following high dose of IV administration of selenite in rabbit using synchrotron-based X-ray absorption spectroscopy provides some useful insights (Liu, 2011). It has been shown that more than 75% of total selenium is in the form of elemental selenium in the plasma following IV injection of selenite. This study reported the highest plasma selenium level of more than 15 μM, corresponding to the level that can be achieved in human without major short-term side effects. Although no selenide was detected in the plasma component, it was one of the major species found in the RBCs along with elemental selenium. The study also reported the presence of selenite in the plasma sample, and its relative abundance was not changed up to 2 hours after selenite administration. Under the assumption that a similar scenario exists in humans, these observations hitherto represent an important question yet to be answered: What are the chemical forms of selenium that are taken up by the tumor following intravenous administration of high doses of selenite?

Does selenite fulfill some of the basic criteria of a potent chemotherapeutic agent with tumor-specific uptake with an acceptable margin of therapeutic index? In the context of tumor-specific uptake, it was previously shown that selenium from [75]Se (as selenite) can be taken up by tumors with a certain degree of specificity following IV administration (Cavalieri et al., 1966; Cavalieri and Scott, 1968). In all these cases, probably tracer doses of radioactive selenite were used. However, it is not well known whether a similar paradigm exists following pharmacological intervention with high doses of selenite.

Selenite exhibits remarkable specificity in inducing cytotoxicity to certain cell lines, and other cell lines are quite resistant as revealed by numerous *in vitro* experiments with cultured tumor cells. In general, selenite uptake is very slow. The fundamental differences in sensitivity rely upon the extracellular redox potential, mainly characterized by a cysteine/cystine redox couple that is a variable characteristic of individual cell lines. We have previously shown that extracellular thiols play critical roles in reducing selenite and subsequent uptake of the reduced form of selenium in inducing cytotoxicity (Olm et al., 2009a). Although it had been previously shown that cysteine (Würmli et al., 1989) and other thiols, such as cysteamine, thioglycolate, and mercaptopyruvate (Scharrer et al., 1992), augment selenium uptake from selenite by several folds, our study was the first to report in the context of a microenvironment of cultured tumor cells. The expression of functional cystine/ glutamate antiporter (SLC7A11, also known as xCT) has been shown as the key determinant of a reducing extracellular environment. SLC7A11 mediates cystine influx in the exchange of glutamate, and subsequently cystine is reduced into cysteine, which is effluxed out of the cells via either the MRP-1 or ASC transport system. In order to ensure an uninterrupted supply of glutamate for exchange, SLC7A11-overexpressing cells require exogenous glutamine supply. In accordance, a recent study indicates overexpression of glutamine transporter SLC1A5 in the A549 cell line known to overexpress SLC7A11 (Hassanein et al., 2013). The same study also reports that glutamine transporter is overexpressed in stage I non-small cell lung cancer patients when compared with matched controls. These observations provide a plausible basis to use selenite for selective targeting of tumors overexpressing SLC7A11 and SLC1A5 (Gln transporter).

6.4.2 GENERATION OF REACTIVE OXYGEN SPECIES

There is considerable controversy over whether ROS scavenging by antioxidant supplementation or induction of ROS production by pro-oxidant drugs is clinically beneficial or detrimental in cancer therapy (Gorrini et al., 2013). In the context of selenite, it can function both as an antioxidant or pro-oxidant, either indirectly or directly, depending on the exposure dose and redox microenvironment. Selenite in trace concentrations functions as a source of selenium, which is an integral part of selenoproteins. Many of these selenoproteins function as redox-active enzymes. It is quite unlikely that antioxidant functions of selenium at trace concentrations may act against the progression of established carcinoma being a constituent of enzymatic antioxidants as these provide crucial growth and survival advantages to tumor cells. In contrast, the characteristic pro-oxidant functions of high doses of selenite have certain distinct advantages as a chemotherapeutic drug. It has long been reported that selenite produces superoxide anion $\left(O_2^{-\bullet}\right)$ following reaction with excess thiols (Yan and Spallholz, 1993; Seko and Imura, 1997). This reaction has two distinct and unique advantages in terms of targeting cancer cells. First, selenite at high doses can oxidize thiols and thereby depletes intracellular GSH and other thiols that provide protection against oxidative insults. In fact, increased GSH biosynthesis is implicated in multidrug and radiation resistance in many cancer cells. An important implication of GSH depletion suggests possible impairment of GSH-based cellular antioxidant machinery in association with compromised ROS scavenging capacity and thereby increased ROS generation. This may be considered as a priming effect of selenite on ROS-induced cellular damage. Second, production of $O_2^{-\bullet}$ following the reaction of selenite and thiols may propagate the deleterious effects of ROS under a thiol-compromised condition. Following the rate laws, the rate of ROS production is of greater magnitude at high doses of selenite under elevated GSH concentration (Yan and Spallholz, 1993). In this context, an obvious question arises whether normal cells can be spared from selenite cytotoxicity under a therapeutic regimen while cancer cells are targeted? To address this, it is important to address the relative basal ROS level in normal cells versus cancer cells. During transformation into invasive carcinoma from normal cells, cancer cells adapt at acquiring nutrients to fulfill the bioenergetic demands associated with increased growth and proliferation (Vander Heiden et al., 2009). Although energetically inefficient aerobic glycolysis prevails over oxidative phosphorylation, such a high metabolic flux is associated with increased basal levels of ROS production in cancer cells. In conjunction, the threshold tolerance for ROS in cancer cells is presumably low compared to normal cells (Gorrini et al., 2013). Therefore, selenite appears to be a unique candidate chemotherapeutic drug with GSH-scavenging property coupled with superoxide production, both of which can be detrimental to the survival of cancer cells with inherent high basal level of ROS. Is that why several studies have reported selenite to be selectively cytotoxic to cancer cells while normal cells are spared at comparable doses (Menter et al., 2000; Husbeck et al., 2006; Nilsonne et al., 2006)?

6.4.3   MODULATION OF IMPORTANT CANCER-ASSOCIATED SIGNALING PATHWAYS BY SELENITE

In recent years, there has been remarkable progress in our understanding of the modulation of important signaling pathways in cancer by selenite. ROS generation by selenite has been implicated in many of the observed effects. Nevertheless, it is important to point out that not all of the observed effects on signaling molecules can be attributed to ROS. There might be involvement of several underlying mechanisms that can exert their effects directly or indirectly. Modifications of critical thiol moieties in proteins by selenite and associated functional change reflect such an example. In this section, we will focus on describing the modulatory roles of selenite on pathways involving cell proliferation and few important transcription factors known to be aberrantly expressed in cancer.

6.4.3.1   Translational Machinery

It has long been known that selenite at pharmacological concentration inhibits amino acid incorporation in protein. Safer et al. (1980) reported selenite (10 μM) inhibition of eukaryotic initiation factor 2 (eIF2), implicated in eIF2-GTP-Met-tRNA ternary complex formation and thus inhibition of ribosomal recruitment of Met-tRNA. As one of the possible mechanisms of inhibition, the authors suggested activation of eIF-2α kinase by selenite, leading to subsequent inhibition of ternary complex formation. In fact, their hypothesis on activation of eIF-2α kinase has been indirectly validated by several independent studies. In HSC-3 cells, increased phosphorylation of eIF-2α was observed upon treatment with 10 μM selenite for 6 hours (Suzuki et al., 2010). Similar effects on eIF-2α have been found in U2OS cells following selenite treatment (1 mM for 1 hour) (Fujimura et al., 2012). Selenite also abrogates eIF4E-binding protein-1 (4EBP1) and its phosphorylation, leading to formation of stress granules without cytoprotective functions. However, it remains to be investigated whether a similar mechanism is involved in pharmacologically achievable selenite concentrations. Transient changes in phosphorylation of eIF-2α have been described in Jurkat cells following selenite treatment (20 μM) (Jiang et al., 2013). The same study also reports selenite-induced induction of phosphorylation of ser209 residue of eIF4E by p38, implicated therein for involvement in autophagic cell death. Notably, ser209 phosphorylation of eIF4E have been implicated in conferring its oncogenic effects, and therefore, its inhibition diminishes tumor progression *in vivo* (Pettersson et al., 2014). The above findings indicate that repression of translational machinery by selenite may have important implications for cancer treatment as a unique chemotherapeutic agent.

6.4.3.2 Cell Cycle

Selenite at pharmacological concentrations inhibits different stages of the cell cycles. It has been shown that selenite induces S-phase cell cycle arrest in tumor cells but not in normal cells (Menter et al., 2000). In the selenite-resistant human acute promyelocytic leukemia NB4 cell line, selenite exposure (20 μM for 36 hours) resulted in cell cycle block at the G_0/G_1 phase (Cao et al., 2006). However, selenite (5 μM for 24 hours) induced S-phase arrest in association with downregulation of p27kip1 and p21cip1 proteins in prostate cancer cell line DU145 (Jiang et al., 2002). In NB4 cells, selenite (20 μM for 0–24 hours) abrogated the expression levels of CDK4, phospho-Rb, cyclin D3, and cyclin A, implicated in cell cycle progression at the G_0/G_1 phase (An et al., 2013). Different from these results, it has recently been shown that selenite (5 μM for 24 hours) induced G_2/M phase arrest in colorectal cancer cells HCT116 and SW620 (Li et al., 2013). In all of these studies, an asynchronous cell population has been used for cell cycle analyses, which makes it difficult to conclude how different stages of cell cycle progression are affected by selenite.

6.4.3.3 p53

Sodium selenite causes ROS-induced DNA damage as reported in several *in vitro* systems. An imminent effect of such damage is the activation of p53, implicated in DNA repair. Both increased expression and phosphorylation of p53 (Ser15) have been reported in a p53-expressing LNCaP cell line following a subtoxic dose of selenite (3.5 μM for 24 hours) treatment (Zhao et al., 2006). Increased p53 expression was associated with upregulation of downstream targets p21[waf1] and Bax, indicating functional activation of p53 by selenite. In the presence of MnTMPyP, a ROS scavenging agent, reversal of some of the effects on p53 suggested an important role of O_2^- in the observed effects. In a p53-null PC3 cell line, forced overexpression of p53 resulted in higher sensitivity to selenite than in mock-transfected cells. The authors argued that selenite-induced DNA damage was not associated with the observed increase in p53 expression although evidence for DNA damage had been presented at a much lower dose (2.5 μM for 24 hours). In human lung cancer cells H1299 with endogenous deletion of both p53 alleles, selenite exposure (10 μM for 5 hours) following overexpression of wild-type p53 resulted in increased expression of total p53 along with phosphorylation of Ser20, Ser37, and Ser46 residues (Smith et al., 2004). A recent study has been shown that ERK2 is involved in the selenite-induced phosphorylation of Ser15 residues of p53 that inhibits MDM2 binding to it (Li et al., 2010). This leads to subsequent p53 transcriptional activation and expression of MnSOD, a downstream target of p53, implicated in the detoxification of O_2^-. Notably, potentiation of selenite cytotoxicity has been related to p53 expression status in several studies. Nevertheless, there are significant differences in selenite sensitivity of different cell lines expressing wild-type p53, including LNCaP (sensitive) and NB4 (resistant) cells. It has also been shown that mutant-type p53 human oral squamous cell carcinoma cell lines (HSC-3 and HSC-4) are more sensitive to selenite than a wild-type p53-expressing human breast cancer MCF-7 cell line (Suzuki et al., 2010). This signifies that there are several other determinants of selenite toxicity other than p53, at least *in vitro*. In view of the above studies, it is clear that at low to moderate doses, the expression of MnSOD may exert cytoprotective effects due to $O_2^{-\bullet}$ scavenging. However, overexpression of a p53-regulated proapoptotic factor, such as Bax, may potentiate the cytotoxic effect of selenite in conjunction with $O_2^{-\bullet}$ spillover at high doses of selenite. These findings have important therapeutic implications as high doses of selenite may augment selenite toxicity irrespective of p53 status under the impaired ROS scavenging capacity of tumor cells.

6.4.3.4   HIf-1α

In the murine melanoma cell line B16F10, exposure to a high concentration of selenite (20 μM for 12 hours) resulted in decreased expression of HIF-1α (Song et al., 2011), which is deregulated in many cancers. In human oral squamous cell carcinoma, selenite (10 μM for 4 hours) inhibited HIF-1α protein level expression and its DNA binding activity (Suzuki et al., 2010). These independent findings implicate that selenite exerts a common inhibition mechanism on HIF-1α. Corroborating these findings, selenite abrogated the expression of VEGF, a downstream target of transcription factor HIF-1α (Pei et al., 2010). Therefore, inhibitory effects of selenite in angiogenesis warrants further investigation.

6.4.3.5   NF-κB

Nuclear factor–kappa B (NF-κB) is one of the major transcription factors involved in the transcription of several important genes in the inflammatory pathways. It has increasingly becoming clear that NF-κB is activated in many types of cancer and plays an important role in evading apoptosis (reviewed in Chaturvedi et al., 2011). It has been shown that selenite inhibits NF-κB DNA binding activity when added to purified nuclear extract (Kim and Stadtman, 1997). However, much higher dose of selenite (100 μM; pharmacologically irrelevant) were required to inhibit NF-κB DNA binding activity. In a later study, selenite (up to 10 μM-mediated) inhibition of NF-κB DNA binding activity could not be reproduced in a purified nuclear extract of DU145 cells (Gasparian et al., 2002). However, the authors found an inhibition of IκBα kinase (IKK) activity by selenite (5 μM for 3 hours). This was associated with a diminished degradation of inhibitory protein IκBα and subsequent abrogation of NF-κB nuclear translocation. Recently, it has been shown that selenite (20 μM for 24 hours) inhibits HSP90 expression and concomitant decrease in NF-κB nuclear translocation in wild-type p53-expressing NB4 cells (Jiang et al., 2011). These findings suggest that a functional p53 status may be an important determinant of selenite-mediated inhibition of NF-κB signaling. It has also been shown that selenite (2 μM for 24 hours) inhibits the expression of TNFα-induced NF-κB downstream targets ICAM-1, VCAM-1, and E-selectin protein levels in HUVEC cells (Zhang et al., 2002).

At pharmacological concentrations, selenite may inhibit NF-κB by a different mechanism than those described above. It has been shown that partial depletion of intracellular GSH and subsequent accumulation of GSSG abrogates the activation and nuclear translocation of NF-κB in a human T-lymphoblast cell line, MOLT-4 cells (Mihm et al., 1995).

In the context of selenite toxicity, it has been shown that selenite depletes intracellular GSH and increases GSSG in a dose-dependent manner (Morgan et al., 2002). Therefore, it is possible that selenite-induced NF-κB inhibition may be partly associated with the accumulation of GSSG following a high dose of selenite exposure.

6.4.3.6 AP-1

Activator protein 1 (AP-1) transcription factor is a homo- or heterodimer of several activating transcription factors, including JUN, FOS, ATF, and MAF families. Both the c-JUN and c-FOS transcription factors play important roles in the upregulation of tumor invasiveness-associated genes (Eferl and Wagner, 2003). It has been shown by us (Spyrou et al., 1995) and others (Handel et al., 1995) that selenite at pharmacological concentrations is a specific inhibitor of AP-1 DNA binding by targeting the critical cysteine residues. Selenite-induced (3 μM for 6 hours) inhibition of AP-1 has been reported to be associated with a decreased expression of its downstream targets MMP-9 and uPA, both of which are important for the degradation of extracellular matrix and thereby promote tumor metastasis (Yoon et al., 2001). The inhibition of NF-κB has been implicated in the inhibition of MMP-2. Context-wise, it is important to note that such inhibitory effects on MMP-2 and MMP-9 were similar in different cell lines of diverse origin, suggesting a similar underlying mechanism. The corresponding increase in TIMP-1 expression is suggestive of potentiation of inhibitory effects on MMP-9. These findings provide a plausible basis for therapeutic application of selenite in inhibiting tumor invasiveness and metastasis as recently suggested (Chen et al., 2013).

6.4.3.7 Epigenetic Signature

Aberrant DNA methylation and decreased histone acetylation are implicated in repression of many antitumor genes expressions in cancer. These alterations present novel targets for inhibition of carcinogenesis and progression of established carcinoma (Azad et al., 2013; Weekley et al., 2014; West and Johnstone, 2014). Selenite has long been recognized to inhibit DNA methyltransferase (DNMT) (Cox and Goorha, 1986; Fiala et al., 1998) in pharmacologically achievable doses. The later study reported that such inhibition was reversible by a reducing agent, such as dithiothreitol. In a subsequent study, it was shown that selenite (1.5 μM for 8 days) inhibited the mRNA expression of DNMT 1 and DNMT3A in LNCaP cells (Xiang et al., 2008). Total genomic methylation data in these cells indicated that the effect of selenite was comparable to 5-aza-2′-deoxycytidine, which is already under clinical evaluation. There is also a report suggesting no effects of selenite on genomic DNA methylation. It has recently been reported that long interspersed nucleotide elements (LINE-1), a marker of global genomic methylation, do not change following prolonged exposure to selenite up to 5 μM concentration (Barrera et al., 2013). However, a study from the same group has reported a time-dependent decrease in LINE-1 methylation and a gene-specific increase of CpG island methylation in normal cell lines with different selenium compounds, including selenite, in the presence of excess folic acid (Charles et al., 2012). Although no comparison has been made in the later study with the former one, it is not clear why no such effects were observed in Caco-2 and HCT116 cells cultured in high folate-containing DMEM media. In view of these studies, it appears that more detailed studies are required to comprehend in-depth understandings of the DNA methylation modulatory effects of selenite in cancer cells.

The study by Xiang et al. (2008) also provides important information on the histone-modifying effects of selenite. They reported decreased activity of total histone deacetylases by selenite treatment (1.5 µM for 7 days). Most importantly, the study revealed that selenite decreased the methylation and increased the acetylation of H3-Lys 9 and thereby positively modulated the common histone covalent modifications involved in epigenetic repression of many genes.

It is interesting to observe that some signaling pathways are modulated by selenite at low doses in certain cell types, and others require high doses. In fact, not all cell types accumulate selenium from selenite in a similar manner, and therefore, differences in intracellular selenium concentration perhaps provide a plausible explanation for such discrepancies. A summary of the observed effects are described below (Figure 6.2) in the context of "hallmarks" of cancer as described by Hanahan and Weinberg (2011).

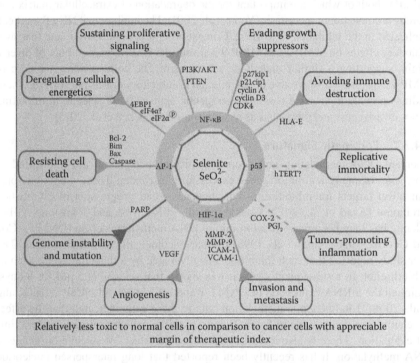

FIGURE 6.2 "Multitarget" nature of selenite in the context of its chemotherapeutic application. The therapeutic targets of hallmarks of cancer (Hanahan and Weinberg, 2011) are shown in the boxes. Transcription factors targeted by selenite are shown inside the circle. The expressions of proteins that are downregulated and upregulated by selenite are depicted in black and gray text, respectively. All of these changes are known to have established tumor-suppressive functions. Question mark indicates that there is not enough evidence. It is important to note that most of these observed effects are reported at pharmacological concentrations of selenite that can be administered in human without any short-term side effects and if any, transient in nature (unpublished data from our clinical trial). The bottom panel of the figure shows a distinct property of selenite based on the available *in vitro* toxicity data.

Multitargeted cancer chemotherapy is gaining momentum in achieving higher efficacy than use of a single drug. This is often achieved by combining several drugs, each targeting their respective pathway(s). In this perspective, selenite represents a unique candidate chemotherapeutic drug in targeting several oncogenic pathways. These multimodal antitumor properties can be exploited in combination therapy with cytostatic drugs in translating into effective therapeutic strategies against multiple tumor types. In the next section, we provide a brief overview on how selenite in combination with other cytostatic drugs can be effective apart from its own effects on cancer cells.

6.4.4 POSSIBLE MECHANISMS AND EVIDENCE FOR POTENTIATION OF CYTOTOXIC EFFECTS OF SELENITE WITH CHEMOTHERAPEUTIC DRUGS

The multitarget nature of selenite in cancer has been discussed above. In this section, we provide the hypothetical basis for the mechanism of how selenite acts in combination with cytostatic drugs. The major classes of anticancer drugs consist of alkylating agents, antimetabolites, antitumor antibiotics, topoisomerase inhibitors, mitotic inhibitors, and corticosteroids (Source: American Cancer Society). In recent years, the evolution of targeted therapy allowed interruption of molecular abnormalities specific to cancer. The target pathways are indicated in Figure 6.2. We argue that the broad category of oncogenic targets of selenite with a variable degree of specificity to cancer cells make it a suitable candidate drug for treatment with either cytostatic drugs or targeted therapeutics, depending on the nature of the tumors.

Previously, we argued for selenite as a potent chemopreventive and chemotherapeutic drug in a therapeutic regimen. Herein, we provide evidence for its superior therapeutic effects against drug-resistant cancer cells. High expression of proteins regulating the synthesis or transport of GSH as a detoxification mechanism (γ-glutamyl synthetase or multidrug resistant proteins) may contribute to the development of resistance against cytostatic drugs in cancer cells. Interestingly, selenite toxicity is attributed to these prosurvival factors when in excess and thereby sensitizes these cells using their own defense mechanisms. In line with this hypothesis, several studies have shown that drug-resistant tumor cells are more sensitive to selenite than their drug-sensitive or benign counterparts (Caffrey and Frenkel, 1994; Nilsonne et al., 2006). In a previous study, we showed that selenite was more toxic to doxorubicin-resistant lung cancer cells expressing MRP than their parental doxorubicin-sensitive cell lines (Björkhem-Bergman et al., 2002; Jönsson-Videsäter et al., 2004). Selenius et al. (2008) compared selenite cytotoxicity in several lung cancer cell lines and showed that the doxorubicin-resistant cell line was the one most sensitive to selenite treatment. In an animal model of cytostatic drug-resistant tumor, selenite was shown to circumvent drug-resistance and thereby tumor growth (Frenkel and Caffrey, 2001). Using a human ovarian tumor xenograft model, it was shown that a pretreatment with selenite prevented development of resistance against carboplatin (Caffrey and Frenkel, 2013). In an *ex vivo* study with patient samples, we also showed that selenite at 5 μM concentration exerted highest toxicity in primary acute myeloid leukemia cells in comparison to conventional antileukemic drugs at clinically relevant concentrations (Olm et al., 2009b).

Apart from its own cytotoxic effects, selenite may interact with cytostatic drugs, resulting in interesting and beneficial effects as summarized below. Selenite may

1. Have an additive or synergistic effect on cancers when combined with conventional cytostatic drugs or targeted therapy
2. Decrease the side effects of the chemotherapeutic treatments
3. Prevent or reverse chemotherapy resistance

The principal factor determining the ultimate effect is the exposure concentration of selenite. Selenite provides protection against the deleterious effects of chemotherapeutic drugs at low concentrations. Vadgama et al. (2000) showed that a low dose of selenium enhanced the chemotherapeutic effect of taxol and doxorubicin in breast, lung, small intestine, and liver cancer cells. Studies on cancer patients have also shown that pretreatment or simultaneous treatment with selenium at low doses and a cytostatic drug may protect against the side effects of chemotherapy without interfering with the effect of the chemotherapy (Hu et al., 1997; Sieja and Talerczyk, 2004). In contrast, at a high dose, selenite may target tumor cells more efficiently than the use of a single cytostatic agent. Schroeder et al. (2004) studied the combined effect of selenite and clinically used cytostatic drugs on different carcinoma cell lines. Out of 11 tested cytostatic drugs, selenite potentiated the cytotoxicity of 5-FU, oxaliplatin, and irinotecan in colon cancer cells (HCT116 and SW620) and increased the inhibitory effect of docetaxel in A549 cells. In a previous study from our lab, a combined treatment of lung cancer cells (H157) with selenite (5–10 µM) with three different cytostatic drugs (cisplatin, docetaxel, and doxorubicin) revealed a potentiating effect of selenite on the cytostatic drugs at low doses (2.5–5 µM of the cytostatic drugs) but had no effect when given individually (Selenius et al., 2008). Similarly, a synergistic growth inhibiting effect of selenite and docetaxel was observed in PC3 prostate cancer cells (Freitas et al., 2011). Recent studies have also shown the efficacy of cotreatment with selenium and cisplatin enhancing the therapeutic effect of treating cancer in mice. Furthermore, treatment of Se-sufficient mice bearing H22 cells with selenosulfate together with cisplatin resulted in a cure rate of 75% compared to 25% by cisplatin treatment alone (Zhang et al., 2008).

6.5 HUMAN STUDIES

6.5.1 RATIONALE FOR USE OF SELENIUM IN TREATMENT OF CANCER PATIENTS

In previous sections, substantial preclinical data are presented on the growth- and survival-modulatory effects of selenite on cancer cells. Although, these studies clearly underline the potential of selenite and other selenium compounds as chemotherapeutic drugs, there are very few studies in humans, and there is no published systematic phase I trial. Several studies confirm the preclinical results, that is, selenium is accumulated by cancer cells. The extent of the specific uptake is demonstrated in a study in which selenium-75 could be used in scintigraphy to visualize tumors (Cavalieri et al., 1966; Cavalieri and Scott, 1968). Blood selenium levels in cancer patients are generally lower compared to healthy individuals (Dennert et al., 2011), indicating that the selective uptake of selenium by cancer cells may modulate its blood level. In accordance, low blood selenium has even been found to be a diagnostic factor in CAT-scan screenings of lung cancer (Grodzki et al., 2013).

The above effects reflect the importance of selenium concentration when low concentrations promote cell growth, a feature harnessed by cancer cells to obtain growth advantage. Low or subtoxic levels may thus potentiate tumor growth, and thereby, supplementation of selenium at nutritional doses to cancer patients could be harmful. However, at higher concentrations of selenium, the efficient uptake will be an Achilles heel that will make selenium specifically cytotoxic to tumor cells in a narrow concentration span.

6.5.2 SELENIUM IN COMBINATION WITH SURGERY, RADIOTHERAPY, AND CHEMOTHERAPY

The majority of the few available human studies focus mainly on the combination of selenium (at subtoxic levels) with routine cytostatic therapy with the aim to prevent side effects or to potentiate the standard treatment. A summary of studies in which medium to high doses of selenium have been administered are presented in Tables 6.2 and 6.3.

In a small study of 50 patients with non-Hodgkin lymphoma (NHL), patients were randomly assigned to daily supplementation with 0.2 mg/kg of selenite over 7 days, together with chemotherapy or only chemotherapy (Asfour et al., 2009). The aim of the study was to find alleviating effects of selenite from the symptoms arising out of chemotherapy treatment. Supplementation with selenite resulted in significant apoptosis in the lymphoma cells in addition to a significant decrease in number of affected lymph nodes. In a similar study with newly diagnosed patients with NHL ($N = 50$), patients were divided into two groups wherein one study group received 0.2 mg/kg/day of selenite for 30 days in addition to chemotherapy, and the other group only received chemotherapy treatment (Asfour et al., 2007). Selenite induced a significant downregulation of the level of Bcl-2 in the lymphoma cells. Furthermore, a more complete response to the treatment was observed in the Se-supplemented group (60%), compared to only chemotherapy treatment (40%).

Beneficial effects of treatment with selenite have been described in most of the studies as exemplified by reduced side effects from cisplatin, reduced leukopenia and anemia after chemotherapy treatment, and reduced lymph edema after radiotherapy in breast cancer patients and in head and neck patients after surgery. However, many of the studies had small sample sizes and were uncontrolled and not randomized. These results are interesting and indicate the need for larger and better-controlled studies in the future. Despite this need, an analysis based on only two randomized studies concludes that "there is insufficient evidence at present that selenium supplementation alleviates the side effects of tumor-specific chemotherapy or radiotherapy treatments" (Dennert and Horneber, 2006).

6.5.3 USE OF SELENIUM AS A CHEMOTHERAPEUTIC DRUG IN HUMANS

To explore the potential of selenite and other selenium compounds, maximum tolerated dose and safety evaluation in systematic phase I studies must be performed. The lack of published phase I data explains the sparse numbers of human trials.

TABLE 6.2
Peroral Treatment with Selenium Compounds in Cancer Populations (NB, not as Cancer Prophylaxis)

Study	Patients	Selenium Dose	Study Design	Endpoint and Results
Weisberger et al. (1956)	$N = 4$; Acute leukemia 2, Chronic myeloid leukemia 2	Seleno-cystine average dose 100 mg/day, ranging from 50–200 mg/day for 10–57 days	Case series	Drop in leukocyte particle concentration (LPK) counts, one patient regained sensitivity to 6-mercaptopurine. Adverse effects reported were severe persistent nausea, vomiting, anorexia, severe alopecia, fingernail destruction. Reversible adverse effects. No overt organ changes related to selenium toxicity (10- and 57-day treatment). No renal or liver function abnormalities reported.
Hu et al. (1997)	$N = 41$; Different carcinomas, (breast, lung and GI cancer)	4000 µg/day (as Kappa-seleno-carrageenan); 4 days before and 4 days after chemotherapy cumulative dose 32 mg selenium	Randomized cross-over-design	Outcome of adverse effects after chemotherapy. LPK, Hb improved and need of GM-CSF less in selenium treated patients. No selenium specific adverse effects reported.
Kasseroller (1998)	$N = 60$; Secondary lymphedema post head and neck surgery	1000 µg Na-selenite/day in week 1300 µg/day for weeks 2 and 3, then 200–300 µg/day for 3 months depending on body weight, cumulative dose selenium approximately 17 mg	Randomized, controlled	Selenium prevented erysipelas. Adverse effects not recorded.

(Continued)

TABLE 6.2 (CONTINUED)

Peroral Treatment with Selenium Compounds in Cancer Populations (NB, not as Cancer Prophylaxis)

Study	Patients	Selenium Dose	Study Design	Endpoint and Results
Micke et al. (2003)	$N = 12$; lymphedema post breast cancer resection	159 µg/m^2/day (corresponding to 350 µg selenite/m^2 for 4–6 weeks, cumulative dose approximately 9.5 mg	Nonrandomized	Design inadequate to correctly assess effect on lymphedema. Statistically significant reduction of pain score, but the study is uncontrolled. No adverse effect of selenium treatment.
Sieja et al. (2004)	$N = 31$; ovarian cancer	200 µg/day (corresponding to 440 µg/day selenite) cumulative dose 16.8 mg selenium, 3 months treatment	Nonrandomized	Adverse effects of cisplatin/cyclophosphamide reduced in selenium treated patients.
Bruns et al. (2004)	$N = 36$; head and neck cancer	159 µg/m^2/day for 4–6 weeks after radiation therapy, (350 µg selenite/m^2/day), cumulative dose approximately 9.5 mg selenium	Nonrandomized uncontrolled	Effect on lymphedema post radiation alone or in combination with surgery was studied. 75% of patients improved their QoL–scores, 65%.
Elango et al. (2006)	$N = 63$; head and neck carcinoma	400 µg/day during 6 months		Selenium level low in nonsupplemented patients, improved selenium levels and antioxidant status in supplemented patients ($p < 0.05$).
Bunzel et al. (2010)	$N = 39$; head and neck carcinomas	500 µg selenite/day of radiotherapy, 300 µg/day without radiotherapy	Randomized phase II	Limited, nonsignificant effect of aguesia (loss of taste) and dysphagia.

TABLE 6.3

Parenteral Administration of Selenium in Cancer Patients

Study	Patients	Selenium Dose	Study Design	Endpoint and Results
Pakdaman (1998)	N = 32; patients with brain tumors and increased intracranial pressure. 70% of patients had subnormal selenium blood levels	454 µg/day IV for 4–8 weeks (Na-selenite 1 mg/day IV). Other symptomatic treatment allowed. Cumulative selenium dose 25 mg	Uncontrolled case series	Symptoms of intracranial pressure reduced in 76% of study population and blood parameters (Hb, thrombocyte particle concentration [TPK]) improved in all patients. Adverse effects of selenium not reported. Steroid medication not shown, but authors say steroid dosage could be reduced (eleven patients received dexamethasone). Additional improved effect in patients receiving oxygen therapy.
Zimmermann et al. (2005)	N = 20; patients with oral tumor surgery with postoperative edema	3000 µg on day of surgery thereafter 1000 µg selenite IV or per oral daily for 3 weeks	Placebo controlled	Inverse correlation between selenium levels, edema and glutathione peroxidase activity, 1.8 µmol/l selenium after two weeks in whole blood.

A few of the published trials are old but are insightful from a therapeutic perspective. However, the observed effects are difficult to interpret because the evaluation did not follow the standard Response Evaluation Criteria In Solid Tumors (RECIST) criteria. One of such study "Preliminary Note on the Treatment of Inoperable Carcinoma with Selenium" was published in which colloidal selenium was administered IV (Todd, 1934). In this study, a total of 91 patients were included. Five milliliters of 0.02% selenium was administered (5–50 injections) either intravenously or intramuscularly or subcutaneously. Of these 91 patients, 66 benefited from the treatment, in most cases as symptomatic relief, but in a few, diminution of the tumor.

In a subsequent study, selenocystine was administered per orally at doses of 50–200 mg/day during 10–57 days to four patients with leukemia of which two patients with acute leukemia and two patients with chronic myeloid leukemia (Weisberger and Suhrland, 1956). A rapid decrease in the total leukocyte count was recorded along with decrease in spleen size in all patients following selenocystine treatment. This effect on immature leukocytes was more consistent and evident in acute leukemia compared to chronic leukemia. Furthermore, a regained sensitivity to 6-mercaptopurine was observed. The treatment resulted in severe side effects, the most prominent of which were severe nausea, vomiting, anorexia, and alopecia. However, there were no signs of organ damage at necropsy.

6.5.4 PHASE I STUDY TO ASSESS SAFETY AND TO DETERMINE THE MAXIMAL TOLERABLE DOSE

We have just finished a long-anticipated phase I trial (Eudra CT Number: 2006-004076-13) on selenium in patients with terminal cancer. The purpose of the study was to determine the MTD following dose escalation and observe side effects (both subjective and clinical), including dose-limiting toxicity, and investigate the pharmacokinetics. The data will be published later this year.

6.6 CONCLUDING REMARKS

There is often a (mis)understanding about the terms "cancer chemoprevention" and "cancer chemotherapy" in the context of selenium and cancer. The classical definition of cancer chemoprevention is the use of natural, synthetic, or biological chemical agents to reverse, suppress, or prevent the initial phase of carcinogenesis or the progression of neoplastic cells to cancer (Sporn et al., 1976). In contrast, chemotherapy is the use of these classes of agents to target established malignancies. After a century of research on selenium and cancer, it is obvious that cancer treatment with selenium must harness the potential of redox-active selenium compounds including selenite. The pro-oxidant properties are central when selenium compounds are used as cancer chemotherapeutics. The applied dose is very important and the determinant of whether selenium will act as an antioxidant or a pro-oxidant. In fact, selenium may promote the growth of dysplastic cells and established malignancies if the concentration is suboptimal to adequate. In modern oncology, efforts are made to find "targeted therapies" that exclusively target single mutations or ligands in tumor cells. However, compensatory mechanisms in cancer may facilitate the development of drug resistance against single-molecule targeted therapy and thereby forming the fundamental basis for "multitargeted therapy." One of the classical examples is p53 mutation in cancer and the multiplicity of the complex nature of gain of functions spanning multiple pathways by mutant p53 (Muller and Vousden, 2013). Context-wise, selenite targets several oncogenic pathways, and the cytotoxicity is remarkably tumor-specific. Selenite is chemically simple, cheap and well-tolerated—properties all of which are favorable for a drug in clinical use. Future effect evaluation studies in humans will reveal whether selenite may be used in clinical practice either as a single drug or in combination with cytostatic drugs.

ACKNOWLEDGMENTS

We are grateful to Cancerfonden (The Swedish Cancer Society), Cancer-och Allergifonden, Stockholm County Council (ALF), Stichting AF Jochnick Foundation, and Radiumhemmets Forskningsfonder for grant support.

REFERENCES

Åkesson, B., T. Bellew, and R. F. Burk. 1994. "Purification of selenoprotein P from human plasma." *Biochim Biophys Acta* no. 1204 (2):243–9.
An, J. J., K. J. Shi, W. Wei et al. 2013. "The ROS/JNK/ATF2 pathway mediates selenite-induced leukemia NB4 cell cycle arrest and apoptosis *in vitro* and *in vivo*." *Cell Death Dis* no. 4:e973.
Asfour, I. A., M. M. El-Tehewi, M. H. Ahmed et al. 2009. "High-dose sodium selenite can induce apoptosis of lymphoma cells in adult patients with non-Hodgkin's lymphoma." *Biol Trace Elem Res* no. 127 (3):200–10.
Asfour, I. A., M. Fayek, S. Raouf et al. 2007. "The impact of high-dose sodium selenite therapy on Bcl-2 expression in adult non-Hodgkin's lymphoma patients: Correlation with response and survival." *Biol Trace Elem Res* no. 120 (1–3):1–10.
Azad, N., C. A. Zahnow, C. M. Rudin, and S. B. Baylin. 2013. "The future of epigenetic therapy in solid tumours—Lessons from the past." *Nat Rev Clin Oncol* no. 10 (5):256–66.
Barrera, L. N., I. T. Johnson, Y. Bao, A. Cassidy, and N. J. Belshaw. 2013. "Colorectal cancer cells Caco-2 and HCT116 resist epigenetic effects of isothiocyanates and selenium *in vitro*." *Eur J Nutr* no. 52 (4):1327–41.
Batist, G., A. G. Katki, R. W. Klecker, Jr., and C. E. Myers. 1986. "Selenium-induced cytotoxicity of human leukemia cells: Interaction with reduced glutathione." *Cancer Res* no. 46 (11):5482–5.
Björkhem-Bergman, L., K. Jönsson, L. C. Eriksson et al. 2002. "Drug-resistant human lung cancer cells are more sensitive to selenium cytotoxicity. Effects on thioredoxin reductase and glutathione reductase." *Biochem Pharmacol* no. 63 (10):1875–84.
Björkhem-Bergman, L., U. B. Torndal, S. Eken et al. 2005. "Selenium prevents tumor development in a rat model for chemical carcinogenesis." *Carcinogenesis* no. 26 (1):125–31.
Björnstedt, M., M. Hamberg, S. Kumar, J. Xue, and A. Holmgren. 1995a. "Human thioredoxin reductase directly reduces lipid hydroperoxides by NADPH and selenocystine strongly stimulates the reaction via catalytically generated selenols." *J Biol Chem* no. 270 (20):11761–4.
Björnstedt, M., S. Kumar, and A. Holmgren. 1995b. "Selenite and selenodiglutathione: Reactions with thioredoxin systems." *Methods Enzymol* no. 252:209–19.
Bruns, F., J. Büntzel, R. Mücke et al. 2004. "Selenium in the treatment of head and neck lymphedema." *Medical Principles and Practice* no. 13 (4):185–90.

Buntzel, J., D. Riesenbeck, M. Glatzel et al. 2010. "Limited effects of selenium substitution in the prevention of radiation-associated toxicities. Results of a randomized study in head and neck cancer patients." *Anticancer Res* no. 30 (5):1829–32.

Caffrey, P. B., and G. D. Frenkel. 1992. "Selenite cytotoxicity in drug resistant and nonresistant human ovarian tumor cells." *Cancer Res* no. 52 (17):4812–6.

Caffrey, P. B., and G. D. Frenkel. 1994. "The development of drug resistance by tumor cells *in vitro* is accompanied by the development of sensitivity to selenite." *Cancer Lett* no. 81 (1):59–65.

Caffrey, P. B., and G. D. Frenkel. 2013. "Prevention of carboplatin-induced resistance in human ovarian tumor xenografts by selenite." *Anticancer Res* no. 33 (10):4249–54.

Cao, T. M., F. Y. Hua, C. M. Xu et al. 2006. "Distinct effects of different concentrations of sodium selenite on apoptosis, cell cycle, and gene expression profile in acute promyeloytic leukemia-derived NB4 cells." *Annals Hematol* no. 85 (7):434–42.

Cavalieri, R. R., and K. G. Scott. 1968. "Sodium selenite Se 75. A more specific agent for scanning tumors." *JAMA* no. 206 (3):591–5.

Cavalieri, R. R., K. G. Scott, and E. Sairenji. 1966. "Selenite (^{75}Se) as a tumor-localizing agent in man." *JAMA* no. 7 (3):197–208.

Charles, M. A., I. T. Johnson, and N. J. Belshaw. 2012. "Supra-physiological folic acid concentrations induce aberrant DNA methylation in normal human cells *in vitro*." *Epigenetics* no. 7 (7):689–94.

Chaturvedi, M. M., B. Sung, V. R. Yadav, R. Kannappan, and B. B. Aggarwal. 2011. "NF-κB addiction and its role in cancer: 'One size does not fit all.'" *Oncogene* no. 30 (14):1615–30.

Chen, Y. C., K. S. Prabhu, and A. M. Mastro. 2013. "Is selenium a potential treatment for cancer metastasis?" *Nutrients* no. 5 (4):1149–68.

Clayton, C. C., and C. A. Baumann. 1949. "Diet and azo dye tumors: Effect of diet during a period when the dye is not fed." *Cancer Res* no. 9 (10):575–82.

Cox, R., and S. Goorha. 1986. "A study of the mechanism of selenite-induced hypomethylated DNA and differentiation of Friend erythroleukemic cells." *Carcinogenesis* no. 7 (12):2015–8.

Cui, S. Y., H. Jin, S. J. Kim, A. P. Kumar, and Y. I. Lee. 2008. "Interaction of glutathione and sodium selenite *in vitro* investigated by electrospray ionization tandem mass spectrometry." *J Bioch* no. 143 (5):685–93.

Dennert, G., and M. Horneber. 2006. "Selenium for alleviating the side effects of chemotherapy, radiotherapy and surgery in cancer patients." *Cochrane Database Syst Rev* no. 3:CD005037.

Dennert, G., M. Zwahlen, M. Brinkman et al. 2011. "Selenium for preventing cancer." *Cochrane Database Syst Rev* no. 5:CD005195.

Eferl, R., and E. F. Wagner. 2003. "AP-1: A double-edged sword in tumorigenesis." *Nature Rev Cancer* no. 3 (11):859–68.

Elango, N., S. Samuel, and P. Chinnakkannu. 2006. "Enzymatic and non-enzymatic antioxidant status in stage (III) human oral squamous cell carcinoma and treated with radical radio therapy: Influence of selenium supplementation." *Clin Chim Acta* no. 373 (1–2):92–8.

Fiala, E. S., M. E. Staretz, G. A. Pandya, K. El-Bayoumy, and S. R. Hamilton. 1998. "Inhibition of DNA cytosine methyltransferase by chemopreventive selenium compounds, determined by an improved assay for DNA cytosine methyltransferase and DNA cytosine methylation." *Carcinogenesis* no. 19 (4):597–604.

Fico, M. E., K. A. Poirier, A. M. Watrach, M. A. Watrach, and J. A. Milner. 1986. "Differential effects of selenium on normal and neoplastic canine mammary cells." *Cancer Res* no. 46 (7):3384–8.

Freitas, M., V. Alves, A. B. Sarmento-Ribeiro, and A. Mota-Pinto. 2011. "Combined effect of sodium selenite and docetaxel on PC3 metastatic prostate cancer cell line." *Biochem Biophys Res Comm* no. 408 (4):713–9.

Frenkel, G. D., and P. B. Caffrey. 2001. "A prevention strategy for circumventing drug resistance in cancer chemotherapy." *Curr Pharm Des* no. 7 (16):1595–614.

Fujimura, K., A. T. Sasaki, and P. Anderson. 2012. "Selenite targets eIF4E-binding protein-1 to inhibit translation initiation and induce the assembly of non-canonical stress granules." *Nucleic Acids Res* no. 40 (16):8099–110.

Galanter, W. L., M. Hakimian, and R. J. Labotka. 1993. "Structural determinants of substrate specificity of the erythrocyte anion transporter." *Am J Physiol* no. 265 (4 Pt 1):C918–26.

Gasparian, A. V., Y. J. Yao, J. Lü et al. 2002. "Selenium compounds inhibit IκB Kinase (IKK) and nuclear factor-κB (NF-κB) in prostate cancer cells." *Mol Cancer Ther* no. 1 (12):1079–87.

Gorrini, C., I. S. Harris, and T. W. Mak. 2013. "Modulation of oxidative stress as an anticancer strategy." *Nature Rev Drug Discov* no. 12 (12):931–47.

Grodzki, T., J. Wójcik, A. Jakubowska et al. 2013. "Low selenium serum level is a good preselection factor for patients invited for low dose chest CT lung cancer screening." Paper read at 15th World Conference on Lung Cancer, at Sydney, Australia.

Hanahan, D., and R. A. Weinberg. 2011. "Hallmarks of cancer: The next generation." *Cell* no. 144 (5):646–74.

Handel, M. L., C. K. Watts, A. deFazio, R. O. Day, and R. L. Sutherland. 1995. "Inhibition of AP-1 binding and transcription by gold and selenium involving conserved cysteine residues in Jun and Fos." *PNAS USA* no. 92 (10):4497–501.

Haratake, M., M. Hongoh, M. Ono, and M. Nakayama. 2009. "Thiol-dependent membrane transport of selenium through an integral protein of the red blood cell membrane." *Inorg Chem* no. 48 (16):7805–11.

Hassanein, M., M. D. Hoeksema, M. Shiota et al. 2013. "SLC1A5 mediates glutamine transport required for lung cancer cell growth and survival." *Clinic Cancer Res* no. 9 (3):560–70.

Hu, Y. J., Y. Chen, Y. Q. Zhang et al. 1997. "The protective role of selenium on the toxicity of cisplatin-contained chemotherapy regimen in cancer patients." *Biol Trace Elem Res* no. 56 (3):331–41.

Husbeck, B., L. Nonn, D. M. Peehl, and S. J. Knox. 2006. "Tumor-selective killing by selenite in patient-matched pairs of normal and malignant prostate cells." *The Prostate* no. 66 (2):218–25.

Husbeck, B., D. M. Peehl, and S. J. Knox. 2005. "Redox modulation of human prostate carcinoma cells by selenite increases radiation-induced cell killing." *Free Radic Biol Med* no. 38 (1):50–7.

Jiang, C., Z. Wang, H. Ganther, and J. Lu. 2002. "Distinct effects of methylseleninic acid versus selenite on apoptosis, cell cycle, and protein kinase pathways in DU145 human prostate cancer cells." *Mol Cancer Ther* no. 1 (12):1059–66.

Jiang, Q., F. Li, K. Shi et al. 2013. "ATF4 activation by the p38MAPK-eIF4E axis mediates apoptosis and autophagy induced by selenite in Jurkat cells." *FEBS Lett* no. 587 (15):2420–9.

Jiang, Q., Y. Wang, T. Li et al. 2011. "Heat shock protein 90–mediated inactivation of nuclear factor-κB switches autophagy to apoptosis through becn1 transcriptional inhibition in selenite-induced NB4 cells." *Mol Biol Cell* no. 22 (8):1167–80.

Jönsson-Videsäter, K., L. Björkhem-Bergman, A. Hossain et al. 2004. "Selenite-induced apoptosis in doxorubicin-resistant cells and effects on the thioredoxin system." *Biochem Pharmacol* no. 67 (3):513–22.

Kasaikina, M. V., M. A. Kravtsova, B. C. Lee et al. 2011. "Dietary selenium affects host selenoproteome expression by influencing the gut microbiota." *FASEB J* no. 25 (7):2492–9.

Kasaikina, M. V., A. A. Turanov, A. Avanesov et al. 2013. "Contrasting roles of dietary selenium and selenoproteins in chemically induced hepatocarcinogenesis." *Carcinogenesis* no. 34 (5):1089–95.

Kasseroller, R. 1998. "Sodium selenite as prophylaxis against erysipelas in secondary lymph-edema." *Anticancer Res* no. 18 (3C):2227–30.

Kim, E. H., S. Sohn, H. J. Kwon et al. 2007. "Sodium selenite induces superoxide-mediated mitochondrial damage and subsequent autophagic cell death in malignant glioma cells." *Cancer Res* no. 67 (13):6314–24.

Kim, I. Y., and T. C. Stadtman. 1997. "Inhibition of NF-κB DNA binding and nitric oxide induction in human T cells and lung adenocarcinoma cells by selenite treatment." *PNAS* no. 94 (24):12904–7.

Kopito, R. R., and H. F. Lodish. 1985. "Primary structure and transmembrane orientation of the murine anion exchange protein." *Nature* no. 316 (6025):234–8.

Krittaphol, W., P. A. Wescombe, C. D. Thomson et al. 2011. "Metabolism of l-selenomethionine and selenite by probiotic bacteria: *In vitro* and *in vivo* studies." *Biol Trace Elem Res* no. 144 (1–3):1358–69.

Kumar, B. S., A. Kunwar, B. G. Singh, A. Ahmad, and K. I. Priyadarsini. 2011. "Anti-hemolytic and peroxyl radical scavenging activity of organoselenium compounds: An *in vitro* study." *Biol Trace Elem Res* no. 140 (2):127–38.

Kumar, S., M. Björnstedt, and A. Holmgren. 1992. "Selenite is a substrate for calf thymus thioredoxin reductase and thioredoxin and elicits a large non-stoichiometric oxidation of NADPH in the presence of oxygen." *Eur J Biochem* no. 207 (2):435–39.

Li, J., L. Zuo, T. Shen, C. Xu, and Z. Zhang. 2003. "Induction of apoptosis by sodium selenite in human acute promyelocytic leukemia NB4 cells: Involvement of oxidative stress and mitochondria." *J Trace Elem Med Biol* no. 17 (1):19–26.

Li, Z., J. Meng, T. J. Xu, X. Y. Qin, and X. D. Zhou. 2013. "Sodium selenite induces apopto-sis in colon cancer cells via Bax-dependent mitochondrial pathway." *Eur Rev Med Pharmacol Sci* no. 17 (16):2166–71.

Li, Z. S., K. J. Shi, L. Y. Guan et al. 2010. "ROS leads to MnSOD upregulation through ERK2 translocation and p53 activation in selenite-induced apoptosis of NB4 cells." *FEBS Lett* no. 584 (11):2291–7.

Lin, Y., and J. E. Spallholz. 1993. "Generation of reactive oxygen species from the reaction of selenium compounds with thiols and mammary tumor cells." *Biochem Pharmacol* no. 45 (2):429–37.

Liu, D. 2011. *Speciation of arsenic and selenium in rabbit using X-ray absorption spectros-copy*, University of Saskatchewan, Saskatoon.

Menter, D. G., A. L. Sabichi, and S. M. Lippman. 2000. "Selenium effects on prostate cell growth." *Cancer Epidemiol Biomarkers Prev* no. 9 (11):1171–82.

Micke, O., F. Bruns, R. Mücke et al. 2003. "Selenium in the treatment of radiation-associated secondary lymphedema." *Int J Radiat Oncol Biol Phys* no. 56 (1):40–9.

Mihm, S., D. Galter, and W. Dröge. 1995. "Modulation of transcription factor NF kappa B activity by intracellular glutathione levels and by variations of the extracellular cysteine supply." *FASEB J* no. 9 (2):246–52.

Milner, J. A., and C. Y. Hsu. 1981. "Inhibitory effects of selenium on the growth of L1210 leukemic cells." *Cancer Res* no. 41 (5):1652–6.

Morgan, K. T., H. Ni, H. R. Brown et al. 2002. "Application of cDNA microarray technology to *in vitro* toxicology and the selection of genes for a real-time RT-PCR-based screen for oxidative stress in Hep-G2 cells." *Toxicol Pathol.* no. 30 (4):435–51.

Moustafa, M. E., B. A. Carlson, M. R. Anver et al. 2013. "Selenium and selenoprotein deficien-cies induce widespread pyogranuloma formation in mice, while high levels of dietary selenium decrease liver tumor size driven by TGFalpha." *PloS One* no. 8 (2):e57389.

Muller, P. A. J., and K. H. Vousden. 2013. "p53 mutations in cancer." *Nat Cell Biol* no. 15 (1):2–8.

Nelson, A. A, O. G. Fitzhugh, and H. O. Calvery. 1943. "Liver tumors following cirrhosis caused by selenium in rats." *Cancer Res* no. 3:230–6.

Nilsonne, G., X. Sun, C. Nyström et al. 2006. "Selenite induces apoptosis in sarcomatoid malignant mesothelioma cells through oxidative stress." *Free Radic Biol Med* no. 41 (6):874–85.

Novoselov, S. V., D. F. Calvisi, V. M. Labunskyy et al. 2005. "Selenoprotein deficiency and high levels of selenium compounds can effectively inhibit hepatocarcinogenesis in transgenic mice." *Oncogene* no. 24 (54):8003–11.

Olm, E., A. P. Fernandes, C. Hebert et al. 2009a. "Extracellular thiol-assisted selenium uptake dependent on the x(c)-cystine transporter explains the cancer-specific cytotoxicity of selenite." *PNAS USA* no. 106 (27):11400–5.

Olm, E., K. Jönsson-Videsäter, I. Ribera-Cortada et al. 2009b. "Selenite is a potent cytotoxic agent for human primary AML cells." *Cancer Lett* no. 282 (1):116–23.

Pakdaman, A. 1998. "Symptomatic treatment of brain tumor patients with sodium selenite, oxygen, and other supportive measures." *Biol Trace Elem Res* no. 62 (1–2):1–6.

Park, S. H., J. H. Kim, G. Y. Chi et al. 2012. "Induction of apoptosis and autophagy by sodium selenite in A549 human lung carcinoma cells through generation of reactive oxygen species." *Toxicol Lett* no. 212 (3):252–61.

Pei, Z., H. Li, Y. Guo, Y. Jin, and D. Lin. 2010. "Sodium selenite inhibits the expression of VEGF, TGFβ1 and IL-6 induced by LPS in human PC3 cells via TLR4-NF-KB signaling blockage." *Int Immunopharmacol* no. 10 (1):50–6.

Pettersson, F., S. V. del Rincon, and W. H. Miller. 2014. "Eukaryotic translation initiation factor 4E as a novel therapeutic target in hematological malignancies and beyond." *Expert Opin Ther Targets* no. 18 (9):1035–48.

Poirier, K. A., and J. A. Milner. 1983. "Factors influencing the antitumorigenic properties of selenium in mice." *J Nutr* no. 113 (11):2147–54.

Round, J. L., and S. K. Mazmanian. 2009. "The gut microbiota shapes intestinal immune responses during health and disease." *Nat Rev Immunol* no. 9 (5):313–23.

Safer, B., R. Jagus, and D. Crouch. 1980. "Indirect inactivation of eukaryotic initiation factor 2 in reticulocyte lysate by selenite." *J Biol Chem* no. 255 (14):6913–7.

Scharrer, E., E. Senn, and S. Wolffram. 1992. "Stimulation of mucosal uptake of selenium from selenite by some thiols at various sites of rat intestine." *Biol Trace Elem Res* no. 33 (1):109–20.

Schroeder, C. P., E. M. Goeldner, K. Schulze-Forster et al. 2004. "Effect of selenite combined with chemotherapeutic agents on the proliferation of human carcinoma cell lines." *Biol Trace Elem Res* no. 99 (1–3):17–25.

Schwarz, K., and C. M. Foltz. 1957. "Selenium as an integral part of factor 3 against dietary necrotic liver degeneration." *J Am Chem Soc* no. 79 (12):3292–3.

Seko, Y., and N. Imura. 1997. "Active oxygen generation as a possible mechanism of selenium toxicity." *Biomed Environ Sci* no. 0 (2–3):333–9.

Selenius, M., A. P. Fernandes, O. Brodin, M. Björnstedt, and A. K. Rundlöf. 2008. "Treatment of lung cancer cells with cytotoxic levels of sodium selenite: Effects on the thioredoxin system." *Biochem Pharmacol* no. 75 (11):2092–9.

Sieja, K., and M. Talerczyk. 2004. "Selenium as an element in the treatment of ovarian cancer in women receiving chemotherapy." *Gynecol Oncol* no. 93 (2):320–7.

Smith, M. L., J. K. Lancia, T. I. Mercer, and C. Ip. 2004. "Selenium compounds regulate p53 by common and distinctive mechanisms." *Anticancer Res* no. 24 (3A):1401–8.

Song, H., J. Kim, H. K. Lee et al. 2011. "Selenium inhibits migration of murine melanoma cells via down-modulation of IL-18 expression." *Int Immunopharmacol* no. 11 (12):2208–13.

Sporn, M. B., N. M. Dunlop, D. L. Newton, and J. M. Smith. 1976. "Prevention of chemical carcinogenesis by vitamin A and its synthetic analogs (retinoids)." *Fed Proc* no. 35 (6):1332–8.

Spyrou, G., M. Björnstedt, S. Kumar, and A. Holmgren. 1995. "Ap-1 DNA-binding activity is inhibited by selenite and selenodiglutathione." *FEBS Lett* no. 368 (1):59–63.

Steinbrenner, H., B. Speckmann, and H. Sies. 2013. "Toward understanding success and failures in the use of selenium for cancer prevention." *Antioxid Redox Signal* no. 19 (2):181–91.

Suzuki, M., M. Endo, F. Shinohara, S. Echigo, and H. Rikiishi. 2010. "Differential apoptotic response of human cancer cells to organoselenium compounds." *Cancer Chemother Pharmacol* no. 66 (3):475–84.

Todd, A. T. 1934. "The selenide treatment of cancer." *Br J Surg* no. 21 (84):619–31.

Turell, L., R. Radi, and B. Alvarez. 2013. "The thiol pool in human plasma: The central contribution of albumin to redox processes." *Free Radic Biol Med* no. 65:244–53.

Vadgama, J. V., Y. Wu, D. Shen, S. Hsia, and J. Block. 2000. "Effect of selenium in combination with adriamycin or taxol on several different cancer cells." *Anticancer Res* no. 20 (3A):1391–414.

Vander Heiden, M. G., L. C. Cantley, and C. B. Thompson. 2009. "Understanding the Warburg effect: The metabolic requirements of cell proliferation." *Science* no. 324 (5930):1029–33.

Wallenberg, M., S. Misra, A. M. Wasik et al. 2014. "Selenium induces a multi-targeted cell death process in addition to ROS formation." *J Cell Mol Med* no. 18 (4):671–84.

Wallenberg, M., E. Olm, C. Hebert, M. Björnstedt, and A. P. Fernandes. 2010. "Selenium compounds are substrates for glutaredoxins: A novel pathway for selenium metabolism and a potential mechanism for selenium mediated cytotoxicity." *Biochem J* no. 429 (1):85–93.

Watrach, A. M., J. A. Milner, M. A. Watrach, and K. A. Poirier. 1984. "Inhibition of human breast cancer cells by selenium." *Cancer Lett* no. 25 (1):41–7.

Watson-Williams, E. 1919. "A preliminary note on the treatment of inoperable carcinoma with selenium." *British Medical J* no. 2 (3067):463–4.

Weekley, C. M., J. B. Aitken, S. Vogt et al. 2011. "Metabolism of selenite in human lung cancer cells: X-ray absorption and fluorescence studies." *J Am Chem Soc* no. 133 (45):18272–9.

Weekley, C. M., G. Jeong, M. E. Tierney et al. 2014. "Selenite-mediated production of superoxide radical anions in A549 cancer cells is accompanied by a selective increase in SOD1 concentration, enhanced apoptosis and Se–Cu bonding." *J Biol Inorg Chem* no. 19 (6):813–28.

Weisberger, A. S., and L. G. Suhrland. 1956. "Studies on analogues of L-cysteine and L-cystine. III. The effect of selenium cystine on leukemia." *Blood* no. 11 (1):19–30.

West, A. C., and R. W. Johnstone. 2014. "New and emerging HDAC inhibitors for cancer treatment." *J Clin Invest* no. 124 (1):30–9.

Würmli, R., S. Wolffram, Y. Stingelin, and E. Scharrer. 1989. "Stimulation of mucosal uptake of selenium from selenite by L-cysteine in sheep small intestine." *Biol Trace Elem Res* no. 20 (1):75–85.

Xiang, N., R. Zhao, G. Song, and W. Zhong. 2008. "Selenite reactivates silenced genes by modifying DNA methylation and histones in prostate cancer cells." *Carcinogenesis* no. 29 (11):2175–81.

Yan, L., and J. E. Spallholz. 1993. "Generation of reactive oxygen species from the reaction of selenium compounds with thiols and mammary tumor cells." *Biochem Pharmacol* no. 45 (2):429–37.

Yoon, S. O., M. M. Kim, and A. S. Chung. 2001. "Inhibitory effect of selenite on invasion of HT1080 tumor cells." *J Biol Chem* no. 276 (23):20085–92.

Zhang, F., W. Yu, J. L. Hargrove et al. 2002. "Inhibition of TNF-α induced ICAM-1, VCAM-1 and E-selectin expression by selenium." *Atherosclerosis* no. 161 (2):381–6.

Zhang, J., D. Peng, H. Lu, and Q. Liu. 2008. "Attenuating the toxicity of cisplatin by using selenosulfate with reduced risk of selenium toxicity as compared with selenite." *Toxicol Appl Pharmacol* no. 226 (3):251–9.

Zhao, R., N. Xiang, F. E. Domann, and W. Zhong. 2006. "Expression of p53 enhances selenite-induced superoxide production and apoptosis in human prostate cancer cells." *Cancer Research* no. 66 (4):2296–304.

Zimmermann, T., H. Leonhardt, S. Kersting et al. 2005. "Reduction of postoperative lymphedema after oral tumor surgery with sodium selenite." *Biol Trace Elem Res* no. 106 (3):193–203.

7 Forms of Selenium in Cancer Prevention

Karam El-Bayoumy, Raghu Sinha,
and John P. Richie, Jr.

CONTENTS

7.1 Introduction 137
7.2 Forms of Selenium in Nature.. 139
7.3 Forms of Selenium in Preclinical Studies 140
    7.3.1 Metabolism and Disposition: Forms That Reach Target Tissues 141
    7.3.2 *In Vitro* Studies.. 141
    7.3.3 *In Vivo* Studies... 147
        7.3.3.1 Carcinogen-Induced and Xenograft Models..................... 147
        7.3.3.2 Transgenic Models... 148
    7.3.4 Mechanism(s) of Action of Various Forms of Selenium 149
7.4 Forms of Selenium in Epidemiological and Clinical Investigations 151
    7.4.1 Epidemiological Studies on Selenium and Cancer........................... 151
    7.4.2 Forms of Selenium in Clinical Trials 152
    7.4.3 Biomarker Studies with Healthy Individuals.................................... 156
7.5 Forms of Selenium in Cancer Prevention: Moving Forward........................ 160
References... 161

7.1 INTRODUCTION

Selenium-containing compounds are known to play an important role in protection against several diseases, including cancer (Fairweather-Tait et al., 2011); however, the mechanisms that can account for cancer prevention remain undefined. Humans have 25 genes coding for selenoproteins by which selenium exerts its biological functions. Selenoproteins contain selenocysteine at their active sites (Papp et al., 2007) (see Chapter 3). Both inorganic and organic forms of selenium compounds can be metabolized in a way that they can supply the requisite selenium for synthesis of these selenoproteins (see Chapter 2). The form as well as optimal levels of selenium for maximal expression of individual selenoproteins and those required for maximal chemoprevention in preclinical animal models have been reported in the literature (El-Bayoumy, 2004; Weekley and Harris, 2013); at present, literature data strongly suggests that, in addition to supplying selenocysteine for synthesis of selenoproteins, different forms of selenium may act as chemopreventive agents through alternative mechanisms (Christensen, 2014).

With the discovery of synthetic chemopreventive organoselenium compounds in the mid-1980s in our laboratories, we realized that not all selenium compounds can be considered equal (El-Bayoumy, 1985, 2004); it was evident to us that different forms of selenium have different biological effects and combined with other factors, including the genetic makeup of the host, collectively can have a major impact on cancer chemopreventive activities. In our quest to understand the relationships between selenium forms and mechanisms of cancer prevention, we continue a journey started back in the thirteenth century by Marco Polo who first observed the "dark side" of the moon element (selenium) on his journey to China (El-Bayoumy, 2001; El-Bayoumy and Sinha, 2005; Ferguson et al., 2012). In our continued efforts to discriminate the toxic from the protective doses of selenium, it is clear that the form of selenium is playing a paramount role in the determination of its biological effects.

Epidemiological studies have suggested that an increased risk for certain human diseases, including cancers of the lung, prostate, breast, and colorectal, is related to insufficient intake of selenium (Brozmanova et al., 2010; Dennert et al., 2011; Hurst et al., 2012). Furthermore, on the basis of meta-analysis of randomized controlled trials of selenium supplements and cancer prevention, Lee et al. (2011) concluded that there was sufficient evidence to support the use of selenium supplements in populations with high risk of cancer and low baseline selenium levels. In these studies, the form(s) of selenium (structure) that are responsible for cancer prevention remain largely undefined, and information on selenium speciation are badly needed in this regard; however, preclinical studies with different selenium forms provide a wealth of knowledge, and certain studies have successfully ranked the chemopreventive efficacy of numerous structurally varied naturally occurring and synthetic forms (Sinha and El-Bayoumy, 2004). In contrast to preclinical studies, limited forms of selenium have been used in human clinical trials conducted in China, India, Italy, New Zealand, and in the United States: namely selenite, selenate, selenomethionine, and selenized yeast. Populations having different risk factors were enrolled in these trials. The outcome of these trials remains inconsistent and even disappointing as demonstrated by the lack of effect of selenomethionine in the Selenium and Vitamin E Cancer Prevention Trial (SELECT), the largest clinical chemoprevention trial ever conducted in highly selenium-proficient men (Lippman et al., 2009). Nevertheless, the results of these trials suggest that the effect of selenium may depend on baseline selenium status prior to the intervention and the form of selenium, populations and genetic makeup, and the target organ and disease under examination (El-Bayoumy, 2001; El-Bayoumy and Sinha, 2005; Ferguson et al., 2012). In fact, as mentioned above, we and others have consistently shown that the form and the dose of selenium are critical factors in cancer prevention (El-Bayoumy, 2001; El-Bayoumy and Sinha, 2005; Ferguson et al., 2012). Clearly, as stated by Lu and his team (Zhang et al., 2010), the quest for developing mechanism-based novel forms of selenium compounds takes on even greater significance and urgency after the SELECT. Furthermore, Marshall et al. (Marshall, Ip et al., 2011) correctly stated that it is of paramount importance not to extrapolate the findings of SELECT using selenomethionine to all selenium compounds and to all populations. In fact, based on baseline selenium levels in the two trials (Clark et al., 1996; Duffield-Lillico et al., 2003; Lippman et al., 2009), Rayman et al. (2009) stated that the results of SELECT are consistent with the outcome of the Nutrition Prevention of Cancer (NPC) trial that demonstrated the chemopreventive effect of selenized yeast (Clark et al., 1996).

The bulk of our knowledge on the mechanisms of action of selenium compounds is primarily based on animal data and studies conducted in *in vitro* systems (El-Bayoumy and Sinha, 2005; Sinha and El-Bayoumy, 2004). How such knowledge is applicable to humans remains largely undefined. Because different chemical forms of selenium can modify cellular processes and molecular targets involved in cancer development, progression, recurrence, and metastasis, it is reasonable to propose that, in addition to its effects on cancer prevention, selenium may also have a place in the arena of cancer therapy (Barger et al., 2012; Sharma and Amin, 2013; Zhang et al., 2010). Clearly, there is a need for small-scale human clinical trials aimed at determining the effects of various forms of selenium on molecular targets that are critical in the multistep carcinogenesis process using contemporary tools, such as selenium speciation and "omics" approaches combined with bioinformatics prior to long-term and costly Phase III clinical chemoprevention trials. Recently, we showed that selenized yeast but not selenomethionine significantly reduced markers of oxidative damage in healthy men (Richie et al., 2014). Identification of selenium compounds (and their metabolites) in human plasma and in commonly consumed foods will assist in better understanding the mechanisms of action of selenium in cancer prevention. About a decade ago, we introduced the term "molecular chemoprevention" as the guiding principle in the design of future clinical chemoprevention trials (El-Bayoumy and Sinha, 2005). Clearly, the ultimate goal is to gain mechanistic insights into how cancer prevention and treatment by selenium become individualized. This chapter is not meant to provide a comprehensive review of the subject but rather to highlight basic mechanistic information on the role of the various forms (structure) of selenium on factors that are critical in cancer chemoprevention; furthermore, potential new approaches for future basic and clinical research are proposed. However, the reader is encouraged and respectfully advised to consult the original publications for more details.

7.2   FORMS OF SELENIUM IN NATURE

Drinking water and air constitute biologically insignificant sources of selenium (El-Bayoumy, 1991). Food is more appealing to the general public when compared to more toxic pharmacological agents and consequently is the major source of selenium intake (Dumont et al., 2006; Thomson et al., 2008; Weekley and Harris, 2013). The selenium content of food is highly dependent on the amount of selenium in the soil, which varies from country to country, from region to region, and it also depends on the ability of plants to take up, accumulate, and metabolize the element. Average dietary intake is within the range of 20–300 µg/day (Patrick, 2004). Intakes are high in the United States (93 µg/day for women; 134 µg/day for men), Canada, Japan, and Venezuela but much lower in Europe (~40 µg/day). Brazil nuts are considered the richest source of selenium with levels as high as 512 mg/kg, but they are not a commonly consumed food. In cereal and grains, the mean values range from 0.02–0.03 mg/kg dry weight in the UK; in high-selenium areas of the United States, levels are as high as 30 mg/kg (Rayman, 2012). Selenite and selenate in the soil are converted in plants to organic forms, including L-selenomethionine and, to a lesser extent, L-selenocysteine and numerous other related compounds (Institute of Medicine [US] Panel on Dietary Antioxidants and Related Compounds, 2000). Selenomethionine represents the major form of selenium in cereals and other plant crops as well as in selenized yeast and other supplements. Selenocysteine can be found in animal foods and Se-methylselenocysteine, and γ-glutamyl-Se-methylselenocysteine can be found in broccoli, garlic, and onions, especially when grown under selenium-rich conditions (Cai et al., 1995; Yang et al., 1997). Sodium selenite and selenate are components of dietary supplements (Schrauzer, 2001). Some selenate is found in fish and plant sources, such as cabbage. The major dietary sources are meat, poultry, grain products, and seafood with the latter responsible for approximately 30% of the dietary selenium intake. Only about 50% of the selenium content in seleniferous wheat was identified as selenomethionine. Selenocystathionine occurs in nuts, and Se-methylselenocysteine and selenocystathionine have been identified in certain species of Astragalus. Selenoneine is the major selenium compound recently discovered in tuna and mackerel (Yamashita and Yamashita, 2010), and tuna contains higher levels of selenium than other selenium-rich foods, such as whole wheat. However, red blood cell GPx activity was significantly lower in rats fed tuna than in rats fed wheat (Douglass et al., 1981). Thus, the chemical forms of selenium in tuna are presumably different from those present in whole wheat, and this is likely to be critical in determining the bioavailability and biological activity of selenium.

Overall, it should be noted that detailed selenium speciation studies have not been performed for most dietary sources of selenium. Therefore, there are likely many other currently undefined forms of selenium that exist in foods that contribute to the overall selenium content of the diet. Even though some of these forms may be relatively minor in concentration compared to selenomethionine, they could still be playing an important protective role based on a higher degree of activity. Indeed, Se-methylselenocysteine, a relatively minor form found in the diet and in selenized yeast, has been found in preclinical studies to be highly active against a variety of cancer-related end points (Unni et al., 2004, 2005).

7.3 FORMS OF SELENIUM IN PRECLINICAL STUDIES

Several naturally occurring and synthetic forms of selenium have been investigated in *in vitro* and *in vivo* systems using a variety of cancer cell types and animal models; it is important to note that not only the dose but also the form of selenium compounds govern their *in vitro* and *in vivo* metabolism and their mechanisms of action that can account for the chemopreventive efficacies (as previously reviewed in Chen et al., 2013; El-Bayoumy and Sinha, 2005; El-Bayoumy et al., 2011; Sinha and El-Bayoumy, 2004; Weekley and Harris, 2013), which mainly differ due to their respective redox properties. Most of the selenium forms alter proliferative potential of epithelial cancers by inducing apoptosis (Sinha and El-Bayoumy, 2004), but selenium may also be considered a double-edged sword in terms of possessing antioxidant as well as prooxidant properties that may be beneficial or harmful depending upon the form and dose being used in a normal or cancer setting (Brozmanova et al., 2010). During the last decade, the concept of the U-shape was introduced to reflect the "good" and the "bad" effects of selenium as a function of dose (Waters et al., 2005).

7.3.1 METABOLISM AND DISPOSITION: FORMS THAT REACH TARGET TISSUES

Selenium enters the body primarily from foods in several forms, including inorganic selenium, selenomethionine, and selenocysteine (see Chapter 2 for metabolism). Study of selenium metabolites in blood and urine can provide a better understanding of the various transformations taking place in the body that produce beneficial or harmful effects. Previous evidence from literature indicates that selenium (selenate, selenite, selenomethionine) is metabolized to hydrogen selenide in biological systems followed by methylation (Ganther, 1986). The major methylated selenium metabolites in this proposed pathway are dimethylselenide, which is excreted by respiration, and trimethylselenonium ion, which is excreted in the urine. However, several selenosugars have also been identified as major urinary metabolites (Francesconi and Pannier, 2004). In a previous study, we showed that the metabolism of a well-tested synthetic chemopreventive organoselenium compound, 1,4-phenylenebis(methylene) selenocyanate (p-XSC) developed by our group, results in the formation of glutathione conjugate (p-XSeSG) followed by aromatic selenol moiety (p-XSeH) that finally converts to tetraselenocyclophane (TSC) (El-Bayoumy et al., 2001). In this pathway, the formation of p-XSeH was thought to be a critical step because the selenol moiety (e.g., methylselenol) may play a significant role in cancer chemoprevention by selenium compounds (Ganther, 1999). Additionally, in an A/J mouse model, we reported that amounts of selenium that reached the target organ (lung) after dietary administration of p-XSC were significantly higher than that from selenized yeast (the major form present as selenomethionine). However, the levels of selenium in plasma from selenized yeast were twofold higher than those from p-XSC (Das et al., 2003). The higher levels of selenium that reach the target organ may, in part, account for the superior chemopreventive efficacy of p-XSC to that of selenized yeast.

When the metabolic pathways for Se-methylselenocysteine, selenomethionine, and selenite were compared in rat tissues and fluids, it was evident that the Se of Se-methylselenocysteine was incorporated into selenoprotein P1 (SePP1) slightly more than or at a comparable level to that of selenomethionine but less than that of selenite (Suzuki et al., 2006). Furthermore, metabolism of selenite in rats was recently reported to require S-adenosylmethionine (SAM)-dependent methylation, and disrupting the conversion of SAM to S-adenosylhomocysteine prevented the conversion of selenite to trimethylselenonium and selenosugars (Jackson et al., 2013).

Because epidemiological investigations (discussed below) support the protective role of selenium, it is critical to investigate commonly consumed selenium-containing food sources rather than employing individual pure chemical forms that are commonly used in preclinical chemopreventive studies (Finley and Davis, 2001) in future clinical chemoprevention studies.

7.3.2 IN VITRO STUDIES

Numerous novel forms of selenium have been synthesized in the past. Table 7.1 summarizes the natural and synthetic forms of selenium studied in a variety of tumor

TABLE 7.1

Effects of Various Forms of Selenium on Cancer and Other Parameters Involved in Carcinogenesis *In Vitro* and *In Vivo*

Forms of Selenium	Model System (*In Vitro/In Vivo*)	Parameters Measured	Refs.
Sodium selenite	a. Mouse mammary cancer cells, human colon carcinoma cells (SW480), human glioma cells (A172, T98G), human hepatoma cells (HEPG2), human ovarian cells (SKVO3), human prostate cancer cells (LNCaP, PC-3, DU145)	DNA fragmentation/apoptosis	Sinha and El-Bayoumy (2004)
	b. Human lung cancer cells (A549)	ROS generation, apoptosis	
Selenodiglutathione	a. Mouse erythro-leukemia cells (MEL), human oral squamous carcinoma cells (SCC), human ovarian cancer cells (A2780), human promyelocytic leukemia cells (HL-60)	DNA fragmentation/apoptosis	Sinha and El-Bayoumy (2004)
	b. Human cervical cells (HeLa)	Apoptosis-like cell death	Wallenberg et al. (2014)
Se-methylselenocysteine	a. Mouse mammary tumor cells (TM6), HEPG2, HL-60	DNA fragmentation/apoptosis	Sinha and El-Bayoumy (2004)
	b. Mouse mammary hyperplasia outgrowth model	Outgrowth reduction	Unni et al. (2004)
	c. Human oral squamous cell carcinoma cells (HSC-3, HSC-4)	Cell growth inhibition/ apoptosis	Suzuki et al. (2006)
Selenomethionine	a. Human breast cancer cells (MCF-7), (HT29, HCT116), human melanoma cells (UACC-375), LNCaP, PC-3, DU145, A549	DNA fragmentation/apoptosis	Sinha and El-Bayoumy (2004)
	b. Human leukemia cells (Jurkat E6-1), PC-3	Cell death	Lunoe et al. (2011)
Methylseleninic acid	a. Mouse mammary tumor cells (TM2H), human breast cancer cells (MFC10dcis. com), HEPG2, DU145, PC-3	DNA fragmentation/apoptosis	Sinha and El-Bayoumy (2004)
	b. PC-3, DU145, PC-3M cells, PC-3M xenograft model	Apoptosis/tumor inhibition	Sinha et al. (2014)
1,4-phenylenebis (methylene) seleno-cyanate (p-XSC)	a. Mouse mammary cancer cells (MOD), rat mammary cancer cells (RBA), human colon cancer cells (Col2), SCC	DNA fragmentation/apoptosis	Sinha and El-Bayoumy (2004)
	b. Human lung cancer cells (NCI-H460, NCI-1299, and A549)	Apoptosis	El-Bayoumy et al. (2006)
	c. LNCaP	Apoptosis	Facompre et al. (2012)

(Continued)

TABLE 7.1 (CONTINUED)

Effects of Various Forms of Selenium on Cancer and Other Parameters Involved in Carcinogenesis *In Vitro* and *In Vivo*

Forms of Selenium	Model System (*In Vitro*/*In Vivo*)	Parameters Measured	Refs.
Selenium nanoparticles	a. Human breast cancer cells (MDA-MB-231), HeLa	Cell growth inhibition	Luo et al. (2012)
Chitosan-stabilized selenium nanoparticles	a. HEPG2	Growth arrest	Estevez et al. (2014)
(+)-1-(2-Deoxy-2-fluoro-4-seleno-D-arabinofuranosyl) cytosine	a. HCT11, A549, T47D, human gastric cancer cells (SNU638), PC3	Cell growth inhibition	Kim, Yu et al. (2014)
2-(phenylselenomethyl) tetrahydrofuran and 2-(phenylselenomethyl) tetrahydropyran	a. HCT-116, MDA-MB-231	Cell growth inhibition	Kosaric et al. (2014)
22-oxo-26-selenocyanocholestane	a. HeLa, human cervical cells (CaSki, ViBo)	Cell growth inhibition	Fernandez-Herrera et al. (2014)
Diaryl selenide	a. Human kidney carcinoma cells (786), HT-29, MCF-7, PC-3	Cell growth inhibition	dos Santos Edos et al. (2013)
1,2-bis (chloropyridazinyl) diselenide	a. MCF-7	Cell growth inhibition	Kim, Lee et al. (2014)
Sucrose selenious ester	a. HeLa, bladder carcinoma cell line (5637), human malignant melanoma cell line (A375), gastric carcinoma cell line (MGC-803)	Apoptosis	Guo et al. (2013)
Selenoester derivatives	a. PC-3, MCF-7, A549, HT-29	Cell growth inhibition	Dominguez-Alvarez et al. (2014)
Methyl N,N′ *bis*(propoxycarbonyl) imido-selenocarbamate	a. Lymphoblastic leukemia cells (CCRF-CEM), human lung cancer cells (HTB-54)	Cell growth inhibition/ apoptosis	Romano et al. (2014)

(Continued)

TABLE 7.1 (CONTINUED)

Effects of Various Forms of Selenium on Cancer and Other Parameters Involved in Carcinogenesis *In Vitro* and *In Vivo*

Forms of Selenium	Model System (*In Vitro*/*In Vivo*)	Parameters Measured	Refs.
α-keto-γ-methylselenobutyrate and β-methylselenopyruvate	a. LNCaP, PC-3, DU145	Cell growth inhibition	Lee et al. (2009)
Selenized milk protein	a. MCF-7 xenograft tumor model	Tumor inhibition	Warrington et al. (2013)
Se-enriched Ziyang green tea	a. Human osteosarcoma cells (U-2OS) xenograft model	Tumor inhibition	Wang et al. (2013)
Selenized yeast	a. Human melanoma cell (K1735) xenograft model	Brain metastasis inhibition	Wrobel et al. (2013)
Bismuth selenide	a. Mouse hepatoma (H22 cells)	Decreased tumor growth in presence of photothermal therapy	Li et al. (2013)
Hyaluronic acid–selenium nanoparticles	a. Mouse hepatic carcinoma	Tumor growth inhibition	Ren et al. (2013)
S,S′-1,4-phenylenebis (1,2-ethanediyl) bis-isoselenourea	a. Malignant melanoma skin construct b. AOM-induced colon cancer model	Apoptosis ACF reduction	Madhunapantula et al. (2008) Janakiram et al. (2013)
Phenylbutyl isoselenocyanate	a. Melanoma cancer cells (UACC 903) xenograft model	Tumor reduction	Sharma et al. (2008)
4-(5-phenyl-3-(seleno cyanatomethyl)-1H-pyrazol-1yl)benzene sulfonamide (Selenocoxib-1)	a. Prostate cancer cells (PAIII, PC-3M), subcutaneous PAIII tumor model	Cell growth inhibition/ apoptosis, tumor inhibition	Desai et al. (2010)
Selenocoxib-1-GSH	a. Melanoma xenograft model	Tumor inhibition	Gowda et al. (2013)
Bis(5-phenylcarbamoyl pentyl) diselenide (SelSA-1) and 5-phenylcarbamoyl pentyl selenocyanide (SelSA-2)	a. Melanoma xenograft model	Tumor inhibition	Gowda et al. (2012)

cell lines *in vitro* as well as *in vivo*, and Figure 7.1 shows the chemical structures of these forms. The structures were drawn using MarvinSketch (ver. 6.3.1, ChemAxon LLC, Cambridge, MA).

Several forms of selenium investigated for their efficacy against numerous tumor cell lines in laboratories around the world include but are not limited to selenite, selenoxide, Se-methylselenocysteine, selenomethionine, ebselen, methylseleninic acid, *p*-XSC, and selenodiglutathione (GS-Se-SG) (Sinha and El-Bayoumy, 2004). In this chapter, studies after 2004 are included (Table 7.1); our intention is to discuss several examples to make the point that the form of selenium is critical in its chemo-preventive activity.

When comparing effects of selenomethionine and sodium selenite on human immortalized keratinocytes (HaCaT) at nontoxic doses, 10 µM selenomethionine was better absorbed than 1 µM selenite, and at toxic doses 100 µM selenomethionine and 5 µM selenite inhibited cell proliferation associated with S-G2 blockage and DNA fragmentation leading to apoptosis (Hazane-Puch et al., 2013).

Selenium nanoparticles are considered as a novel form of selenium with potent antioxidant properties as measured by increased glutathione peroxidase, thioredoxin reductase, and glutathione S-transferase activities (Bai et al., 2011) and have shown superior growth inhibitory effects in HeLa and MDA-MB-231 when compared to selenoxide (Luo et al., 2012) and in HEPG2 cells when compared to selenomethionine, Se-methylselenocysteine, and selenocystine (Estevez et al., 2014).

Based on a potent anticancer agent, D-arabino-configured cytosine nucleoside (ara-C), several novel 2′-substituted-4′-selenoarabinofuranosyl pyrimidines were synthesized (Kim, Yu et al., 2014), and among the compounds tested, the 2′-fluoro derivative was found to be the more potent anticancer agent as compared to control ara-C in all the human cell lines tested (HCT11, A549, T47D, SNU638, and PC-3) except leukemia cell lines (K562).

A set of structurally similar selenium analogues, 2-(phenylselenomethyl) tetra-hydrofuran and 2-(phenylselenomethyl) tetrahydropyran, were successful in reducing reactive oxygen species as a means to inhibit growth of colon (HCT11) and breast cancer (MDA-MB-231) cell lines (Kosaric et al., 2014). More recently, a novel strat-egy for synthesizing selenium-containing steroidal compounds showed potential antitumor effects of 22-oxo-26-selenocyanocholestanic steroids in cervico-uterine cancer cells (Fernandez-Herrera et al., 2014).

A selenium analog of combretastatin A-4 (diaryl selenide) was reported to inhibit tubulin polymerization and cell growth of several cancer cells, including 786, HT-29, MCF-7, and PC-3 (dos Santos Edos et al., 2013). Similarly, a new series of bis (aryl or aralkyl) diselenides was synthesized recently for development of new anticancer agents, and among the 17 synthesized compounds in this scheme, 1,2-bis(chloropyridazinyl) diselenide was the most potent against MCF-7 cells (Kim, Lee et al., 2014).

A sucrose-selenious ester is able to inhibit proliferation of a variety of tumor cells lines, including HeLa, bladder carcinoma cell line 5637, human malignant melanoma cell line A375, and gastric carcinoma cell line MGC-803 (Guo et al., 2013). Furthermore, another set of selenoester derivatives demonstrates strong

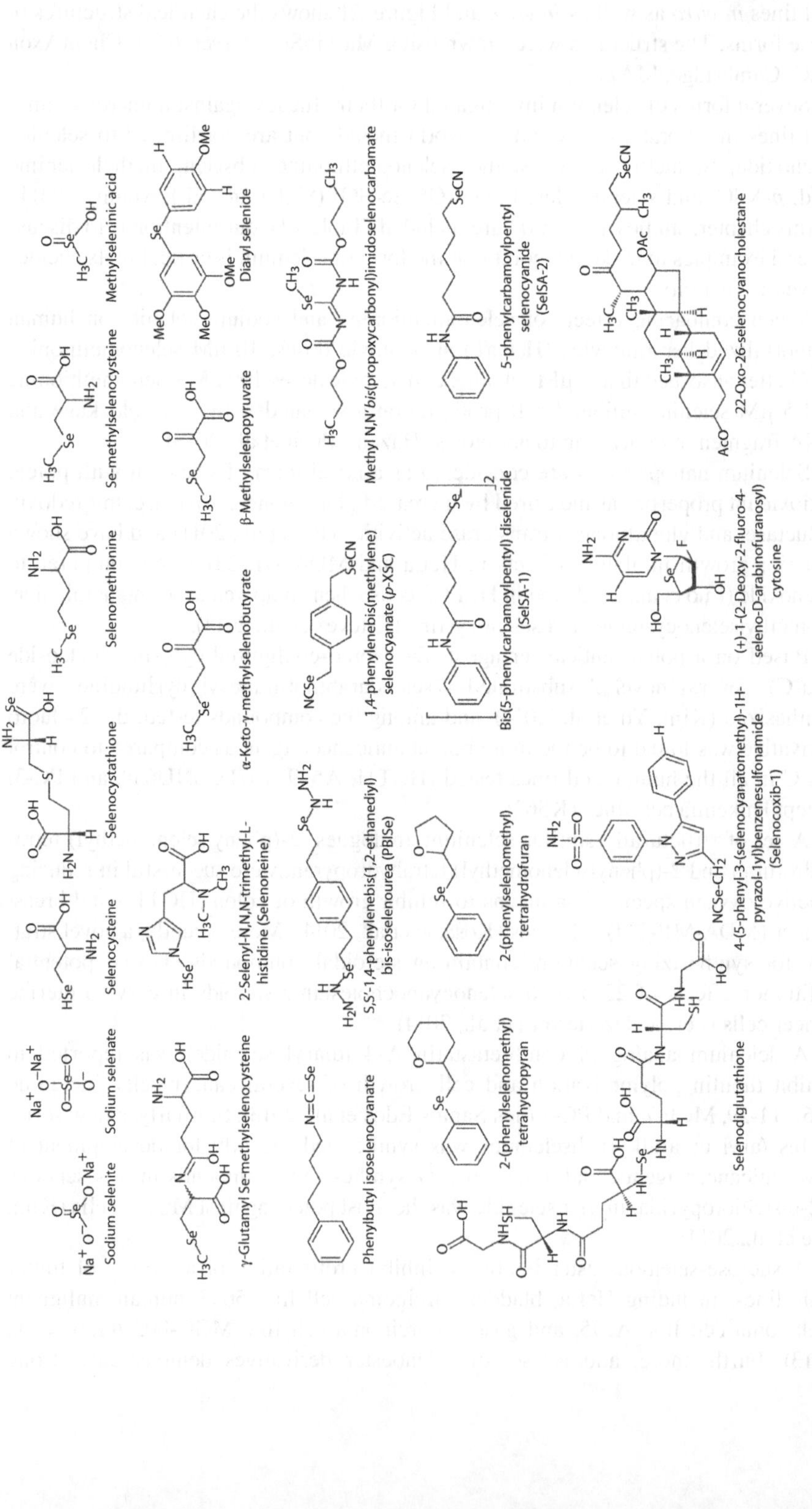

FIGURE 7.1   Selenium forms and their structures.

redox modulation and growth inhibition in PC-3, MCF-7, A549, and HT-29 cells (Dominguez-Alvarez et al., 2014).

Another series of new aliphatic, aromatic, and heteroaromatic carbamate derivatives that contain a methylseleno-moiety were synthesized, and the most effective form, with a propyl-group, showed promising antitumor effects against a panel of human cell lines, including lymphoblastic leukemia and lung carcinoma cells *in vitro* (Romano et al., 2014).

A class of recently discovered α-keto acid metabolites of Se-methylselenocysteine and selenomethionine; β-methylselenopyruvate and α-keto-γ-methylselenobutyrate, respectively, have demonstrated greater cell growth inhibition in prostate and colon cancer cells than their parent compounds (Lee et al., 2009; Pinto et al., 2011).

Although most of the experimental data in the literature are convincing and provide leads to potential agents that need to be examined further in animal models, the reader is cautioned about the fact that the choice of cells, the metabolizing enzymes existing within the cells, the form of selenium compound being tested, and the corresponding doses will dictate the outcomes.

7.3.3 IN VIVO STUDIES

A variety of selenium forms have shown remarkable efficacy against tumor development in several animal models, thus providing convincing preclinical data to potentially promote these compounds for human trials. More importantly, the dose and form of selenium play an important part in determining their chemopreventive properties, which also take into account the desired reduced or no toxic side effects. Major forms of selenium existing in nature and synthesized in laboratories around the world that have successfully shown significant inhibition of tumor formation in several animal models are also listed in Table 7.1.

7.3.3.1 Carcinogen-Induced and Xenograft Models

A detailed comparison of the chemopreventive efficacy of structurally different selenium compounds in the 7,12-dimethylbenz[a]anthracene (DMBA)-induced rat mammary tumor model in terms of effective dose (ED_{50}), maximum tolerated dose, chemopreventive index, and apoptosis has been documented earlier for sodium selenite, potassium selenocyanate, selenomethionine, Se-methylselenocysteine, selenobetaine, methylseleninic acid, benzyl selenocyanate, p-XSC, p-XSeSG, triphenylselenonium chloride, and allylselenocysteine (Sinha and El-Bayoumy, 2004). These studies clearly demonstrate that the form (structure) of the selenium-containing compounds is a determining factor in their chemopreventive efficacy.

Selenized milk protein proves to be an effective antiproliferative supplement for MCF-7 tumor inhibition (Warrington et al., 2013) as well as for colorectal cancer inhibition (Hu et al., 2008) when tested in xenograft models. Selenium-polysaccharide from Ziyang green tea showed remarkable tumor regression in a xenograft model for human osteosarcoma U2OS cells (Wang et al., 2013). Selenized yeast attenuated brain metastasis in a xenograft model using K1735 melanoma cells (Wrobel et al., 2013). More recently, bismuth selenide has been successfully used for its theranostic properties *in vitro* and *in vivo* against mouse hepatoma (Li et al., 2013) and oral hyaluronic acid–selenium nanoparticles were effective in reducing mouse hepatic carcinoma *in vivo* (Ren et al., 2013).

Using a similar approach that was introduced by our group in the mid-1980s (El-Bayoumy, 1985), a new category of potent chemopreventive/chemotherapeutic compounds is being synthesized by isosterically replacing sulfur by selenium in existing agents with antitumor properties (as previously reviewed in Sharma and Amin, 2013). One such example of this selenium form is S,S'-1,4-phenylenebis(1,2-ethanediyl)bis-isoselenourea (PBISe), a selenium analog of S,S'-1,4-phenylenebis(1,2-ethanediyl)bis-isothiourea (PBIT) (well-known iNOS inhibitor), which showed chemopreventive properties against AOM-induced aberrant crypt foci (Janakiram et al., 2013) and growth inhibition of malignant melanoma in a skin construct model (Madhunapantula et al., 2008). Within the same context, the potency of chemopreventive isothiocyanates in a malignant melanoma xenograft model (Sharma et al., 2008) can be enhanced by replacing sulfur in isothiocyanates (R-N = C = S) by selenium to give isoselenocyanates (R-N = C = Se).

Similarly, selenocoxibs have recently been synthesized by incorporation of selenium moiety into Celecoxib, a nonsteroidal anti-inflammatory drug that selectively targets COX-2. Selenocoxib-1 has been shown to be effective against subcutaneous prostate tumors (Desai et al., 2010) and a glutathione conjugated selenocoxib-1 form was effective against melanoma xenografted mice (Gowda et al., 2013). More recently, selenium derivatives (SelSA-1, SelSA-2) (see Figure 7.1) of well-known histone deacetylase (HDAC) inhibitor suberoylanilide hydroxamic acid were found to be very effective in a melanoma model (Gowda et al., 2012).

Selenized yeast contains a large proportion of selenium in the form of selenomethionine. When dogs supplemented with selenomethionine or selenized yeast were compared based on the equivalent intraprostatic selenium concentration following supplementation, there was no significant differences in potency of either selenium form on any of the six parameters studied—intraprostatic dihydrotestosterone (DHT), testosterone (T), DHT:T, and epithelial cell DNA damage, proliferation, and apoptosis—over three different ranges of target tissue selenium concentration (Waters et al., 2012). In contrast, a genomics approach revealed that different forms of selenium (selenite, selenomethionine, selenized yeast) resulted in clear differences in gene expression profiles in rodents (Barger et al., 2012).

7.3.3.2 Transgenic Models

Selenium compounds with effective chemopreventive efficacy and low toxicity, including *p*-XSC, benzyl selenocyanate (BSC), diphenylmethyl selenocyanate (DPMSC), as well as selenium-enriched plant foods such as garlic and broccoli, have been investigated in an azoxymethane (AOM)-induced colorectal cancer rat model as well as in the APC (min/+) mouse model (El-Bayoumy, 2001; Finley, 2003; Hu et al., 2012; Nayini et al., 1991).

Compounds such as Se-methylselenocysteine and methylseleninic acid (both are precursors of the active intermediate methylselenol) are equally efficacious in their chemopreventive properties against prostate cancer development in the TRAMP model. Methylseleninic acid mainly affected proteins related to prostate functional differentiation, androgen receptor signaling, protein (mis)folding, and endoplasmic reticulum-stress responses whereas Se-methylselenocysteine impacted proteins involved in phase II detoxification or cytoprotection and in stromal cells (Zhang et al., 2010). Clearly, although both compounds can lead to the same intermediate, the varied protective profiles suggest different mechanisms that can account for their efficacy in this animal model.

Considering the several promising selenium forms that have been tested in pre-clinical studies, it is intriguing that selenomethionine, an organic form with no or minimal efficacy (depending upon dose used) in animal models (Ozten et al., 2014) was selected for a Phase III Clinical Trial (SELECT) (El-Bayoumy, 2009). An important lesson learned from this huge and expensive trial is that selection of the form of selenium is absolutely critical prior to initiation of any future Phase III clinical trials. In order to determine the form of selenium that may provide beneficial effects in the prevention and/or therapeutic setting, it is essential to develop programs that utilize novel tools to perform selenium speciation initially in commonly consumed selenium containing foods, in preclinical studies (in tissues and in blood), and finally in small clinical studies utilizing different forms of selenium that will assist in identification and quantification of selenium metabolites that may play a critical role in cancer chemoprevention.

7.3.4 MECHANISM(S) OF ACTION OF VARIOUS FORMS OF SELENIUM

One of the determining factors in the selection of potential chemopreventive agents is the toxicity of the compound being studied. Depending upon the form and dose, the selenium compounds alter the expression and/or activities of a number of cell cycle regulatory proteins, interactions between cyclins and their partnering kinases, signaling molecules, mitochondrial-associated factors, proteases, transcriptional factors, tumor-suppressor genes, polyamine and glutathione levels, and finally induce apoptosis in a variety of cancer cell types depending upon the dose and time of treatments (Ferguson et al., 2012; Sinha and El-Bayoumy, 2004).

Previous reports indicate that the role of selenium in cancer prevention is mainly due to ROS-induced apoptosis (Stewart et al., 1999). In this study, selenite and selenocystamine but not selenomethionine induced apoptosis in mouse keratinocytes. Recently, when selenite and selenodiglutathione (GS-Se-SG) were compared in HeLa cells, the cell death mode of selenite, known to mainly induce ROS formation, was predominantly necroptosis, and selenodiglutathione was able to glutathionylate protein thiols, which might lead to activation of the extrinsic pathway of cell death (Wallenberg et al., 2014). Furthermore, upon comparing noncancerous lung and colon fibroblasts with prostate and colon cancer cells to treatments with sodium selenite, methylseleninic acid, and Se-methylselenocysteine, Wu et al. (2010) demonstrated that selenium is able to induce an ATM-dependent senescence response via redox regulation in noncancerous but not in cancerous cells regardless of the selenium form used. Moreover, inorganic selenium form can sensitize prostate cancer cells to TRAIL-induced apoptosis, and this phenomenon may involve superoxide/p53/Bax-mediated activation of the mitochondrial pathway (Hu et al., 2006).

Several selenium compounds have also been shown to impact other important biological targets and processes, including HDAC activity (Gowda et al., 2012;

Kassam et al., 2011; Lee et al., 2005), estrogen receptor (Lee et al., 2005), androgen receptor (Dong et al., 2004; Facompre et al., 2010; Lee et al., 2006), 14-3-3 isoforms (Bortner et al., 2009), mammalian target of rapamycin (mTOR) (Facompre et al., 2012), protein kinases (Sharma et al., 2009; Singh et al., 2008), osteopontin (Unni et al., 2004), hypoxia inducible factor (HIF-1α) (Sinha et al., 2012), and regulated in development and DNA damage responses (REDD1) (Sinha et al., 2014).

Methylseleninic acid inhibits HDAC activity in diffuse large B-cell lymphoma cells, which results in the acetylation of histone H3 and α-tubulin. However, cellular metabolism of methylseleninic acid to methylselenol is required for this effect (Kassam et al., 2011). The α-keto acid metabolites of Se-methylselenocysteine and selenomethionine have shown to alter HDAC activity and histone acetylation status in human prostate cancer and colon cancer cells (Lee et al., 2009; Pinto et al., 2011).

Selenium in the form of methylseleninic acid can differentially influence the ERα or ERβ mRNA expression, depending upon the cell type, and may present yet another potential for use of selenium in future breast cancer trials (Lee et al., 2005). Similarly, androgen receptor expression has been shown to be reduced by methylseleninic acid (Dong et al., 2004), Se-methylselenocysteine (Lee et al., 2006), and p-XSC, but not selenomethionine (Facompre et al., 2010) in prostate cancer cells.

Several forms of 14-3-3 proteins were rescued by p-XSC in lung tissues of the tobacco-specific lung carcinogen 4-(methylnitrosamino)-1-(3-pyridyl)-1-butanone (NNK)-treated A/J mice, and this may provide a novel mechanism involving chaperone proteins in cancer prevention by selenium (Bortner et al., 2009). Furthermore, p-XSC also exhibited remarkable inhibition of mTORC2 signaling in prostate cancer cells, and in combination with Rapamycin (mTORC1 inhibitor), it provided a significantly greater growth inhibition (Facompre et al., 2012).

Selenium can influence a variety of protein kinases in mammary tumors (Singh et al., 2008), and it would be interesting to examine the network of these kinases, which when targeted in proper combination may provide a novel chemoprevention strategy using the most effective form of selenium. Another form of selenium, such as isoselenocyanate, has been implicated in specifically decreasing Akt3 signaling in cultured melanoma cells and tumors (Sharma et al., 2009).

In mammary hyperplasia, osteopontin, an extracellular structure protein present in bones but also playing a role in tumor progression, was markedly reduced by Se-methylselenocysteine supplementation in diet (Unni et al., 2004) and may help explain the impact of this form on PI3K-AKT pathway disruption leading to apoptosis in mammary hyperplasia cells *in vitro* (Unni et al., 2005).

More recently, methylseleninic acid has been documented to induce REDD1 protein expression, which promotes apoptosis in invasive prostate cancer cells in hypoxic conditions. This event is accompanied by downregulation of hypoxia inducible factor (HIF)-1α and activation of AKT and mTORC1 signaling (Sinha et al., 2014).

As noted above, several molecular pathways that interact with or are disrupted by various forms of selenium may converge onto apoptosis, but the key information would be to delineate these molecular events in the target organs *in vivo* as a function of the dose and the form of selenium. Such studies are limited, and highly promising selenium forms proven effective *in vitro* need to be thoroughly investigated in the target organs *in vivo* prior to being translated to possible human trials.

7.4   FORMS OF SELENIUM IN EPIDEMIOLOGICAL AND CLINICAL INVESTIGATIONS

A substantial body of preclinical data supports a potential role for selenium in the prevention of cancer with efficacy being particularly dependent on the form of selenium. However, the ultimate utility of selenium-derived agents for cancer chemoprevention relies on the results of epidemiological investigations and clinical trials aimed at elucidating the efficacy of different forms of selenium on cancer development and cancer-related intermediate end points. Below, we have summarized the selenium-related cancer epidemiologic studies as well as the disease end point clinical trials and shorter term biomarker studies that have been reported.

7.4.1   EPIDEMIOLOGICAL STUDIES ON SELENIUM AND CANCER

Many epidemiological and laboratory investigations show a protective effect of selenium against the development of cancer at numerous sites, including prostate, colon, and lung, as reviewed previously (Brozmanova et al., 2010; Dennert et al., 2011; El-Bayoumy, 2001; El-Bayoumy and Sinha, 2005; Hurst et al., 2012; Vinceti et al., 2013) with studies as early as 1969 first linking dietary selenium intake with reduced cancer risk (Shamberger and Frost, 1969). The selenium content of foods is tightly linked to regional differences in soil selenium, and throughout the world, cancer mortality rates tend to be greater in regions with low soil concentrations of selenium (Schrauzer et al., 1977; Shamberger and Frost, 1969; Shamberger et al., 1976). Both prospective and case-control studies have shown associations between low blood or toenail selenium and increased risk for cancer, particularly of the prostate (Platz and Helzlsouer, 2001; Vogt et al., 2003). In a cohort study, men with the highest levels of selenium intake were found to have half the risk for developing prostate cancer (Yoshizawa et al., 1998). In a nested case-control study of ovarian cancer, increased serum selenium levels were associated with decreased cancer risk (Helzlsouer et al., 1996). In a study of selenium intake and colorectal cancer, individuals with low selenium intake were at four times the risk compared to those with high intake (Russo et al., 1997). Risk for uterine cervix cancer were also inversely associated with selenium intake (Bhuvarahamurthy et al., 1996) and serum selenium levels (Guo et al., 1994).

Not all studies have observed a protective effect for selenium. In a report that used data from the prospective Alpha-Tocopherol, Beta-Carotene Cancer Prevention Study (ATBC) (n = 29,133), no associations between baseline selenium levels and overall cancer risk were observed (Hartman et al., 1998). In a case-control study of prostate cancer (358 cases; 679 controls) no association was observed between intake of selenium and prostate cancer (West et al., 1991). However, recent meta-analyses of observational studies conclude that high selenium levels in plasma, serum, or toenails may be associated with reduced risk for prostate cancer (Dennert et al., 2011; Holben and Smith, 1999; Hurst et al., 2012; Vinceti et al., 2014; Yang et al., 2013). Indeed, the outcomes of these studies are consistent with those reported in preclinical investigations (El-Bayoumy, 2001).

Altogether, results from these observational studies provide support, with some exceptions, for a potential protective role of selenium against cancers at different

sites. Many studies suffered from a variety of weaknesses inherent in epidemiologic study design, including the use of dietary assessment instruments that are often
fraught with inaccuracies due to the large variation in the selenium content of foods
based upon geographic differences in selenium concentrations in soil (Holben and
Smith, 1999), lack of individual data in geographic distribution studies, and difficul-
ties in interpreting diet recall data and blood or toenail selenium levels as they relate
to long-term selenium intake in case-control designs. Further, it is of paramount
importance to emphasize that these studies offer little information regarding the
specific form(s) of selenium that may be most effective because dietary sources of
selenium contain many different selenium-containing compounds that can contribute to blood and tissue selenium status to varying degrees.

7.4.2 Forms of Selenium in Clinical Trials

Driven by results from preclinical studies and epidemiologic investigations, several
disease end point clinical trials have been conducted to test the impact of selenium
supplementation on cancer risk. These trials have used both inorganic and organic
forms of selenium and have focused on a variety of end points, including cancers of
the prostate, lung, colon, skin, and liver (Table 7.2). Results of these trials have been
the subject of several reviews and commentaries (Christensen, 2014; El-Bayoumy,
2001; Steinbrenner et al., 2013; Vinceti et al., 2013; Weekley and Harris, 2013; Yang
et al., 2013). Below, we have summarized the results of these trials with particular
emphasis on the relative efficacy of the forms of selenium used in the intervention.

Earlier studies conducted in China focused on the use of either selenized yeast or
selenite supplementation in regions in which low concentrations of selenium in the
soil have led to low selenium intake and plasma levels in the inhabitants (Table 7.2,
Studies 1 and 2). Although selenium supplementation reduced the incidence of liver
cancer, it has been noted that these studies suffered from a variety of methodological
and quality control problems (Rayman, 2004).

Because preclinical studies have suggested that certain organoselenium agents
tend to be more effective and less toxic than inorganic forms (e.g., selenite), more
recent clinical studies have focused primarily on the use of organoselenium agents.
In this regard, trials conducted to date have either used selenomethionine or selenized yeast (Baker's yeast grown in a selenium-enriched medium), which are common sources of selenium for supplementation. Selenized yeast contains as much as
2 mg Se per g, primarily in the form of organoselenium. Among the different forms
identified in selenized yeast, selenomethionine is typically the most abundant, but
in addition, more than 15 other forms are present (Dernovics et al., 2009; Far et al.,
2010). Although only a few have been studied in laboratory models to date, some,
such as Se-methylselenocysteine, may be particularly active as described earlier.

Perhaps the most influential trial conducted to date was the NPC Trial (Table 7.2,
Study 8) (Clark et al., 1996). In this randomized, double-blind placebo-controlled
trial, selenized yeast supplementation was tested to determine if it could prevent
recurrence of nonmelanoma skin cancer in 1312 subjects. Although no association
between treatment and the primary end point was observed, striking effects were
noted for a variety of secondary end points, including a 25% decrease in total cancer

incidence, 52% decrease in prostate cancer incidence, 26% decrease in lung cancer incidence, 54% decrease in colorectal cancer incidence, and 41% decrease in total cancer mortality. In a follow-up analysis, it was observed that subjects with low baseline levels of selenium in plasma (<122 ng/ml) showed the greatest benefit of selenium supplementation (Duffield-Lillico et al., 2002).

The favorable results of the NPC sparked a number of cancer site-specific trials of selenium supplementation, some with selenized yeast and others with selenomethionine, a major component of selenized yeast. One of these trials, SELECT (Table 7.2, Study 9), the largest cancer prevention trial ever conducted in the United States, was designed to test the protective effects of selenomethionine and vitamin E individually and in combination against prostate cancer (Lippman et al., 2009). Although SELECT was highly successful from the standpoint of recruitment and compliance, results demonstrated a lack of protection by selenomethionine, supporting the notion that selenomethionine is a form of selenium that is not highly active against prostate cancer (Lippman et al., 2009). This is consistent with laboratory studies that have consistently demonstrated that, although organic forms of selenium in general have greater anticancer activities, among those forms tested, selenomethionine is relatively inactive. These results provide further evidence that, although selenomethionine represents a major form of selenium in selenized yeast, it is not likely the form responsible for the chemopreventive properties of selenized yeast.

In the Negative Biopsy Trial (NBT) (Table 7.2, Study 11), 200 μg/day selenized yeast was tested for its effectiveness at inhibiting the development of prostate cancer in men at high risk for the disease (PSA >4 ng/ml, suspicious digital rectal exam or PSA velocity >0.75 ng/ml/year); baseline plasma selenium levels >122 ng/ml (Algotar et al., 2013). The ineffectiveness of selenized yeast in this trial suggests that supplementation may need to be started early and may not be effective when initiated in individuals at high risk when the stage of the disease process is already well underway.

Baseline levels of selenium may also represent an important factor in determining the efficacy of selenium. In the NPC, low baseline levels were associated with enhanced protection. Also, baseline levels of selenium in NPC were significantly lower than those observed in either SELECT or NBT. The importance of baseline selenium levels in predicting the efficacy of selenium was highlighted by a recent report of selenized yeast supplementation in healthy men in which beneficial effects of selenized yeast on oxidative damage biomarkers and antioxidants were only observed in individuals with low baseline blood levels (Karunasinghe et al., 2013). Altogether, these findings combined with the results of epidemiological investigations and extensive preclinical studies suggest that the effectiveness of selenium supplementation at preventing cancer development may only be apparent in individuals with low baseline selenium levels. We should point out that 23%–25% of subjects in both the SELECT and the National Health and Nutrition Examination Survey (NHANES) had baseline selenium levels <123 μg/L (Laclaustra et al., 2010; Lippman et al., 2009). It remains undefined whether these ~20% of the American men with low selenium can benefit from selenium supplementation.

Overall, among the few clinical trials that have examined the impact of selenium supplementation on cancer outcomes, only a limited number of selenium forms

TABLE 7.2
Clinical Trials Employing Selenium Alone or in Combination with Other Minerals and Vitamins

Study #	Study Design	Selenium Form (Dose)	Country	Population	Type of Cancer/Outcome	Refs.
1	Randomized placebo-controlled trial (n = 226)[a]	Selenized yeast (200 μg Se/day)	China	Hepatitis surface antigen carriers	Liver cancer/inhibition	Yu et al. (1991, 1997)
2	Population-based intervention trial (n = 130,471)[a]	Selenite in table salt (15 mg/kg)	China	General population	Preneoplastic liver lesions/inhibition	Yu et al. (1991, 1997)
3	Randomized placebo-controlled trial (n = 29,584)	Selenized yeast (50 μg Se/day); β-carotene (15 mg/day); α-tocopherol (30 mg/day)	China	General population	Stomach cancer/inhibition	Blot et al. (1993)
4	Randomized placebo-controlled trial (n = 3318)	Selenized yeast (50 μg Se/day) in combination with 14 vitamins and 11 other minerals	China	Esophageal dysplasia	Esophageal cancer/no effect	Li et al. (1993)
5	Randomized placebo-controlled trial (n = 298)	Selenomethionine (50–100 μg/day); vitamin A (10,000–25,000 IU), riboflavin (15–50 mg/day), and zinc (12.5–25 mg/day)	India	Reverse smokers	Oral preneoplastic lesions/inhibition	Prasad et al. (1995)
6	Randomized, double-blind, placebo-controlled trial (n = 411)	Selenomethionine (200 Se μg/day); zinc (30 mg), vitamin A (2 mg), vitamin C (180 mg), and vitamin E (30 mg)	Italy	Patients with resected adenomatous polyps	New adenomatous polyps/inhibition	Bonelli et al. (2013)
7	Randomized, double-blind, placebo-controlled trial (n = 238)	Selenomethionine (200 Se μg/day)	China	Esophageal dysplasia	Esophageal cancer/no effect Dysplasia progression/protective effect among individuals with mild dysplasia at baseline	Limburg et al. (2005)

(Continued)

TABLE 7.2 (CONTINUED)

Clinical Trials Employing Selenium Alone or in Combination with Other Minerals and Vitamins

Study #	Study Design	Selenium Form (Dose)	Country	Population	Type of Cancer/Outcome	Refs.
8	Randomized, double-blind, placebo-controlled trial Nutrition Prevention of Cancer Trial (NPC) (n = 1312)	Selenized yeast (200 Se μg/day)	US	Patients with prior skin cancer	Basal cell or squamous cell carcinoma/no effect Lung cancer/inhibition Colon cancer/inhibition Prostate cancer/inhibition	Clark et al. (1996, 1998)
9	Randomized, double-blind, placebo-controlled trial Selenium and Vitamin E Prostate Cancer Prevention Trial (SELECT) (n = 35,533)	Selenomethionine (200 Se μg/day)	US	Healthy males	Prostate cancer/no effect	Lippman et al. (2009)
10	Randomized, double-blind, placebo-controlled trial SWOG S9917 (n = 423)	Selenomethionine (200 Se μg/day)	US	Men with high-grade PIN	Prostate cancer/no effect	Marshall, Tangen et al. (2011)
11	Randomized, double-blind, placebo-controlled trial Negative Biopsy Trial (NBT) (n = 699)	Selenized yeast (200 Se μg/day)	US	Men with high PSA (>4 ng/ml), PSA velocity (>0.75 ng/ml/year), or suspicious digital rectal exam	Prostate cancer/no effect	Algotar et al. (2013)
12	Randomized, double-blind, placebo-controlled trial ECOG 5597 (n = 1561)	Selenized yeast (200 Se μg/day)	US	Patients with resected stage I NSCLC, RDBPC Trial	Second primary lung tumors/no effect	Karp et al. (2013)

a Studies suffered from methodological and quality control problems.

have been investigated. As for the organoselenium forms, only selenomethionine and selenized yeast have been tested with the latter containing >15 different forms. Although the results of these trials have yielded both positive and negative results regarding the potential role of selenium-containing compounds in cancer prevention, no definitive recommendations regarding an effective chemoprevention strategy can be made. However, these trials have provided some important information regarding the potentially effective forms of selenium for future study. The striking results of the NPC trial with selenized yeast suggest that one or more of its many forms of organoselenium, including relatively minor forms, may be playing a key role in the prevention of cancer. Inconsistent results with selenized yeast may reflect differences in the content of the active form(s) in the yeast used or may be due to other factors, such as differences in the patient populations under study. Nonetheless, identification and testing of these other forms of organoselenium in selenized yeast is warranted.

7.4.3 BIOMARKER STUDIES WITH HEALTHY INDIVIDUALS

Given the expense and lack of success of Phase III clinical trials, the use of smaller scale clinical studies involving cancer-related biomarkers represents a more feasible and effective approach for future investigations. One of the major problems in the design of these studies is the lack of suitable and reliable intermediate end point disease biomarkers for most cancers. Even with prostate cancer, the use of PSA levels or velocity as a marker for early detection has recently been questioned, and there remains widespread disagreement over its utility (Hayes and Barry, 2014). Thus, in the absence of relevant disease biomarkers, attention has been given to earlier events in the carcinogenic process and the analysis of biomarkers that reflect earlier end points, such as oxidative damage, inflammation, DNA damage, and antioxidant levels. Despite the plethora of positive preclinical data and disappointing outcomes in Phase III trials described above, there have been surprisingly few such small-scale biomarker studies with selenium performed to date. A number of these trials are summarized in Table 7.3.

Based on the success of the NPC trial with selenized yeast, a randomized, double-blind, placebo-controlled clinical biomarker trial was performed in healthy men to identify the effects of selenized yeast supplementation on early biomarkers, including blood and urine oxidative stress markers and levels of PSA and relevant hormones in serum (Table 7.3, Study 1) (El-Bayoumy et al., 2002). After 9 months, selenium supplementation (247 µg/day) was associated with a reduction in oxidative stress as assessed by decreases in blood oxidized to reduced glutathione levels, which are likely driven by increases in total glutathione levels. Interestingly, plasma PSA levels, which were within normal range in all groups, were nonetheless decreased by selenium supplementation. Proteomic profiling of plasma from study subjects revealed numerous proteins differentially expressed by selenium supplementation, including a number of redox-sensitive proteins and the inflammatory and possible prostate cancer marker α-1-antitrypsin (Sinha et al., 2010). In this same study, it was also noted that the efficacy of selenized yeast at enhancing blood selenium levels and oxidized and free glutathione levels occurred primarily in White participants but not in Blacks (Richie et al., 2011). These relationships between plasma selenium and glutathione levels in Blacks and Whites were also observed in a study of healthy unsupplemented adults (Richie et al., 2011). Plasma selenium concentrations were positively associated with levels of GSH and activity of its rate-limiting biosynthetic enzyme glutamyl cysteine ligase in Caucasians but much less so in African Americans. These results are of particular interest given the observed racial differences in incidence rates noted for numerous cancers, including prostate cancer (American Cancer Society, 2009).

Selenized yeast supplementation in prostate cancer patients demonstrated a dose-responsive increase in prostate selenium levels after 4–6 weeks of supplementation (Table 7.3, Study 7) (Algotar et al., 2011) whereas no effects were observed in PSA velocity after 5 years (Table 7.3, Study 3) (Stratton et al., 2010). In healthy individuals, selenized yeast supplementation was associated with changes in thioredoxin reductase activity and peroxide-induced DNA damage, which were greatest in individuals with the highest baseline DNA damage (Table 7.3, Study 8) (Karunasinghe et al., 2013). Supplementation with selenized yeast (300 µg/day) also resulted in changes in T-cell and NK cell activation and related cellular pathways in men after 1 year (Table 7.3, Studies 9 and 10) (Hawkes et al., 2009, 2013). Short-term studies of selenomethionine in combination with other agents revealed no effects on biomarkers of oxidative stress in prostate cancer patients (Table 7.3, Study 4) (Vidlar et al., 2010) but a reduction in plasma inflammatory and oxidative stress markers in colorectal adenoma patients (Table 7.3, Study 5) (Hopkins et al., 2010).

In our recent randomized, double-blind, placebo-controlled, biomarker trial, the two forms of organoselenium previously tested in Phase III clinical trials, selenized yeast (200 or 285 µg Se/day) and selenomethionine (200 µg Se/day), were directly compared for the first time (Table 7.3, Study 11) (Richie et al., 2014). The trial was designed so that the selenomethionine dose would be equivalent to the selenomethionine group in the high-dose selenized yeast group. Results indicated that urinary levels of oxidative stress biomarkers 8-OHdG and 8-iso-PGF$_{2\alpha}$ were decreased after 9 months in the high-dose selenized yeast group but not the selenomethionine group, despite comparable increases in plasma selenium levels in both groups. These decreases in oxidative stress markers were greatest in individuals in the lowest plasma selenium quartile. These findings support the notion that selenium-containing compounds in selenized yeast other than selenomethionine may account for the decrease in oxidative stress.

There have been few clinical studies involving other forms of selenium. Se-methylselenocysteine, a common form of selenium found in foods and one of the organoselenium compounds in selenized yeast was found to be bioavailable from a single oral dose in a Phase 1 pharmacokinetic study in healthy adults (Table 7.3, Study 6) (Marshall, Ip et al., 2011). In a recent study of blood and urine selenium speciation, effects of supplementation with Se-methylselenocysteine, selenomethionine, selenized yeast, selenite, or selenate were compared at various times after a single oral dose (Table 7.3, Study 12) (Kokarnig et al., 2015). Although preliminary in nature, results indicate differences between treatments for specific low molecular weight metabolites including selenosugar 1 and trimethylselenonium. Larger scale studies of this type will be required to fully elucidate the metabolic profiles of different selenium supplements.

TABLE 7.3
Clinical Biomarkers Studies Employing Selenium Alone or in Combination with Other Minerals and Vitamins

Study #	Selenium Form (Dose)	Population	Study Design	End Points/Outcomes	Refs.
1	Selenized yeast (247 μg Se/day)	Healthy men (n = 36)	Randomized, double-blind, placebo controlled trial 9 months supplementation, 3 months wash out	Blood glutathione/increased Glutathionylated proteins in blood/ decreased Plasma PSA/decreased	El-Bayoumy et al. (2002)
2	Selenite (200 or 500 μg Se/day)	Patients with coronary artery disease (n = 465)	Randomized, placebo-controlled trial 12 weeks	Blood GPx1/increase Oxidative stress and inflammation biomarkers/no change	Schnabel et al. (2008)
3	Selenized yeast (200 or 800 μg Se/ day)	Nonmetastatic prostate cancer patients (n = 53)	Randomized, double-blind, placebo controlled Phase 2 trial 5 years	PSA velocity/no effect (except for increase at high dose)	Stratton et al. (2010)
4	Selenomethionine (240 μg Se/day) in combination with silymarin (570 mg)	Prostate cancer patients after radical prostatectomy (n = 37)	Randomized, double-blind, placebo controlled Phase 2 trial 6 months	Blood antioxidant status and biomarkers of oxidative stress/no effect	Vidlar et al. (2010)
5	Selenomethionine (200 μg Se/day) in combination with seven other vitamins and minerals	Patients with history of sporadic colorectal adenoma (n = 47)	Randomized, double-blind, placebo controlled Phase 2 trial 4 months	Plasma TNF-α/decreased Plasma cystine/decreased	Hopkins et al. (2010)
6	Se-methylselenocysteine (400, 800, 1200 μg Se)	Healthy men (n = 18)	Phase I single-dose, dose-escalation pharmaco-kinetic/toxicity study; Randomized, placebo controlled	Plasma Se/increased Urine Se/increased	Marshall, Ip et al. (2011)
7	Selenized yeast (200 or 400 μg Se/ day)	Nonmetastatic prostate cancer patients (n = 53)	Randomized, double-blind, placebo controlled Phase 2 trial 4–6 weeks	Prostate Se levels/dose responsive increase	Algotar et al. (2011)

(Continued)

TABLE 7.3 (CONTINUED)
Clinical Biomarkers Studies Employing Selenium Alone or in Combination with Other Minerals and Vitamins

Study #	Selenium Form (Dose)	Population	Study Design	End Points/Outcomes	Refs.
8	Selenized yeast (200 μg Se/day)	Healthy men with serum Se <200 ng/ml (n = 425)	Supplementation trial, 6 months	TR activity/increased Peroxide-induced DNA damage/increased Greatest benefit from Se supplementation was in individuals with highest baseline DNA damage	Karunasinghe et al. (2013)
9	Selenized yeast (300 μg Se/day)	Healthy men (n = 42)	Randomized, controlled trial, 48 weeks	Suppression of T-cell and NK cell activation	Hawkes et al. (2009)
10	Selenized yeast (300 μg Se/day)	Healthy men (n = 16)	Randomized, controlled trial, 1 year	Whole blood gene expression/changes in genes involved in T-cell and NK cell cytotoxicity consistent with an anti-inflammatory effect	Hawkes et al. (2013)
11	Selenized yeast (200 or 285 μg Se/day) or Selenomethionine (200 μg Se/day)	Healthy men (n = 69)	Randomized, double-blind, placebo controlled trial 9 months supplementation, 3 months washout	Urinary oxidative stress biomarkers (8-OHdG and 8-iso-PGF$_{2\alpha}$)/decreased (selenized yeast group only)	Richie et al. (2014)
12	Selenate, Selenite, Selenized yeast, Selenomethionine, Se-methylselenocysteine	Healthy adults	Metabolism and disposition study 24 h	Comparison of serum and urinary low molecular weight selenium species profiles by form/differences between forms (e.g., selenized yeast and selenomethionine)	Kokarnig et al. (2015)

Note: 8-OHdG, 8-hydroxy-2′-deoxyguanosine; 8-iso-PGF$_{2\alpha}$, 8-iso-prostaglandin-F$_{2\alpha}$.

Given the many forms of organoselenium observed in commonly ingested foods, it is clear that we have only scratched the surface in our clinical investigation to date. Thus, there is a distinct need for small-scale clinical studies to test and compare the potential effectiveness of these different selenium forms on critical and relevant biomarkers of the disease being studied.

7.5   FORMS OF SELENIUM IN CANCER PREVENTION: MOVING FORWARD

Epidemiological studies and preclinical investigations support the protective role of selenium in cancer prevention. These previous preclinical studies support the notion that not only the dose but also the form of selenium is critical in cancer prevention. Based on a wealth of mechanistic information derived solely from preclinical studies, it is apparent that depending on the form, selenium compounds can modify cellular processes (cell proliferation, apoptosis, cell cycle, angiogenesis, immune response) and molecular targets (AKT, BCl-2, HDAC, NF-κB, mTOR, HIF-1α, REDD1, VEGF, to mention a few) involved in cancer development, progression, recurrence, and metastasis.

Why then, after decades of research, has selenium not established itself as an effective chemopreventive agent in the clinic? Unfortunately, our knowledge of the structures of selenium compounds in our commonly consumed selenium-containing foods, selenized yeast, as well as in tissues and in circulation of rodents and humans, remains largely undefined. Scheme 7.1 refers to our proposed strategic plan for future research on selenium and cancer chemoprevention. Obviously, there is a need to develop and optimize analytical methods for selenium speciation; such methods can provide information on quantification and structural characterization of various forms of selenium in foods, and selenized yeast (Stage A, Scheme 7.1). The focus in Stage B will be on speciation of biological samples that can be easily obtained from studies in animals and humans supplemented with various forms of selenium (selenium-enriched foods, selenized yeast); this step will lead to identification and quantification of selenium metabolites in tissues and/or in circulation. Furthermore, human archival samples can be obtained from previous clinical trials that employed selenized-yeast and selenomethionine. Information obtained from A and B on the forms of selenium will allow the design of comparative chemopreventive efficacy studies in preclinical systems of multiple forms and explore their potential mechanisms (Stage C). Following the identification of the most effective form(s) of selenium and the least toxic, small-scale clinical trials can be proposed (Stage D) to determine the impact in humans of these forms on relevant biomarkers of carcinogenesis (inflammation, oxidative stress, etc.). In addition, using contemporary omics approaches can lead to the discovery of targets at the gene, protein, and metabolite levels that may provide insights into the mechanisms of action that can account for chemoprevention. The ultimate goal of this exciting area of research is to determine the form and the optimal level of selenium to achieve remarkable chemopreventive efficacy and gain better mechanistic understanding on how cancer chemoprevention and treatment by specific form(s) of selenium become

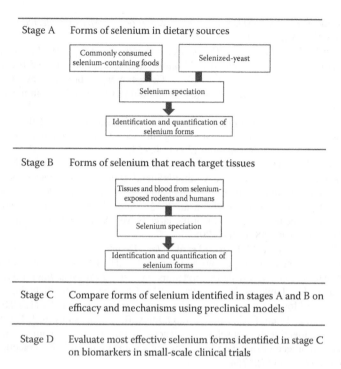

Stage A   Forms of selenium in dietary sources

Commonly consumed selenium-containing foods

Selenized-yeast

Selenium speciation

Identification and quantification of selenium forms

Stage B   Forms of selenium that reach target tissues

Tissues and blood from selenium-exposed rodents and humans

Selenium speciation

Identification and quantification of selenium forms

Stage C   Compare forms of selenium identified in stages A and B on efficacy and mechanisms using preclinical models

Stage D   Evaluate most effective selenium forms identified in stage C on biomarkers in small-scale clinical trials

SCHEME 7.1   Proposed strategic plan for future research on selenium and cancer chemoprevention.

individualized. Therefore, future studies (Scheme 7.1) are needed to answer the following questions: Do we have sensitive and relevant biomarkers to monitor disease (cancer) progression being examined? What is the most effective form of selenium that can alter relevant disease biomarkers? What is the appropriate population for chemoprevention studies (race, gender, country)? Do African Americans need more selenium than Whites? Do subjects of different ages require different levels of selenium? We believe that answers to these questions are urgently needed prior to the initiation of long-term and costly Phase III clinical trials using selenium. In summary, our goal is to enhance the bright side of the moon element (selenium) and to move the cancer chemoprevention arena by selenium to a higher level, that is, precision chemoprevention by selenium.

REFERENCES

Algotar, A. M., M. S. Stratton, M. J. Xu et al. 2011. "Dose-dependent effects of selenized yeast on total selenium levels in prostatic tissue of men with prostate cancer." *Nutr Cancer* no. 63 (1):1–5.

Algotar, A. M., M. S. Stratton, F. R. Ahmann et al. 2013. "Phase 3 clinical trial investigating the effect of selenium supplementation in men at high-risk for prostate cancer." *Prostate* no. 73 (3):328–35.

American Cancer Society. 2009. *Cancer facts & figures for African Americans 2009–2010.* Atlanta, GA: American Cancer Society.

Bai, Y., B. Qin, Y. Zhou et al. 2011. "Preparation and antioxidant capacity of element selenium nanoparticles sol-gel compounds." *J Nanosci Nanotechnol* no. 11 (6):5012–7.

Barger, J. L., T. Kayo, T. D. Pugh et al. 2012. "Gene expression profiling reveals differential effects of sodium selenite, selenomethionine, and yeast-derived selenium in the mouse." *Genes Nutr* no. 7 (2):155–65.

Bhuvarahamurthy, V., N. Balasubramanian, and S. Govindasamy. 1996. "Effect of radiotherapy and chemoradiotherapy on circulating antioxidant system of human uterine cervical carcinoma." *Mol Cell Biochem* no.158 (1):17–23.

Blot, W. J., J. Y. Li, P. R. Taylor et al. 1993. "Nutrition intervention trials in Linxian, China: Supplementation with specific vitamin/mineral combinations, cancer incidence, and disease-specific mortality in the general population." *J Natl Cancer Inst* no. (18):1483–92.

Bonelli, L., M. Puntoni, B. Gatteschi et al. 2013. "Antioxidant supplement and long-term reduction of recurrent adenomas of the large bowel. A double-blind randomized trial." *J Gastroenterol* no. 48 (6):698–705.

Bortner, J. D., Jr., A. Das, T. M. Umstead et al. 2009. "Down-regulation of 14-3-3 isoforms and annexin A5 proteins in lung adenocarcinoma induced by the tobacco-specific nitrosamine NNK in the A/J mouse revealed by proteomic analysis." *J Proteome Res* no. 8 (8):4050–61.

Brozmanova, J., D. Manikova, V. Vlckova, and M. Chovanec. 2010. "Selenium: A double-edged sword for defense and offense in cancer." *Arch Toxicol* no. 84 (12):919–38.

Cai, X. J., Block, E. Uden, P. C., Zhang, X., Quimby, B. D., and Sullivan, J. J. 1995. "Alium chemistry: Identification of selenoamino acids in ordinary and selenium-enriched garlic, onion, and broccoli using gas chromatography with atomic emission detection." *J Agric Food Chem* no. 43 (7):1754–7.

Chen, Y. C., K. S. Prabhu, A. Das, and A. M. Mastro. 2013. "Dietary selenium supplementation modifies breast tumor growth and metastasis." *Int J Cancer* no. 133 (9):2054–64.

Christensen, M. J. 2014. "Selenium and prostate cancer prevention: What next-if anything?" *Cancer Prev Res (Phila)* no. 7 (8):781–5.

Clark, L. C., G. F. Combs, Jr., B. W. Turnbull et al. 1996. "Effects of selenium supplementation for cancer prevention in patients with carcinoma of the skin. A randomized controlled trial. Nutritional Prevention of Cancer Study Group." *JAMA* no. (24):1957–63.

Clark, L. C., B. Dalkin, A. Krongrad et al. 1998. "Decreased incidence of prostate cancer with selenium supplementation: Results of a double-blind cancer prevention trial." *Br J Urol* no. 81 (5):730–4.

Das, A., D. Desai, B. Pittman, S. Amin, and K. El-Bayoumy. 2003. "Comparison of the chemopreventive efficacies of 1,4-phenylenebis(methylene)selenocyanate and selenium-enriched yeast on 4-(methylnitrosamino)-1-(3-pyridyl)-1-butanone induced lung tumorigenesis in A/J mouse." *Nutr Cancer* no. 46 (2):179–85.

Dennert, G., M. Zwahlen, M. Brinkman et al. 2011. "Selenium for preventing cancer." *Cochrane Database Syst Rev* no. (5):CD005195.

Dernovics, M., J. Far, and R. Lobinski. 2009. "Identification of anionic selenium species in Se-rich yeast by electrospray QTOF MS/MS and hybrid linear ion trap/orbitrap MSn." *Metallomics* no. 1 (4):317–29.

Desai, D., I. Sinha, K. Null et al. 2010. "Synthesis and antitumor properties of selenocoxib-1 against rat prostate adenocarcinoma cells." *Int J Cancer* no. 127 (1):230–8.

Dominguez-Alvarez, E., D. Plano, M. Font et al. 2014. "Synthesis and antiproliferative activity of novel selenoester derivatives." *Eur J Med Chem* no. 73:153–66.

Dong, Y., S. O. Lee, H. Zhang et al. 2004. "Prostate specific antigen expression is down-regulated by selenium through disruption of androgen receptor signaling." *Cancer Res* no. 64 (1):19–22.

dos Santos Edos, A., E. Hamel, R. Bai et al. 2013. "Synthesis and evaluation of diaryl sulfides and diaryl selenide compounds for antitubulin and cytotoxic activity." *Bioorg Med Chem Lett* no. 23 (16):4669–73.

Douglass, J. S., V. C. Morris, J. H. Soares, Jr., and O. A. Levander. 1981. "Nutritional availability to rats of selenium in tuna, beef kidney, and wheat." *J Nutr* no. 111 (12):2180–7.

Duffield-Lillico, A. J., M. E. Reid, B. W. Turnbull et al. 2002. "Baseline characteristics and the effect of selenium supplementation on cancer incidence in a randomized clinical trial: A summary report of the Nutritional Prevention of Cancer Trial." *Cancer Epidemiol Biomarkers Prev* no. 11 (7):630–9.

Duffield-Lillico, A. J., B. L. Dalkin, M. E. Reid et al. 2003. "Selenium supplementation, baseline plasma selenium status and incidence of prostate cancer: An analysis of the complete treatment period of the Nutritional Prevention of Cancer Trial." *BJU Int* no. 91 (7):608–12.

Dumont, E., F. Vanhaecke, and R. Cornelis. 2006. "Selenium speciation from food source to metabolites: A critical review." *Anal Bioanal Chem* no. 385 (7):1304–23.

El-Bayoumy, K. 1985. "Effects of organoselenium compounds on induction of mouse forestomach tumors by benzo(a)pyrene." *Cancer Res* no. 45 (8):3631–5.

El-Bayoumy, K. 1991. The role of selenium in cancer prevention. In *Cancer Principles and Practice of Oncology*, edited by V. T. J. DeVita, Hellman, S., Rosenberg, S. A. Philadelphia, PA: J. P. Lippincott Company.

El-Bayoumy, K. 2001. "The protective role of selenium on genetic damage and on cancer." *Mutat Res* no. 475 (1–2):123–39.

El-Bayoumy, K., C. V. Rao, and B. S. Reddy. 2001. "Multiorgan sensitivity to anticarcinogenesis by the organoselenium 1,4-phenylenebis(methylene)selenocyanate." *Nutr Cancer* no. 40 (1):18–27.

El-Bayoumy, K., J. P. Richie, Jr., T. Boyiri et al. 2002. "Influence of selenium-enriched yeast supplementation on biomarkers of oxidative damage and hormone status in healthy adult males: A clinical pilot study." *Cancer Epidemiol Biomarkers Prev* no. 11 (11):1459–65.

El-Bayoumy, K. 2004. Not all chemopreventive selenium compounds are created equal. In *Cancer Chemoprevention, Vol. 1: Promising Cancer Chemopreventive Agents* edited by G. J. Kelloff, Hawk, E. T., Segman, C. C. Totowa, NJ: Humana Press, Inc.

El-Bayoumy, K., and R. Sinha. 2005. "Molecular chemoprevention by selenium: A genomic approach." *Mutat Res* no. 591 (1–2):224–36.

El-Bayoumy, K., A. Das, B. Narayanan et al. 2006. "Molecular targets of the chemopreventive agent 1,4-phenylenebis (methylene)-selenocyanate in human non-small cell lung cancer." *Carcinogenesis* no. 27 (7):1369–76.

El-Bayoumy, K. 2009. "The negative results of the SELECT study do not necessarily discredit the selenium-cancer prevention hypothesis." *Nutr Cancer* no. 61 (3):285–6.

El-Bayoumy, K., Sinha, R., Cooper, A. J. L., and Pinto, J. T. 2011. "The role of alliums and their sulfur and selenium constituents in cancer prevention." In *Vegetables Whole Grains, and Their Derivatives in Cancer Prevention*, edited by M. Mutanen, A.M. Pajari, Springer, Netherlands.

Estevez, H., J. C. Garcia-Lidon, J. L. Luque-Garcia, and C. Camara. 2014. "Effects of chitosan-stabilized selenium nanoparticles on cell proliferation, apoptosis and cell cycle pattern in HepG2 cells: Comparison with other selenospecies." *Colloids Surf B Biointerfaces* no. 122C:184–93.

Facompre, N. D., K. El-Bayoumy, Y. W. Sun, J. T. Pinto, and R. Sinha. 2010. "1,4-phenylenebis (methylene)selenocyanate, but not selenomethionine, inhibits androgen receptor and Akt signaling in human prostate cancer cells." *Cancer Prev Res (Phila)* no. 3 (8):975–84.

Facompre, N. D., I. Sinha, K. El-Bayoumy, J. T. Pinto, and R. Sinha. 2012. "Remarkable inhibition of mTOR signaling by the combination of rapamycin and 1,4-phenylenebis(methylene) selenocyanate in human prostate cancer cells." *Int J Cancer* no. 131 (9):2134–42.

Fairweather-Tait, S. J., Y. Bao, M. R. Broadley et al. 2011. "Selenium in human health and disease." *Antioxid Redox Signal* no. 14 (7):1337–83.

Far, J., H. Preud'homme, and R. Lobinski. 2010. "Detection and identification of hydrophilic selenium compounds in selenium-rich yeast by size exclusion-microbore normal-phase HPLC with the on-line ICP-MS and electrospray Q-TOF-MS detection." *Anal Chim Acta* no. 657 (2):175–90.

Ferguson, L. R., N. Karunasinghe, S. Zhu, and A. H. Wang. 2012. "Selenium and its' role in the maintenance of genomic stability." *Mutat Res* no. 733 (1–2):100–10.

Fernandez-Herrera, M. A., J. Sandoval-Ramirez, L. Sanchez-Sanchez, H. Lopez-Munoz, and M. L. Escobar-Sanchez. 2014. "Probing the selective antitumor activity of 22-oxo-26-selenocyanocholestane derivatives." *Eur J Med Chem* no. 74:451–60.

Finley, J. W., and C. D. Davis. 2001. "Selenium (Se) from high-selenium broccoli is utilized differently than selenite, selenate and selenomethionine, but is more effective in inhibiting colon carcinogenesis." *Biofactors* no. 14 (1–4):191–6.

Finley, J. W. 2003. "Reduction of cancer risk by consumption of selenium-enriched plants: Enrichment of broccoli with selenium increases the anticarcinogenic properties of broccoli." *J Med Food* no. 6 (1):19–26.

Francesconi, K. A., and F. Pannier. 2004. "Selenium metabolites in urine: A critical overview of past work and current status." *Clin Chem* no. 50 (12):2240–53.

Ganther, H. E. 1986. "Pathways of selenium metabolism including respiratory excretory products." *J Am Coll Toxicol* no. 5:1–5.

Ganther, H. E. 1999. "Selenium metabolism, selenoproteins and mechanisms of cancer prevention: Complexities with thioredoxin reductase." *Carcinogenesis* no. 20 (9):1657–66.

Gowda, R., S. V. Madhunapantula, D. Desai, S. Amin, and G. P. Robertson. 2012. "Selenium-containing histone deacetylase inhibitors for melanoma management." *Cancer Biol Ther* no. 13 (9):756–65.

Gowda, R., S. V. Madhunapantula, D. Desai, S. Amin, and G. P. Robertson. 2013. "Simultaneous targeting of COX-2 and AKT using selenocoxib-1-GSH to inhibit melanoma." *Mol Cancer Ther* no. 12 (1):3–15.

Guo, P., P. Zhao, J. Liu et al. 2013. "Preparation of a novel organoselenium compound and its anticancer effects on cervical cancer cell line HeLa." *Biol Trace Elem Res* no. 151 (2):301–6.

Guo, W. D., A. W. Hsing, J. Y. Li et al. 1994. "Correlation of cervical cancer mortality with reproductive and dietary factors, and serum markers in China." *Int J Epidemiol* no. 23 (6):1127–32.

Hartman, T. J., D. Albanes, P. Pietinen et al. 1998. "The association between baseline vitamin E, selenium, and prostate cancer in the alpha-tocopherol, beta-carotene cancer prevention study." *Cancer Epidemiol Biomarkers Prev* no. 7 (4):335–40.

Hawkes, W. C., A. Hwang, and Z. Alkan. 2009. "The effect of selenium supplementation on DTH skin responses in healthy North American men." *J Trace Elem Med Biol* no. 23 (4):272–80.

Hawkes, W. C., D. Richter, and Z. Alkan. 2013. "Dietary selenium supplementation and whole blood gene expression in healthy North American men." *Biol Trace Elem Res* no. 155 (2):201–8.

Hayes, J. H., and M. J. Barry. 2014. "Screening for prostate cancer with the prostate-specific antigen test: A review of current evidence." *JAMA* no. 311 (11):1143–9.

Hazane-Puch, F., P. Champelovier, J. Arnaud et al. 2013. "Long-term selenium supplementation in HaCaT cells: Importance of chemical form for antagonist (protective versus toxic) activities." *Biol Trace Elem Res* no. 154 (2):288–98.

Helzlsouer, K. J., A. J. Alberg, E. P. Norkus et al. 1996. "Prospective study of serum micronutrients and ovarian cancer." *J Natl Cancer Inst* no. 88 (1):32–7.

Holben, D. H., and A. M. Smith. 1999. "The diverse role of selenium within selenoproteins: A review." *J Am Diet Assoc* no. 99 (7):836–43.

Hopkins, M. H., V. Fedirko, D. P. Jones, P. D. Terry, and R. M. Bostick. 2010. "Antioxidant micronutrients and biomarkers of oxidative stress and inflammation in colorectal adenoma patients: Results from a randomized, controlled clinical trial." *Cancer Epidemiol Biomarkers Prev* no. 19 (3):850–8.

Hu, H., C. Jiang, T. Schuster et al. 2006. "Inorganic selenium sensitizes prostate cancer cells to TRAIL-induced apoptosis through superoxide/p53/Bax-mediated activation of mitochondrial pathway." *Mol Cancer Ther* no. 5 (7):1873–82.

Hu, Y., G. H. McIntosh, R. K. Le Leu, R. Woodman, and G. P. Young. 2008. "Suppression of colorectal oncogenesis by selenium-enriched milk proteins: Apoptosis and K-ras mutations." *Cancer Res* no. 68 (12):4936–44.

Hu, Y., G. H. McIntosh, and G. P. Young. 2012. "Selenium-rich foods: A promising approach to colorectal cancer prevention." *Curr Pharm Biotechnol* no. 13 (1):165–72.

Hurst, R., L. Hooper, T. Norat et al. 2012. "Selenium and prostate cancer: Systematic review and meta-analysis." *Am J Clin Nutr* no. 96 (1):111–22.

Institute of Medicine (US). Panel on Dietary Antioxidants and Related Compounds. 2000. *Dietary reference intakes for vitamin C, vitamin E, selenium, and carotenoids: A report of the Panel on Dietary Antioxidants and Related Compounds, Subcommittees on Upper Reference Levels of Nutrients and of Interpretation and Use of Dietary Reference Intakes, and the Standing Committee on the Scientific Evaluation of Dietary Reference Intakes, Food and Nutrition Board, Institute of Medicine.* Washington, D.C.: National Academy Press.

Jackson, M. I., K. Lunoe, C. Gabel-Jensen, B. Gammelgaard, and G. F. Combs, Jr. 2013. "Metabolism of selenite to selenosugar and trimethylselenonium *in vivo*: Tissue dependency and requirement for S-adenosylmethionine-dependent methylation." *J Nutr Biochem* no. 24 (12):2023–30.

Janakiram, N. B., A. Mohammed, D. Ravillah et al. 2013. "Chemopreventive effects of PBI-Se, a selenium-containing analog of PBIT, on AOM-induced aberrant crypt foci in F344 rats." *Oncol Rep* no. 30 (2):952–60.

Karp, D. D., S. J. Lee, S. M. Keller et al. 2013. "Randomized, double-blind, placebo-controlled, phase III chemoprevention trial of selenium supplementation in patients with resected stage I non-small-cell lung cancer: ECOG 5597." *J Clin Oncol* no. 31 (33):4179–87.

Karunasinghe, N., D. Y. Han, S. Zhu et al. 2013. "Effects of supplementation with selenium, as selenized yeast, in a healthy male population from New Zealand." *Nutr Cancer* no. 65 (3):355–66.

Kassam, S., H. Goenaga-Infante, L. Maharaj et al. 2011. "Methylseleninic acid inhibits HDAC activity in diffuse large B-cell lymphoma cell lines." *Cancer Chemother Pharmacol* no. 68 (3):815–21.

Kim, C., J. Lee, and M. S. Park. 2014. "Synthesis of new diorganodiselenides from organic halides: Their antiproliferative effects against human breast cancer MCF-7 cells." *Arch Pharm Res*. May 27. doi: 10.1007/s12272-014-0407-4.

Kim, J. H., J. Yu, V. Alexander et al. 2014. "Structure-activity relationships of 2′-modified-4′-selenoarabinofuranosyl-pyrimidines as anticancer agents." *Eur J Med Chem* no. 83:208–25.

Kokarnig, S., A. Tsirigotaki, T. Wiesenhofer et al. 2015. "Concurrent quantitative HPLC-mass spectrometry profiling of small selenium species in human serum and urine after ingestion of selenium supplements." *J Trace Elem Med Biol* no. 29:83–90.

Kosaric, J. V., D. M. Cvetkovic, M. N. Zivanovic et al. 2014. "Antioxidative and antiproliferative evaluation of 2-(phenylselenomethyl)tetrahydrofuran and 2-(phenylselenomethyl) tetrahydropyran." *J BUON* no. 19 (1):283–90.

Laclaustra, M., S. Stranges, A. Navas-Acien, J. M. Ordovas, and E. Guallar. 2010. "Serum selenium and serum lipids in US adults: National Health and Nutrition Examination Survey (NHANES) 2003–2004." *Atherosclerosis* no. 210 (2):643–8.

Lee, E. H., S. K. Myung, Y. J. Jeon et al. 2011. "Effects of selenium supplements on cancer prevention: Meta-analysis of randomized controlled trials." *Nutr Cancer* no. 63 (8):1185–95.

Lee, J. I., H. Nian, A. J. Cooper et al. 2009. "Alpha-keto acid metabolites of naturally occurring organoselenium compounds as inhibitors of histone deacetylase in human prostate cancer cells." *Cancer Prev Res (Phila)* no. 2 (7):683–93.

Lee, S. O., N. Nadiminty, X. X. Wu et al. 2005. "Selenium disrupts estrogen signaling by altering estrogen receptor expression and ligand binding in human breast cancer cells." *Cancer Res* no. 65 (8):3487–92.

Lee, S. O., J. Yeon Chun, N. Nadiminty et al. 2006. "Monomethylated selenium inhibits growth of LNCaP human prostate cancer xenograft accompanied by a decrease in the expression of androgen receptor and prostate-specific antigen (PSA)." *Prostate* no. 66 (10):1070–5.

Li, J. Y., P. R. Taylor, B. Li et al. 1993. "Nutrition intervention trials in Linxian, China: Multiple vitamin/mineral supplementation, cancer incidence, and disease-specific mortality among adults with esophageal dysplasia." *J Natl Cancer Inst* no. 85 (18):1492–8.

Li, J., F. Jiang, B. Yang et al. 2013. "Topological insulator bismuth selenide as a theranostic platform for simultaneous cancer imaging and therapy." *Sci Rep* no. 3:1998.

Limburg, P. J., W. Wei, D. J. Ahnen et al. 2005. "Randomized, placebo-controlled, esophageal squamous cell cancer chemoprevention trial of selenomethionine and celecoxib." *Gastroenterology* no. 129 (3):863–73.

Lippman, S. M., E. A. Klein, P. J. Goodman et al. 2009. "Effect of selenium and vitamin E on risk of prostate cancer and other cancers: The Selenium and Vitamin E Cancer Prevention Trial (SELECT)." *JAMA* no. 301 (1):39–51.

Lunoe, K., C. Gabel-Jensen, S. Sturup et al. 2011. "Investigation of the selenium metabolism in cancer cell lines." *Metallomics* no. 3 (2):162–8.

Luo, H., F. Wang, Y. Bai, T. Chen, and W. Zheng. 2012. "Selenium nanoparticles inhibit the growth of HeLa and MDA-MB-231 cells through induction of S phase arrest." *Colloids Surf B Biointerfaces* no. 94:304–8.

Madhunapantula, S. V., D. Desai, A. Sharma et al. 2008. "PBISe, a novel selenium-containing drug for the treatment of malignant melanoma." *Mol Cancer Ther* no. 7 (5):1297–308.

Marshall, J. R., C. Ip, K. Romano et al. 2011. "Methyl selenocysteine: Single-dose pharmacokinetics in men." *Cancer Prev Res (Phila)* no. 4 (11):1938–44.

Marshall, J. R., C. M. Tangen, W. A. Sakr et al. 2011. "Phase III trial of selenium to prevent prostate cancer in men with high-grade prostatic intraepithelial neoplasia: SWOG S9917." *Cancer Prev Res (Phila)* no. 4 (11):1761–9.

Nayini, J. R., S. Sugie, K. El-Bayoumy et al. 1991. "Effect of dietary benzylselenocyanate on azoxymethane-induced colon carcinogenesis in male F344 rats." *Nutr Cancer* no. 15 (2):129–39.

Ozten, N., M. Schlicht, A. M. Diamond, and M. C. Bosland. 2014. "L-selenomethionine does not protect against testosterone plus 17beta-estradiol-induced oxidative stress and preneoplastic lesions in the prostate of NBL rats." *Nutr Cancer* no. 66 (5):825–34.

Papp, L. V., J. Lu, A. Holmgren, and K. K. Khanna. 2007. "From selenium to selenoproteins: Synthesis, identity, and their role in human health." *Antioxid Redox Signal* no. 9 (7):775–806.

Patrick, L. 2004. "Selenium biochemistry and cancer: A review of the literature." *Altern Med Rev* no. 9 (3):239–58.

Pinto, J. T., J. I. Lee, R. Sinha, M. E. MacEwan, and A. J. Cooper. 2011. "Chemopreventive mechanisms of alpha-keto acid metabolites of naturally occurring organoselenium compounds." *Amino Acids* no. 41 (1):29–41.

Platz, E. A., and K. J. Helzlsouer. 2001. "Selenium, zinc, and prostate cancer." *Epidemiol Rev* no. 23 (1):93–101.

Prasad, M. P., M. A. Mukundan, and K. Krishnaswamy. 1995. "Micronuclei and carcinogen DNA adducts as intermediate end points in nutrient intervention trial of precancerous lesions in the oral cavity." *Eur J Cancer B Oral Oncol* no. 31B (3):155–9.

Rayman, M. P. 2004. "The use of high-selenium yeast to raise selenium status: How does it measure up?" *Br J Nutr* no. 92 (4):557–73.

Rayman, M. P. 2012. "Selenium and human health." *Lancet* no. 379 (9822):1256–68.

Rayman, M. P., G. F. Combs, Jr., and D. J. Waters. 2009. "Selenium and vitamin E supplementation for cancer prevention." *JAMA* no. 301 (18):1876; author reply 1877.

Ren, Y., T. Zhao, G. Mao et al. 2013. "Antitumor activity of hyaluronic acid-selenium nanoparticles in Heps tumor mice models." *Int J Biol Macromol* no. 57:57–62.

Richie, J. P., Jr., J. E. Muscat, I. Ellison et al. 2011. "Association of selenium status and blood glutathione concentrations in blacks and whites." *Nutr Cancer* no. 63 (3):367–75.

Richie, J. P., Jr., A. Das, A. M. Calcagnotto et al. 2014. "Comparative effects of two different forms of selenium on oxidative stress biomarkers in healthy men: A randomized clinical trial." *Cancer Prev Res (Phila)* no. 7 (8):796–804.

Romano, B., M. Font, I. Encio, J. A. Palop, and C. Sanmartin. 2014. "Synthesis and antiproliferative activity of novel methylselenocarbamates." *Eur J Med Chem* no. 83:674–84.

Russo, M. W., S. C. Murray, J. I. Wurzelmann, J. T. Woosley, and R. S. Sandler. 1997. "Plasma selenium levels and the risk of colorectal adenomas." *Nutr Cancer* no. 28 (2):125–9.

Schnabel, R., E. Lubos, C. M. Messow et al. 2008. "Selenium supplementation improves antioxidant capacity *in vitro* and *in vivo* in patients with coronary artery disease The Selenium Therapy in Coronary Artery disease Patients (SETCAP) Study." *Am Heart J* no. 156 (6):1201 e1–11.

Schrauzer, G. N., D. A. White, and C. J. Schneider. 1977. "Cancer mortality correlation studies—III: Statistical associations with dietary selenium intakes." *Bioinorg Chem* no. 7 (1):23–31.

Schrauzer, G. N. 2001. "Nutritional selenium supplements: Product types, quality, and safety." *J Am Coll Nutr* no. 20 (1):1–4.

Shamberger, R. J., and D. V. Frost. 1969. "Possible protective effect of selenium against human cancer." *Can Med Assoc J* no. 100 (14):682.

Shamberger, R. J., S. A. Tytko, and C. E. Willis. 1976. "Antioxidants and cancer. Part VI. Selenium and age-adjusted human cancer mortality." *Arch Environ Health* no. 31 (5):231–5.

Sharma, A. K., A. Sharma, D. Desai et al. 2008. "Synthesis and anticancer activity comparison of phenylalkyl isoselenocyanates with corresponding naturally occurring and synthetic isothiocyanates." *J Med Chem* no. 51 (24):7820–6.

Sharma, A., A. K. Sharma, S. V. Madhunapantula et al. 2009. "Targeting Akt3 signaling in malignant melanoma using isoselenocyanates." *Clin Cancer Res* no. 15 (5):1674–85.

Sharma, A. K., and S. Amin. 2013. "Post SELECT: Selenium on trial." *Future Med Chem* no. 5 (2):163–74.

Singh, U., K. Null, and R. Sinha. 2008. "*In vitro* growth inhibition of mouse mammary epithelial tumor cells by methylseleninic acid: Involvement of protein kinases." *Mol Nutr Food Res* no. 52 (11):1281–8.

Sinha, I., K. Null, W. Wolter et al. 2012. "Methylseleninic acid downregulates hypoxia-inducible factor-1alpha in invasive prostate cancer." *Int J Cancer* no. 130 (6):1430–9.

Sinha, I., J. E. Allen, J. T. Pinto, and R. Sinha. 2014. "Methylseleninic acid elevates REDD1 and inhibits prostate cancer cell growth despite AKT activation and mTOR dysregulation in hypoxia." *Cancer Med* no. 3 (2):252–64.

Sinha, R., and K. El-Bayoumy. 2004. "Apoptosis is a critical cellular event in cancer chemoprevention and chemotherapy by selenium compounds." *Curr Cancer Drug Targets* no. 4 (1):13–28.

Sinha, R., I. Sinha, N. Facompre et al. 2010. "Selenium-responsive proteins in the sera of selenium-enriched yeast-supplemented healthy African American and Caucasian men." *Cancer Epidemiol Biomarkers Prev* no. 19 (9):2332–40.

Steinbrenner, H., B. Speckmann, and H. Sies. 2013. "Toward understanding success and failures in the use of selenium for cancer prevention." *Antioxid Redox Signal* no. 19 (2):181–91.

Stewart, M. S., J. E. Spallholz, K. H. Neldner, and B. C. Pence. 1999. "Selenium compounds have disparate abilities to impose oxidative stress and induce apoptosis." *Free Radic Biol Med* no. 26 (1–2):42–8.

Stratton, M. S., A. M. Algotar, J. Ranger-Moore et al. 2010. "Oral selenium supplementation has no effect on prostate-specific antigen velocity in men undergoing active surveillance for localized prostate cancer." *Cancer Prev Res (Phila)* no. 3 (8):1035–43.

Suzuki, K. T., C. Doi, and N. Suzuki. 2006. "Metabolism of 76Se-methylselenocysteine compared with that of 77Se-selenomethionine and 82Se-selenite." *Toxicol Appl Pharmacol* no. 217 (2):185–95.

Thomson, C. D., A. Chisholm, S. K. McLachlan, and J. M. Campbell. 2008. "Brazil nuts: An effective way to improve selenium status." *Am J Clin Nutr* no. 87 (2):379–84.

Unni, E., F. S. Kittrell, U. Singh, and R. Sinha. 2004. "Osteopontin is a potential target gene in mouse mammary cancer chemoprevention by Se-methylselenocysteine." *Breast Cancer Res* no. 6 (5):R586–92.

Unni, E., D. Koul, W. K. Yung, and R. Sinha. 2005. "Se-methylselenocysteine inhibits phosphatidylinositol 3-kinase activity of mouse mammary epithelial tumor cells *in vitro*." *Breast Cancer Res* no. 7 (5):R699–707.

Vidlar, A., J. Vostalova, J. Ulrichova et al. 2010. "The safety and efficacy of a silymarin and selenium combination in men after radical prostatectomy—A six month placebo-controlled double-blind clinical trial." *Biomed Pap Med Fac Univ Palacky Olomouc Czech Repub* no. 154 (3):239–44.

Vinceti, M., C. M. Crespi, C. Malagoli, C. Del Giovane, and V. Krogh. 2013. "Friend or foe? The current epidemiologic evidence on selenium and human cancer risk." *J Environ Sci Health C Environ Carcinog Ecotoxicol Rev* no. 31 (4):305–41.

Vinceti, M., G. Dennert, C. M. Crespi et al. 2014. "Selenium for preventing cancer." *Cochrane Database Syst Rev* no. 3:CD005195.

Vogt, T. M., R. G. Ziegler, B. I. Graubard et al. 2003. "Serum selenium and risk of prostate cancer in US blacks and whites." *Int J Cancer* no. 103 (5):664–70.

Wallenberg, M., S. Misra, A. M. Wasik et al. 2014. "Selenium induces a multi-targeted cell death process in addition to ROS formation." *J Cell Mol Med* no. 18 (4):671–84.

Wang, Y., J. Chen, D. Zhang et al. 2013. "Tumoricidal effects of a selenium (Se)-polysaccharide from Ziyang green tea on human osteosarcoma U-2 OS cells." *Carbohydr Polym* no. 98 (1):1186–90.

Warrington, J. M., J. J. Kim, P. Stahel et al. 2013. "Selenized milk casein in the diet of BALB/c nude mice reduces growth of intramammary MCF-7 tumors." *BMC Cancer* no. 13:492.

Waters, D. J., S. Shen, L. T. Glickman et al. 2005. "Prostate cancer risk and DNA damage: Translational significance of selenium supplementation in a canine model." *Carcinogenesis* no. 26 (7):1256–62.

Waters, D. J., S. Shen, S. S. Kengeri et al. 2012. "Prostatic response to supranutritional sele-nium supplementation: Comparison of the target tissue potency of selenomethionine vs. selenium-yeast on markers of prostatic homeostasis." *Nutrients* no. 4 (11):1650–63.

Weekley, C. M., and H. H. Harris. 2013. "Which form is that? The importance of selenium speciation and metabolism in the prevention and treatment of disease." *Chem Soc Rev* no.42 (23):8870–94.

West, D. W., M. L. Slattery, L. M. Robison, T. K. French, and A. W. Mahoney. 1991. "Adult dietary intake and prostate cancer risk in Utah: A case-control study with special empha-sis on aggressive tumors." *Cancer Causes Control* no. 2 (2):85–94.

Wrobel, J. K., M. J. Seelbach, L. Chen, R. F. Power, and M. Toborek. 2013. "Supplementation with selenium-enriched yeast attenuates brain metastatic growth." *Nutr Cancer* no. 65 (4):563–70.

Wu, M., M. M. Kang, N. W. Schoene, and W. H. Cheng. 2010. "Selenium compounds activate early barriers of tumorigenesis." *J Biol Chem* no. 285 (16):12055–62.

Yamashita, Y., and M. Yamashita. 2010. "Identification of a novel selenium-containing com-pound, selenoneine, as the predominant chemical form of organic selenium in the blood of bluefin tuna." *J Biol Chem* no. 285 (24):18134–8.

Yang, L., M. Pascal, and X. H. Wu. 2013. "Review of selenium and prostate cancer preven-tion." *Asian Pac J Cancer Prev* no. 14 (4):2181–4.

Yang, X., Y. Tian, P. Ha, and L. Gu. 1997. "[Determination of the selenomethionine content in grain and human blood]." *Wei Sheng Yan Jiu* no. 26 (2):113–6.

Yoshizawa, K., W. C. Willett, S. J. Morris et al. 1998. "Study of prediagnostic selenium level in toenails and the risk of advanced prostate cancer." *J Natl Cancer Inst* no. 90 (16):1219–24.

Yu, S. Y., Y. J. Zhu, W. G. Li et al. 1991. "A preliminary report on the intervention trials of pri-mary liver cancer in high-risk populations with nutritional supplementation of selenium in China." *Biol Trace Elem Res* no. 29 (3):289–94.

Yu, S. Y., Y. J. Zhu, and W. G. Li. 1997. "Protective role of selenium against hepatitis B virus and primary liver cancer in Qidong." *Biol Trace Elem Res* no. 56 (1):117–24.

Zhang, J., L. Wang, L. B. Anderson et al. 2010. "Proteomic profiling of potential molecu-lar targets of methyl-selenium compounds in the transgenic adenocarcinoma of mouse prostate model." *Cancer Prev Res (Phila)* no. 3 (8):994–1006.

Section IV

Dual Functions
of Selenoproteins in Cancer

8 Thioredoxin Reductase 1

Bradley A. Carlson

CONTENTS

8.1 Introduction 173
8.2 TrxR1 Maintains Normal Cells in Homeostasis 175
8.3 TrxR1 Tumor-Promoting Effects and Their Possible Mechanisms 175
    8.3.1 Deletion of TrxR1 Protects against Cancer 176
    8.3.2 Deletion of TrxR1 in Promoting/Sustaining Liver Cancer 176
    8.3.3 Effect of Deletion of TrxR1 in Other Cancers 178
    8.3.4 TrxR1 Role in Promoting/Sustaining Cancer Varies in Different
        Tissues Providing Possible Avenues for Cancer Therapy 178
8.4 The Trx System and Hallmarks of Cancer 178
8.5 TrxR1 and Selenium in Cancer 179
    8.5.1 Is the Sec Residue Essential in All Functions of TrxR1
        and Could Selenium-Free Forms of TrxR1 Play a Role in Cancer? 180
8.6 Conclusions 181
Acknowledgments 183
References 183

8.1 INTRODUCTION

The thioredoxin system, which is an oxidoreductase system, occurs in virtually all living cells and consists of three vital components, thioredoxin (Trx), thioredoxin reductase (TrxR), and NADPH (Arnér, 2009; Arnér and Holmgren, 2000; Lu and Holmgren, 2014). In most organisms, this flavoprotein system performs thiol-dependent/thiol-disulfide exchange reactions to maintain cellular redox homeostasis that play essential roles in many cellular processes. These cellular processes encompass a wide range of roles that include cell growth, transcription, DNA synthesis and repair, redox regulation of signal transduction, protection against oxidative damage, and apoptosis (Arnér, 2009; Gromer et al., 2004; Labunskyy et al., 2014; Lu and Holmgren, 2014). Thioredoxin reductases (TrxRs) occur in two known classes, wherein one class has been characterized in prokaryotes and lower eukaryotes with proteins of ~35,000 molecular weight, and another class has been characterized in animals with proteins of ~55,000 molecular weight that share a common ancestry with glutathione reductase (Arnér, 2009; Arnér and Holmgren, 2000; Lu and Holmgren, 2014). These two classes evolved independently of each other, but all TrxRs are flavoproteins and function as homodimers although the specific details of their reaction mechanisms differ.

In mammalian cells, there are three TrxR isozymes, thioredoxin reductase 1 (designated TrxR1), mitochondrial thioredoxin reductase 2 (designated TrxR2), and thioredoxin glutathione reductase (designated TGR), wherein each isozyme contains a catalytic selenocysteine (Sec) residue as the C-terminal penultimate amino acid, and the –Gly-Cys-Sec-Gly-COOH motif is conserved in all mammalian TrxRs (Arnér, 2009; Gladyshev et al., 1996; Labunskyy et al., 2014; Lothrop et al., 2014a; Lu and Holmgren, 2014). The Sec residue is essential for maximum reducing activity of this selenoenzyme (Cunniff et al., 2014; Lothrop et al., 2009, 2014a; Peng et al., 2014). TrxR1 is the cytoplasmic form that is also present in the nucleus, and its major function in normal cells is to maintain Trx1 in the reduced state (Arnér, 2009; Rundlof and Arnér, 2004; Rundlof et al., 2004; Sun et al., 2001b). This selenoenzyme also serves important functions apart from Trx1 reduction, has a wide variety of substrates, and reduces a number of low molecular weight compounds with antioxidant functions (Arnér, 2009; Arnér and Holmgren, 2000; Lu and Holmgren, 2014). It should be noted that at least six isoforms of this selenoenzyme have been identified in mammals, and they arise by alternative splicing and amino-terminal protein extension due to different transcription initiation sites (Arnér, 2009; Cebula et al., 2013; Damdimopoulou et al., 2009; Dammeyer et al., 2008; Su and Gladyshev, 2004). The transcription of these variants results in proteins with different N-terminal sequences, cell-specific expression patterns, and unique functions (Arnér, 2009; Cebula et al., 2013; Damdimopoulou et al., 2009; Dammeyer et al., 2008; Su and Gladyshev, 2004). The details regarding the nature and mechanism of these splice variants is beyond the scope of this chapter, but it will be important going forward to gain a better understanding of the specific molecular functions of these isoforms and also which isoforms of TrxR1 are expressed in different cancers or cell types. For example, a study using lung cancer tissues has shown that the TrxR1-v.2,3,5 isoforms show a significant correlation with the degree of differentiation, suggesting that one of these variants may have a tumorigenic function (Fernandes et al., 2009).

TrxR2 is the mitochondrial form of thioredoxin reductase, and its major function is to keep mitochondrial thioredoxin, Trx2, in the reduced state as well as having a role in the reduction of glutaredoxin (Grx) 2. TrxR1 and TrxR2 occur in all vertebrates and have been shown to be essential selenoenzymes because their targeted removal is embryonic lethal (Bondareva et al., 2007; Conrad et al., 2004; Jakupoglu et al., 2005; Matsui et al., 1996; Nonn et al., 2003). TGR is a thioredoxin/glutathione reductase, and it is quite different from TrxR1 and TrxR2 as it also has an additional Grx domain. Thus, this selenoenzyme appears to have roles in the two major antioxidant systems in mammals, the Trx and glutathione systems (Su et al., 2005; Sun et al., 2001a, 2005). TGR is highly expressed in testis following puberty, and its role is still poorly understood (Su et al., 2005).

Because TrxR1 is the major focus of this chapter, its roles in normal and malignant cells with special emphasis on the dual personality of this selenoenzyme in protecting cells from cancer and in promoting cancer are examined in much greater detail. There are two other selenoproteins that are known to exhibit such dual personalities, glutathione peroxidase 2 (GPx2) and selenoprotein 15 (Sep15), that are discussed in Chapters 9 and 10, respectively.

8.2 TRxR1 MAINTAINS NORMAL CELLS IN HOMEOSTASIS

TrxR1 is one of most studied selenoproteins in mammals because, as noted above, it plays a key role in the Trx system being the enzyme that controls the redox state and thus the function of Trx1 in normal cells (Arnér and Holmgren, 2000; Rundlof and Arnér, 2004; Rundlof et al., 2004; Sun et al., 2001b). It is expressed in all cell types, tissues, and organs, and by virtue of keeping Trx1 in the reduced state, it has roles in angiogenesis, cell proliferation, DNA replication and repair, transcription, and in resistance to oxidative stress and apoptosis (Arnér, 2009; Arnér and Holmgren, 2000; Biaglow and Miller, 2005; Fujino et al., 2006; Holmgren, 2006; Holmgren and Lu, 2010; Rundlof and Arnér, 2004). As also noted above, this selenoenzyme is regarded as a major antioxidant and redox regulator in mammalian cells and is an essential cellular protein with a highly significant role in embryogenesis. Its Sec moiety is located near the C-terminus as the penultimate amino acid (Gladyshev et al., 1996), which is required for catalytic activity (see Nordberg et al., 1998; Zhong et al., 1998; Zhong and Holmgren, 2000, and above). For the role of TrxR1 in metabolism, see also Chapter 13.

There are other cellular phenomena that TrxR1 is involved in, providing further evidence of this protein's function in cancer prevention and maintaining cells in a normal state. For example, TrxR1 is known to activate the p53 tumor suppressor (Merrill et al., 1999; Moos et al., 2003) and to exhibit other tumor-suppressing activities (Arnér and Holmgren, 2006).

8.3 TRxR1 TUMOR-PROMOTING EFFECTS
AND THEIR POSSIBLE MECHANISMS

TrxR1 also plays a prominent role in promoting and/or sustaining cancer. For example, this selenoenzyme is overexpressed in many tumors and cancer cell lines (Arnér, 2009; Arnér and Holmgren, 2006; Biaglow and Miller, 2005; Fujino et al., 2006; Gundimeda et al., 2009; Hedstrom et al., 2009; Rundlof et al., 2004; Yoo et al., 2006), and this overexpression may be associated with tumor aggressiveness (Berggren et al., 1996; Lincoln et al., 2003). Because many types of cancer cells suffer from oxidative stress, the upregulation of TrxR1 in cancer certainly provides a stronger antioxidant defense system, allowing cancer cells to escape cell death. Further, reduced Trx1 is required to inhibit apoptosis signal-regulating kinase (ASK1), and overexpression of TrxR1 has been associated with resistance to apoptosis (Marzano et al., 2007). The elevated level of TrxR1 observed in many cancer cells and tumors has prompted numerous investigators to target the down-regulation of the *TrxR1* transcript or inhibition of the TrxR1 protein as a means of reducing the antioxidant potential in malignant cells. Targeting the expression of TrxR1 with any of a number of anticancer drugs and other inhibitors has indeed resulted in an alteration of some of the cancer-related properties in malignant cells, further suggesting that this selenoenzyme has a role in driving malignancy and is therefore a target for cancer therapy (Arnér and Holmgren, 2006; Biaglow and Miller, 2005; Fujino et al., 2006; Hatfield et al., 2009; Lu et al., 2007; Moos et al.,

2003; Nguyen et al., 2006; Powis and Kirkpatrick, 2007; Rundlof and Arnér, 2004; Yoo et al., 2006, 2007).

8.3.1 DELETION OF TrxR1 PROTECTS AGAINST CANCER

One of the initial studies that provided the strongest evidence that TrxR1 is a target for cancer therapy demonstrated that the specific downregulation of this procancer protein in mouse lung cancer, LLC1, cells resulted in the reversal of several of the malignant properties that included morphology, growth rate, anchorage-independent growth, and in the reduced expression of two cancer-characteristic mRNAs, hepatocyte growth factor (Hgf) and osteopontin (Opn1), which play important roles in metastasis and tumor growth (Yoo et al., 2006). Furthermore, TrxR1-deficient and control LLC1 cells were injected into compatible mice to assess both solid tumor production and metastasis. The level of tumor formation and metastasis was dramatically reduced, and the tumor sizes were much smaller in mice injected with the TrxR1 knockdown cells. Most interestingly, the tumors recovered from mice injected with the TrxR1-deficient cells expressed this selenoenzyme but had lost the TrxR1 knockdown construct demonstrating that TrxR1 expression was essential for tumor formation and to sustain tumor growth (Yoo et al., 2006). Although TrxR1 expression, and most certainly TrxR1 overexpression, is required in this mouse model to drive the malignancy, the underlying mechanism of how this critical protein becomes elevated in the malignancy process is poorly understood. As discussed above, TrxR1 has an important role in maintaining normal cells in a healthy state, but then, at least in mouse lung cancer cells, its enhancement is required to drive the cancer. Importantly, its disruption upsets the malignancy process and can reverse several cancer-related properties, such as morphology, growth rate, and anchorage-independent growth, resulting in the cells becoming more like normal cells.

Arnér and collaborators examined human lung adenocarcinoma A549 cells to assess whether TrxR1 deficiency would reverse some of its cancer-related properties in human lung cancer cells (Eriksson et al., 2009). The knockdown of TrxR1 had only a minor effect on the growth of A549 cells but no other apparent phenotypic affects. The authors concluded that the excessively high levels of TrxR1 activity in A549 cells far exceeded its requirements to maintain a much lower Trx activity in the reduced state, and even though the knockdown of this selenoenzyme was approximately 90%, there was sufficient activity to cause continued reduction of Trx for the cell to retain its malignant properties. Interestingly, there were drug-specific alterations, wherein TrxR1-deficient A549 cells manifested highly sensitive responses to 1-chloro-2,4-dinitrobenzene (DNCB) or menadione but became less susceptible to cisplatin compared to the control cells suggesting roles of TrxR1 in cellular functions other than those involving Trx (Eriksson et al., 2009).

8.3.2 DELETION OF TrxR1 IN PROMOTING/SUSTAINING LIVER CANCER

The role of TrxR1 in hepatocarcinogenesis was examined in liver of mice carrying a tissue-specific deletion of TrxR1, and the mice were exposed to the liver carcinogen diethylnitrosamine (Carlson et al., 2012). Approximately 90% of the TrxR1$^{-/-}$ livers had developed tumors while only ~16% of the corresponding wild-type livers had tumors, indicating that TrxR1 has a protective role in hepatocarcinogenesis. Interestingly, another selenoenzyme that has been implicated in cancer promotion, GPx2 (see Chapter 9), was enriched in the TrxR1$^{-/-}$ liver tumors as were several components in the GSH system that included enzymes involved in GSH synthesis and maintaining GSH in the reduced state (Carlson et al., 2012). The results suggested that, in the absence of TrxR1 as in the absence of GPx2, other antioxidant enzymes are enhanced, providing a sufficient redox background to sustain tumor growth. Nrf2, the nuclear factor (erythroid-derived 2)-like 2 transcription factor, which upregulates the expression of numerous antioxidant proteins, including TrxR1, GPx2, and those in the GSH system, by binding to the antioxidant responsive element (ARE) in the promoter of target enzymes (Kensler and Wakabayashi, 2010), has been reported to be activated in TrxR1-deficient mouse hepatocytes (Carlson et al., 2012; Suvorova et al., 2009). TrxR1-deficient hepatocytes were able to proliferate, indicating a TrxR1-independent means of supporting normal growth, which may be due to the increase in other Nrf2-regulated genes (Rollins et al., 2010). The full details of the mechanism by which Nrf2 is activated in the absence of TrxR1 are still unknown. Nrf2 is maintained in the cytosol by kelch-like ECH-associated protein 1 (Keap1), and dissociation of the complex occurs when thiol residues are modified, leading to nuclear translocation of Nrf2. Trx has been shown to reduce the Cys151 residue of Keap 1 (Fourquet et al., 2010). Thus, it has been proposed that TrxR1 may inhibit Nrf2 translocation by means of keeping Trx in a reduced state (Brigelius-Flohé et al., 2012). Decreased activity or inhibition of TrxR1 may prevent Keap1 restoration, permitting newly synthesized Nrf2 to enter the nucleus (Brigelius-Flohé and Kipp, 2013). It has recently been shown that simultaneous inhibition of TrxR1 and the phosphoinositide 3-kinase/serine/threonine kinase (PI3K/AKT) pathway in lung cancer cells led to increased reactive oxygen species (ROS) production and cell apoptosis via the Keap1/Nrf2 pathway and required a functional Keap1 gene (Dai et al., 2013; see Chapter 13). Whether TrxR1 plays a similar role in protein disulfide formation in the procarcinogenic Wnt/β-catenin pathway remains to be determined (Brigelius-Flohé and Kipp, 2013).

Enhanced activation of Nrf2 most likely accounts for the upregulation of several other antioxidant proteins in hepatocarcinogenesis (Carlson et al., 2012; Suvorova et al., 2009) as well as in other cell types (Locy et al., 2012; Mandal et al., 2010). However, the role of TrxR1 in supporting tumor growth in this latter investigation was likely minimal as this selenoenzyme did not appear to be highly enriched in tumors of control livers compared to surrounding normal tissues, and its role appeared to be more of a protective nature in liver (Carlson et al., 2012). Interestingly, the only examined protein that appeared to be highly elevated in the TrxR1-containing tumor compared to normal, surrounding tissue, which might play a role in driving the tumor progression, was GPx4 (Carlson et al., 2012). GPx4 has recently been shown to have a prominent role in skin cancer development (Yang et al., 2014), but its possible function in hepatocarcinogenesis requires further study. It should be noted that tumors in control animals in this study were not sufficient in numbers or size to carry out a thorough analysis of the levels of the major antioxidants and antioxidant systems to fully assess their roles in driving the malignancy process.

8.3.3 EFFECT OF DELETION OF TRxR1 IN OTHER CANCERS

Other studies have used different approaches to examine the role of TrxR1 in cancer. For example, the targeted removal of TrxR1 in mouse embryonic fibroblasts (MEFs) that had been transformed with the oncogenes *c-myc* and *H-ras* resulted in no effect on the tumorigenic properties of the cells as assessed by *in vitro* and *in vivo* studies (Mandal et al., 2010). Also, a *c-myc*-driven B-cell lymphoma mouse model showed no effect of loss of TrxR1 on lymphoma incidence and mouse survival (Mandal et al., 2010). However, TrxR1-deficient tumors were exceedingly susceptible to glutathione depletion, demonstrating that tumor progression was dependent on both of these antioxidant systems.

8.3.4 TRxR1 ROLE IN PROMOTING/SUSTAINING CANCER VARIES IN DIFFERENT TISSUES PROVIDING POSSIBLE AVENUES FOR CANCER THERAPY

The fact that TrxR1 has major cancer preventing as well as cancer promoting characteristics raises an important question as to which of these properties has the greater effect on governing cell destiny. The above studies suggest that TrxR1 has a major role in promoting malignancies in some cell types as the cells appear to be highly dependent on this antioxidant for sustaining the disease (e.g., mouse lung cancer cells, Section 8.3.1). However, in other cells, reducing TrxR1 expression appears to enhance the levels of other antioxidants that, in turn, promote and sustain the disorder (e.g., mouse hepatocytes Section 8.3.2). The role of TrxR1 in cancer may depend on the model used, the type of cancer, and the cancer stage as well as various other factors as reported for other selenoproteins, such as GPx2 (Chapter 9), Sep15 (Chapter 10), and others (Brigelius-Flohé, 2008; Davis et al., 2012). An elucidation in understanding the switch in TrxR1's role from one of protection against malignancy in normal cells to one of promotion in some cell types and/or of promoting other antioxidants to sustain the malignancy in other cell types would certainly seem to provide important avenues in cancer prevention, treatment, and/or therapy. Therefore, examination of TrxR1 in the malignancy process is further considered below.

8.4 THE TRx SYSTEM AND HALLMARKS OF CANCER

Arnér and Holmgren examined Hanahan's and Weinberg's initial proposal of the six hallmarks of cancer (Hanahan and Weinberg, 2000) that must occur for cancer to develop in relation to the Trx system and noted that, if the Trx system played a major role in cancer development, then the Trx system must be involved in most of these hallmarks (Arnér and Holmgren, 2006). The hallmarks are (1) self-sufficiency in growth signals, (2) evading apoptosis, (3) insensitivity to antigrowth signals, (4) limitless replicative potential, (5) tissue invasion and metastasis, and (6) sustained angiogenesis. Examples of how the Trx system was involved in each of these criteria for cancer development were provided with the possible exception of tumor cell tissue invasion (Arnér and Holmgren, 2006). These investigators also noted that additional studies had been undertaken to assess the underlying molecular mechanisms of how these proposed hallmarks occur and to elucidate the relative importance of the Trx system in cancer development.

Other investigators have also mentioned that the molecular details of TrxR1 and Trx1 in cancer development are indeed poorly understood and have pursued additional avenues to elucidate possible mechanisms and the role of this antioxidant system in the hallmarks of cancer. For example, TrxR1 was targeted for downregulation in DT cells (Yoo et al., 2007), which is the oncogenic *k-ras* cell line generated from the parental NIH 3T3 cells (Noda et al., 1983). NIH 3T3 cells are a normal fibroblast cell line. The morphological changes in the resulting TrxR1-deficient cells were more characteristic of the parental cells, and those in the corresponding DT cells transfected with the control vector were unchanged. Most interestingly, however, were the observations that TrxR1-deficient DT cells had lost their self-sufficiency growth properties, manifest decreased expression of DNA polymerase α, which is important in DNA replication, and decreased progression through the S growth phase compared to DT control cells (Yoo et al., 2007). Thus, TrxR1 appeared to have a role in overriding self-sufficiency of growth signals. It should also be noted that two important features of using DT cells in this study were that TrxR1 appeared to be uniquely overexpressed in the cancer cell line compared to the parental cells, and unlike the vast majority of studies carried out with cancer cells that have examined the role of antioxidants in the malignancy process, this study had a parental, normal cell line for comparison.

Another study that examined the role of the Trx system in an additional hallmark of cancer, the evasion of apoptosis, involved TrxR1 or Trx1 deficiencies in the mouse mammary cancer cell line EMT6 (Yoo et al., 2013). The relative sensitivities of these cells to tumor necrosis factor-α (TNF-α)–induced cell death in TrxR1 or Trx1 knockdown cells showed that deficiencies in either component caused greater sensitivity to TNF-α–induced cell death, but the enhanced sensitivity was far more pronounced in Trx1-deficient cells. In both TrxR1- and Trx1-deficient cells, TNF-α–induced nuclear localization of phosphorylated ERK 1/2 correlated with increased apoptosis and was found to require PI3K activity (Yoo et al., 2013). The data provided strong evidence that the Trx system has an important role in protecting cancer cells against TNF-α–induced apoptosis and, thus, may support cancer progression.

8.5 TRXR1 AND SELENIUM IN CANCER

Even though selenium is an essential trace element in mammals and is known to have many health benefits, including being a cancer chemopreventive agent, it is toxic to cells if absorbed in high levels. In fact, this element was first described as a toxin causing numerous disorders, primarily in livestock that ate seleniferous plants (Franke et al., 1934). Interestingly, however, in regard to the relative sensitivities to the toxicity of selenium, cancer cells have been shown to be more sensitive than normal cells, and selenium sensitivities vary in different cancers (Husbeck et al., 2006; Menter et al., 2000; see Chapters 5 and 6). TrxR1 has been shown to reduce various selenium compounds and can reduce selenite to selenide, which is the redox state of selenium required for selenoprotein synthesis (Arnér and Holmgren, 2006; see Chapter 3). It has also been shown that toxic doses of selenite lead to reduced

TrxR1 protein levels despite an increase in mRNA levels (Selenius et al., 2008). These observations prompted us to assess the role of TrxR1 and Trx1 in the sensitivity of cancer cells to selenite (Tobe et al., 2012). The interplay in the deficiencies between these two proteins and selenite toxicity was examined in DT and NIH 3T3 cells. The advantage of this study, like that discussed above in the self sufficiency of growth signals, was that the malignant DT cells had a parental, normal control cell line, NIH 3T3 cells, for comparison.

The comparison of the sensitivities of selenite toxicity in DT and NIH 3T3 cells was carried out using 5 µM sodium selenite because selenite was toxic to DT cells at this level but had very little effect on the parental cell line (Tobe et al., 2012). TrxR1- and Trx1-deficient NIH 3T3 cells were sensitive to selenite treatment, and the corresponding DT cells were even more sensitive, demonstrating that the Trx system has a role in protecting both normal and malignant cells against selenium toxicity. TrxR1-deficient NIH 3T3 and DT cells, however, were more sensitive than Trx1-deficient cells, but interestingly, TrxR1-deficient DT cells were by far the most sensitive to selenite toxicity. The latter data show that the downregulation of TrxR1 in the malignant cells results in an extremely higher sensitivity to selenite than the corresponding Trx1 cells. The data, thus, suggest that TrxR1 had a new role in cancer, which is independent of reducing Trx1, the major function of this selenoenzyme in normal cells (Arnér and Holmgren, 2000; Rundlof and Arnér, 2004; Rundlof et al., 2004; Sun et al., 2001b). An examination of other parameters of the unexpectedly greater sensitivity of TrxR1-deficient DT cells to selenite revealed that exposing these cells to hydrogen peroxide, which was used as a positive control for oxidative stress, manifested no differences between these cells and control and Trx1-deficient cells, and thus, the increased sensitivity to selenite was not the result of hydrogen peroxide-induced oxidative stress. The effect of hydrogen peroxide and selenite on glutathione (GSH) metabolism in each of the DT cell lines was also examined to elucidate the role of the GSH system in TrxR1-deficient DT cells and in selenite toxicity. Enzymes involved in the GSH system were more enriched in TrxR1-deficient cells treated with either hydrogen peroxide or selenite than control cells, and Trx1-deficient cells manifested the greatest enrichment in selenite-treated cells. Furthermore, only TrxR1-deficient DT cells generated and secreted enhanced levels of GSH connected with selenite toxicity, suggesting greater dependence on the GSH system to compensate for the loss of TrxR1 in control or Trx1-deficient cells (Tobe et al., 2012). Thus, the enhanced toxicity of cancer cells to selenite coupled with TrxR1 inhibition may provide a new approach in cancer therapy.

8.5.1 IS THE SEC RESIDUE ESSENTIAL IN ALL FUNCTIONS OF TRXR1 AND COULD SELENIUM-FREE FORMS OF TRXR1 PLAY A ROLE IN CANCER?

Cysteine (Cys) has been reported to exist in place of the Sec residue in TrxR1, resulting in a TrxR1-Sec minus protein (Lu et al., 2009; Xu et al., 2010). The occurrence of Cys at the catalytic site in this selenoenzyme can arise naturally by Sec/Cys replacement, wherein Cys is biosynthesized on Sec tRNA (designated tRNA[Ser]Sec]) using the Sec biosynthetic machinery (Xu et al., 2010), and by mistranslation, wherein antibiotics can induce this amino acid and other amino acids to replace Sec (Tobe et al., 2013; see Chapter 4).

A question arises whether this selenium-free form of TrxR1 may have a cellular role. In fact, the question of whether the various forms of Sec-less TrxR1, including the amino acid penultimate truncated form, have cellular functions has been a topic of considerable interest for a number of years. Deletion of Sec or mutation of Sec to Cys in TrxR1 leads to greatly reduced activity toward many small molecule substrates as well as Trx1 (Zhong and Holmgren, 2000). Among TrxRs, this appears to be unique to TrxR1, as C-terminal mutant TrxR2 has been shown to be less dependent upon Sec for reduction of substrates (Lothrop et al., 2014a). This may be due to the presence of a "guiding bar" (amino acids 407–422) that is present in TrxR1 but not TrxR2 that restricts motion of the C-terminal tail (Lothrop et al., 2014b). TrxR1 knockout MEF cells were subjected to glucose-induced hydrogen peroxide that resulted in cell death (Peng et al., 2014). The cells could be rescued by introducing TrxR1 but not by mutant TrxR1s with Cys or serine (Ser) in place of Sec or by TrxR1 truncated at the 3′-penultimate site. Furthermore, neither TrxR1 with a Ser replacement of Sec or the truncated form of TrxR1 rescued TrxR1 knockout MEF cells. This study clearly demonstrated that neither Cys nor Ser would replace Sec at the catalytic site in TrxR1, nor would the truncated form of TrxR1, in preventing "oxidative stress-triggered cell death" (Peng et al., 2014). However, because Cys is synthesized by the Sec biosynthetic pathway under normal conditions in mammals, yielding a Cys in place of Sec TrxR1 form (Xu et al., 2010), and the major form of this enzyme in nature contains Cys at the catalytic site, it is tempting to propose that the Sec-less, Cys-containing mammalian form has other cellular functions.

Modified TrxR1s, in which the C-terminal region is truncated or in which the Sec residue is compromised by targeting with electrophiles, have been given the designation "SecTRAPs" (selenium-compromised thioredoxin reductase-derived apoptotic proteins) (Anestal and Arnér, 2003; Anestal et al., 2008). SecTRAPs lack TrxR activity but could induce rapid cell death in cancer cells through what appears to be both apoptotic and necrotic mechanisms and an extensive increase in ROS production (Anestal et al., 2008). SecTRAPs were unable to reduce Trx1 but efficiently reduced the quinone substrate juglone, indicating that selenium-compromised TrxR1 can still function as an oxidoreductase with substrates other than Trx1. Recently, it has been proposed that SecTRAPs undergo a conformational change that allows for greater accessibility of a substrate to the N-terminal redox domain of TrxR1 (Lothrop et al., 2014b). It will be important to determine what role, if any, these selenium-free forms of TrxR1 play in cancer initiation and/or progression.

8.6 CONCLUSIONS

Clearly, TrxR1 plays a major role in normal cells of preventing cancer and in malignant cells of promoting and/or sustaining cancer. Because this selenoenzyme is one of the most important antioxidants and redox regulators in mammals, it undoubtedly provides a highly significant function in both normal and cancer cells of protecting them from oxidative stress (Figure 8.1). Because a characteristic of most, if not all, cancer cells is that they suffer from oxidative stress, and because normal cells

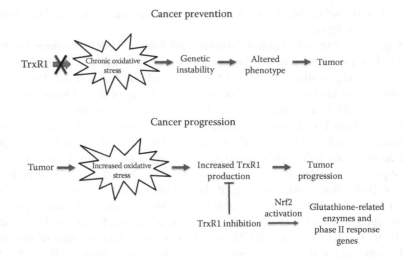

FIGURE 8.1 TrxR1 involvement in cancer prevention and progression. In cancer prevention (upper panel), TrxR1 functions as a major redox regulator, protecting normal cells from oxidative damage as shown by the blocked arrow. Changes in TrxR1 status could lead to chronic oxidative stress, resulting in genetic instability, altered cell phenotype and tumor development. In cancer progression (lower panel), upregulation of TrxR1 may enable cancer cells to overcome the burden of increased oxidative stress resulting in cell proliferation. Inhibition of TrxR1 by specific inhibitors or gene knockdown in some cell types leads to a compensatory upregulation of other redox-active enzymes dependent on Nrf2 activation (see text). This must be taken into account when pursuing TrxR1 inhibition as a means of cancer therapy.

require a balanced and strong antioxidant system, the dual personality of TrxR1 in preventing and promoting cancer may be much the same, which is, in large part, to protect them from oxidative stress. However, in most cancer cells, it appears that TrxR1 protection is required at an elevated level and that TrxR1 may take on additional roles that it did not possess in normal cells. Understanding how TrxR1 activity is elevated and its apparent novel roles in malignant cells would undoubtedly provide new and important avenues for protecting and preventing malignancy as well as avenues for cancer therapy.

In fact, one of the major focuses of elucidating TrxR1's role in cancer is to provide new means of treating this horrific family of diseases. However, an important caution in carrying out human clinical trials involving the inhibition of TrxR1 in cancer therapy is that Nrf2-regulated antioxidant systems other than the Trx system appear to be enhanced in some cancers (e.g., liver cancer) when TrxR1 activity is reduced (see Figure 8.1, lower panel). Thus, inhibiting TrxR1 may have little effect on hepatocarcinogenesis and may even promote tumor growth or make the tumors resistant against chemotherapy. In other cancers (e.g., lung cancer), inhibiting TrxR1 may provide greater potential for reducing the malignancy process. TrxR1's role in malignancy most likely depends on factors such as the model employed, the type of cancer, the stage of cancer, and the time of intervention (e.g., initiation vs. promotion stage). Although there is considerable effort in numerous laboratories to elucidate the underlying molecular mechanisms of TrxR1 in cancer, including exploring the roles of this selenoenzyme and the TrxR1 system in the hallmarks of cancer and in cancer therapy, there is still much to be done.

ACKNOWLEDGMENTS

I thank Drs. Dolph Hatfield, Vadim Gladyshev, and Petra Tsuji for their careful review of this chapter and for their many helpful and insightful comments. The work was supported by the Intramural Research Program of the National Institutes of Health, National Cancer Institute, Center for Cancer Research.

REFERENCES

Anestal, K., and E. S. Arnér. 2003. "Rapid induction of cell death by selenium-compromised thioredoxin reductase 1 but not by the fully active enzyme containing selenocysteine." *J Biol Chem* no. 278 (18):15966–72. doi: 10.1074/jbc.M210733200.

Anestal, K., S. Prast-Nielsen, N. Cenas, and E. S. Arnér. 2008. "Cell death by SecTRAPs: Thioredoxin reductase as a prooxidant killer of cells." *PLoS One* no. 3 (4):e1846. doi: 10.1371/journal.pone.0001846.

Arnér, E. S. J. 2009. "Focus on mammalian thioredoxin reductases—Important selenoproteins with versatile functions." *Biochimica Et Biophysica Acta-General Subjects* no. 1790 (6):495–526. doi: 10.1016/j.bbagen.2009.01.014.

Arnér, E. S., and A. Holmgren. 2000. "Physiological functions of thioredoxin and thioredoxin reductase." *Eur J Biochem* no. 267 (20):6102–9.

Arnér, E. S., and A. Holmgren. 2006. "The thioredoxin system in cancer." *Semin Cancer Biol* no. 16 (6):420–6. doi: 10.1016/j.semcancer.2006.10.009.

Berggren, M., A. Gallegos, J. R. Gasdaska et al. 1996. "Thioredoxin and thioredoxin reductase gene expression in human tumors and cell lines, and the effects of serum stimulation and hypoxia." *Anticancer Res* no. 16 (6B):3459–66.

Biaglow, J. E., and R. A. Miller. 2005. "The thioredoxin reductase/thioredoxin system: Novel redox targets for cancer therapy." *Cancer Biol Ther* no. 4 (1):6–13.

Bondareva, A. A., M. R. Capecchi, S. V. Iverson et al. 2007. "Effects of thioredoxin reductase-1 deletion on embryogenesis and transcriptome." *Free Radic Biol Med* no. 43 (6):911–23. doi: 10.1016/j.freeradbiomed.2007.05.026.

Brigelius-Flohé, R. 2008. "Selenium compounds and selenoproteins in cancer." *Chem Biodivers* no. 5 (3):389–95. doi: 10.1002/cbdv.200890039.

Brigelius-Flohé, R., and A. P. Kipp. 2013. "Selenium in the redox regulation of the Nrf2 and the Wnt pathway." *Methods Enzymol* no. 527:65–86. doi: 10.1016/B978-0-12-405882-8.00004-0.

Brigelius-Flohé, R., M. Muller, D. Lippmann, and A. P. Kipp. 2012. "The yin and yang of nrf2-regulated selenoproteins in carcinogenesis." *Int J Cell Biol* no. 2012:486147. doi: 10.1155/2012/486147.

Carlson, B. A., M. H. Yoo, R. Tobe et al. 2012. "Thioredoxin reductase 1 protects against chemically induced hepatocarcinogenesis via control of cellular redox homeostasis." *Carcinogenesis* no. 33 (9):1806–13. doi: 10.1093/carcin/bgs230.

Cebula, M., N. Moolla, A. Capovilla, and E. S. Arnér. 2013. "The rare TXNRD1_v3 ("v3") splice variant of human thioredoxin reductase 1 protein is targeted to membrane rafts by N-acylation and induces filopodia independently of its redox active site integrity." *J Biol Chem* no. 288 (14):10002–11. doi: 10.1074/jbc.M112.445932.

Conrad, M., C. Jakupoglu, S. G. Moreno et al. 2004. "Essential role for mitochondrial thioredoxin reductase in hematopoiesis, heart development, and heart function." *Mol Cell Biol* no. 24 (21):9414–23. doi: 10.1128/MCB.24.21.9414-9423.2004.

Cunniff, B., G. W. Snider, N. Fredette et al. 2014. "Resolution of oxidative stress by thioredoxin reductase: Cysteine versus selenocysteine." *Redox Biol* no. 2:475–84. doi: 10.1016/j.redox.2014.01.021.

Dai, B., S. Y. Yoo, G. Bartholomeusz et al. 2013. "KEAP1-dependent synthetic lethality induced by AKT and TXNRD1 inhibitors in lung cancer." *Cancer Res* no. 73 (17):5532–43. doi: 10.1158/0008-5472.CAN-13-0712.

Damdimopoulou, P. E., A. Miranda-Vizuete, E. S. Arnér, J. A. Gustafsson, and A. E. Damdimopoulos. 2009. "The human thioredoxin reductase-1 splice variant TXNRD1_v3 is an atypical inducer of cytoplasmic filaments and cell membrane filopodia." *Biochim Biophys Acta* no. 1793 (10):1588–96. doi: 10.1016/j.bbamcr.2009.07.007.

Dammeyer, P., A. E. Damdimopoulos, T. Nordman et al. 2008. "Induction of cell membrane protrusions by the N-terminal glutaredoxin domain of a rare splice variant of human thioredoxin reductase 1." *J Biol Chem* no. 283 (5):2814–21. doi: 10.1074/jbc.M708939200.

Davis, C. D., P. A. Tsuji, and J. A. Milner. 2012. "Selenoproteins and cancer prevention." *Annu Rev Nutr* no. 32:73–95. doi: 10.1146/annurev-nutr-071811-150740.

Eriksson, S. E., S. Prast-Nielsen, E. Flaberg, L. Szekely, and E. S. Arnér. 2009. "High levels of thioredoxin reductase 1 modulate drug-specific cytotoxic efficacy." *Free Radic Biol Med* no. 47 (11):1661–71. doi: 10.1016/j.freeradbiomed.2009.09.016.

Fernandes, A. P., A. Capitanio, M. Selenius et al. 2009. "Expression profiles of thioredoxin family proteins in human lung cancer tissue: Correlation with proliferation and differentiation." *Histopathology* no. 55 (3):313–20. doi: 10.1111/j.1365-2559.2009.03381.x.

Fourquet, S., R. Guerois, D. Biard, and M. B. Toledano. 2010. "Activation of NRF2 by nitrosative agents and H2O2 involves KEAP1 disulfide formation." *J Biol Chem* no. 285 (11):8463–71. doi: 10.1074/jbc.M109.051714.

Franke, K. W. 1934. "A new toxicant occurring naturally in certain samples of plant foodstuffs I. Results obtained in preliminary feeding trials." *J Nutr* no. 8:597–608.

Fujino, G., T. Noguchi, K. Takeda, and H. Ichijo. 2006. "Thioredoxin and protein kinases in redox signaling." *Semin Cancer Biol* no. 16 (6):427–35. doi: 10.1016/j.semcancer.2006.09.003.

Gladyshev, V. N., K. T. Jeang, and T. C. Stadtman. 1996. "Selenocysteine, identified as the penultimate C-terminal residue in human T-cell thioredoxin reductase, corresponds to TGA in the human placental gene." *Proc Natl Acad Sci U S A* no. 93 (12):6146–51.

Gromer, S., S. Urig, and K. Becker. 2004. "The thioredoxin system—From science to clinic." *Med Res Rev* no. 24 (1):40–89. doi: 10.1002/med.10051.

Gundimeda, U., J. E. Schiffman, S. N. Gottlieb, B. I. Roth, and R. Gopalakrishna. 2009. "Negation of the cancer-preventive actions of selenium by over-expression of protein kinase Cepsilon and selenoprotein thioredoxin reductase." *Carcinogenesis* no. 30 (9):1553–61. doi: 10.1093/carcin/bgp164.

Hanahan, D., and R. A. Weinberg. 2000. "The hallmarks of cancer." *Cell* no. 100 (1):57–70.

Hatfield, D. L., M. H. Yoo, B. A. Carlson, and V. N. Gladyshev. 2009. "Selenoproteins that function in cancer prevention and promotion." *Biochimica Et Biophysica Acta-General Subjects* no. 1790 (11):1541–5. doi: 10.1016/j.bbagen.2009.03.001.

Hedstrom, E., S. Eriksson, J. Zawacka-Pankau, E. S. Arnér, and G. Selivanova. 2009. "p53-dependent inhibition of TrxR1 contributes to the tumor-specific induction of apoptosis by RITA." *Cell Cycle* no. 8 (21):3576–83.

Holmgren, A. 2006. "Selenoproteins of the thioredoxin system." In *Selenium: Its molecular biology and role in human health (2nd Edition)*, 183–94. New York: Springer Science+Business Media, LLC.

Holmgren, A., and J. Lu. 2010. "Thioredoxin and thioredoxin reductase: Current research with special reference to human disease." *Biochem Biophys Res Commun* no. 396 (1):120–4. doi: 10.1016/j.bbrc.2010.03.083.

Husbeck, B., L. Nonn, D. M. Peehl, and S. J. Knox. 2006. "Tumor-selective killing by selenite in patient-matched pairs of normal and malignant prostate cells." *Prostate* no. 66 (2):218–25. doi: 10.1002/pros.20337.

Jakupoglu, C., G. K. Przemeck, M. Schneider et al. 2005. "Cytoplasmic thioredoxin reductase is essential for embryogenesis but dispensable for cardiac development." *Mol Cell Biol* no. 25 (5):1980–8. doi: 10.1128/MCB.25.5.1980-1988.2005.

Kensler, T. W., and N. Wakabayashi. 2010. "Nrf2: Friend or foe for chemoprevention?" *Carcinogenesis* no. 31 (1):90–9. doi: 10.1093/carcin/bgp231.

Labunskyy, V. M., D. L. Hatfield, and V. N. Gladyshev. 2014. "Selenoproteins: Molecular Pathways and Physiological Roles." *Physiol Rev* no. 94 (3):739–77. doi: 10.1152/physrev .00039.2013.

Lincoln, D. T., E. M. Ali Emadi, K. F. Tonissen, and F. M. Clarke. 2003. "The thioredoxin-thioredoxin reductase system: Over-expression in human cancer." *Anticancer Res* no. 23 (3B):2425–33.

Locy, M. L., L. K. Rogers, J. R. Prigge et al. 2012. "Thioredoxin reductase inhibition elicits Nrf2-mediated responses in clara cells: Implications for oxidant-induced lung injury." *Antioxid Redox Signal* no. 17 (10):1407–16. doi: 10.1089/ars.2011.4377.

Lothrop, A. P., E. L. Ruggles, and R. J. Hondal. 2009. "No selenium required: Reactions catalyzed by mammalian thioredoxin reductase that are independent of a selenocysteine residue." *Biochemistry* no. 48 (26):6213–23. doi: 10.1021/bi802146w.

Lothrop, A. P., G. W. Snider, S. Flemer, Jr. et al. 2014a. "Compensating for the absence of selenocysteine in high-molecular weight thioredoxin reductases: The electrophilic activation hypothesis." *Biochemistry* no. 53 (4):664–74. doi: 10.1021/bi4007258.

Lothrop, A. P., G. W. Snider, E. L. Ruggles, and R. J. Hondal. 2014b. "Why is mammalian thioredoxin reductase 1 so dependent upon the use of selenium?" *Biochemistry* no. 53 (3):554–65. doi: 10.1021/bi400651x.

Lu, J., and A. Holmgren. 2014. "The thioredoxin antioxidant system." *Free Radic Biol Med* no. 66:75–87. doi: 10.1016/j.freeradbiomed.2013.07.036.

Lu, J., E. H. Chew, and A. Holmgren. 2007. "Targeting thioredoxin reductase is a basis for cancer therapy by arsenic trioxide." *Proc Natl Acad Sci U S A* no. 104 (30):12288–93. doi: 10.1073/pnas.0701549104.

Lu, J., L. Zhong, M. E. Lonn et al. 2009. "Penultimate selenocysteine residue replaced by cysteine in thioredoxin reductase from selenium-deficient rat liver." *FASEB J* no. 23 (8):2394–402. doi: 10.1096/fj.08-127662.

Mandal, P. K., M. Schneider, P. Kolle et al. 2010. "Loss of thioredoxin reductase 1 renders tumors highly susceptible to pharmacologic glutathione deprivation." *Cancer Res* no. 70 (22):9505–14. doi: 10.1158/0008-5472.CAN-10-1509.

Marzano, C., V. Gandin, A. Folda et al. 2007. "Inhibition of thioredoxin reductase by auranofin induces apoptosis in cisplatin-resistant human ovarian cancer cells." *Free Radic Biol Med* no. 42 (6):872–81. doi: 10.1016/j.freeradbiomed.2006.12.021.

Matsui, M., M. Oshima, H. Oshima et al. 1996. "Early embryonic lethality caused by targeted disruption of the mouse thioredoxin gene." *Dev Biol* no. 178 (1):179–85. doi: 10.1006 /dbio.1996.0208.

Menter, D. G., A. L. Sabichi, and S. M. Lippman. 2000. "Selenium effects on prostate cell growth." *Cancer Epidemiol Biomarkers Prev* no. 9 (11):1171–82.

Merrill, G. F., P. Dowell, and G. D. Pearson. 1999. "The human p53 negative regulatory domain mediates inhibition of reporter gene transactivation in yeast lacking thioredoxin reductase." *Cancer Res* no. 59 (13):3175–9.

Moos, P. J., K. Edes, P. Cassidy, E. Massuda, and F. A. Fitzpatrick. 2003. "Electrophilic prostaglandins and lipid aldehydes repress redox-sensitive transcription factors p53 and hypoxia-inducible factor by impairing the selenoprotein thioredoxin reductase." *J Biol Chem* no. 278 (2):745–50. doi: 10.1074/jbc.M211134200.

Nguyen, P., R. T. Awwad, D. D. Smart, D. R. Spitz, and D. Gius. 2006. "Thioredoxin reductase as a novel molecular target for cancer therapy." *Cancer Lett* no. 236 (2):164–74. doi: 10.1016/j.canlet.2005.04.028.

Noda, M., Z. Selinger, E. M. Scolnick, and R. H. Bassin. 1983. "Flat revertants isolated from Kirsten sarcoma virus-transformed cells are resistant to the action of specific oncogenes." *Proc Natl Acad Sci U S A* no. 80 (18):5602–6.

Nonn, L., R. R. Williams, R. P. Erickson, and G. Powis. 2003. "The absence of mitochondrial thioredoxin 2 causes massive apoptosis, exencephaly, and early embryonic lethality in homozygous mice." *Mol Cell Biol* no. 23 (3):916–22.

Nordberg, J., L. Zhong, A. Holmgren, and E. S. Arnér. 1998. "Mammalian thioredoxin reductase is irreversibly inhibited by dinitrohalobenzenes by alkylation of both the redox active selenocysteine and its neighboring cysteine residue." *J Biol Chem* no. 273 (18):10835–42.

Peng, X., P. K. Mandal, V. O. Kaminskyy et al. 2014. "Sec-containing TrxR1 is essential for self-sufficiency of cells by control of glucose-derived H_2O_2." *Cell Death Dis* no. 5:e1235. doi: 10.1038/cddis.2014.209.

Powis, G., and D. L. Kirkpatrick. 2007. "Thioredoxin signaling as a target for cancer therapy." *Current Opinion in Pharmacology* no. 7 (4):392–7. doi: 10.1016/j.coph.2007.04.003.

Rollins, M. F., D. M. van der Heide, C. M. Weisend et al. 2010. "Hepatocytes lacking thioredoxin reductase 1 have normal replicative potential during development and regeneration." *J Cell Sci* no. 123 (Pt 14):2402–12. doi: 10.1242/jcs.068106.

Rundlof, A. K., and E. S. Arnér. 2004. "Regulation of the mammalian selenoprotein thioredoxin reductase 1 in relation to cellular phenotype, growth, and signaling events." *Antioxid Redox Signal* no. 6 (1):41–52. doi: 10.1089/152308604771978336.

Rundlof, A. K., M. Janard, A. Miranda-Vizuete, and E. S. Arnér. 2004. "Evidence for intriguingly complex transcription of human thioredoxin reductase 1." *Free Radic Biol Med* no. 36 (5):641–56. doi: 10.1016/j.freeradbiomed.2003.12.004.

Selenius, M., A. P. Fernandes, O. Brodin, M. Bjornstedt, and A. K. Rundlof. 2008. "Treatment of lung cancer cells with cytotoxic levels of sodium selenite: Effects on the thioredoxin system." *Biochem Pharmacol* no. 75 (11):2092–9. doi: 10.1016/j.bcp.2008.02.028.

Su, D., and V. N. Gladyshev. 2004. "Alternative splicing involving the thioredoxin reductase module in mammals: A glutaredoxin-containing thioredoxin reductase 1." *Biochemistry* no. 43 (38):12177–88. doi: 10.1021/bi048478t.

Su, D., S. V. Novoselov, Q. A. Sun et al. 2005. "Mammalian selenoprotein thioredoxin-glutathione reductase. Roles in disulfide bond formation and sperm maturation." *J Biol Chem* no. 280 (28):26491–8. doi: 10.1074/jbc.M503638200.

Sun, Q. A., L. Kirnarsky, S. Sherman, and V. N. Gladyshev. 2001a. "Selenoprotein oxidoreductase with specificity for thioredoxin and glutathione systems." *Proc Natl Acad Sci U S A* no. 98 (7):3673–8. doi: 10.1073/pnas.051454398.

Sun, Q. A., F. Zappacosta, V. M. Factor et al. 2001b. "Heterogeneity within animal thioredoxin reductases. Evidence for alternative first exon splicing." *J Biol Chem* no. 276 (5):3106–14. doi: 10.1074/jbc.M004750200.

Sun, Q. A., D. Su, S. V. Novoselov et al. 2005. "Reaction mechanism and regulation of mammalian thioredoxin/glutathione reductase." *Biochemistry* no. 44 (44):14528–37. doi: 10.1021/bi051321w.

Suvorova, E. S., O. Lucas, C. M. Weisend et al. 2009. "Cytoprotective Nrf2 pathway is induced in chronically txnrd 1-deficient hepatocytes." *PLoS One* no. 4 (7):e6158. doi: 10.1371/journal.pone.0006158.

Tobe, R., M. H. Yoo, N. Fradejas et al. 2012. "Thioredoxin reductase 1 deficiency enhances selenite toxicity in cancer cells via a thioredoxin-independent mechanism." *Biochem J* no. 445 (3):423–30. doi: 10.1042/BJ20120618.

Tobe, R., S. Naranjo-Suarez, R. A. Everley et al. 2013. "High error rates in selenocysteine insertion in mammalian cells treated with the antibiotic doxycycline, chloramphenicol, or geneticin." *J Biol Chem* no. 288 (21):14709–15. doi: 10.1074/jbc.M112.446666.

Xu, X. M., A. A. Turanov, B. A. Carlson et al. 2010. "Targeted insertion of cysteine by decoding UGA codons with mammalian selenocysteine machinery." *Proc Natl Acad Sci U S A* no. 107 (50):21430–4. doi: 10.1073/pnas.1009947107.

Yang, W. S., R. SriRamaratnam, M. E. Welsch et al. 2014. "Regulation of ferroptotic cancer cell death by GPX4." *Cell* no. 156 (1–2):317–31. doi: 10.1016/j.cell.2013.12.010.

Yoo, M. H., X. M. Xu, B. A. Carlson, V. N. Gladyshev, and D. L. Hatfield. 2006. "Thioredoxin reductase 1 deficiency reverses tumor phenotype and tumorigenicity of lung carcinoma cells." *J Biol Chem* no. 281 (19):13005–8. doi: 10.1074/jbc.C600012200.

Yoo, M. H., X. M. Xu, B. A. Carlson et al. 2007. "Targeting thioredoxin reductase 1 reduction in cancer cells inhibits self-sufficient growth and DNA replication." *PLoS One* no. 2 (10):e1112. doi: 10.1371/journal.pone.0001112.

Yoo, M. H., B. A. Carlson, V. N. Gladyshev, and D. L. Hatfield. 2013. "Abrogated thioredoxin system causes increased sensitivity to TNF-alpha-induced apoptosis via enrichment of p-ERK 1/2 in the nucleus." *PLoS One* no. 8 (9):e71427. doi: 10.1371/journal .pone.0071427.

Zhong, L., and A. Holmgren. 2000. "Essential role of selenium in the catalytic activities of mammalian thioredoxin reductase revealed by characterization of recombinant enzymes with selenocysteine mutations." *J Biol Chem* no. 275 (24):18121–8. doi: 10.1074/jbc .M000690200.

Zhong, L., E. S. Arnér, J. Ljung, F. Aslund, and A. Holmgren. 1998. "Rat and calf thioredoxin reductase are homologous to glutathione reductase with a carboxyl-terminal elongation containing a conserved catalytically active penultimate selenocysteine residue." *J Biol Chem* no. 273 (15):8581–91.

9 Glutathione Peroxidase 2

Anna P. Kipp and Mike F. Müller

CONTENTS

9.1 Introduction 189
9.2 Dual Role of Factors Regulating GPx2 Expression.. 190
    9.2.1 Nrf2.. 190
    9.2.2 Wnt... 192
9.3 Tumor Stage–Specific Functions of GPx2 193
    9.3.1 Redox Regulation 193
    9.3.2 Modulation of Proliferation and Apoptosis 194
    9.3.3 Inhibition of Inflammation 196
9.4 Conclusion 197
Acknowledgments... 197
References... 197

9.1 INTRODUCTION

Glutathione peroxidase 2 (GPx2) was first described in 1993 (Chu et al., 1993) and belongs to a family of hydroperoxide-reducing thiolperoxidases. The whole family consists of eight isoforms of which five are known to be selenoproteins in humans (For an overview, see Brigelius-Flohé and Maiorino, 2013). GPx2 has a high sequence homology with GPx1. Both use glutathione (GSH) to reduce H_2O_2 and soluble organic hydroperoxides, but GPx2 is supposed to have a stronger preference for organic hydroperoxides (Chu et al., 1993). So far, no purified GPx2 has been available to convincingly identify its specific substrates or kinetic constants. However, unlike the ubiquitously expressed GPx1, GPx2 is epithelium-specific with a predominant expression in epithelial cells lining the whole gastrointestinal tract as well as in breast, bladder, and lung epithelium (Cho et al., 2002; Chu et al., 1993, 1999; Florian et al., 2001; Komatsu et al., 2001). In the intestinal epithelium, GPx2 is not uniformly expressed but rather shows a gradient with high expression at the base of intestinal crypts and lower expression in villi or the upper crypt region of the small intestine or colon, respectively (Florian et al., 2001). In addition, GPx2 is upregulated under conditions of inflammation, for example, during colitis (Florian et al., 2010), chronic hepatitis (Suzuki et al., 2013), or in inflamed lung in response to cigarette smoke (Singh et al., 2006). A high GPx2 expression is also detectable in various tumors of epithelial origin. Colorectal adenomas (Mörk et al., 2000), colorectal carcinomas (Murawaki et al., 2008), and moderately differentiated colorectal tumors (Banning et al., 2008a; Florian et al., 2001) show enhanced GPx2 expression in comparison to nontransformed tissue. Also extraintestinal tumors of epithelial origin, such as squamous cell carcinomas of the skin (Serewko et al., 2002; Walshe et al., 2007), adenocarcinomas of the lung (Wönckhaus et al., 2006), and ductal mammary carcinomas (Naiki-Ito et al., 2007) display an increased GPx2 expression. Furthermore, GPx2 is highly expressed in hepatocellular carcinomas (Suzuki et al., 2013) and prostate cancer (Naiki et al., 2014). Based on this specific expression pattern, which is not observed for any other GPx, it was supposed that GPx2 plays an important role during inflammation and cancer development. Under selenium shortage, GPx2 expression is kept relatively constant while GPx1 levels decline rapidly. This indicates that GPx2 has a higher priority over other selenoproteins, a phenomenon that is also described as the hierarchy of selenoproteins (Brigelius-Flohé, 1999; Brigelius-Flohé et al., 2001; Wingler et al., 1999).

The concept of U-shaped effects of different factors during carcinogenesis is based on the observation that there is an optimal concentration of a protein, transcription factor, or substance in suppressing tumor development, which means that deficiency as well as excess can have similar effects. Regarding selenium, it has been discussed for a long time that a low selenium supply is associated with an increased risk of developing different types of cancer (Medina and Shepherd, 1981; Shamberger and Frost, 1969) but that also overexpression of certain selenoproteins can support cancer cell growth. According to a recent study in rats, a low selenium status supports carcinogenesis in early stages of tumor development whereas selenium supplementation had no effect in later stages (Yang et al., 2012). Thus, before selenium can be claimed to have anticarcinogenic properties, it is of essential interest to elucidate the functions of selenoproteins mediating the observed effects of selenium. At least three selenoproteins, namely thioredoxin reductase 1 (Chapter 8), selenoprotein 15 (Chapter 10), and GPx2, are known to unify tumor-promoting as well as tumor-suppressive functions in themselves. The dual role of GPx2 during carcinogenesis has been extensively analyzed using knockout mice as well as cancer cell lines with GPx2 knockdown (kd) or overexpression, most often focusing on colorectal cancer. The actual role of GPx2 in tumorigenesis appears to depend on the tumor stage and the involvement of inflammation. We will, thus, summarize and discuss recent findings about the regulation and function of GPx2 during different stages of tumor development and during inflammation.

9.2   DUAL ROLE OF FACTORS REGULATING GPx2 EXPRESSION

9.2.1   Nrf2

The transcription factor nuclear factor E2-related factor 2 (Nrf2) is known to regulate cell survival and defense against endogenous and exogenous stress. Under basal conditions, Nrf2 is kept in the cytosol by binding to Keap1. In this complex, Nrf2 becomes ubiquitinated and degraded via the proteasome. There are multiple ways to activate the pathway of which the thiol modification of Keap1 has been most extensively studied. Modified Keap1 continues to bind Nrf2, which however is not degraded anymore. Thus, due to space limit at Keap1, newly synthesized Nrf2 can enter the nucleus and activate transcription (see overview in recent reviews Jaramillo and Zhang, 2013; Leinonen et al., 2014). Its target genes include detoxifying enzymes, such as glutathione S-transferase (GST); enzymes related to glutathione synthesis, such as glutamate-cysteine ligase (GCL) (McWalter et al., 2004); and antioxidant enzymes, such as peroxiredoxin 1 (Ishii et al., 2004) or sulfiredoxin 1 (Singh et al., 2009). In addition, GPx2 is a target gene of the Nrf2 pathway, which was first discovered in cancer cell lines, and by identifying a functional antioxidant responsive element (ARE) within the human GPx2 promoter (Banning et al., 2005). *In vivo*, GPx2 expression is downregulated in Nrf2 KO mice (Singh et al., 2006) and is upregulated in response to stimulation with Nrf2 activators, such as sulforaphane (Krehl et al., 2012; Lippmann et al., 2014).

The Nrf2 pathway was, for a long time, believed to mainly protect from tumorigenesis. This was obvious from its ability to inactivate potential carcinogens via induction of phase II enzymes. In addition, the induction of antioxidant enzymes appears to be protective in terms of reducing potential DNA-harming substances (see also Section 9.3.1). This idea is also supported by studies using Nrf2 KO mice. These mice are more susceptible to DNA adduct or tumor formation after exposure to different chemicals (Cho et al., 2006; Enomoto et al., 2001; Ramos-Gomez et al., 2001). For example, they develop a more severe inflammation in response to the colitis-inducing substance dextran sodium sulphate (DSS) (Khor et al., 2006; Osburn et al., 2007) and accordingly develop more tumors when colitis is combined with the carcinogen azoxymethane (AOM) (Khor et al., 2008). Recently, it was shown that Nrf2 ablation accelerated tumor growth in a mouse model of Braf^{V600E}-induced lung cancer (Strohecker et al., 2013). In a urethane-induced multistep model of lung carcinogenesis, Nrf2 KO mice exhibited an increase in tumor foci 8 weeks after urethane administration. However, after 16 weeks, tumors in Nrf2 KO mice showed less advanced malignancy (Satoh et al., 2013). This clearly indicates that Nrf2 activation appears to be protective during initiation and early tumor stages but also has the potential to support tumor cells when they reach later stages. In line with that, an shRNA-mediated downregulation of Nrf2 in non-small cell lung cancer cells inhibited tumor growth and resulted in a greater sensitivity to chemotherapeutic drugs (Singh et al., 2008). In many cancer types, Nrf2 is constitutively active, which either results from inactivating mutations in Keap1 or from gain-of-function mutations in Nrf2. Accordingly, Nrf2 target genes, such as NADPH quinone oxidoreductase 1 (NQO1) and GST isoenzymes, as well as GPx2 are also most often upregulated in tumors (see overview in Leinonen et al., 2014).

Genomic characterization of common human cancers identified the mutation of the Nrf2 encoding gene *NFE2L2* as one of the 140 driver mutations in tumorigenesis (Vogelstein et al., 2013). In non-small cell lung carcinoma patients, Nrf2 expression was associated with poor survival (Solis et al., 2010), which was also the case in patients with pancreatic adenocarcinomas (Soini et al., 2014). Keap1 KO mice clearly confirm that Nrf2 activation is supportive for tumor cells. However, Nrf2 overactivation was not sufficient to induce spontaneous tumor initiation (Taguchi et al., 2010). Thus, the Nrf2 system is an adaptive defense system, which can protect against tumorigenic substances but at the same time is used by established tumor cells to ensure their survival.

9.2.2 Wnt

The majority of colorectal tumors are caused by mutations in key components of the Wnt signaling pathway. In 85% of all cases of sporadic intestinal tumorigenesis, the gene for adenomatous polyposis coli (APC) is functionally inactivated (Moser et al., 1990; Nishisho et al., 1991). APC acts as a scaffold for a protein complex that is essential to phosphorylate and degrade β-catenin under unstimulated conditions. Once APC is mutated, β-catenin is constitutively activated. If APC is not mutated, β-catenin itself most often harbors an activating mutation (Sparks et al., 1998). Thus, the Wnt pathway obviously is oncogenic, but according to recent studies, this only holds true for some types of tumors. In contrast to, for example, colorectal and liver cancer, it is associated with tumor suppression in melanomas (see overview in Serio, 2014). The reason for this difference appears to be that Wnt rather enhances oncogenic Ras in colorectal cancer while it negatively regulates Ras in melanomas (Jeong et al., 2012). It is, therefore, discussed that the downstream effects of Wnt activity are largely determined by the context in which signaling occurs. For example, an inflammatory environment aggravates loss of functional Wnt signaling, indicated by the transformation of differentiated intestinal epithelial cells into tumor-initiating cells after activation of both β-catenin and NF-κB (Schwitalla et al., 2013).

In the healthy intestinal epithelium, the Wnt pathway is responsible for building a niche at the crypt base that enables stem cells to continuously divide and to maintain pluripotency (Pinto et al., 2003; Scoville et al., 2008). However, Wnt gradients not only maintain proliferation, but also directly affect differentiation into the different lineages. Ring and colleagues proposed that these antagonistic effects are regulated by binding of different transcriptional coactivators to β-catenin. According to this hypothesis, CBP/β-catenin-mediated transcription drives proliferation whereas p300/β-catenin-mediated transcription initiates differentiation (Ring et al., 2014). As discussed for the Nrf2 pathway, also the Wnt pathway appears to have tumor-supportive as well as antagonizing effects, which highly depend on the tumor environment and also on the involvement of inflammation.

The highest expression of GPx2 can be detected at proliferative crypt bases of the intestinal epithelium (Chu et al., 1993; Florian et al., 2001), an area also characterized by active Wnt signaling. Based on the colocalization of GPx2 and active Wnt at the crypt base, a causal relationship was hypothesized. The human GPx2 promoter was subsequently shown to harbor a functional TCF binding element (Kipp et al., 2007). In cells isolated from the crypt base of inducible β-catenin KO mice, β-catenin deletion led to a loss of GPx2 expression (Kipp et al., 2012), indicating that Wnt signaling ensures the specific localization of GPx2 in the intestinal epithelium. Moreover, ΔNp63γ, another transcription factor that is specific for basal, proliferating epithelial cells and often overexpressed in tumors, can induce GPx2 expression (Yan and Chen, 2006). On the contrary, differentiation of embryonic stem cells (Saretzki et al., 2008) and CaCo2 cells (Speckmann et al., 2011) led to a reduced mRNA expression of GPx2. This specific localization and transcriptional regulation of GPx2 suggests a specific role in proliferating cells (see Section 9.3.2). Further evidence for this comes from the enhanced intestinal GPx2 expression not only in regenerating intestinal epithelium after irradiation and in growth periods of young mice (Chu et al., 2004b; Esworthy et al., 2000), but also during carcinogenesis.

9.3 TUMOR STAGE–SPECIFIC FUNCTIONS OF GPx2

9.3.1 REDOX REGULATION

Being a member of the GPx family, GPx2 is considered to have a relevant hydroperoxide-reducing activity. Based on this assumption, GPx2 was supposed to protect the gastrointestinal epithelium from hydroperoxide absorption and consecutive damage (Chu et al., 1993; Wingler et al., 2000). Hydroperoxides belong to a multitude of substances that are able to damage DNA and thus have the potential to initiate tumorigenesis. According to the stem cell theory, intestinal stem cells that reside at the base of each intestinal crypt and give rise to progenitor cells that further differentiate into the mature epithelial cell types are also the cells of origin for tumor stem cells (Barker et al., 2009). In order to inhibit intestinal tumor development, it is, thus, of high importance to protect the stem cells from oxidative damage.

Apart from their damaging properties, hydroperoxides act as second messengers in many signaling pathways, mostly by modifying redox-sensitive cysteines of key proteins (see overview in Brigelius-Flohé and Flohé, 2011). One of those redox-sensitive pathways is the Wnt pathway, which is essential for the homeostasis of the intestinal epithelium (see Section 9.2.2). Oxidation of critical cysteines of nucleoredoxin (NRX) was shown to enhance β-catenin-mediated transactivation (Funato et al., 2006). Under reducing conditions, NRX binds and sequesters disheveled (Dvl), which is important for the stabilization of β-catenin. It has been shown recently that NADPH oxidase 1 (NOX1) provides H_2O_2 for modulating the Wnt pathway (Kajla et al., 2012). NOX1 is the isoform that is specifically expressed in intestinal epithelial cells, and it is further upregulated in colon cancer and during colitis (Szanto et al., 2005). In line with that, the intestinal epithelium of NOX1 KO mice is characterized by disturbed Wnt and Notch signaling and by changes in cell fate of progenitor cells (Coant et al., 2010). Thus, it is of great importance to tightly regulate the redox balance of intestinal stem cells in order to maintain a balanced level of H_2O_2 to ensure proliferation but to avoid DNA damage or hyperproliferation. GPx2 obviously is one of the enzymes involved in this regulation. However, removal of hydroperoxides, for example by GPx2, helps especially damaged cells to evade apoptosis (Song et al., 2009). Supporting proliferation, which, on the one hand, is necessary for the continuous renewal of the intestinal epithelium, will, on the other hand, enhance tumor growth. Hydroperoxides, and thus also hydroperoxide-reducing enzymes, obviously have a dual role during carcinogenesis, depending on whether cells are just recently damaged or already established tumor cells. Many recent reviews are focusing on this topic in more detail, which is beyond the scope of this chapter (Brigelius-Flohé and Kipp, 2009; Gupta et al., 2012; Holmström and Finkel, 2014; Kardeh et al., 2014).

Different cell lines with shRNA-induced GPx2 knockdown were characterized for their redox status and stress resistance. Prostate cancer cell lines, such as PC3 or PCai1 (a castration-resistant rat prostate cancer cell line), show higher dichlorodihydrofluorescein (DCF) fluorescence upon downregulation of GPx2 expression (Naiki et al., 2014). Lower GPx2 expression is supposed to result in H_2O_2 accumulation and in higher production of OH-radicals. Overexpression of GPx2 decreased the sensitivity of MCF7 cells to H_2O_2-induced cell death, which was p53-dependent (Yan et al., 2006). However, in liver and colon-derived cancer cell lines with shRNA-mediated knockdown of GPx2, no difference in the basal redox status was observed (Banning et al., 2008a, Suzuki et al., 2013). Obviously, other enzymes have compensated for the loss of GPx2, which appears to depend on the cellular system used (see Section 8.3.2 in Chapter 8 on TrxR1).

Similarly, GPx2 KO mice do not suffer from oxidative stress under basal conditions either, which can be explained by a compensatory upregulation of GPx1 and a higher total GPx activity (Florian et al., 2010). Interestingly, the upregulation of GPx1 specifically occurs at the crypt base where GPx1 is only faintly expressed in WT mice and where GPx2 normally resides. At first glance, GPx2 KO mice behaved phenotypically normal in the unstressed state (Esworthy et al., 2000), especially in comparison to GPx1/GPx2 double knockout (DKO) mice. Deletion of both isoenzymes resulted in the spontaneous development of ileocolitis and cancer. These effects could be alleviated by a single allele of GPx2, but not of GPx1, indicating that GPx2 might have a crucial role in preventing colitis and colitis-associated colorectal cancer (Chu et al., 2004a; Esworthy et al., 2001). At that point, the major source of H_2O_2 that needs to be reduced by GPx1 and 2 was still unknown, but it was supposed to be produced by NOX1. Therefore, Chu and Esworthy decided to generate triple knockout mice (TKO) for GPx1, GPx2, and NOX1. As hypothesized, loss of NOX1 rescued the phenotype of the DKO mice (Esworthy et al., 2014). Taken together, this suggests that NOX1-derived H_2O_2 promotes inflammation and tumor development in the DKO intestine.

In the meantime, single GPx2 KO mice were further characterized and were found to have an intestinal phenotype, which includes aberrant apoptotic cell death at the crypt base (Florian et al., 2010), an enlargement of the proliferation zone, and higher numbers of intraepithelial inflammatory cells (Müller et al., 2013). As these effects were detectable, although total GPx activity was increased, the hypotheses were established that GPx2 might have a specific substrate that cannot be eliminated by other enzymes or that GPx2 might have functions independent of its hydroperoxide-scavenging activity.

9.3.2 MODULATION OF PROLIFERATION AND APOPTOSIS

As already mentioned, GPx2 KO mice were characterized by high numbers of apoptotic cells at the crypt base while mitotic cell numbers were unchanged in this area (Florian et al., 2010). However, the length of colonic crypts was enhanced together with the PCNA-positive proliferation zone (Müller et al., 2013). Under physiological conditions, GPx2 thus may inhibit apoptosis as well as proliferation. To test whether this also holds true during tumor development, different cancer cell lines were analyzed. In contrast to the healthy intestine, a reduced proliferation capacity associated with a cyclin B1-dependent G_2/M arrest was observed in prostate cancer cells transfected with shGPx. In these cells, apoptotic signaling was not modified by GPx2 expression (Naiki et al., 2014). An shRNA-mediated knockdown of GPx2 in hepatocarcinoma cells (HCC) did not only reduce cellular proliferation rate, but also the metastatic potential as indicated by a suppressed invasion and migration, which was accompanied by reduction of matrix metalloprotease 9 (MMP-9) secretion (Suzuki et al., 2013). The knockdown of GPx2 expression also caused the inhibition of proliferation in both rat and human mammary carcinoma cell lines with wild-type p53, but not with an inactivating mutation of p53 (Naiki-Ito et al., 2007). Because human HT-29 colorectal cancer cells also have a mutant p53, it is not surprising that the GPx2 kd did not affect proliferation rates in those cells (Banning et al., 2008b). It, however, resulted in decreased colony formation in soft agar and smaller tumors in a xenograft model, which clearly shows that certain growth properties were reduced upon loss of GPx2 (Banning et al., 2008b). In line with that, also prostate cancer cells with GPx2 knockdown produced smaller tumors when injected into nude mice (Naiki et al., 2014). It is hypothesized, but not yet proven, that GPx2 kd cells are more prone to anoikis, a type of apoptosis induced by loss of contact to the basement membrane or cell culture dish. Without such contact, GPx2 kd cells would be more susceptible to death and thus develop smaller tumors when injected into nude mice. The mechanism via which GPx2 affects apoptosis needs to be evaluated in the future. For GPx4, it was convincingly shown that cells with GPx4 KO are more prone to 12/15-lipoxygenase-mediated lipid peroxidation, which induces AIF-mediated cell death (Seiler et al., 2008).

About 65%–85% of colorectal cancers occur spontaneously (Munkholm, 2003), and spontaneous colorectal carcinogenesis can be mimicked in mice by repeated application of the colon-specific carcinogen AOM. In this model, GPx2 KO mice developed fewer tumors and preneoplastic lesions than WT mice (Figure 9.1a), which fits with the results obtained using xenografts. It is well established that within 6–8 hours after AOM application, massive apoptotic cell death of intestinal epithelial cells takes place. Accordingly, AOM-induced apoptotic cells in the midcrypt region

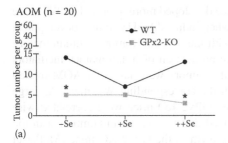 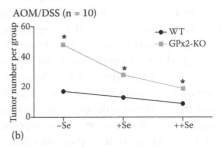

FIGURE 9.1 Tumor numbers in the colon of WT and GPx2 KO mice treated with AOM or AOM/DSS. Tumor development was induced by injecting the carcinogen azoxymethane (AOM) (Müller et al., 2013) six times (a) or by a combined application of AOM and dextran sulfate sodium (DSS) (Krehl et al., 2012) (b). The selenium status was modified by feeding an adequate (+Se, 0.15 mg/kg diet), supplemented (++Se, 0.6 mg/kg diet) or a suboptimal (−Se, 0.05–0.08 mg/kg diet) selenium diet. Graphs indicate the total tumor number per group. *$p <$ 0.05 versus respective WT groups using the U test. Graphs were recalculated from own data of the quoted references.

were higher in GPx2 KO mice than in WT mice (Müller et al., 2013). Thus, higher levels of basal and AOM-induced apoptosis in GPx2-KO mice are supposed to efficiently eliminate AOM-initiated cells resulting in lower numbers of damaged cells that have the potential to develop into tumors. The hypothesis was supported by increased GPx2 expression in dysplastic crypts as well as in mucin-depleted foci. In both cases, GPx2 expression was only detectable in highly proliferating ki-67-positive cells (Müller et al., 2013). These results indicate that, depending on the model, GPx2 can support proliferation and growth of tumor cells. In addition, GPx2 appears to help damaged cells to evade apoptotic cell death and thus to support tumor development.

9.3.3 INHIBITION OF INFLAMMATION

Mice with a deletion of both GPx1 and GPx2 are characterized by a spontaneous microbiota-related ileocolitis (Esworthy et al., 2001). Also unstressed GPx2 KO mice display signs of inflammation indicated by higher numbers of F4/80-positive myeloid cells within the epithelium (Müller et al., 2013). In a mouse model of allergic airway inflammation, GPx2 KO mice developed a more severe inflammation and stronger infiltration of immune cells into the lung epithelium after sensitization and challenge with antigen (Dittrich et al., 2010). It is well established that inflammation is a risk factor for tumor development because inflammatory cells help to create a tumor-promoting environment (Coussens and Werb, 2002; Reuter et al., 2010). Colitis-associated colon carcinogenesis can be mimicked in mice by a single AOM application followed by induction of colitis by DSS. In this AOM/DSS model, GPx2 acted protectively by inhibition of inflammation as obvious from a more severe inflammation in GPx2 KO mice (Krehl et al., 2012). In contrast to the lower tumor numbers in the AOM model (Figure 9.1a) but in line with the higher severity of colitis, GPx2 KO mice displayed a higher tumor load in the AOM/DSS model (Figure 9.1b) (Krehl et al., 2012). Also in the tumor model for UV-induced squamous cell carcinoma (SCC) of the skin, GPx2 KO mice developed more tumors than WT mice (Walshe et al., 2007). In SCC, especially when induced by UV, tumor development is accompanied by an immune response with massive infiltration of inflammatory cells, and consequently, local and systemic production of inflammatory mediators (Sluyter and Halliday, 2000). Unexpectedly, tumors were larger in AOM-treated selenium-adequate GPx2 KO mice than in the corresponding WT mice. As large tumors in selenium-deficient and -adequate GPx2 KO mice were correlated with markers of inflammation, it was hypothesized that a subacute inflammation in the AOM model might have promoted tumor growth in the GPx2 KO mice (Müller et al., 2013).

Interestingly, HT-29 cells with a GPx2 knockdown showed a massive basal and IL-1β induced upregulation of cyclooxygenase 2 (COX-2) in comparison to control cells. In immunohistochemical analysis of human colitis patients, GPx2 and COX-2 were colocalized in areas of inflammatory cell infiltration (Banning et al., 2008a). Based on that, GPx2 was discussed to inhibit COX-2 activity by lowering the local hydroperoxide tone, which is needed for proper COX-2 activity. Whether or not higher COX-2 activity is causing the inflammatory phenotype of GPx2 KO mice needs to be further evaluated. Additional analysis of the HT-29 shGPx2 model revealed that GPx2 kd cells showed a higher invasiveness and ability to migrate (Banning et al., 2008b), indicating that GPx2 appears to inhibit two important characteristics of tumor cells. Both, enhanced invasion and migration were COX-2 dependent because GPx2 kd effects were lost after treatment with the COX-2 inhibitor Celecoxib. However, in hepatocarcinoma cells with GPx2 kd, invasion and migration were suppressed (Suzuki et al., 2013). Obviously, a GPx2 kd without COX-2 upregulation affects invasion and migration. Based on these results, it can be concluded that GPx2 inhibits inflammation differently. Thus, it inhibits tumor development when inflammation is causally involved.

9.4 CONCLUSION

GPx2 clearly has dual properties during cancer development. Both protective and detrimental effects depend on the model, the tumor stage, the tumor environment, and on the involvement of inflammation. Further studies are needed to decipher the effect of upregulating selenoproteins such as GPx2 more precisely and, thus, to evaluate the effect of selenium supplementation during tumor development. Based on the current knowledge, it is rather difficult to give general recommendations.

ACKNOWLEDGMENTS

Work providing the basis for the review was funded by the German Research Council (DFG), grant Br778/8-1 and KI1590/2-1.

REFERENCES

Banning, A., S. Deubel, D. Kluth, Z. Zhou, and R. Brigelius-Flohé. 2005. "The GI-GPx gene is a target for Nrf2." *Mol Cell Biol* no. 25 (12):4914–23.

Banning, A., S. Florian, S. Deubel et al. 2008a. "GPx2 counteracts PGE2 production by dampening COX-2 and mPGES-1 expression in human colon cancer cells." *Antioxid Redox Signal* no. 10 (9):1491–500.

Banning, A., A. Kipp, S. Schmitmeier et al. 2008b. "Glutathione peroxidase 2 inhibits cyclooxygenase-2-mediated migration and invasion of HT-29 adenocarcinoma cells but supports their growth as tumors in nude mice." *Cancer Res* no. 68 (23):9746–53.

Barker, N., R. A. Ridgway, J. H. van Es et al. 2009. "Crypt stem cells as the cells-of-origin of intestinal cancer." *Nature* no. 457 (7229):608–11.

Brigelius-Flohé, R. 1999. "Tissue-specific functions of individual glutathione peroxidases." *Free Radic Biol Med* no. 27 (9–10):951–65.

Brigelius-Flohé, R., and A. Kipp. 2009. "Glutathione peroxidases in different stages of carcinogenesis." *Biochim Biophys Acta* no. 1790 (11):1555–68.

Brigelius-Flohé, R., and L. Flohé. 2011. "Basic principles and emerging concepts in the redox control of transcription factors." *Antioxid Redox Signal* no. 15 (8):2335–81.

Brigelius-Flohé, R., and M. Maiorino. 2013. "Glutathione peroxidases." *Biochim Biophys Acta* no. 1830 (5):3289–303. doi: 10.1016/j.bbagen.2012.11.020.

Brigelius-Flohé, R., M. Maiorino, F. Ursini, and L. Flohé. 2001. "Selenium: An antioxidant?" In *Handbook of Antioxidants*, 2nd Edition. Cadenas, E., Parker, L., eds. (New York, Basel, Marcel Dekker) pp. 633–64.

Cho, H. Y., A. E. Jedlicka, S. P. Reddy et al. 2002. "Role of NRF2 in protection against hyperoxic lung injury in mice." *Am J Respir Cell Mol Biol* no. 26 (2):175–82.

Cho, H. Y., S. P. Reddy, and S. R. Kleeberger. 2006. "Nrf2 defends the lung from oxidative stress." *Antioxid Redox Signal* no. 8 (1–2):76–87. doi: 10.1089/ars.2006.8.76.

Chu, F.F., J. H. Doroshow, and R. S. Esworthy. 1993. "Expression, characterization, and tissue distribution of a new cellular selenium-dependent glutathione peroxidase, GSHPx-GI." *J Biol Chem* no. 268 (4):2571–6.

Chu, F.F., R.S. Esworthy, L. Lee, and S. Wilczynski. 1999. "Retinoic acid induces Gpx2 gene expression in MCF-7 human breast cancer cells." *J Nutr* no. 129 (10):1846-54.

Chu, F.F., R. S. Esworthy, P. G. Chu et al. 2004a. "Bacteria-induced intestinal cancer in mice with disrupted Gpx1 and Gpx2 genes." *Cancer Res* no. 64 (3):962–8.

Chu, F.F., R.S. Esworthy, and J.H. Doroshow. 2004b. "Role of Se-dependent glutathione peroxidases in gastrointestinal inflammation and cancer." *Free Radic Biol Med* no. 36 (12):1481-95.

Coant, N., S. Ben Mkaddem, E. Pedruzzi et al. 2010. "NADPH oxidase 1 modulates WNT and NOTCH1 signaling to control the fate of proliferative progenitor cells in the colon." *Mol Cell Biol* no. 30 (11):2636–50.

Coussens, L. M., and Z. Werb. 2002. "Inflammation and cancer." *Nature* no. 420 (6917):860–7.

Dittrich, A. M., H. A. Meyer, M. Krokowski et al. 2010. "Glutathione peroxidase-2 protects from allergen-induced airway inflammation in mice." *Eur Respir J* no. 35 (5):1148–54.

Enomoto, A., K. Itoh, E. Nagayoshi et al. 2001. "High sensitivity of Nrf2 knockout mice to acetaminophen hepatotoxicity associated with decreased expression of ARE-regulated drug metabolizing enzymes and antioxidant genes." *Toxicol Sci* no. 59 (1):169–77.

Esworthy, R. S., J. R. Mann, M. Sam, and F. F. Chu. 2000. "Low glutathione peroxidase activity in Gpx1 knockout mice protects jejunum crypts from gamma-irradiation damage." *Am J Physiol Gastrointest Liver Physiol* no. 279 (2):G426–36.

Esworthy, R. S., R. Aranda, M. G. Martin et al. 2001. "Mice with combined disruption of Gpx1 and Gpx2 genes have colitis." *Am J Physiol Gastrointest Liver Physiol* no. 281 (3):G848–G55.

Esworthy, R. S., B. W. Kim, J. Chow et al. 2014. "Nox1 causes ileocolitis in mice deficient in glutathione peroxidase-1 and -2." *Free Radic Biol Med* no. 68:315–25. doi: 10.1016/j.freeradbiomed.2013.12.018.

Florian, S., K. Wingler, K. Schmehl et al. 2001. "Cellular and subcellular localization of gastrointestinal glutathione peroxidase in normal and malignant human intestinal tissue." *Free Radic Res* no. 35 (6):655–63.

Florian, S., S. Krehl, M. Loewinger et al. 2010. "Loss of GPx2 increases apoptosis, mitosis, and GPx1 expression in the intestine of mice." *Free Radic Biol Med* no. 49 (11):1694–702.

Funato, Y., T. Michiue, M. Asashima, and H. Miki. 2006. "The thioredoxin-related redox-regulating protein nucleoredoxin inhibits Wnt-beta-catenin signalling through dishevelled." *Nat Cell Biol* no. 8 (5):501–8.

Gupta, S. C., D. Hevia, S. Patchva et al. 2012. "Upsides and downsides of reactive oxygen species for cancer: The roles of reactive oxygen species in tumorigenesis, prevention, and therapy." *Antioxid Redox Signal* no. 16 (11):1295–322. doi: 10.1089/ars.2011.4414.

Holmström, K. M., and T. Finkel. 2014. "Cellular mechanisms and physiological consequences of redox-dependent signalling." *Nat Rev Mol Cell Biol* no. 15 (6):411–21. doi: 10.1038/nrm3801.

Ishii, T., K. Itoh, E. Ruiz et al. 2004. "Role of Nrf2 in the regulation of CD36 and stress protein expression in murine macrophages: Activation by oxidatively modified LDL and 4-hydroxynonenal." *Circ Res* no. 94 (5):609–16. doi: 10.1161/01.RES.0000119171.44657.45.

Jaramillo, M. C., and D. D. Zhang. 2013. "The emerging role of the Nrf2-Keap1 signaling pathway in cancer." *Genes Dev* no. 27 (20):2179–91. doi: 10.1101/gad.225680.113.

Jeong, W. J., J. Yoon, J. C. Park et al. 2012. "Ras stabilization through aberrant activation of Wnt/beta-catenin signaling promotes intestinal tumorigenesis." *Sci Signal* no. 5 (219):ra30. doi: 10.1126/scisignal.2002242.

Kajla, S., A. S. Mondol, A. Nagasawa et al. 2012. "A crucial role for Nox 1 in redox-dependent regulation of Wnt-beta-catenin signaling." *FASEB J* no. 26 (5):2049–59. doi: 10.1096 /fj.11-196360.

Kardeh, S., S. Ashkani-Esfahani, and A. M. Alizadeh. 2014. "Paradoxical action of reactive oxygen species in creation and therapy of cancer." *Eur J Pharmacol* no. 735:150–68. doi: 10.1016/j.ejphar.2014.04.023.

Khor, T. O., M. T. Huang, K. H. Kwon et al. 2006. "Nrf2-deficient mice have an increased susceptibility to dextran sulfate sodium-induced colitis." *Cancer Res* no. 66 (24):11580–4.

Khor, T. O., M. T. Huang, A. Prawan et al. 2008. "Increased susceptibility of Nrf2 knockout mice to colitis-associated colorectal cancer." *Cancer Prev Res* no. 1 (3):187–91.

Kipp, A., A. Banning, and R. Brigelius-Flohé. 2007. "Activation of the glutathione peroxidase 2 (GPx2) promoter by beta-catenin." *Biol Chem* no. 388 (10):1027–33.

Kipp, A. P., M. F. Müller, E. M. Göken, S. Deubel, and R. Brigelius-Flohé. 2012. "The selenoproteins GPx2, TrxR2 and TrxR3 are regulated by Wnt signalling in the intestinal epithelium." *Biochim Biophys Acta* no. 1820 (10):1588–96. doi: 10.1016/j.bbagen.2012.05.016.

Komatsu, H., I. Okayasu, H. Mitomi et al. 2001. "Immunohistochemical detection of human gastrointestinal glutathione peroxidase in normal tissues and cultured cells with novel mouse monoclonal antibodies." *J Histochem Cytochem* no. 49 (6):759–66.

Krehl, S., M. Loewinger, S. Florian et al. 2012. "Glutathione peroxidase-2 and selenium decreased inflammation and tumors in a mouse model of inflammation-associated carcinogenesis whereas sulforaphane effects differed with selenium supply." *Carcinogenesis* no. 33 (3):620–8.

Leinonen, H. M., E. Kansanen, P. Polonen, M. Heinaniemi, and A. L. Levonen. 2014. "Role of the Keap1-Nrf2 pathway in cancer." *Adv Cancer Res* no. 122:281–320. doi: 10.1016 /B978-0-12-420117-0.00008-6.

Lippmann, D., C. Lehmann, S. Florian et al. 2014. "Glucosinolates from pak choi and broccoli induce enzymes and inhibit inflammation and colon cancer differently." *Food Funct* no. 5 (6):1073–81. doi: 10.1039/c3fo60676g.

McWalter, G. K., L. G. Higgins, L. I. McLellan et al. 2004. "Transcription factor Nrf2 is essential for induction of NAD(P)H:quinone oxidoreductase 1, glutathione S-transferases, and glutamate cysteine ligase by broccoli seeds and isothiocyanates." *J Nutr* no. 134 (12 Suppl):3499S–506S.

Medina, D., and F. Shepherd. 1981. "Selenium-mediated inhibition of 7,12-dimethylbenz[a] anthracene-induced mouse mammary tumorigenesis." *Carcinogenesis* no. 2 (5):451–5.

Mörk, H., O. H. al-Taie, K. Bahr et al. 2000. "Inverse mRNA expression of the selenocysteine-containing proteins GI-GPx and SeP in colorectal adenomas compared with adjacent normal mucosa." *Nutr Cancer* no. 37 (1):108–16.

Moser, A. R., H. C. Pitot, and W. F. Dove. 1990. "A dominant mutation that predisposes to multiple intestinal neoplasia in the mouse." *Science* no. 247 (4940):322–4.

Müller, M. F., S. Florian, S. Pommer et al. 2013. "Deletion of glutathione peroxidase-2 inhibits azoxymethane-induced colon cancer development." *PLoS One* no. 8 (8):e72055. doi: 10.1371/journal.pone.0072055.

Munkholm, P. 2003. "Review article: The incidence and prevalence of colorectal cancer in inflammatory bowel disease." *Aliment Pharmacol Ther* no. 18 Suppl 2:1–5.

Murawaki, Y., H. Tsuchiya, T. Kanbe et al. 2008. "Aberrant expression of selenoproteins in the progression of colorectal cancer." *Cancer Lett* no. 259 (2):218–30.

Naiki, T., A. Naiki-Ito, M. Asamoto et al. 2014. "GPX2 overexpression is involved in cell proliferation and prognosis of castration-resistant prostate cancer." *Carcinogenesis* no. 35 (9):1962–7. doi: 10.1093/carcin/bgu048.

Naiki-Ito, A., M. Asamoto, N. Hokaiwado et al. 2007. "Gpx2 is an overexpressed gene in rat breast cancers induced by three different chemical carcinogens." *Cancer Res* no. 67 (23):11353–8.

Nishisho, I., Y. Nakamura, Y. Miyoshi et al. 1991. "Mutations of chromosome 5q21 genes in FAP and colorectal cancer patients." *Science* no. 253 (5020):665–9.

Osburn, W. O., B. Karim, P. M. Dolan et al. 2007. "Increased colonic inflammatory injury and formation of aberrant crypt foci in Nrf2-deficient mice upon dextran sulfate treatment." *Int J Cancer* no. 121 (9):1883–91.

Pinto, D., A. Gregorieff, H. Begthel, and H. Clevers. 2003. "Canonical Wnt signals are essential for homeostasis of the intestinal epithelium." *Genes Dev* no. 17 (14):1709–13.

Ramos-Gomez, M., M. K. Kwak, P. M. Dolan et al. 2001. "Sensitivity to carcinogenesis is increased and chemoprotective efficacy of enzyme inducers is lost in nrf2 transcription factor-deficient mice." *Proc Natl Acad Sci U S A* no. 98 (6):3410–5.

Reuter, S., S. C. Gupta, M. M. Chaturvedi, and B. B. Aggarwal. 2010. "Oxidative stress, inflammation, and cancer: How are they linked?" *Free Radic Biol Med* no. 49 (11):1603–16. doi: 10.1016/j.freeradbiomed.2010.09.006.

Ring, A., Y. M. Kim, and M. Kahn. 2014. "Wnt/catenin signaling in adult stem cell physiology and disease." *Stem Cell Rev* no. 10 (4):512–25. doi: 10.1007/s12015-014-9515-2.

Saretzki, G., T. Walter, S. Atkinson et al. 2008. "Downregulation of multiple stress defense mechanisms during differentiation of human embryonic stem cells." *Stem Cells* no. 26 (2):455–64. doi: 10.1634/stemcells.2007-0628.

Satoh, H., T. Moriguchi, J. Takai, M. Ebina, and M. Yamamoto. 2013. "Nrf2 prevents initiation but accelerates progression through the Kras signaling pathway during lung carcinogenesis." *Cancer Res* no. 73 (13):4158–68. doi: 10.1158/0008-5472.CAN-12-4499.

Schwitalla, S., A. A. Fingerle, P. Cammareri et al. 2013. "Intestinal tumorigenesis initiated by dedifferentiation and acquisition of stem-cell-like properties." *Cell* no. 152 (1–2):25–38. doi: 10.1016/j.cell.2012.12.012.

Scoville, D. H., T. Sato, X. C. He, and L. Li. 2008. "Current view: Intestinal stem cells and signaling." *Gastroenterology* no. 134 (3):849–64.

Seiler, A., M. Schneider, H. Forster et al. 2008. "Glutathione peroxidase 4 senses and translates oxidative stress into 12/15-lipoxygenase dependent- and AIF-mediated cell death." *Cell Metab* no. 8 (3):237–48.

Serewko, M. M., C. Popa, A. L. Dahler et al. 2002. "Alterations in gene expression and activity during squamous cell carcinoma development." *Cancer Res* no. 62 (13):3759–65.

Serio, R. N. 2014. "Wnt of the two horizons: Putting stem cell self-renewal and cell fate determination into context." *Stem Cells Dev* no. 23 (17):1975–90. doi: 10.1089/scd.2014.0055.

Shamberger, R. J., and D. V. Frost. 1969. "Possible protective effect of selenium against human cancer." *Can Med Assoc J* no. 100 (14):682.

Singh, A., T. Rangasamy, R. K. Thimmulappa et al. 2006. "Glutathione peroxidase 2, the major cigarette smoke-inducible isoform of GPX in lungs, is regulated by Nrf2." *Am J Respir Cell Mol Biol* no. 35 (6):639–50.

Singh, A., S. Boldin-Adamsky, R. K. Thimmulappa et al. 2008. "RNAi-mediated silencing of nuclear factor erythroid-2-related factor 2 gene expression in non-small cell lung cancer inhibits tumor growth and increases efficacy of chemotherapy." *Cancer Res* no. 68 (19):7975–84. doi: 10.1158/0008-5472.CAN-08-1401.

Singh, A., G. Ling, A. N. Suhasini et al. 2009. "Nrf2-dependent sulfiredoxin-1 expression protects against cigarette smoke-induced oxidative stress in lungs." *Free Radic Biol Med* no. 46 (3):376–86. doi: 10.1016/j.freeradbiomed.2008.10.026.

Sluyter, R., and G. M. Halliday. 2000. "Enhanced tumor growth in UV-irradiated skin is associated with an influx of inflammatory cells into the epidermis." *Carcinogenesis* no. 21 (10):1801–7.

Soini, Y., M. Eskelinen, P. Juvonen et al. 2014. "Nuclear Nrf2 expression is related to a poor survival in pancreatic adenocarcinoma." *Pathol Res Pract* no. 210 (1):35–9. doi: 10.1016/j.prp.2013.10.001.

Solis, L. M., C. Behrens, W. Dong et al. 2010. "Nrf2 and Keap1 abnormalities in non-small cell lung carcinoma and association with clinicopathologic features." *Clin Cancer Res* no. 16 (14):3743–53. doi: 10.1158/1078-0432.CCR-09-3352.

Song, I. S., S. U. Kim, N. S. Oh et al. 2009. "Peroxiredoxin I contributes to TRAIL resistance through suppression of redox-sensitive caspase activation in human hepatoma cells." *Carcinogenesis* no. 30 (7):1106–14. doi: 10.1093/carcin/bgp104.

Sparks, A. B., P. J. Morin, B. Vogelstein, and K. W. Kinzler. 1998. "Mutational analysis of the APC/beta-catenin/Tcf pathway in colorectal cancer." *Cancer Res* no. 58 (6):1130–4.

Speckmann, B., H. J. Bidmon, A. Pinto et al. 2011. "Induction of glutathione peroxidase 4 expression during enterocytic cell differentiation." *J Biol Chem* no. 286 (12):10764–72.

Strohecker, A. M., J. Y. Guo, G. Karsli-Uzunbas et al. 2013. "Autophagy sustains mitochondrial glutamine metabolism and growth of BrafV600E-driven lung tumors." *Cancer Discov* no. 3 (11):1272–85. doi: 10.1158/2159-8290.CD-13-0397.

Suzuki, S., P. Pitchakarn, K. Ogawa et al. 2013. "Expression of glutathione peroxidase 2 is associated with not only early hepatocarcinogenesis but also late stage metastasis." *Toxicology* no. 311 (3):115–23. doi: 10.1016/j.tox.2013.07.005.

Szanto, I., L. Rubbia-Brandt, P. Kiss et al. 2005. "Expression of NOX1, a superoxide-generating NADPH oxidase, in colon cancer and inflammatory bowel disease." *J Pathol* no. 207 (2):164–76. doi: 10.1002/path.1824.

Taguchi, K., J. M. Maher, T. Suzuki et al. 2010. "Genetic analysis of cytoprotective functions supported by graded expression of Keap1." *Mol Cell Biol* no. 30 (12):3016–26.

Vogelstein, B., N. Papadopoulos, V. E. Velculescu et al. 2013. "Cancer genome landscapes." *Science* no. 339 (6127):1546–58. doi: 10.1126/science.1235122.

Walshe, J., M. M. Serewko-Auret, N. Teakle et al. 2007. "Inactivation of glutathione peroxidase activity contributes to UV-induced squamous cell carcinoma formation." *Cancer Res* no. 67 (10):4751–8.

Wingler, K., M. Böcher, L. Flohé, H. Kollmus, and R. Brigelius-Flohé. 1999. "mRNA stability and selenocysteine insertion sequence efficiency rank gastrointestinal glutathione peroxidase high in the hierarchy of selenoproteins." *Eur J Biochem* no. 259 (1–2):149–57.

Wingler, K., C. Müller, K. Schmehl, S. Florian, and R. Brigelius-Flohé. 2000. "Gastrointestinal glutathione peroxidase prevents transport of lipid hydroperoxides in CaCo-2 cells." *Gastroenterology* no. 119 (2):420–30.

Wönckhaus, M., L. Klein-Hitpass, U. Grepmeier et al. 2006. "Smoking and cancer-related gene expression in bronchial epithelium and non-small-cell lung cancers." *J Pathol* no. 210 (2):192–204.

Yan, W., and X. Chen. 2006. "GPX2, a direct target of p63, inhibits oxidative stress-induced apoptosis in a p53-dependent manner." *J Biol Chem* no. 281 (12):7856–62.

Yang, H., X. Jia, X. Chen, C. S. Yang, and N. Li. 2012. "Time-selective chemoprevention of vitamin E and selenium on esophageal carcinogenesis in rats: The possible role of nuclear factor kappaB signaling pathway." *Int J Cancer* no. 131 (7):1517–27. doi: 10.1002/ijc.27423.

10 The 15-kDa Selenoprotein (SEP15)

Petra A. Tsuji and Cindy D. Davis

CONTENTS

10.1 Introduction 203
10.2 SEP15 and Cancer 205
    10.2.1 Gene Expression 205
    10.2.2 Genetic Polymorphisms 206
    10.2.3 Mechanistic Studies 208
10.3 Comparison with Other Selenoproteins 210
10.4 Combined Knockdown of Both SEP15 and TrxR1 210
10.5 Conclusions 211
Acknowledgments 211
References 211

10.1 INTRODUCTION

Selenium has been implicated in cancer prevention in numerous animal models and some human studies, as previously reviewed (Davis et al., 2012). As discussed in Chapter 3, selenium is incorporated into proteins in the form of selenocysteine, which is typically present in the active centers of selenium-containing proteins, termed selenoproteins. At least three of these selenoproteins are known to play dual roles in the cancer process, including both cancer prevention and promotion. This chapter describes one such protein, designated the 15-kDa Selenoprotein (SEP15). In 1998, SEP15 was purified from human T-cells, sequenced, and molecularly characterized (Gladyshev et al., 1998). In subsequent studies, the highest levels of *Sep15* expression were reported to occur in human, rat, and mouse tissues from liver, kidney, testes, thyroid, and prostate (Hu et al., 2001). *Sep15* orthologs have been described in many animals, including mice, cows, dogs, chickens (Liu et al., 2014), zebra fish (Thisse et al., 2003), the Western clawed frog, the nematode *Caenorhabditis elegans*, fruit fly (Ferguson et al., 2006), and others. The endoplasmic reticulum-residing fish SEP15-like protein (Fep15) likely evolved by gene duplication and is homologous to mammalian SEP15, but its function is not known (Labunskyy et al., 2014). A selenoprotein of similar molecular weight, Selenoprotein M (SelM), is also found in many species and is thought to be a distant SEP15 homolog; however, its function and implications to human health remain to be elucidated, too (Ferguson et al., 2006).

In humans, *SEP15* is located on chromosome locus 1p31 (Gladyshev et al., 1998), a locus implicated in tumor suppression and commonly deleted or mutated in human cancer (Nasr et al., 2004). In the mammalian *Sep15* gene, the selenocysteine-encoding UGA codon is found at position 93 in exon 3 with the SECIS element present, as occurs in other eukaryotes, in the gene's 3'-untranslated region. The existence of two alternatively spliced transcript variants has been proposed, which would encode for two distinct *Sep15* isoforms: the isoform 1 precursor consisting of five exons translated into 165 amino acids and the isoform 2 precursor translated into 124 amino acids. This second isoform, with a shorter and distinct C-terminus, presumably would lack an exon in the 3' coding region, resulting in a translational frame shift (Jablonska et al., 2011). However, because this would disrupt the gene's C-terminal thioredoxin domain, it is possible that this second isoform may not result in a functional protein, and this finding remains to be elucidated.

Studies have suggested that SEP15 may be involved in the quality control of post-translational protein folding. During the initial purification from a human T-cell line, SEP15 was isolated in the denatured state; however, when native SEP15 was isolated from rat prostate and mouse liver, it copurified with a protein of 160-240 kDa. SEP15 was found to form a 1:1 complex with the UDP-glucose:glycoprotein glucosyltransferase (UGGT), an enzyme that is responsible for quality control in the endoplasmic reticulum, primarily recognizing structural maturation of N-glycosylated proteins (Gladyshev et al., 1998). Recently, the effect of a mutant SEP15, in which the selenocysteine was replaced with cysteine, on the catalytic activity of UGGT has been reported. When coexpressed with SEP15, the glucosyltransferase activities of UGGT1 and UGGT2 were enhanced about twofold and fourfold, respectively (Takeda et al., 2014). Thus, considering that selenoproteins with known functions are oxidoreductases, one potential role for SEP15 is its involvement with disulfide bond formation in the endoplasmic reticulum, contributing to the quality of protein folding or secretion of specific glycoproteins (Ferguson et al., 2006).

Sequence analysis of the SEP15 family members identified two distinct domains within this selenoprotein: a C-terminal thioredoxin-like domain that contains the redox active Cys-x-Sec motif and a novel cysteine-rich domain located in the N-terminal part of the protein. Thus, like other functionally described selenoproteins, SEP15 belongs to the group of thiol-oxidoreductase-like selenoproteins with a thioredoxin-like fold in which selenocysteine is the functional residue. SEP15 resides in the endoplasmic reticulum, and SEP15 expression was found to be upregulated in response to tunicamycin and brefeldin A-induced adaptive endoplasmic reticulum stress. Furthermore, dithiothreitol-induced endoplasmic reticulum stress was found to induce rapid proteasomal degradation of SEP15 (Labunskyy et al., 2009). Thus, it seems likely that SEP15 is indeed regulated by certain forms of endoplasmic reticulum stress. Recent *in vivo* studies in mice demonstrated a role for SEP15 in the visual function, wherein SEP15 was found to be enriched during lens development (Kasaikina et al., 2011). Additionally, mice lacking SEP15 showed increased levels of both lipid and protein oxidation, which may have caused or contributed to the observed lens cataract formation. However, the ratio of reduced to oxidized glutathione and levels of thiols in lenses were not different from controls, and thus it remains to be determined whether general oxidative stress contributed to cataract formation in SEP15 knockout mice or if other pathways were involved (Kasaikina et al., 2011, 2012). For example, the role of SEP15 in quality control of protein folding would also be supported by the evidence of SEP15-dependent nuclear cataract development in developing eyes in mice lacking SEP15 systemically (Kasaikina et al., 2011) as many lens proteins are expressed only once during embryonic development, and early mis-folded lens proteins would thus be maintained into adulthood.

10.2   SEP15 AND CANCER

In addition to its chromosomal location, there are several lines of evidence that have implicated a role for SEP15 in both cancer prevention and tumor promotion. These include (1) mRNA and protein expression patterns in normal and malignant cells; (2) identification of polymorphic sites that regulate SEP15 expression, differentially respond to nutritional selenium supplementation, and are linked to cancer risk; and (3) mechanistic studies using either targeted gene downregulation or knockout technology.

10.2.1   Gene Expression

The expression of SEP15 in malignant versus normal tissue appears to be tissue-/ organ-specific and may also be dependent on the cancer stage. *SEP15* mRNA is expressed in high amounts in normal liver and prostate, but is reduced in the cor-responding malignant tissues (Kumaraswamy et al., 2000). Similarly, low expression of *SEP15* mRNA has been observed in malignant tissues of lung, breast, prostate, and liver (Wright and Diamond, 2012). *SEP15* mRNA was also shown to be down-regulated by twofold or greater in approximately 60% (14 of 23) of malignant mesothelioma-derived cell lines compared with the average of three normal meso-thelial cell samples and in at least five malignant mesothelioma tumor specimens whereas *SEP15* expression was not detectable in normal mesothelial cells (Apostolou et al., 2004). Likewise, in an epidemiological study conducted in Poland, *SEP15* mRNA expression was found to be generally decreased in leukocytes of male patients with bladder cancer compared to controls. However, among these bladder cancer patients, the mRNA expression of *SEP15* gradually increased with tumor grade (Reszka et al., 2009), and patients with high-grade, poorly differentiated tumors expressed a greater amount of *SEP15* than those with lower-grade tumors. Regardless of tumor grade, no association or correlation was found between *SEP15* mRNA levels in leukocytes, dietary selenium status, and selenoprotein genetic poly-morphisms in these bladder cancer patients. Interestingly, *SEP15* mRNA expres-sion was found to be significantly higher in smokers than nonsmokers in the control group (Reszka et al., 2009), but reasons for such elevated levels of *SEP15* remain unclear. In contrast to these findings with bladder cancer patients, mRNA expression of *SEP15* in malignant tissue was unchanged in non-small cell lung cancer (Gresner et al., 2009) and in lung and laryngeal cancer patients in Poland (Jaworska et al., 2013). Although decreased expression of *SEP15* has been observed in liver, blad-der, prostate, and breast cancer compared to nonmalignant tissues, the expression of *SEP15* in colon cancer is less clear. Analysis of the National Cancer Institute's

Developmental Therapeutics Program database (http://dtp.nci.nih.gov/mtweb/index .jsp), which screens 60 human tumor cell lines for molecular targets, demonstrated a higher *SEP15* expression in human colon cancer cell lines than in other cancer cell lines as well as compared to other selenoproteins. In combination, these data suggest that *SEP15* expression is downregulated in many but not all tumors.

10.2.2 GENETIC POLYMORPHISMS

Two polymorphisms, which are located in the 3′UTR in the *SEP15* gene (rs5859 and rs5845) are in linkage disequilibrium and appear to have functional consequences. These polymorphisms cause a C-T substitution at position 811 and a G-A substitution at position 1125, respectively. They were found to decrease the efficiency of the SECIS element at higher concentrations of selenium via measurement of reporter gene activity (Hu et al., 2001). However, it remains to be elucidated how SEP15 protein expression correlates to these polymorphisms. Additionally, malignant meso-thelium cells lines with the A allele substituted at position 1125 were less responsive to the growth-inhibiting and apoptotic effects of added selenium than those with the homozygous G genotype (Apostolou et al., 2004). Only the two haplotypes, with either a T at position 811 and an A at 1125, or a C at 811 and a G at 1125, respectively, were observed in 700 DNA samples (Hu et al., 2001). Moreover, the TA haplotype homozygosity is relatively rare, occurring in only 7% of Caucasians, albeit in 31% of African-Americans (Hu et al., 2001).

A number of epidemiological studies have suggested a relationship between these polymorphisms and cancer risk. Investigating breast cancer, Hu et al. found a statistically significant difference in *SEP15* allelic distribution for rs5859 between African-Americans with breast cancer (DNA obtained from tumors, n = 60) and cancer-free controls (DNA obtained from lymphocytes, n = 490), and a loss of heterozygosity in particular was evident (Hu et al., 2001). Studies conducted by Nasr et al. also suggested a possible role of *SEP15* in breast cancer development among African-American women (Nasr et al., 2004). The authors used four highly polymorphic microsatellite markers on the chromosome 1 region that included the *SEP15* gene to assess the difference in heterozygosity index at studied loci between DNA obtained from breast cancer tumors (n = 61) and DNA obtained from lymphocytes of cancer-free individuals (n = 50). A significant reduction of heterozygosity was found for the locus that was most tightly linked to *SEP15* (Nasr et al., 2004).

In a Polish population, an association between *SEP15* genotype and small cell or non-small cell lung cancer as a function of selenium status revealed that there was an increased risk of lung cancer in those individuals with an AA genotype at position 1125 and low selenium status. Intriguingly, among those with higher selenium status (above 80 ng/mL), the opposite was true, indicating that GA and GG genotypes showed increased risk but AA genotype benefitted from higher serum selenium concentrations (Jablonska et al., 2008; Figure 10.1). The results from this study suggested that those with the AA genotype benefitted most from increasing dietary selenium whereas in those with the GG or GA genotype, higher selenium status may increase the risk for lung cancer. Therefore, genetic *SEP15* polymorphisms

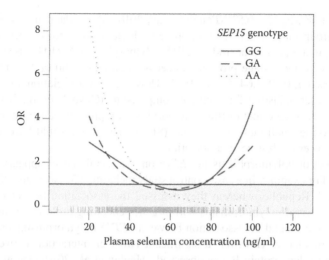

FIGURE 10.1 Joint effect of plasma selenium concentration and *SEP15* 1125G/A polymorphism for odds ratio (OR) of lung cancer development. Test for trends in *SEP15* genotypes: p = 0.038 for GG, p = 0.035 for GA, p = 0.030 or AA. Test for trend differences: AA versus GG: p = 0.049, AA versus GA: p = 0.025. These results suggested that for those with the GG or GA genotype, higher selenium status may increase the risk for lung cancer whereas those with the *SEP15* AA genotype benefitted most from increasing dietary selenium. (Reprinted with the kind permission from *Eur. J. Nutr.* and W. Wasowicz.)

can interact with selenium nutrient status to influence cancer risk, and both must be considered when evaluating individual susceptibility.

Polymorphisms in *SEP15* may also be related to prostate cancer stage or cancer mortality. In a study conducted in New Zealand with 275 cases of malignant disease, 135 cases of benign prostate disease, and 441 healthy controls, the rs5845 minor T allele was associated with a higher risk for benign prostate disease compared to controls (odds ratio [OR] = 1.98 [95%CI 1.40–2.81], p = 0.0001), but a lower risk of developing malignant disease compared to benign disease (OR = 0.62 [95% CI 0.44–0.87], p = 0.005) (Karunasinghe et al., 2012). However, this association was not apparent in other studies investigating the relationship between polymorphisms in *SEP15* and prostate cancer. In a nested case-control study involving 1286 cases and 1267 controls to assess the relationship between 5 *SEP15* polymorphisms (rs5859, rs479341, rs561104, rs527281, rs1407131), selenium status, prostate cancer risk and survival, the authors did not find any association between *SEP15* polymorphisms and prostate cancer risk (Penney et al., 2010). However, they did observe that three variants (rs479341, rs1407131, and rs561104) were significantly associated with prostate cancer mortality, and one of the polymorphisms (rs561104) was shown to modify prostate cancer survival in association with selenium status. An inverse association between plasma selenium and prostate cancer mortality was only apparent among individuals carrying at least one copy of the high risk or minor T allele (hazard ratio [HR] = 0.82 [95% CI, 0.67–1.00], P_{trend} = 0.05) whereas the beneficial association was completely absent in those homozygous for the major allele (HR = 1.08 [95%

CI = 0.85–1.38], P_{trend} = 0.53) (Penney et al., 2010). These data suggest that only a small subgroup may benefit from improved selenium status for cancer prevention. In contrast, in a more recent study, *SEP15* polymorphism rs561104 was significantly associated with local stage prostate cancer with an odds ratio of 1.28 for GG versus AA (95% CI, 0.99–1.64, P_{trend} = 0.03). However, this association did not remain significant after adjusting for multiple comparisons (Geybels et al., 2013). In this study, the lack of an effect of this polymorphism on prostate cancer mortality may be linked to the small number of events (81 deaths), which would have limited the statistical power to observe an association.

The effect of polymorphisms in *SEP15* on colorectal cancer appears much less consistent. For example, in a case-control study including 832 cases and 705 controls in the Czech Republic, wherein they assessed the association between colorectal cancer and several polymorphisms within genes encoding proteins, including selenoproteins, they found no association between *SEP15* polymorphisms and colorectal cancer incidence. However, a significant two-loci interaction between *SEP15* (rs5859) and selenoprotein P was observed (Méplan et al., 2010) (see also Chapter 15). Similarly, in a study in a Korean patient population, no association was found for *SEP15* and colon cancer. However, the minor alleles for either rs5845 (GG-GA) or rs5859 (CC-CT) in the *SEP15* gene were associated with a borderline increased risk of rectal cancer in men (OR 2.47; [95% CI 0.99–6.19], and OR 2.51; [95% CI 1.00–6.28], p = 0.052 and p = 0.049, respectively) but not women (Sutherland et al., 2010). More recently, no association between polymorphisms in *SEP15* and colon or rectal cancer in the United States was found, but an association between a different *SEP15* polymorphism (rs9433110) and survival after diagnosis with colon and rectal cancer was observed (OR = 1.45; [95% CI 1.07–1.45], p = 0.004) (Slattery et al., 2012).

In summary, the literature investigating the relationship between polymorphisms in *SEP15* and cancer incidence and mortality is rather inconsistent. Part of these inconsistencies may relate to the specific ethnic groups being studied, variations in selenium status, the type of cancer and the specific tissue affected, and/or the cancer stage. Additionally, *SEP15* polymorphisms may have different effects on cancer incidence versus mortality thus affecting subpopulations differently. Clearly, a better understanding of the functional consequences of many of the recently identified polymorphisms is needed. For further SNPs and related functions in *SEP15*, see Chapter 15.

10.2.3 MECHANISTIC STUDIES

Both targeted gene downregulation through RNA interference in mammalian cells in culture and development of systemic *Sep15* knockout mice have demonstrated tissue specificity in the biological effects associated with decreased expression of this gene. For example, decreased expression of *Sep15* has been associated with protection against colon cancer but not lung cancer in mammalian cells as further described below.

Mouse colon carcinoma CT26 cells that were stably transfected with shRNA constructs targeting *Sep15* displayed decreased growth abilities both under anchorage-dependent and anchorage-independent conditions (Irons et al., 2010) compared to empty plasmid-transfected control cells. Moreover, whereas most (14/15) mice sub-cutaneously injected with control cells developed solid tumors at the injection site, few (3/30) mice injected with *Sep15*-deficient cells developed any tumors. The ability to form pulmonary metastases had similar results. Mice intravenously injected with the empty plasmid-transfected control cells developed >250 lung metastases/mouse as expected. In contrast, mice injected with cells with downregulation of *Sep15* only developed 7.8 ± 5.4 metastases. This influence on cellular proliferation was likely mediated, at least in part, by the observed G_2/M cell cycle arrest in mouse colon can-cer cells that lacked SEP15 expression (Irons et al., 2010). Additionally, functional analyses in these mouse colon cancer cell lines using microarrays suggested a pos-sible link between SEP15 expression and interferon-γ–regulated proteins.

Targeted downregulation of *SEP15* in two human colon cancer cell lines, HCT116 and HT29, also resulted in a reversal of the cancer phenotype (Tsuji et al., 2012c). Both cell lines, upon downregulation of SEP15, demonstrated significantly reduced growth abilities under both anchorage-dependent and -independent conditions com-pared to their respective controls. This effect also appeared to be mediated via cell cycle arrest. However, in contrast to CT26 cells, which exhibited a G_2/M cell cycle arrest, the human colon cancer cells had a delayed release from G_0/G_1 phase. These observations indicate that a loss of *Sep15* expression leads to alterations in the cell cycle in both mouse and human colorectal cells whereas the specific molecular mechanism remained unclear.

Knockout of *Sep15* in mice has also been shown to influence colon cancer sus-ceptibility *in vivo* (Tsuji et al., 2012a). Similarly to what has been described in gluta-thione peroxidase 2 (GPx2) knockout mice (Müller et al., 2013), the total number of carcinogen-induced aberrant crypt foci per colon and the number of aberrant crypts per focus were significantly lower in mice systemically lacking SEP15 expression compared to wild-type and heterozygous littermate controls. Because the develop-ment of aberrant crypt foci in the intestinal lining serves as a surrogate biomarker for colon cancer risk in humans, these results indicate that a lack of SEP15 expres-sion may be protective against colon tumor initiation *in vivo*. This effect appears to likely be mediated through altered inflammatory pathways. Molecular analysis indicated that guanylate binding protein-1 (Gbp-1) mRNA and protein expression were strongly upregulated in SEP15 knockout mice. Gbp-1, which is expressed in response to interferon-γ, is considered to be an activation marker during inflam-matory diseases, and upregulation of *GBP-1* in humans has been associated with a highly significant, increased five-year survival rate in colorectal cancer patients (Naschberger et al., 2008). In agreement with these studies, an almost twofold higher protein expression of interferon-γ in plasma of untreated SEP15 knockout mice was observed (Tsuji et al., 2012a).

Whereas targeted downregulation of *Sep15* is protective against tumorigenesis in colon cancer cells, it does not appear to be protective against various forms of lung can-cer. *Sep15*-targeted downregulation in mouse Lewis lung carcinoma (LLC1) cells did not affect anchorage-dependent or -independent cell growth (Irons et al., 2010). Even though the basal expression of SEP15 is lower in LLC1 cells than in the colon cancer cell lines described above, the inability to affect lung carcinoma cells by removal of SEP15 is more likely due to SEP15 not being as important for tumorigenesis in lung tissue. Furthermore, when a malignant mesothelioma cell line, Meso 6, was transiently transfected with small interfering RNA against *SEP15*, the cells exhibited increased cell proliferation following exposure to pharmacologic concentrations of selenium (1–25 μmol/L) compared with Meso 6 cells that possessed the wild-type *SEP15* (Apostolou et al., 2004). These data suggest that lung cancer/mesothelioma cells responded differently to RNAi-induced targeted downregulation of SEP15 than mouse and human colon cancer cells, suggesting that there might be strong tissue specificity in the response.

10.3   COMPARISON WITH OTHER SELENOPROTEINS

In addition to SEP15, there are at least two other selenoproteins, glutathione peroxidase 2 (GPx2) and thioredoxin reductase 1 (TrxR1), that appear to be important regulators of the cellular redox state and are likely involved in both cancer prevention and tumor promotion (Hatfield et al., 2014). GPx2, which is preferentially expressed in the healthy intestine, plays a major role in the prevention of inflammation and may contribute to the detoxification of carcinogens because it is regulated by the transcription factor Nrf2/Keap1system (Brigelius-Flohé et al., 2012). GPx2 is also upregulated in malignancies of epithelial origin and appears to help cells escape apoptosis. Similar to knockout of SEP15 in mice, systemic knockout of GPx2 lacks an obvious phenotype. Interestingly, in animal models, either lack of SEP15 (Tsuji et al., 2012a) or GPx2 (Müller et al., 2013) appeared to be protective against azoxymethane-induced preneoplastic lesions whereas lack of GPx2 predisposed mice to cancer cell formation when exposed to UV (Walshe et al., 2007). This is further explained in detail in Chapter 9.

Thioredoxin and TrxR1 function in thiol-disulfide exchange reactions and are crucial for the control of the intracellular redox environment, cellular growth, and defense against oxidative stress as described in Chapter 8. By maintaining protein thiols in the reduced state, thioredoxin and TrxR1 help to maintain a reduced cellular environment and support the activity of transcription factors, including the p53 tumor suppressor, hypoxia-inducible factor-1 (HIF-1) and NFκB (Yoo et al., 2007). The observations that TrxR1 activates the p53 tumor suppressor, manifests other tumor suppressor activities, and that it is specifically targeted by carcinogenic electrophilic compounds, suggest that TrxR1 plays a major role in cancer prevention. However, because TrxR1 also regulates redox homeostasis in malignant cells, this selenoprotein can also serve a role as a cancer-promoting protein once malignancy is initiated. Removal of TrxR1 using RNA interference technology in both a murine lung cancer cell line (Yoo et al., 2006) and in a mouse cancer cell line driven by oncogenic *k-ras* (Yoo et al., 2007), inhibited both anchorage-dependent and -independent cell growth as well as tumor progression and metastasis when injected into mice. Therefore, much like what we described for SEP15 in this chapter, TrxR1 as well as GPx2 is thought to harbor a dual personality in that it plays important roles in both maintaining and preventing cancer as detailed in Chapters 8 and 9.

10.4   COMBINED KNOCKDOWN OF BOTH SEP15 AND TrxR1

Because targeted downregulation of either *Sep15* or *TrxR1* has been shown to reverse the malignant phenotype of cancer cells, it seemed likely that combined knockdown may have additive or synergistic effects. As expected, shRNA knockdown of either *Sep15* or *Txnrd* in CT26 cells inhibited these cells' anchorage-dependent and -independent cell growth, primary tumor growth, and lung metastasis. Surprisingly, a double knockdown of both *Sep15* and *Txnrd* in mouse colon cancer cells reversed the effects seen in single knockdowns (Tsuji et al., 2012b). Cell cycle analysis revealed that the significant G_2/M cell cycle arrest in shSEP15 cells was no longer observed in cells lacking both SEP15 and TrxR1. Microarray analysis demonstrated that the expression of *Gbp-1*, the top upregulated gene in shSEP15 cells, was not affected in colon cancer cells lacking both SEP15 and TrxR1. The results suggest that the independent down-regulation of either SEP15 or TrxR1, but not the double-knockdown, resulted in changes in the cancer phenotype of mouse colon carcinoma cells. The mechanisms behind this unexpected interplay between SEP15 and TrxR1 remain to be investigated.

10.5 CONCLUSIONS

Collectively, the role of SEP15 in cancer, similar to TrxR1 and GPx2, very likely depends on tissue type, selenium status, cancer stage, and the cancer model used for the investigation. GPx2 and TrxR1 appear to serve a protective function in most healthy tissues and in the very early phase of cancer initiation but may have a tumor maintaining or promoting one during cancer progression when cells have already transformed and a tumor has developed. Similarly, SEP15 may possibly serve a protective function against lung cancer. In contrast, lack of SEP15 is protective against chemically induced preneoplastic colonic lesions, suggesting that SEP15 expression may promote cancer is this tissue. A better understanding of tissue specificity and interactions among selenoproteins is needed.

ACKNOWLEDGMENTS

We thank Dolph Hatfield, Brad Carlson, and Vadim Gladyshev for their helpful comments, and for their critical review of this chapter. The work was supported by Towson University's Fisher College of Science and Mathematics and the NIH Office of Dietary Supplements.

REFERENCES

Apostolou, S., J. O. Klein, Y. Mitsuuchi et al. 2004. "Growth inhibition and induction of apoptosis in mesothelioma cells by selenium and dependence on selenoprotein SEP15 genotype." *Oncogene* no. 23 (29):5032–40.

Brigelius-Flohé, R., M. Müller, D. Lippmann, and A. P. Kipp. 2012. "The yin and yang of nrf2-regulated selenoproteins in carcinogenesis." *Int J Cell Biol* no. 2012:486147.

Davis, C. D., P. A. Tsuji, and J. A. Milner. 2012. "Selenoproteins and cancer prevention." *Annu Rev Nutr* no. 32:73–95.

Ferguson, A. D., V. M. Labunskyy, D. E. Fomenko et al. 2006. "NMR structure of the seleno-protein Sep15 and SelM reveal redox activity of a new thioredoxin-like family." *J Biol Chem* no. 281 (6):3536–43.

Geybels, M. S., C. M. Hutter, E. M. Kwon et al. 2013. "Variation in selenoenzyme genes and prostate cancer risk and survival." *Prostate* no. 73 (7):734–42.

Gladyshev, V. N., K. T. Jeang, J. C. Wootton, and D. L. Hatfield. 1998. "A new human selenium-containing protein. Purification, characterization, and cDNA sequence." *J Biol Chem* no. 273 (15):8910–5.

Gresner, P., J. Gromadzinska, E. Jablonska, J. Kaczmarski, and W. Wasowicz. 2009. "Expression of selenoprotein-coding genes SEPP1, SEP15, and hGPX1 in non-small cell lung cancer." *Lung Cancer* no. 65:34–40.

Hatfield, D. L., P. A. Tsuji, B. A. Carlson, and V. N. Gladyshev. 2014. "Selenium and sele-nocysteine: Roles in cancer, health, and development." *Trends Biochem Sci* no. 39 (3):112–20.

Hu, Y. J., K. V. Korotkov, R. Mehta et al. 2001. "Distribution and functional consequences of nucleotide polymorphisms in the 3′-untranslated region of the human Sep15 gene." *Cancer Res* no. 61 (5):2307–10.

Irons, R., P. A. Tsuji, B. A. Carlson et al. 2010. "Deficiency in the 15 kDa selenoprotein inhibits tumorigenicity and metastasis of colon cancer cells." *Cancer Prev Res* no. 3 (5):630–9.

Jablonska, E., J. Gromadzinska, W. Sobala, E. Reszka, and W. Wasowicz. 2008. "Lung cancer risk associated with selenium status is modified in smoking individuals by Sep15 poly-morphism." *Eur J Cancer* no. 47:47–54.

Jablonska, E., J. Gromadzinska, E. Reszka, and W. Wasowicz. 2011. "SEP15 (15-kDa seleno-protein)." *Atlas Genet Cytogenet Oncol Haematol* no. 15 (6).

Jaworska, K., S. Gupta, K. Surda et al. 2013. "A low selenium level is associated with lung and laryngeal cancer." *PLoS One* no. 8 (3):e59051. doi: 10.1371/journal.pone.0059051.

Karunasinghe, N., D. Y. Han, M. Goudie et al. 2012. "Prostate disease risk factors among a New Zealand cohort." *J Nutrigenet Nutrigenomics* no. 5 (6):339–51.

Kasaikina, M. V., D. E. Fomenko, V. M. Labunskyy et al. 2011. "Roles of the 15-kDa seleno-protein (Sep15) in redox homeostasis and cataract development revealed by the analysis of Sep 15 knockout mice." *J Biol Chem* no. 286 (38):33203–12.

Kasaikina, M. V., D. L. Hatfield, and V. N. Gladyshev. 2012. "Understanding selenoprotein function and regulation through the use of rodent models." *Biochim Biophys Acta* no. 1823 (9):1633–42.

Kumaraswamy, E., A. Malykh, K. V. Korotkov et al. 2000. "Structure-expression relationships of the 15-kDa selenoprotein gene. Possible role of the protein in cancer etiology." *J Biol Chem* no. 275 (45):35540–7.

Labunskyy, V. M., M. H. Yoo, D. L. Hatfield, and V. N. Gladyshev. 2009. "Sep15, a thioredoxin-like selenoprotein, is involved in the unfolded protein response and differentially regu-lated by adaptive and acute ER stresses." *Biochemistry* no. 48 (35):8458–65.

Labunskyy, V. M., D. L. Hatfield, and V. N. Gladyshev. 2014. "Selenoproteins: Molecular pathways and physiological roles." *Physiol Rev* no. 94 (3):739–77.

Liu, C. P., J. Fu, S. L. Lin, X. S. Wang, and S. Li. 2014. "Effects of dietary selenium deficiency on mRNA levels of twenty-one selenoprotein genes in the liver of layer chicken." *Biol Trace Elem Res* no. 159 (1–3):192–8.

Méplan, C., D. J. Hughes, B. Pardini et al. 2010. "Genetic variants in selenoprotein genes increase risk of colorectal cancer." *Carcinogenesis* no. 31 (6):1074–9.

Müller, M. F., S. Florian, S. Pommer et al. 2013. "Deletion of glutathione peroxidase-2 inhibits azoxymethane-induced colon cancer development." *PLoS One* no. 8 (8):e72055.

Naschberger, E., R. S. Croner, S. Merkel et al. 2008. "Angiostatic immune reaction in colorec-tal carcinoma: Impact on survival and perspectives for antiangiogenic therapy." *Int J Cancer* no. 123 (9):2120–9.

Nasr, M. A., Y. J. Hu, and A. M. Diamond. 2004. "Allelic loss of the Sep15 locus in breast cancer." *Cancer Ther* no. 1:293–8.

Penney, K. L., F. R. Schumacher, H. Li et al. 2010. "A large prospective study of SEP15 genetic variation, interaction with plasma selenium levels, and prostate cancer risk and survival." *Cancer Prev Res* no. 3:604–10.

Reszka, E., J. Gromadzinska, E. Jablonska, W. Wasowicz, Z. Jablonowski, and M. Sosnowski. 2009. "Level of selenoprotein transcripts in peripheral leukocytes of patients with bladder cancer and healthy individuals." *Clin Chem Lab Med* no. 47 (9):1125–32.

Slattery, M. L., A. Lundgreen, B. Welbourn, C. Corcoran, and R. K. Wolff. 2012. "Genetic variation in selenoprotein genes, lifestyle, and risk of colon and rectal cancer." *PLoS One* no. 7 (5):e37312.

Sutherland, A., D. H. Kim, C. Relton, Y. O. Ahn, and J. Hesketh. 2010. "Polymorphisms in the selenoprotein S and 15-kDa selenoprotein genes are associated with altered susceptibility to colorectal cancer." *Genes Nutr* no. 5 (3):215–23.

Takeda, Y., A. Seko, M. Hachisu et al. 2014. "Both isoforms of human UDP-glucose:glycoprotein glucosyltransferase are enzymatically active." *Glycobiology* no. 24 (4):344–50.

Thisse, C., A. Degrave, G. V. Kryukov et al. 2003. "Spatial and temporal expression patterns of selenoprotein genes during embryogenesis in zebrafish." *Gene Expr Patterns* no. 3 (4):525–32.

Tsuji, P. A., B. A. Carlson, S. Naranjo-Suarez et al. 2012a. "Knockout of the 15 kDa selenoprotein protects against chemically-induced aberrant crypt formation in mice." *PLoS One* no. 7 (12):e50574. doi: 10.1371/journal.pone.0050574.

Tsuji, P. A., B. A. Carlson, M. H. Yoo et al. 2012b. "Independent down-regulation of Sep15 and TR1, but not deficiency in both genes, affects cancer phenotypes of mouse colon carcinoma cells." *FASEB J* no. 26:253.1.

Tsuji, P. A., S. Naranjo-Suarez, B. A. Carlson, R. Tobe, M. H. Yoo, and C. D. Davis. 2012c. "Deficiency in the 15 kDa selenoprotein inhibits colon cancer cell growth." *Nutrients* no. 3 (9):805–17.

Walshe, J., M. M. Serewko-Auret, N. Teakle et al. 2007. "Inactivation of glutathione peroxidase activity contributes to UV-induced squamous cell carcinoma formation." *Cancer Res* no. 67 (10):4751–8.

Wright, M. E., and A. M. Diamond. 2012. "Polymorphisms in selenoprotein genes and cancer." In *Selenium—Its molecular biology and role in human health*, edited by D. L. Hatfield, M. J. Berry and V. N. Gladyshev. New York: Springer Science + Business Media, LLC.

Yoo, M. H., X. M. Xu, B. A. Carlson, V. N. Gladyshev, and D. L. Hatfield. 2006. "Thioredoxin reductase 1 deficiency reverses tumor phenotypes and tumorigenicity of lung carcinoma cells." *J Biol Chem* no. 281:13005–8.

Yoo, M. H., X. M. Xu, B. A. Carlson, A. D. Patterson, V. N. Gladyshev, and D. L. Hatfield. 2007. "Targeting thioredoxin reductase 1 reduction in cancer cells inhibits self-sufficient growth and DNA replication." *PLOS One* no. 2:e1112.

Section V

Unexpected Links

11 Multifaceted and Intriguing Effects of Selenium and Selenoproteins on Glucose Metabolism and Diabetes

Ji-Chang Zhou, Holger Steinbrenner,
Margaret P. Rayman, and Xin Gen Lei

CONTENTS

11.1 Introduction 218
11.2 Redox Control of Insulin Secretion and Signaling 218
11.3 Dual Roles of Se in Glucose Metabolism and Diabetes 220
    11.3.1 Attempted Use of Se Compounds for Prevention and Treatment
    of Diabetes 220
    11.3.2 Supranutritional Se Intake as a Potential Risk Factor for T2DM 221
        11.3.2.1 Animal Studies 221
        11.3.2.2 Cross-Sectional Studies 222
        11.3.2.3 Randomized Controlled Trials (RCTs) 223
11.4 GPx1 and Diabetes 223
    11.4.1 Paradoxical Roles of GPx1 in Diabetes 223
    11.4.2 Regulatory Mechanisms of GPx1 on Insulin-Regulated Glucose
    Metabolism 226
        11.4.2.1 Cellular Protection through the Hydrogen
        Peroxide-Degrading Activity of GPx1 226
        11.4.2.2 Outcome of Overproduction and Knockout of GPx1
        Alone or in Combination with SOD1 226
11.5 Sepp1 and Diabetes 228
    11.5.1 Functions of Sepp1 228
    11.5.2 Sepp1 as a Negative Regulator of Insulin Signaling 229
    11.5.3 Elevation of Circulating Sepp1 Levels in T2DM Patients 230
11.6 Other Selenoproteins, Glucose Metabolism, and Diabetes 231

11.7 Concluding Remarks and Perspective 233
Acknowledgments 235
References 235

11.1 INTRODUCTION

Selenium (Se) is an essential nutrient for both humans and animals. This trace element has received attention for a plethora of (assumed) beneficial effects on health as well as for its unique biochemistry. Compared with other micronutrients, Se has a rather narrow "safe window" and a U-shaped dose-response curve (Fairweather-Tait et al., 2011; Rayman, 2012). Although European and US health authorities recommend daily intakes of around 55 μg Se (Fairweather-Tait et al., 2011) and suggest 300–450 μg Se/day as the "tolerable upper intake level" (Division for Nutrition, 2006), excessive Se intake, even below the toxic level, might still trigger adverse effects. The main chemical forms of Se are selenomethionine (SeMet) in plant foods with selenocysteine (Sec) additionally coming from meats (Rayman et al., 2008). Nutritional supplements of Se for humans and animals include SeMet, sodium selenite, sodium selenate, and Se-enriched yeast (Rayman, 2012). Metabolism of various forms of Se converges at hydrogen selenide (Rayman et al., 2008) that supplies Se for the cellular biosynthesis of selenoproteins. Expression of selenoprotein genes in tissues is highly regulated by the body Se status (Sunde and Raines, 2011).

The human genome contains 25 selenoprotein genes encoding glutathione peroxidases (GPxs), thioredoxin reductases (TrxRs or Txnrds), iodothyronine deiodinases (DIOs), selenoprotein P (Sepp1 or SelP), and other selenoproteins (Kryukov et al., 2003). In the ribosomal translation of selenoproteins, specific *cis*-acting elements and *trans*-acting factors are required to help the UGA triplet(s) in the open reading frame of the selenoprotein mRNAs in directing the incorporation of Sec into the growing polypeptide chain (see Chapters 3 and 4). Sec is more reactive in nucleophilic reactions, and it is more readily oxidized than its sulfur analogue, cysteine (Cys) (Johansson et al., 2005; Byun and Kang, 2011). These biochemical properties underlie the effectiveness of Sec in the active center of selenoproteins (Squires and Berry, 2008; Allmang et al., 2009; Turanov et al., 2011) that include multiple oxidoreductases (Steinbrenner and Sies, 2009; Hawkes and Alkan, 2010; Skaff et al., 2012; Lenart and Pawlowski, 2013). Thus, nutritionally or genetically altering the production of GPx1, Sepp1, and other selenoproteins may shift the cellular and/or extracellular steady-state reduction-oxidation (redox) status and modulate metabolic functions. This provides a putative mechanistic basis for the potential of Se to prevent diabetic complications related to oxidative stress and for a link between dietary Se oversupply and an enhanced risk of developing type-2 diabetes mellitus (T2DM) (Steinbrenner, 2013).

11.2 REDOX CONTROL OF INSULIN SECRETION AND SIGNALING

The intracellular and extracellular redox states affect cellular responses to intrinsic and extrinsic stimuli. Disturbing the redox balance leads to either oxidative or reductive stress that may result in the development of metabolic disorders such as T2DM (Trachootham et al., 2008). As a major component of the redox balance, reactive oxygen species (ROS) (Ray et al., 2012) are mainly generated as by-products of the mitochondrial respiratory chain and by enzymes including NAD(P)H oxidases (Noxs) and xanthine oxidase located in the endoplasmic reticulum (ER), in lysosomes, and at the plasma membrane (Thannickal and Fanburg, 2000; Bartosz, 2009). Although excessive ROS cause oxidative damage to macromolecules, metabolic dysfunction, and cell death, physiological levels of ROS are important for many biological activities (Hensley et al., 2000; Thannickal and Fanburg, 2000; Castro and Freeman, 2001; Alfadda and Sallam, 2012; Brieger et al., 2012) and serve as second messengers in the signal transduction of various growth factors and hormones, including insulin (Goldstein et al., 2005; Loh et al., 2009; Tiganis, 2011).

The elevated plasma glucose after a carbohydrate-rich meal in humans and animals initiates a series of signaling and metabolic events that in turn stimulate the release of insulin from the pancreatic β-cells (MacDonald et al., 2005). This process is termed glucose-stimulated insulin secretion (GSIS). Insulin is transported to peripheral tissues through the blood circulation, captured by its cell-membrane receptors on target tissues, and then triggers downstream insulin-signaling cascades (Saltiel and Kahn, 2001; Kadowaki et al., 2012). Among the proteins in these cascades, phosphoinositide-3-kinase (PI3K) and protein kinase B (PKB, a serine/threonine kinase that is also known as Akt), are key factors (Taniguchi et al., 2006; Manning, 2010) that integrate many metabolic actions of insulin by regulating gene expression, phosphorylation, stability, and/or cellular localization of downstream effector proteins. A general scheme of insulin signaling and the potential sites affected by the redox balance or ROS status can be found in a previous review (Bashan et al., 2009).

Being one of the etiological factors of insulin resistance (Houstis et al., 2006), excessive ROS are involved in the pathogenesis of the three major types of diabetes, that is, type 1 diabetes mellitus (T1DM), T2DM, and gestational diabetes mellitus (GDM). By activating transcription factors, such as nuclear factor kappa-light-chain-enhancer of activated B-cells (NF-κB) (Gloire et al., 2006), ROS can alter gene expression and protein production linked to insulin synthesis, secretion, and function. In diabetic subjects, sustained high blood glucose may induce oxidation and nonenzymatic glycation of proteins, thus elevating ROS (Brownlee et al., 1988) and causing further deterioration of the diabetic condition and late complications (Roberts and Sindhu, 2009). Excessive ROS have been shown to inhibit the insulin-induced tyrosine phosphorylation of the insulin receptor (INSR) β subunit in NIH-B cells (Hansen et al., 1999), Akt activation (Gardner et al., 2003), and K(ATP) channel activation in vascular smooth-muscle cells (Yasui et al., 2012).

Meanwhile, many studies suggest that low levels of ROS, notably hydrogen peroxide (H_2O_2), can play a positive, insulin-mimetic role (Czech et al., 1974). In hepatocytes and adipocytes, binding of insulin to its receptor stimulates a burst of H_2O_2 that temporarily inactivates protein-tyrosine phosphatase 1B (PTP1B) and other phosphatases and extends the phosphorylation of INSR, insulin receptor substrate (IRS), and Akt (Mahadev et al., 2001). This H_2O_2-mediated fine regulation of the phosphorylation states of key molecules in the insulin-signaling cascade is vitally important for the control of insulin action and sensitivity. H_2O_2 also mediates GSIS in pancreatic β-cells. Short-term exposure to high glucose stimulates H_2O_2 production, resulting in increased intracellular calcium concentrations to stimulate insulin secretion (Pi et al., 2007).

The dual role of ROS in insulin signaling and secretion might explain the poor success of clinical approaches that have attempted to use antioxidants for the treatment of diabetes (Rains and Jain, 2011). Moreover, an oxidative environment is required in the ER to oxidize thiol groups of Cys residues in newly synthesized proteins to disulphide bonds as insufficient production of ROS might result in failure in protein folding in the ER and contribute to the development of T2DM (Watson, 2014).

11.3 DUAL ROLES OF SE IN GLUCOSE METABOLISM AND DIABETES

11.3.1 ATTEMPTED USE OF SE COMPOUNDS FOR PREVENTION AND TREATMENT OF DIABETES

Se compounds have been found to improve pancreatic β-cell function and to mimic the action of insulin on its major target cells. In the 1960s and 1970s, pancreatic atrophy was observed in Se-deficient chicks, which was prevented by Se supplementation (Thompson and Scott, 1970; Bunk and Combs, 1981). Sodium selenite (30 nmol/L) stimulated insulin synthesis and secretion in insulinoma cells and rat islets (Campbell et al., 2008). High doses of sodium selenate (1 mmol/L) stimulated glucose uptake in rat adipocytes (Ezaki, 1990). This insulin-mimetic effect of selenate was associated with increased phosphorylation of the INSR β-subunit, endogenous proteins of 60–170 kDa, and ribosomal S6 kinase. Similar results were obtained with other cell types *in vitro* (Pillay and Makgoba, 1992; Stapleton et al., 1997; Hei et al., 1998). Meanwhile, Ezaki (1990) demonstrated in a cell-free system that selenate itself did not affect protein tyrosine kinase or phosphatase activity, suggesting that its effect on the phosphorylation status of the INSR in cells was indirectly mediated by metabolites.

Sodium selenate, sodium selenite, and SeMet were used at high dosages (e.g., 10–15 μmol/kg of body weight [BW]/day, by intraperitoneal injection) to alleviate or improve T1DM-like phenotypes induced by treatment with the β-cell toxin streptozotocin in animals (McNeill et al., 1991; Berg et al., 1995; Douillet et al., 1998; Mukherjee et al., 1998; Mueller et al., 2003; Sheng et al., 2004; Ozdemir et al., 2005; Ulusu and Turan, 2005; Mueller and Pallauf, 2006; Agbor et al., 2007; Hwang et al., 2007; Aydemir-Koksoy and Turan, 2008). In *db/db* mice (an animal model of T2DM), selenate reduced hepatic insulin resistance by inhibiting (through intermediary metabolites) the activity of protein tyrosine phosphatases (Mueller and Pallauf, 2006). It should be noted that high doses of Se compounds can easily act as pro-oxidants and produce large amounts of ROS that can damage the thiol redox system (Mézes and Balogh, 2009; see also Chapter 5). To ensure the physiological needs of rodents, 0.15 (up to 0.4 in pregnancy and lactation) mg Se/kg of feed is adequate (NRC, 1995), and 5.0 mg Se/kg of feed slows the growth rate and causes deterioration of biochemical and molecular markers (Sunde and Raines, 2011). It has been suggested that the beneficial effects of Se on glucose regulation and diabetes prevention are similar to the insulin-mimetic effects of many other elements and merely represent a manifestation of the stress response (Steinbrenner et al., 2011). Moreover, prevention or treatment of diabetes in humans is not practical with nearly toxic doses of Se.

Several epidemiological studies reported inverse associations between blood/ toenail Se concentrations and diabetes incidence along with beneficial effects of dietary Se supplementation. Compared with healthy subjects, T1DM and/or T2DM patients had lower circulating Se (Navarro-Alarcon et al., 1999; Kljai and Runje, 2001; Whiting et al., 2008) and toenail Se concentrations (Rajpathak et al., 2005). Similarly, pregnant women with impaired glucose tolerance (IGT) or GDM had much lower serum Se concentrations than women with normal pregnancies (Tan et al., 2001; Kilinc et al., 2008) and displayed an inverse relationship between serum Se and blood-glucose concentration (Kilinc et al., 2008). Interestingly, Se concentration in whole blood, plasma, or serum decreases over the course of pregnancy (Ferrer et al., 1999; Tan et al., 2001; Hawkes et al. 2004; Molnar et al., 2008; Rayman et al., 2014). In a small group of pregnant women, the increase in fasting glucose as pregnancy advanced was inversely correlated with plasma Se concentration as was the increase in plasma glucose following an oral glucose tolerance test administered at 12 weeks of gestation (Hawkes et al., 2004). Compared with healthy subjects, T1DM and/or T2DM patients had higher levels of thiobarbituric acid reactive substances in plasma (Faure et al., 2004) and urine (Whiting et al., 2008). Supplementing the T2DM patients with sodium selenite lowered the activation of NF-κB in peripheral blood monocytes (960 μg Se/day for 3 months) (Faure et al., 2004), and a dose of 100 μg Se/day lowered serum concentrations of thiobarbituric acid reactive substances and urinary albumin excretion (Kahler et al., 1993).

A number of epidemiological studies have shown a benefit of higher Se status on lowering T2DM risk. In a cohort study of 1162 French elderly, the risk of developing impaired fasting plasma glucose (FPG)/T2DM over 9 years of follow-up was significantly lower in 574 men with plasma Se in the highest (1.19–1.94 μmol/L) than in the lowest (0.18–1.00 μmol/L) tertile at baseline (hazard ratio [HR] = 0.48, 95% confidence interval [CI] = 0.25–0.92). However, higher Se did not appear to benefit women (Akbaraly et al., 2010). In a joint US study that combined results from 3535 men of the Health Professionals Follow-up Study and 3630 women of the Nurses' Health Study, 780 cases of T2DM developed over 142,550 person-years of follow-up (Park et al., 2012). After multivariable adjustment, the risk of T2DM was lower with increasing quintiles of toenail Se ($P = 0.01$).

11.3.2 SUPRANUTRITIONAL SE INTAKE AS A POTENTIAL RISK FACTOR FOR T2DM

11.3.2.1 Animal Studies

Four weeks after treatment, rats that received 5.0 μmol of Se/kg BW (sodium selenite) via intraperitoneal injection had 16% higher FPG than control rats (127 vs. 109 mg/dL) (Ulusu and Turan, 2005). In comparison with mice fed with a Torula-yeast Se-deficient diet containing 0.1 mg Se/kg (as sodium selenite), mice fed with 0.4 mg Se/kg diet for 3 months had hyperinsulinemia and higher plasma glucose

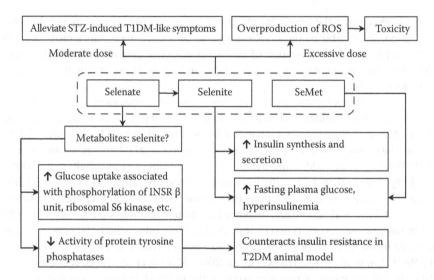

FIGURE 11.1 Role and mechanism of different levels of Se supplementation in glucose metabolism. "Moderate dose" refers to supplements close to recommended dietary intakes whereas "excessive dose" refers to supplements above tolerable upper intake levels. Arrows linking the text boxes indicate the cause–result (arrow direction) relationship whereas in the text boxes upward arrows mean "increase," and downward arrows mean "decrease." Selenate, selenite, and SeMet embraced with the dashed box sometimes had similar effects. INSR: insulin receptor; ROS: reactive oxygen species; SeMet: selenomethionine; STZ: streptozotocin; T1DM: type 1 diabetes mellitus; T2DM: type 2 diabetes mellitus.

after glucose challenge (Labunskyy et al., 2011). Dietary supplementation of gestating rats with Se-enriched yeast (3 mg Se/kg) impaired body insulin sensitivity and resulted in hyperinsulinemia (Zeng et al., 2012). Elevated plasma insulin levels were also observed in pigs supplemented for 16 weeks with 0.5 mg Se/kg (as Se-enriched yeast) compared to pigs fed a Se-adequate (0.17 mg Se/kg) diet (Pinto et al., 2012) and in pigs fed 3.0 mg Se/kg compared to 0–0.3 mg Se/kg (Liu et al., 2012). Figure 11.1 highlights the effects of Se in its three commonly used chemical forms on glucose metabolism and diabetes reported by animal and/or *in vitro* experiments.

11.3.2.2   Cross-Sectional Studies

Children suffering from T1DM were found to have higher circulating Se concentrations than their healthy controls (Gebre-Medhin et al., 1984; Cser et al., 1993; Wang et al., 1995) although those differences were interpreted as resulting from metabolic abnormalities, insulin therapy (Gebre-Medhin et al., 1985), or elevated food intake in the afflicted subjects (Wang et al., 1995). Five out of eight cross-sectional studies in adults have shown positive associations between serum/plasma Se and T2DM or FPG (Rayman and Stranges, 2013). These include large studies, such as the French SU.VI.MAX (Czernichow et al., 2006) and the American NHANES III and NHANES 2003–2004 (Bleys et al., 2007; Laclaustra et al., 2009). In SU.VI.MAX, baseline FPG was positively ($P < 0.0001$) associated with plasma Se (Czernichow et al., 2006). NHANES III (8876 adults aged ≥ 20 years) found that diabetes prevalence in quintile 5 of serum Se was 1.57-fold higher than in quintile 1 and that the risk of diabetes was significantly increased with serum Se concentration > 130 µg/L (Bleys et al., 2007). NHANES 2003–2004 (917 adults aged ≥ 40 years) reported a positive correlation between serum Se concentration and diabetes prevalence, FPG, and glycosylated hemoglobin A1c (HbA1c) levels (Laclaustra et al., 2009). Similar results were found in smaller studies of 398 adult Lebanese (Obeid et al., 2008) and 200 elderly Taiwanese subjects (Yang et al., 2010) and in much smaller studies of T2DM patients ($n = 53$) (Ekmekcioglu et al., 2001) and obese GDM pregnant women ($n = 11$) (Al-Saleh et al., 2007).

11.3.2.3 Randomized Controlled Trials (RCTs)

There have been four RCTs with Se as a single agent plus one that used a hypocaloric diet rich in legumes in both Se and placebo groups (Alizadeh et al., 2012; Rayman and Stranges, 2013). Of these five trials, Rayman and Stranges (2013) concluded that Se supplementation had no effect in three, and one showed significantly lowered fasting serum insulin and homeostasis model assessment of insulin resistance (HOMA-IR). The only trial that found an increased risk of T2DM was the Nutritional Prevention of Cancer (NPC) trial (Stranges et al., 2007). In that trial, 600 out of 1202 dermatology outpatients without T2DM at baseline received Se-enriched yeast (200 µg Se/day), and 602 received placebo. After 7.7 years, the incidence of T2DM was higher in the Se-supplemented group than in the placebo group (HR = 1.55, 95% CI = 1.03–2.33). In fact, the increased risk was only significant in those in the highest Se tertile at baseline (HR = 2.70, 95% CI = 1.30–5.61).

Although the Se and Vitamin E Cancer Prevention Trial (SELECT) in 32,400 male volunteers (Lippman et al., 2009) is often described as causing an increased risk of T2DM among men given 200 µg Se/day (as SeMet), the effect never reached statistical significance (relative risk [RR] = 1.07, 99% CI = 0.94–1.22, $P = 0.16$ (Lippman et al., 2009) and further diminished after an additional follow-up of 2 years and 9 months (RR = 1.04, 99% CI = 0.93–1.17, $P = 0.34$) (Klein et al., 2011). However, a recent RCT in 60 T2DM patients found that supplementation with 200 µg Se/day (as sodium selenite) versus placebo for 3 months caused elevated FPG, HbA1c, and high-density lipoprotein-cholesterol (HDL-C), suggesting adverse effects of supranutritional Se on blood glucose and lipid homeostasis in T2DM patients (Faghihi et al., 2013).

11.4 GPx1 AND DIABETES

11.4.1 PARADOXICAL ROLES OF GPx1 IN DIABETES

Diabetes has been found to be associated with systemic decreases in both low-molecular-weight and enzymatic antioxidants. Decreased blood or tissue concentrations of GPx1 (L'Abbé and Trick, 1994; Dominguez et al., 1998; Martin-Gallan et al., 2003; Memisogullari et al., 2003; Whiting et al., 2008), glutathione (L'Abbé and Trick, 1994; Dominguez et al., 1998; Martin-Gallan et al., 2003; Memisogullari et al., 2003), β-carotene (Dominguez et al., 1998; Martin-Gallan et al., 2003), vitamin E, and vitamin C (Salonen et al., 1995; Sundaram et al., 1996; Vijayalingam et al., 1996) have been measured in diabetic humans and animals. GPx activity was lower in erythrocytes and in the plasma of patients with 6–8 years of diabetes history than in healthy controls and patients with ≤2 years of diabetes history (Whiting et al., 2008). Some of the above studies also determined superoxide dismutase (SOD) activity as SOD catalyzes the conversion of O_2^- to H_2O_2, which is a substrate of GPx. However, the results were not as consistent as for GPx. Erythrocyte SOD activity was elevated in T1DM (Martin-Gallan et al., 2003) or T2DM (Dominguez et al., 1998) subjects compared with nondiabetic subjects in some studies although others found no change (Memisogullari et al., 2003) or observed a decrease in SOD activity within the first 2 years after diagnosis of T2DM (Sundaram et al., 1996). The Cu/Zn-SOD (cytoplasmic SOD or SOD1) and total SOD activities in the pancreas of prediabetic BioBreeding rats were increased (L'Abbé and Trick, 1994).

Overexpression of GPx1 in β-cells protected against β-cell loss induced by oxidative stress and attenuated hyperglycemia in *db/db* mice treated with streptozotocin (Harmon et al., 2009). On the other hand, pancreatic islets isolated from transgenic mice overexpressing GPx1 were not able to survive pro-oxidant (hypoxanthine/xanthine oxidase) treatment whereas combined overexpression of SOD1, extracellular SOD (SOD3), and GPx1 protected the cells from death induced by oxidative stress. It is probable that the combined activities of SOD1, SOD3, and GPx1 were necessary to retain the functional homeostasis of the islets. Combined overexpression of SOD1, SOD3, and GPx1 also enabled transplanted islets to function better in the regulation of blood glucose in recipient mice as compared to transplanted islets isolated from wild-type mice (Mysore et al., 2005). SOD1$^{-/-}$ mice had more apparent pancreatitis than GPx1$^{-/-}$ mice and knockdown of SOD1 impaired islet function, pancreatic integrity, and body glucose homeostasis more than knockdown of GPx1 (Wang et al., 2011). Mechanistically, knockdown of SOD1 impaired glucose homeostasis through the p53-AMPK (adenosine monophosphate-activated protein kinase) pathway and lipid metabolism via the PTP1B-SREBP1 (sterol regulatory binding protein 1) pathway whereas knockdown of GPx1 had little effect on those pathways (Wang et al., 2012).

Intriguingly, both human and animal studies showed elevated biosynthesis/activity of GPx as a risk factor for hyperglycemia and diabetes. In pregnant women without GDM, erythrocyte GPx1 activity increased with trimester of pregnancy and was positively associated with FPG, plasma insulin, C-peptide, and the HOMA-IR index (Chen et al., 2003). Children with T1DM had higher Se levels in whole blood, plasma, and erythrocytes and higher plasma GPx3 activities than healthy children, and their blood glucose was positively correlated with GPx activity in plasma and erythrocytes (Cser et al., 1993). Whole-blood GPx activity and 8-hydroxy-2'-deoxyguanosine (8-OHdG) levels in T2DM patients ($n = 100$) were 1.3 and 1.5 times those in healthy controls ($n = 50$), respectively. Both in T2DM patients and in healthy controls, FPG and HbA1c were positively associated with blood GPx activity, plasma lipid peroxides, and plasma 8-OHdG, but negatively correlated with blood SOD activity (Bolajoko et al., 2008).

Global GPx1 overexpression induced T2DM-like phenotypes in aged mice, manifested as hyperinsulinemia, hyperglycemia, obesity, and insulin resistance (McClung et al., 2004). Diet restriction could prevent all these symptoms except hyperinsulinemia and hypersecretion of insulin after glucose stimulation. The GPx1 overproduction dysregulates the expression of transcription factors, signaling proteins, and functional enzymes that are key regulators or effectors of glucose homeostasis. These changes include hyperacetylation of H3 and H4 histones in the approximate region of pancreatic duodenal homeobox 1 (PDX1) promoter (Wang et al., 2008), elevated mRNA, and/or protein or activity of PDX1 (Wang et al., 2008; Pepper et al., 2011; Yan et al., 2012), acetyl-CoA carboxylase 1 (ACC1) (Yan et al., 2012), phosphoenolpyruvate carboxykinase (PEPCK or PCK) (Yan et al., 2012) and glucokinase (GK or GK1) (Pepper et al., 2011; Yan et al., 2012), and decreased mRNA and/or protein concentration of uncoupling protein 2 (UCP2) (Wang et al., 2008). PDX1 is one of the most important transcription factors required for pancreatic development and β-cell differentiation and maturation (Offield et al., 1996), ACC1 is a rate-limiting enzyme for lipogenesis (Scott et al., 2012), PEPCK is a rate-limiting enzyme for gluconeogenesis (Granner and Pilkis, 1990), GK plays a key role in the catabolism of glucose (Winzell et al., 2011), and UCP2 mainly functions in the control of mitochondrial-derived generation of ROS and may affect the mitochondrial potential (Arsenijevic et al., 2000). Notably, dietary Se deficiency essentially precludes GPx1 overproduction in GPx1-overexpressing mice and could thus restore or minimize most dysregulations, resulting in alleviation of the phenotypes (Pepper et al., 2011; Yan et al., 2012).

Supplementation of rat myocytes cultivated in Se-poor (serum-free) medium with both inorganic (sodium selenite and sodium selenate) and organic (SeMet and methylseleninic acid) Se compounds at a dose of 1 μmol/L increased the gene expression and activity of GPx1 (Pinto et al., 2011). Selenite exhibited a more pronounced effect on GPx1 than the other Se compounds and attenuated intracellular ROS generation as well as insulin-induced glucose transporter 4 (GLUT4) gene expression, glucose uptake, and phosphorylation of Akt, forkhead box O1 (FoxO1a or FoxO1), and FoxO3. FoxO1a is a transcription factor for multiple genes including two gluconeogenic enzymes (PEPCK and glucose-6-phosphatase) (Cheng and White, 2012) and Sepp1 (Speckmann et al., 2008; Walter et al., 2008). Dephosphorylated FoxO1a is the active form that promotes expression of its target genes (Cheng and White, 2012). Prolonged feeding from weaning of first-parity female rats with 3.0 mg Se/kg of feed (as Se-rich yeast) resulted in moderate gestational diabetes and insulin resistance in their offspring that were fed the same high Se diet (Zeng et al., 2012). The high-Se diet upregulated gene and/or protein expression of GPx1 and several other selenoproteins in pancreas, liver, and erythrocytes of the dams and their offspring (Zeng et al., 2012). Among the three key insulin target tissues (liver, skeletal muscle, and visceral adipose tissue), GPx activity was elevated only in the skeletal muscle of pigs fed a supranutritional Se diet (0.5 against 0.17 mg Se/kg), which was accompanied by elevation of mRNA abundance of FoxO1a and peroxisomal proliferator-activated receptor-γ coactivator 1α (PGC1α) and nonsignificant increases in fasting plasma insulin and cholesterol levels (Pinto et al., 2012). Zhou et al. (2013) have reviewed recent studies conducted in mice, rats, and pigs that illustrate the diabetogenic potential of dietary Se intake above nutritional requirements (from 0.4 to 3.0 mg/kg of diet).

11.4.2 REGULATORY MECHANISMS OF GPx1 ON INSULIN-REGULATED GLUCOSE METABOLISM

11.4.2.1 Cellular Protection through the Hydrogen Peroxide-Degrading Activity of GPx1

GPx1 catalyzes the reduction of H_2O_2 and low molecular weight organic hydroperoxides including peroxides of cholesterol, fatty acids, and other lipids (Ladenstein et al., 1979; Arthur, 2000). As pancreatic β-cells produce relatively low levels of GPx, SOD, and catalase (Grankvist et al., 1981), excessive ROS or other reactive oxygen (and nitrogen) species are likely to cause damage of their structure and function, resulting in insufficient insulin production/secretion (Rosen et al., 2001; Evans et al., 2002). Although prolonged exposure to high glucose and ribose induces oxidative stress, cellular damage can be prevented by overexpression of GPx1 in the islets (Tanaka et al., 2002). In insulin target tissues, insulin resistance is positively correlated with ROS production in mitochondria. For example, overproduction of ROS induced by a high-fat diet caused insulin resistance in skeletal muscle, but the insulin-mediated glucose uptake was improved by elevating mitochondrial catalase or by treatment with antioxidants (Anderson et al., 2009). As body Se status represents one of the most important regulators of selenoprotein biosynthesis and GPx1 is rapidly depleted in Se deficiency, adequate Se supply is required to maintain physiological levels of GPx1 activity in the pancreas, liver, skeletal muscle, and other tissues to counteract excessive ROS production (Faure et al., 2004; Sheng et al., 2004).

11.4.2.2 Outcome of Overproduction and Knockout of GPx1 Alone or in Combination with SOD1

Because of the dual role of H_2O_2 in generating oxidative destruction of functional molecules and serving as a second messenger for redox signaling (Veal et al., 2007; Gough and Cotter, 2011), GPx1 can be either metabolically beneficial at its physiological expression level when it prevents excessive H_2O_2-induced oxidative stress or detrimental when it dysregulates (blocks or shifts) H_2O_2-mediated signaling (McClung et al., 2004; Lei et al., 2007; Fourquet et al., 2008; Wang et al., 2008; Steinbrenner et al., 2011; Wang et al., 2011). Our current understanding of the roles and mechanisms of GPx1 in insulin signaling and glucose metabolism is summarized in Figure 11.2. The physiological importance of maintaining the balance between intracellular GPx1 activity and ROS production and the potential redox-related signaling mechanisms underlying the development of a T2DM-like phenotype in GPx1-overexpressing mice (Finley et al., 2011; Steinbrenner et al., 2011) are elaborated in the following four paragraphs.

When insulin binds the INSR on the plasma membrane, Nox4 catalyzes the production of superoxide $O_2^{\cdot-}$ and H_2O_2. H_2O_2 or other hydroperoxides transiently inactivate PTP1B, thus prolonging insulin-induced phosphorylation of INSR and

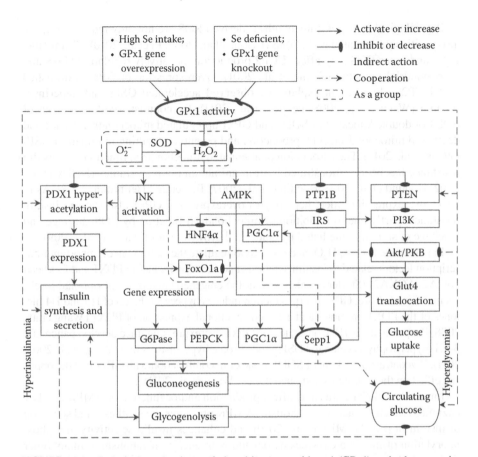

FIGURE 11.2   Role and mechanisms of glutathione peroxidase-1 (GPx1) and selenoprotein P (Sepp1) in insulin signaling and glucose metabolism. Akt/PKB: protein kinase B; AMPK: adenosine monophosphate-activated protein kinase; FoxO1a: forkhead box O1; G6Pase: glucose-6-phosphatase; Glut4: glucose transporter 4; HNF4α: hepatocyte nuclear factor 4α; IRS: insulin receptor substrate; PDX1: pancreatic duodenal homeobox 1; PEPCK: phospho-enolpyruvate carboxykinase; PGC1α: peroxisomal proliferator-activated receptor-γ coactivator 1α; PI3K: phosphoinositide 3-kinase; PTEN: phosphatase and tensin homolog; PTP1B: protein-tyrosine phosphatase 1B; SOD: superoxide dismutase. GPx1 overexpression was defined in Chapter 9, Section 9.2.

downstream target signal proteins (Mahadev et al., 2004). Apparently, overproduction of GPx1 diminishes intracellular H_2O_2 and thus attenuates or blocks this important function of ROS. In fact, the insulin-stimulated phosphorylation of INSR and Akt are suppressed in the liver and muscle of the GPx1-overexpressing mice (McClung et al., 2004). Likewise, H_2O_2 inhibits phosphatase and tensin homolog (PTEN) preventing its inhibition of PI3K, a key signaling molecule in the regulation of insulin sensitivity (Loh et al., 2009). By contrast, knockout of GPx1 promotes the ROS-mediated inactivation of PTEN and enhances the insulin-induced PI3K/Akt phosphorylation and glucose uptake in muscle (Loh et al., 2009).

Physiological levels of mitochondrial-derived ROS are not only required for, but also positively correlated with, GSIS (Leloup et al., 2009; Pi et al., 2010). Meanwhile, H_2O_2 and other ROS may affect UCP2 production, negatively regulating GSIS and mitochondrial potential (Zhang et al., 2001). Overexpression of GPx1 diminished islet UCP2, which might explain the observed accelerated GSIS and hyperinsulinemia (Wang et al., 2008; Pepper et al., 2011). By contrast, single knockout of SOD1 or double knockout of SOD1 and GPx1 upregulated mitochondrial UCP2 and decreased mitochondrial ATP production and potential, resulting in impaired GSIS (Wang et al., 2011). The association of attenuated mitochondrial function and insulin resistance has been demonstrated both in animal and human studies (Kelley et al., 2002; Hojlund et al., 2003; Boudina et al., 2007; Turner and Heilbronn, 2008). As in GPx1 overexpressing mice, hepatocytes overexpressing GPx1 also showed insulin resistance, partly due to inhibition of mitochondrial function by ROS shortage and blockage of signaling mediated by epidermal growth factor (Handy et al., 2009).

In pancreatic β-cells, H_2O_2 and other ROS regulate the expression of important transcription factors related to insulin synthesis and function including PDX1 and forkhead box A2 (FOXA2) (Stoffers et al., 1998) at the epigenetic and/or transcriptional levels. As mentioned above, GPx1 overproduction induces hyperacetylation of H3 and H4 histones of the PDX1 promoter and elevated functional expression of PDX1 (Wang et al., 2008; Pepper et al., 2011). Consequently, GPx1-overexpressing mice develop pancreatic β-cell hypertrophy, accelerated GSIS, and chronic hyperinsulinemia (Wang et al., 2008, 2011). By contrast, knockout of SOD1 alone or in combination with GPx1 had the opposite effect on these proteins and phenotypes (Wang et al., 2011).

Intracellular ROS status may affect production and/or function of AMPK and key enzymes related to energy metabolism. AMPK promotes the synthesis and secretion of insulin in β-cells (Misu et al., 2010) and enhances insulin sensitivity and phosphorylation of multiple enzymes associated with energy metabolism in many other tissues (Kahn et al., 2005). Upregulation of AMPKα1 expression and phosphorylation in the liver of SOD1$^{-/-}$ mice was associated with decreased PEPCK activity and increased GK activity, resulting in lowered gluconeogenesis and glycogen storage (Wang et al., 2012). AMPK also mediates the translocation of intracellular GLUT4 to the cell membrane, thereby facilitating the uptake of glucose (Kurth-Kraczek et al., 1999; Uchizono et al., 2006). H_2O_2 can modify the α- and β-subunits of AMPK by oxidation. In fact, oxidative modification of Cys299 and Cys304 in the α-subunit is an essential step in kinase activation of the entire AMPKαβγ complex, thus affecting the uptake of glucose (Uchizono et al., 2006; Zmijewski et al., 2010). As mentioned above, GPx1 overproduction resulted in upregulation of ACC1 (Yan et al., 2012), PEPCK (Yan et al., 2012), and GK (Pepper et al., 2011; Yan et al., 2012) although the redox or metabolic regulatory mechanisms remain unclear.

11.5 SEPP1 AND DIABETES

11.5.1 FUNCTIONS OF SEPP1

Sepp1 is the major Se-containing protein in plasma. Full-length human Sepp1 contains 10 Sec residues (Stoytcheva et al., 2006). One Sec is located at the N-terminus and accounts for the presumed antioxidant properties of Sepp1, and the remaining Sec residues are distributed within the C-terminus and are used for Se transport and Se supply of peripheral tissues (Saito et al., 2004). The Sepp1 circulating in plasma is mainly secreted by the liver and has recently been postulated to be a so-called "hepatokine" (Misu et al., 2010). The concentration of Sepp1 in plasma is a good indicator of Se nutritional status as plasma Sepp1 reaches a plateau only at a rather high intake of ~105 µg Se/day (Burk and Hill, 2009). Liver-derived Sepp1 may play a role in glucose metabolism and insulin sensitivity (Misu et al., 2010).

In addition to their presence in liver, Sepp1 mRNA and protein have been detected in many other organs, but the function of extrahepatic Sepp1 remains largely elusive. Abundant expression of Sepp1 has been found in white adipocytes; knockdown of Sepp1 resulted in impaired adipogenesis, attenuated insulin signaling, and dysregulated production of adipokines (Zhang and Chen, 2011). In the pancreas, Sepp1 is exclusively expressed in the endocrine part, that is, in glucagon-producing α-cells and in insulin-producing β-cells of the islets (Steinbrenner et al., 2013). Sepp1 expression in isolated islets and in INS-1 insulinoma cells decreased in the presence of high-glucose concentrations. One week after intraperitoneal injection of streptozotocin, Sepp1 expression in the pancreas was significantly increased, suggesting that it may contribute to antioxidant protection of pancreatic islets (Steinbrenner et al., 2013).

11.5.2 SEPP1 AS A NEGATIVE REGULATOR OF INSULIN SIGNALING

A high (supraphysiological) concentration (10 µg/mL) of SEPP1 purified from human plasma counteracted the insulin-induced phosphorylation of INSR and Akt, and the insulin-stimulated downregulation of two key enzymes in gluconeogenesis, Pck1 (one of the isoforms of PEPCK) and glucose-6-phosphatase in cultured hepatocytes. As a result, hepatocyte-produced glucose increased by 30% in the presence of insulin. Without insulin, SEPP1 did not affect gene expression of gluconeogenic enzymes or the rate of glucose production, suggesting that SEPP1 is involved in the fine-tuning of insulin signaling processes. In cultured myocytes, SEPP1 lowered insulin-stimulated glucose uptake. Intraperitoneal injection of mice with SEPP1 (1 mg/kg BW) resulted in impaired glucose tolerance and insulin resistance. Although their plasma insulin levels were significantly increased, following the injection of SEPP1, mice had lowered insulin-stimulated serine phosphorylation of Akt in the liver and skeletal muscle (Misu et al., 2010). Phosphorylation of AMPKα and ACC was decreased by SEPP1 only in hepatocytes but not in myocytes, resulting in suppression of fatty acid β-oxidation in the liver (Misu et al., 2010). It remains to be elucidated whether these insulin-inhibitory actions of high doses of SEPP1 as used in the experiments described above are relevant under physiological or pathophysiological conditions. Furthermore, the molecular mechanisms underlying these autocrine and endocrine actions of SEPP1 as a postulated "hepatokine" are not yet understood. Upregulation of Sepp1 and GPx1 production through sodium selenite supplementation was also found to attenuate insulin signaling in hepatocytes and in the liver of rats (Wang et al., 2014).

Conversely, Sepp1 knockout ($Sepp1^{-/-}$) mice had lower postprandial plasma insulin levels and higher glucose tolerance compared to their wild-type littermates.

In liver and skeletal muscle of the *Sepp1⁻/⁻* mice and in hepatocytes isolated from these mice, phosphorylation of INSR and Akt were both increased. AMPKα and ACC phosphorylation were increased in the liver of the *Sepp1⁻/⁻* mice. Moreover, Sepp1 deficiency attenuated adipocyte hypertrophy, glucose intolerance, and insulin resistance in diet-induced obese mice (Misu et al., 2010). A mechanistic scheme for the action of Sepp1 in insulin signaling and glucose metabolism is shown in Figure 11.2.

11.5.3 ELEVATION OF CIRCULATING SEPP1 LEVELS IN T2DM PATIENTS

Compared with nondiabetic individuals, T2DM patients had elevated SEPP1 concentrations that were positively correlated with FPG (Misu et al., 2010, 2012; Yang et al., 2011) and HbA1c (Misu et al., 2010; Yang et al., 2011). Serum SEPP1 concentration was also higher in overweight and obese subjects than in those who were lean (Yang et al., 2011). Similarly, hepatic Sepp1 mRNA levels and serum Sepp1 concentrations were elevated in rodent models of T2DM (Misu et al., 2010).

A potential explanation for these observations is provided by the concept that hepatic Sepp1 is regulated virtually like a gluconeogenic enzyme (Steinbrenner et al., 2011). Hepatic Sepp1 mRNA levels of mice were higher in the fasting state than in the postprandial state (Misu et al., 2010). Insulin downregulated, but the synthetic glucocorticoid dexamethasone upregulated, Sepp1 biosynthesis in cultured hepatocytes (Speckmann et al., 2008; Walter et al., 2008; Misu et al., 2010). Moreover, Sepp1 mRNA and protein levels were upregulated by glucose or palmitate (Misu et al., 2010). Gene expression and secretion of Sepp1 were increased in hepatocytes cultured in high-glucose medium and decreased by metformin, a drug that is widely prescribed to T2DM patients to suppress hepatic gluconeogenesis and lower their blood-glucose levels (Speckmann et al., 2009).

Mechanistically, the regulation of Sepp1 biosynthesis through the abovementioned factors may be attributed to the presence of a combined binding element for the transcription factors FoxO1a and hepatocyte nuclear factor 4α (HNF4α) within the human and mouse *Sepp1* promoter (Figure 11.2). High-level Sepp1 transcription in liver is ensured by the concerted action of FoxO1a and HNF4α with their coactivator PGC1α (Speckmann et al., 2008; Walter et al., 2008). Key hormones of glucose metabolism (e.g., insulin, glucagon, and glucocorticoids) as well as glucose itself regulate hepatic gluconeogenesis via PGC1α (Yoon et al., 2001; Puigserver et al., 2003). PGC1α activity can be inhibited at physiologically increased insulin levels via PI3K/Akt/FoxO1a (Southgate et al., 2005; Hong et al., 2011), resulting in the suppression of Sepp1 transcription (Walter et al., 2008). Dexamethasone increased the expression and secretion of Sepp1 by activating PGC1α in cultured rat hepatocytes (Speckmann et al., 2008) as well as by its direct interaction with the glucocorticoid-receptor binding domain in the *Sepp1* promoter in human embryonic kidney 293 cells (Rock and Moos, 2009). On the other hand, hepatic PGC1α is dysregulated (strongly increased) at chronically elevated blood glucose concentrations in T2DM, resulting in the upregulation of gluconeogenic enzymes and Sepp1. Therefore, it has been proposed that PGC1α links glucose metabolism and Sepp1 expression (Speckmann et al., 2008; Steinbrenner et al., 2011).

11.6   OTHER SELENOPROTEINS, GLUCOSE METABOLISM, AND DIABETES

Other selenoproteins, including iodothyronine deiodinase 2 (DIO2), selenoproteins R (SelR or SelX or Sepx1, i.e., methionine-R-sulfoxide reductase 1 [MsrB1]), selenoprotein S (SelS or Seps1 or TANIS or VIMP), and selenoprotein T (SelT) have also been linked to glucose homeostasis and/or diabetes (Figures 11.3 and 11.4). These selenoproteins are discussed in more detail in Chapter 12, which highlights particularly the use of transgenic mouse models to elucidate the role of individual selenoproteins in energy metabolism.

SelS is located at the plasma membrane (Gao et al., 2003) and the ER membrane (Ye et al., 2004) and has been postulated to be a receptor of serum amyloid A (SAA), an inflammatory factor (Walder et al., 2002). SelS mRNA levels in the liver of the T2DM animal model, *Psammomys obesus*, were inversely correlated with blood glucose and insulin levels (Walder et al., 2002). In cultured HepG2 hepatocytes, 3T3-L1 adipocytes, and C2C12 myotubes, SelS gene expression was inhibited by glucose or insulin (Walder et al., 2002), and its mRNA and protein levels were increased by glucose deficiency in HepG2 cells (Gao et al., 2004). Hepatic mRNA levels of SelS and SAA were significantly elevated in rats that developed insulin resistance on a high-fat diet, compared to lean control rats and rosiglitazone-treated insulin-resistant rats; the two measures were positively correlated in all three groups of rats (Liu et al., 2009). In human studies, SelS mRNA levels were increased in adipocytes isolated

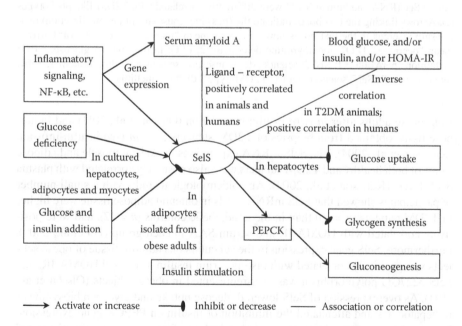

FIGURE 11.3   Effects and potential mechanisms of selenoprotein S (SelS) on insulin signaling and glucose metabolism. HOMA-IR: homeostasis model assessment of insulin resistance; PEPCK: phosphoenolpyruvate carboxykinase.

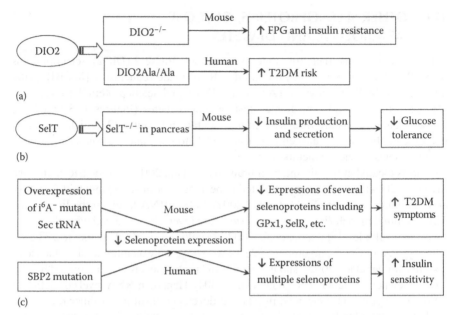

FIGURE 11.4 Effects of selected selenoproteins on insulin signaling and glucose metabolism. Gene knockout or Ala homozygote of DIO2 (a), and SelT knockout (b) increased T2DM risk. By contrast, decreased selenoprotein expression caused by gene mutations of mouse Sec tRNA and human SBP2 were differently correlated with T2DM-like phenotypes (c). Arrows linking the text boxes indicate the (possible) cause–result (arrow direction) relationship whereas in the text boxes upward arrows mean "increase," and downward arrows mean "decrease." DIO2: iodothyronine deiodinase 2; FPG: fasting plasma glucose; GPx1: glutathione peroxidase 1; SBP2: selenocysteine insertion sequence-binding protein 2; SelR: selenoprotein R; SelT: selenoprotein T; T2DM: type 2 diabetes mellitus.

from obese adults after *in vitro* insulin stimulation (Olsson et al., 2011) and in adipose tissue obtained from biopsies of T2DM subjects after *in vivo* insulin infusion (Karlsson et al., 2004). The SelS mRNA expressions in skeletal muscle and adipose tissue of both healthy and T2DM subjects were also positively correlated with plasma SAA levels (Karlsson et al., 2004). An epidemiological study with a small number of participants showed that SelS mRNA levels in omental adipose tissue were higher in T2DM patients ($n = 10$) than in the nondiabetic controls ($n = 12$), and were positively correlated with HOMA-IR and serum SAA in each group (Du et al., 2008). Furthermore, SelS gene expression in the subcutaneous adipose tissue of obese subjects was positively correlated with fasting serum insulin levels and HOMA-IR, and SelS 5227GG polymorphism was over-represented in obese subjects (Olsson et al., 2011). As overexpression of SelS lowered glucose uptake and glycogen biosynthesis in hepatocytes and attenuated the inhibition of insulin on PEPCK gene expression, it seemed to promote gluconeogenesis but had no effect on the insulin-stimulated phosphorylation of INSR (Gao et al., 2003). Being a target of the transcription factor NF-κB (Gao et al., 2003; Rock and Moos, 2009) and a putative receptor for the acute phase protein SAA (Walder et al., 2002), SelS has been proposed to be associated with diabetes risk, possibly through inflammatory signaling pathways (Walder et al., 2002; Curran et al., 2005; Gao et al., 2006; Zhang et al., 2011) (Figure 11.3).

Residing at the ER membrane, DIO2 catalyzes the conversion of the thyroid hormone precursor thyroxine (T4) to the active hormone triiodothyronine (T3) (Arrojo et al., 2011). In humans, a polymorphism in the coding region of the DIO2 gene (Thr92Ala) has been linked to insulin resistance, as Ala/Ala homozygotes exhibited a higher risk for T2DM (Mentuccia et al., 2002; Canani et al., 2005; Dora et al., 2010). Elevated FPG and insulin resistance was reported in DIO2 knockout mice. These mice were more susceptible to obesity induced by a high fat-diet than were wild-type mice (Marsili et al., 2011) (Figure 11.4a). Hypothyroidism has been associated with insulin resistance that can be improved with thyroid hormone treatment (Rochon et al., 2003). However, serum T3 was not lower in DIO2 knockout mice than in wild-type mice (Marsili et al., 2011). Thus, the mechanism of DIO2 deletion on glucose metabolism dysregulation requires further exploration.

Abundant expression of another ER-located selenoprotein, SelT, has recently been detected in human and murine pancreatic β-cells. Mice with targeted deletion of SelT in the pancreas showed impaired glucose tolerance due to a defect in insulin production/secretion (Prevost et al., 2013) (Figure 11.4b).

Overexpression of GPx1 has been reported to affect protein levels of SelS, SelT, and SelR in the liver of mice and in cultured hepatocytes whereas overexpression of an i^6A^- mutant Sec tRNA in mice (see Chapter 4) decreased several selenoproteins, including GPx1 and SelR. Both types of transgenic mice had increased FPG, blood glucose after glucose loading, and plasma insulin in the fed state (Labunskyy et al., 2011). Interestingly, humans with a mutation in the gene for the Sec insertion sequence-binding protein 2 (SBP2) resulting in the very rare phenotype of severely decreased synthesis of multiple selenoproteins and a multisystem disorder exhibited an increase in systemic and cellular insulin sensitivity (Schoenmakers et al., 2010) (Figure 11.4c) and Chapter 16.

It is noteworthy that SelS, DIO2, and SelT are all located in the ER, and their alterations are potentially able to shift the redox status or functional integrity of the ER. Loss of ER homeostasis, including redox homeostasis and accumulation of misfolded proteins, may activate the ER-stress or unfolded-protein response. Under conditions of obesity and T2DM, the unfolded-protein response affects both the major insulin target tissues and pancreatic β-cells, contributing to the development of inflammation, insulin resistance, and impaired insulin secretion (Cnop et al., 2012).

11.7   CONCLUDING REMARKS AND PERSPECTIVE

The relationship between Se and the risk of developing T2DM can be depicted as a U-shaped curve. Although both Se deficiency and oversupply appear to dysregulate glucose metabolism and potentiate the risk for T2DM in a number of animal studies, current data from human studies in support of this association are much less convincing. Although various mechanistic insights have been obtained from the overexpression and/or knockdown of several selenoproteins in mice and in cultured cells that may explain the impact of Se and selenoproteins on glucose metabolism and diabetes, redox regulation of insulin signaling, and oxidative modifications of selected protein phosphatases or kinases remain the main, if not the sole, mode of

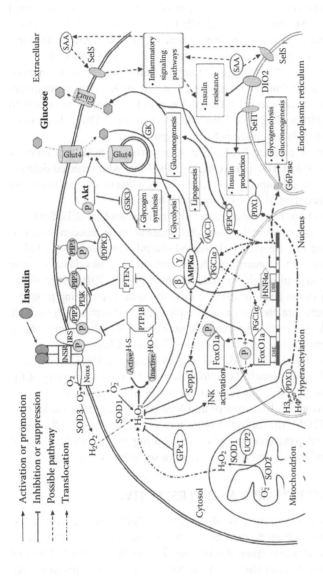

FIGURE 11.5 Pathways and coordinated mechanisms for multiple selenoproteins involved in insulin signaling and glucose metabolism. ACC1: acetyl-CoA carboxylase 1; Akt: protein kinase B; AMPK: adenosine monophosphate-activated protein kinase; DBE: DNA-binding element; DIO2: iodo-thyronine deiodinase 2; FoxO1a: forkhead box O1; G6Pase: glucose-6-phosphatase; GK: glucokinase; Glut: glucose transporter; GPx1: glutathione peroxidase 1; GSK3: glycogen synthase kinase 3; INSR: insulin receptor; IRS: insulin receptor substrate; Noxs: NAD(P)H oxidases; PIP2: phos-phatidylinositol 4,5-bisphosphate; PIP3: phosphatidylinositol (3,4,5)-trisphosphate; PDPK1: phosphoinositide-dependent kinase 1; PDX1: pancreatic duodenal homeobox 1; PEPCK: phosphoenolpyruvate carboxykinase; PGC1α: peroxisomal proliferator-activated receptor-γ coactivator 1α; PI3K: phos-phoinositide 3-kinase; PTEN: phosphatase and tensin homolog; PTP1B: protein-tyrosine phosphatase 1B; SAA: serum amyloid A; SelS: selenoprotein S; SelT: selenoprotein T; Sepp1: selenoprotein P; SOD: superoxide dismutase; UCP2: uncoupling protein 2.

action (Figure 11.5). Future research should therefore be directed toward more global or coordinated effects of dietary Se and various functional levels of selenoproteins on whole energy, rather than just glucose, metabolism on multiple hormones, rather than just insulin, involved in the regulation of the energy metabolism and on ER stress in addition to oxidative stress.

It is essential that we develop a full and accurate understanding of the role and mechanism(s) of Se and selenoproteins in the regulation of energy metabolism and the development of insulin resistance, diabetes, and the metabolic syndrome. Overlaid upon that is the need to take account of geographical locations and individual differences in selenoprotein polymorphisms that affect the ability to utilize or tolerate Se (see also Chapter 15). Only then will we be in a position to advise the public on appropriate Se intakes for optimal health.

ACKNOWLEDGMENTS

Research in the authors' laboratories was supported in part by NIH DK53018 and NSFC Projects 30628019, 30700585, 30871844, 31320103920, and 81372993. The work of H. S. was supported by a grant (STE 1782/2) of the Deutsche Forschungsgemeinschaft (DFG).

REFERENCES

Agbor, G. A., J. A. Vinson, S. Patel et al. 2007. "Effect of selenium- and glutathione-enriched yeast supplementation on a combined atherosclerosis and diabetes hamster model." *J Agric Food Chem* no. 55 (21):8731–6.

Akbaraly, T. N., J. Arnaud, M. P. Rayman et al. 2010. "Plasma selenium and risk of dysglycemia in an elderly French population: Results from the prospective Epidemiology of Vascular Ageing Study." *Nutr Metab (Lond)* no. 7:21.

Alfadda, A. A., and R. M. Sallam. 2012. "Reactive oxygen species in health and disease." *J Biomed Biotechnol* no. 2012:936486.

Alizadeh, M., A. Safaeiyan, A. Ostadrahimi et al. 2012. "Effect of L-arginine and selenium added to a hypocaloric diet enriched with legumes on cardiovascular disease risk factors in women with central obesity: A randomized, double-blind, placebo-controlled trial." *Ann Nutr Metab* no. 60 (2):157–68.

Allmang, C., L. Wurth, and A. Krol. 2009. "The selenium to selenoprotein pathway in eukaryotes: More molecular partners than anticipated." *Biochim Biophys Acta* no. 1790 (11):1415–23.

Al-Saleh, E., M. Nandakumaran, I. Al-Rashdan, J. Al-Harmi, and M. Al-Shammari. 2007. "Maternal-foetal status of copper, iron, molybdenum, selenium and zinc in obese gestational diabetic pregnancies." *Acta Diabetol* no. 44 (3):106–13.

Anderson, E. J., M. E. Lustig, K. E. Boyle et al. 2009. "Mitochondrial H2O2 emission and cellular redox state link excess fat intake to insulin resistance in both rodents and humans." *J Clin Invest* no. 119 (3):573–81.

Arrojo, E., R. Drigo, and A. C. Bianco. 2011. "Type 2 deiodinase at the crossroads of thyroid hormone action." *Int J Biochem Cell Biol* no. 43 (10):1432–41.

Arsenijevic, D., H. Onuma, C. Pecqueur et al. 2000. "Disruption of the uncoupling protein-2 gene in mice reveals a role in immunity and reactive oxygen species production." *Nat Genet* no. 26 (4):435–9.

Arthur, J. R. 2000. "The glutathione peroxidases." *Cell Mol Life Sci* no. 57 (13–14):1825–35.

Aydemir-Koksoy, A., and B. Turan. 2008. "Selenium inhibits proliferation signaling and restores sodium/potassium pump function of diabetic rat aorta." *Biol Trace Elem Res* no. 126 (1–3):237–45.

Bartosz, G. 2009. "Reactive oxygen species: Destroyers or messengers?" *Biochem Pharmacol* no. 77 (8):1303–15.

Bashan, N., J. Kovsan, I. Kachko, H. Ovadia, and A. Rudich. 2009. "Positive and negative regulation of insulin signaling by reactive oxygen and nitrogen species." *Physiol Rev* no. 89 (1):27–71.

Berg, E. A., J. Y. Wu, L. Campbell, M. Kagey, and S. R. Stapleton. 1995. "Insulin-like effects of vanadate and selenate on the expression of glucose-6-phosphate dehydrogenase and fatty acid synthase in diabetic rats." *Biochimie* no. 77 (12):919–24.

Bleys, J., A. Navas-Acien, and E. Guallar. 2007. "Serum selenium and diabetes in US adults." *Diabetes Care* no. 30 (4):829–34.

Bolajoko, E. B., K. S. Mossanda, F. Adeniyi, O. Akinosun, A. Fasanmade, and M. Moropane. 2008. "Antioxidant and oxidative stress status in type 2 diabetes and diabetic foot ulcer." *S Afr Med J* no. 98 (8):614–7.

Boudina, S., S. Sena, H. Theobald et al. 2007. "Mitochondrial energetics in the heart in obesity-related diabetes: Direct evidence for increased uncoupled respiration and activation of uncoupling proteins." *Diabetes* no. 56 (10):2457–66.

Brieger, K., S. Schiavone, F. J. Miller, Jr., and K. H. Krause. 2012. "Reactive oxygen species: From health to disease." *Swiss Med Wkly* no. 142:w13659.

Brownlee, M., A. Cerami, and H. Vlassara. 1988. "Advanced glycosylation end products in tissue and the biochemical basis of diabetic complications." *N Engl J Med* no. 318 (20):1315–21.

Bunk, M. J., and G. F. Combs, Jr. 1981. "Relationship of selenium-dependent glutathione peroxidase activity and nutritional pancreatic atrophy in selenium-deficient chicks." *J Nutr* no. 111 (9):1611–20.

Burk, R. F., and K. E. Hill. 2009. "Selenoprotein P-expression, functions, and roles in mammals." *Biochim Biophys Acta* no. 1790 (11):1441–7.

Byun, B. J., and Y. K. Kang. 2011. "Conformational preferences and pK(a) value of selenocysteine residue." *Biopolymers* no. 95 (5):345–53.

Campbell, S. C., A. Aldibbiat, C. E. Marriott et al. 2008. "Selenium stimulates pancreatic beta-cell gene expression and enhances islet function." *FEBS Lett* no. 582 (15):2333–7.

Canani, L. H., C. Capp, J. M. Dora et al. 2005. "The type 2 deiodinase A/G (Thr92Ala) polymorphism is associated with decreased enzyme velocity and increased insulin resistance in patients with type 2 diabetes mellitus." *J Clin Endocrinol Metab* no. 90 (6):3472–8.

Castro, L., and B. A. Freeman. 2001. "Reactive oxygen species in human health and disease." *Nutrition* no. 17 (2):161, 163–5.

Chen, X., T. O. Scholl, M. J. Leskiw, M. R. Donaldson, and T. P. Stein. 2003. "Association of glutathione peroxidase activity with insulin resistance and dietary fat intake during normal pregnancy." *J Clin Endocrinol Metab* no. 88 (12):5963–8.

Cheng, Z., and M. F. White. 2012. "The AKTion in non-canonical insulin signaling." *Nat Med* no. 18 (3):351–3.

Cnop, M., F. Foufelle, and L. A. Velloso. 2012. "Endoplasmic reticulum stress, obesity and diabetes." *Trends Mol Med* no. 18 (1):59–68.

Cser, A., I. Sziklai-Laszlo, H. Menzel, and I. Lombeck. 1993. "Selenium status and lipoproteins in healthy and diabetic children." *J Trace Elem Electrolytes Health Dis* no. 7 (4):205–10.

Curran, J. E., J. B. Jowett, K. S. Elliott et al. 2005. "Genetic variation in selenoprotein S influences inflammatory response." *Nat Genet* no. 37 (11):1234–41.

Czech, M. P., J. C. Lawrence, Jr., and W. S. Lynn. 1974. "Evidence for electron transfer reactions involved in the Cu^{2+}-dependent thiol activation of fat cell glucose utilization." *J Biol Chem* no. 249 (4):1001–6.

Czernichow, S., A. Couthouis, S. Bertrais et al. 2006. "Antioxidant supplementation does not affect fasting plasma glucose in the Supplementation with Antioxidant Vitamins and Minerals (SU.VI.MAX) study in France: Association with dietary intake and plasma concentrations." *Am J Clin Nutr* no. 84 (2):395–9.

Division for Nutrition, Danish Veterinary and Food Administration. 2006. Safe upper intake levels for vitamins and minerals. http://ec.europa.eu/food/food/labellingnutrition/supplements /documents/denmark_annex2.pdf.

Dominguez, C., E. Ruiz, M. Gussinye, and A. Carrascosa. 1998. "Oxidative stress at onset and in early stages of type 1 diabetes in children and adolescents." *Diabetes Care* no. 21 (10):1736–42.

Dora, J. M., W. E. Machado, J. Rheinheimer, D. Crispim, and A. L. Maia. 2010. "Association of the type 2 deiodinase Thr92Ala polymorphism with type 2 diabetes: Case-control study and meta-analysis." *Eur J Endocrinol* no. 163 (3):427–34.

Douillet, C., M. Bost, M. Accominotti, F. Borson-Chazot, and M. Ciavatti. 1998. "Effect of selenium and vitamin E supplements on tissue lipids, peroxides, and fatty acid distribution in experimental diabetes." *Lipids* no. 33 (4):393–9.

Du, J. L., C. K. Sun, B. Lu et al. 2008. "Association of SelS mRNA expression in omental adipose tissue with Homa-IR and serum amyloid A in patients with type 2 diabetes mellitus." *Chin Med J (Engl)* no. 121 (13):1165–8.

Ekmekcioglu, C., C. Prohaska, K. Pomazal, I. Steffan, G. Schernthaner, and W. Marktl. 2001. "Concentrations of seven trace elements in different hematological matrices in patients with type 2 diabetes as compared to healthy controls." *Biol Trace Elem Res* no. 79 (3):205–19.

Evans, J. L., I. D. Goldfine, B. A. Maddux, and G. M. Grodsky. 2002. "Oxidative stress and stress-activated signaling pathways: A unifying hypothesis of type 2 diabetes." *Endocr Rev* no. 23 (5):599–622.

Ezaki, O. 1990. "The insulin-like effects of selenate in rat adipocytes." *J Biol Chem* no. 265 (2):1124–8.

Faghihi, T., M. Radfar, M. Barmal et al. 2013. "A randomized, placebo-controlled trial of selenium supplementation in patients with type 2 diabetes: Effects on glucose homeostasis, oxidative stress, and lipid profile." *Am J Ther* no. 21 (6):491–5.

Fairweather-Tait, S. J., Y. Bao, M. R. Broadley et al. 2011. "Selenium in human health and disease." *Antioxid Redox Signal* no. 14 (7):1337–83.

Faure, P., O. Ramon, A. Favier, and S. Halimi. 2004. "Selenium supplementation decreases nuclear factor-kappa B activity in peripheral blood mononuclear cells from type 2 diabetic patients." *Eur J Clin Invest* no. 34 (7):475–81.

Ferrer, E., A. Alegria, R. Barbera, R. Farre, M. J. Lagarda, and J. Monleon. 1999. "Whole blood selenium content in pregnant women." *Sci Total Environ* no. 227 (2–3):139–43.

Finley, J. W., A. N. Kong, K. J. Hintze, E. H. Jeffery, L. L. Ji, and X. G. Lei. 2011. "Antioxidants in foods: State of the science important to the food industry." *J Agric Food Chem* no. 59 (13):6837–46.

Fourquet, S., M. E. Huang, B. D'Autreaux, and M. B. Toledano. 2008. "The dual functions of thiol-based peroxidases in H2O2 scavenging and signaling." *Antioxid Redox Signal* no. 10 (9):1565–76.

Gao, Y., K. Walder, T. Sunderland et al. 2003. "Elevation in Tanis expression alters glucose metabolism and insulin sensitivity in H4IIE cells." *Diabetes* no. 52 (4):929–34.

Gao, Y., H. C. Feng, K. Walder et al. 2004. "Regulation of the selenoprotein SelS by glucose deprivation and endoplasmic reticulum stress—SelS is a novel glucose-regulated protein." *FEBS Lett* no. 563 (1–3):185–90.

Gao, Y., N. R. Hannan, S. Wanyonyi et al. 2006. "Activation of the selenoprotein SEPS1 gene expression by pro-inflammatory cytokines in HepG2 cells." *Cytokine* no. 33 (5):246–51.

Gardner, C. D., S. Eguchi, C. M. Reynolds, K. Eguchi, G. D. Frank, and E. D. Motley. 2003. "Hydrogen peroxide inhibits insulin signaling in vascular smooth muscle cells." *Exp Biol Med (Maywood)* no. 228 (7):836–42.

Gebre-Medhin, M., U. Ewald, L. O. Plantin, and T. Tuvemo. 1984. "Elevated serum selenium in diabetic children." *Acta Paediatr Scand* no. 73 (1):109–14.

Gebre-Medhin, M., E. Kylberg, U. Ewald, and T. Tuvemo. 1985. "Dietary intake, trace elements and serum protein status in young diabetics." *Acta Paediatr Scand Suppl* no. 320:38–43.

Gloire, G., S. Legrand-Poels, and J. Piette. 2006. "NF-kappaB activation by reactive oxygen species: Fifteen years later." *Biochem Pharmacol* no. 72 (11):1493–505.

Goldstein, B. J., K. Mahadev, X. Wu, L. Zhu, and H. Motoshima. 2005. "Role of insulin-induced reactive oxygen species in the insulin signaling pathway." *Antioxid Redox Signal* no. 7 (7–8):1021–31.

Gough, D. R., and T. G. Cotter. 2011. "Hydrogen peroxide: A Jekyll and Hyde signalling molecule." *Cell Death Dis* no. 2:e213.

Grankvist, K., S. L. Marklund, and I. B. Taljedal. 1981. "CuZn-superoxide dismutase, Mn-superoxide dismutase, catalase and glutathione peroxidase in pancreatic islets and other tissues in the mouse." *Biochem J* no. 199 (2):393–8.

Granner, D., and S. Pilkis. 1990. "The genes of hepatic glucose metabolism." *J Biol Chem* no. 265 (18):10173–6.

Handy, D. E., E. Lubos, Y. Yang et al. 2009. "Glutathione peroxidase-1 regulates mitochondrial function to modulate redox-dependent cellular responses." *J Biol Chem* no. 284 (18):11913–21.

Hansen, L. L., Y. Ikeda, G. S. Olsen, A. K. Busch, and L. Mosthaf. 1999. "Insulin signaling is inhibited by micromolar concentrations of H(2)O(2). Evidence for a role of H(2)O(2) in tumor necrosis factor alpha-mediated insulin resistance." *J Biol Chem* no. 274 (35):25078–84.

Harmon, J. S., M. Bogdani, S. D. Parazzoli et al. 2009. "beta-Cell-specific overexpression of glutathione peroxidase preserves intranuclear MafA and reverses diabetes in db/db mice." *Endocrinology* no. 150 (11):4855–62.

Hawkes, W. C., and Z. Alkan. 2010. "Regulation of redox signaling by selenoproteins." *Biol Trace Elem Res* no. 134 (3):235–51.

Hawkes, W. C., Z. Alkan, K. Lang, and J. C. King. 2004. "Plasma selenium decrease during pregnancy is associated with glucose intolerance." *Biol Trace Elem Res* no. 100 (1):19–29.

Hei, Y. J., S. Farahbakhshian, X. Chen, M. L. Battell, and J. H. McNeill. 1998. "Stimulation of MAP kinase and S6 kinase by vanadium and selenium in rat adipocytes." *Mol Cell Biochem* no. 178 (1–2):367–75.

Hensley, K., K. A. Robinson, S. P. Gabbita, S. Salsman, and R. A. Floyd. 2000. "Reactive oxygen species, cell signaling, and cell injury." *Free Radic Biol Med* no. 28 (10):1456–62.

Hojlund, K., K. Wrzesinski, P. M. Larsen et al. 2003. "Proteome analysis reveals phosphorylation of ATP synthase beta -subunit in human skeletal muscle and proteins with potential roles in type 2 diabetes." *J Biol Chem* no. 278 (12):10436–42.

Hong, T., J. Ning, X. Yang et al. 2011. "Fine-tuned regulation of the PGC-1alpha gene transcription by different intracellular signaling pathways." *Am J Physiol Endocrinol Metab* no. 300 (3):E500–7.

Houstis, N., E. D. Rosen, and E. S. Lander. 2006. "Reactive oxygen species have a causal role in multiple forms of insulin resistance." *Nature* no. 440 (7086):944–8.

Hwang, D., S. Seo, Y. Kim et al. 2007. "Selenium acts as an insulin-like molecule for the down-regulation of diabetic symptoms via endoplasmic reticulum stress and insulin signalling proteins in diabetes-induced non-obese diabetic mice." *J Biosci* no. 32 (4):723–35.

Johansson, L., G. Gafvelin, and E. S. Arner. 2005. "Selenocysteine in proteins-properties and biotechnological use." *Biochim Biophys Acta* no. 1726 (1):1–13.

Kadowaki, T., N. Kubota, K. Ueki, and T. Yamauchi. 2012. "SnapShot: Physiology of insulin signaling." *Cell* no. 148 (4):834–834 e1.

Kahler, W., B. Kuklinski, C. Ruhlmann, and C. Plotz. 1993. "[Diabetes mellitus—A free radical-associated disease. Results of adjuvant antioxidant supplementation]." *Z Gesamte Inn Med* no. 48 (5):223–32.

Kahn, B. B., T. Alquier, D. Carling, and D. G. Hardie. 2005. "AMP-activated protein kinase: Ancient energy gauge provides clues to modern understanding of metabolism." *Cell Metab* no. 1 (1):15–25.

Karlsson, H. K., H. Tsuchida, S. Lake, H. A. Koistinen, and A. Krook. 2004. "Relationship between serum amyloid A level and Tanis/SelS mRNA expression in skeletal muscle and adipose tissue from healthy and type 2 diabetic subjects." *Diabetes* no. 53 (6):1424–8.

Kelley, D. E., J. He, E. V. Menshikova, and V. B. Ritov. 2002. "Dysfunction of mitochondria in human skeletal muscle in type 2 diabetes." *Diabetes* no. 51 (10):2944–50.

Kilinc, M., M. A. Guven, M. Ezer, I. E. Ertas, and A. Coskun. 2008. "Evaluation of serum selenium levels in Turkish women with gestational diabetes mellitus, glucose intolerants, and normal controls." *Biol Trace Elem Res* no. 123 (1–3):35–40.

Klein, E. A., I. M. Thompson, Jr., C. M. Tangen et al. 2011. "Vitamin E and the risk of prostate cancer: The Selenium and Vitamin E Cancer Prevention Trial (SELECT)." *JAMA* no. 306 (14):1549–56.

Kljai, K., and R. Runje. 2001. "Selenium and glycogen levels in diabetic patients." *Biol Trace Elem Res* no. 83 (3):223–9.

Kryukov, G. V., S. Castellano, S. V. Novoselov et al. 2003. "Characterization of mammalian selenoproteomes." *Science* no. 300 (5624):1439–43.

Kurth-Kraczek, E. J., M. F. Hirshman, L. J. Goodyear, and W. W. Winder. 1999. "5' AMP-activated protein kinase activation causes GLUT4 translocation in skeletal muscle." *Diabetes* no. 48 (8):1667–71.

L'Abbé, M. R., and K. D. Trick. 1994. "Changes in pancreatic glutathione peroxidase and superoxide dismutase activities in the prediabetic diabetes-prone BB rat." *Proc Soc Exp Biol Med* no. 207 (2):206–12.

Labunskyy, V. M., B. C. Lee, D. E. Handy, J. Loscalzo, D. L. Hatfield, and V. N. Gladyshev. 2011. "Both maximal expression of selenoproteins and selenoprotein deficiency can promote development of type 2 diabetes-like phenotype in mice." *Antioxid Redox Signal* no. 14 (12):2327–36.

Laclaustra, M., A. Navas-Acien, S. Stranges, J. M. Ordovas, and E. Guallar. 2009. "Serum selenium concentrations and diabetes in US adults: National Health and Nutrition Examination Survey (NHANES) 2003-2004." *Environ Health Perspect* no. 117 (9):1409–13.

Ladenstein, R., O. Epp, K. Bartels, A. Jones, R. Huber, and A. Wendel. 1979. "Structure analysis and molecular model of the selenoenzyme glutathione peroxidase at 2.8 A resolution." *J Mol Biol* no. 134 (2):199–218.

Lei, X. G., W. H. Cheng, and J. P. McClung. 2007. "Metabolic regulation and function of glutathione peroxidase-1." *Annu Rev Nutr* no. 27:41–61.

Leloup, C., C. Tourrel-Cuzin, C. Magnan et al. 2009. "Mitochondrial reactive oxygen species are obligatory signals for glucose-induced insulin secretion." *Diabetes* no. 58 (3):673–81.

Lenart, A., and K. Pawlowski. 2013. "Intersection of selenoproteins and kinase signalling." *Biochim Biophys Acta.* no. 1834 (7):1279–84.

Lippman, S. M., E. A. Klein, P. J. Goodman et al. 2009. "Effect of selenium and vitamin E on risk of prostate cancer and other cancers: The Selenium and Vitamin E Cancer Prevention Trial (SELECT)." *JAMA* no. 301 (1):39–51.

Liu, J., H. Tang, L. Niu, and Y. Xu. 2009. "Upregulation of Tanis mRNA expression in the liver is associated with insulin resistance in rats." *Tohoku J Exp Med* no. 219 (4):307–10.

Liu, Y., H. Zhao, Q. Zhang et al. 2012. "Prolonged dietary selenium deficiency or excess does not globally affect selenoprotein gene expression and/or protein production in various tissues of pigs." *J Nutr* no. 142 (8):1410–6.

Loh, K., H. Deng, A. Fukushima et al. 2009. "Reactive oxygen species enhance insulin sensitivity." *Cell Metab* no. 10 (4):260–72.

MacDonald, P. E., J. W. Joseph, and P. Rorsman. 2005. "Glucose-sensing mechanisms in pancreatic beta-cells." *Philos Trans R Soc Lond B Biol Sci* no. 360 (1464):2211–25.

Mahadev, K., A. Zilbering, L. Zhu, and B. J. Goldstein. 2001. "Insulin-stimulated hydrogen peroxide reversibly inhibits protein-tyrosine phosphatase 1b *in vivo* and enhances the early insulin action cascade." *J Biol Chem* no. 276 (24):21938–42.

Mahadev, K., H. Motoshima, X. Wu et al. 2004. "The NAD(P)H oxidase homolog Nox4 modulates insulin-stimulated generation of H2O2 and plays an integral role in insulin signal transduction." *Mol Cell Biol* no. 24 (5):1844–54.

Manning, B. D. 2010. "Insulin signaling: Inositol phosphates get into the Akt." *Cell* no. 143 (6):861–3.

Marsili, A., C. Aguayo-Mazzucato, T. Chen et al. 2011. "Mice with a targeted deletion of the type 2 deiodinase are insulin resistant and susceptible to diet induced obesity." *PLoS One* no. 6 (6):e20832.

Martin-Gallan, P., A. Carrascosa, M. Gussinye, and C. Dominguez. 2003. "Biomarkers of diabetes-associated oxidative stress and antioxidant status in young diabetic patients with or without subclinical complications." *Free Radic Biol Med* no. 34 (12):1563–74.

McClung, J. P., C. A. Roneker, W. Mu et al. 2004. "Development of insulin resistance and obesity in mice overexpressing cellular glutathione peroxidase." *Proc Natl Acad Sci U S A* no. 101 (24):8852–7.

McNeill, J. H., H. L. Delgatty, and M. L. Battell. 1991. "Insulin like effects of sodium selenate in streptozocin-induced diabetic rats." *Diabetes* no. 40 (12):1675–8.

Memisogullari, R., S. Taysi, E. Bakan, and I. Capoglu. 2003. "Antioxidant status and lipid peroxidation in type II diabetes mellitus." *Cell Biochem Funct* no. 21 (3):291–6.

Mentuccia, D., L. Proietti-Pannunzi, K. Tanner et al. 2002. "Association between a novel variant of the human type 2 deiodinase gene Thr92Ala and insulin resistance: Evidence of interaction with the Trp64Arg variant of the beta-3-adrenergic receptor." *Diabetes* no. 51 (3):880–3.

Mézes, M., and K. Balogh. 2009. "Prooxidant mechanisms of selenium toxicity—A review." *Acta Biol Szeged* no. 53 (S1):4.

Misu, H., T. Takamura, H. Takayama et al. 2010. "A liver-derived secretory protein, selenoprotein P, causes insulin resistance." *Cell Metab* no. 12 (5):483–95.

Misu, H., K. Ishikura, S. Kurita et al. 2012. "Inverse correlation between serum levels of selenoprotein P and adiponectin in patients with type 2 diabetes." *PLoS One* no. 7 (4):e34952.

Molnar, J., Z. Garamvolgyi, M. Herold, N. Adanyi, A. Somogyi, and J. Rigo, Jr. 2008. "Serum selenium concentrations correlate significantly with inflammatory biomarker high-sensitive CRP levels in Hungarian gestational diabetic and healthy pregnant women at mid-pregnancy." *Biol Trace Elem Res* no. 121 (1):16–22.

Mueller, A. S., and J. Pallauf. 2006. "Compendium of the antidiabetic effects of supranutritional selenate doses: *In vivo* and *in vitro* investigations with type II diabetic db/db mice." *J Nutr Biochem* no. 17 (8):548–60.

Mueller, A. S., J. Pallauf, and J. Rafael. 2003. "The chemical form of selenium affects insulin-omimetic properties of the trace element: Investigations in type II diabetic dbdb mice." *J Nutr Biochem* no. 14 (11):637–47.

Mukherjee, B., S. Anbazhagan, A. Roy, R. Ghosh, and M. Chatterjee. 1998. "Novel implications of the potential role of selenium on antioxidant status in streptozotocin-induced diabetic mice." *Biomed Pharmacother* no. 52 (2):89–95.

Mysore, T. B., T. A. Shinkel, J. Collins et al. 2005. "Overexpression of glutathione peroxidase with two isoforms of superoxide dismutase protects mouse islets from oxidative injury and improves islet graft function." *Diabetes* no. 54 (7):2109–16.

Navarro-Alarcon, M., H. Lopez-Garcia de la Serrana, V. Perez-Valero, and C. Lopez-Martinez. 1999. "Serum and urine selenium concentrations in patients with cardiovascular diseases and relationship to other nutritional indexes." *Ann Nutr Metab* no. 43 (1):30–6.

NRC. 1995. *Nutrient requirements of laboratory animals*. 4th ed. Washington, DC: National Academy Press.

Obeid, O., M. Elfakhani, S. Hlais et al. 2008. "Plasma copper, zinc, and selenium levels and correlates with metabolic syndrome components of lebanese adults." *Biol Trace Elem Res* no. 123 (1–3):58–65.

Offield, M. F., T. L. Jetton, P. A. Labosky et al. 1996. "PDX-1 is required for pancreatic outgrowth and differentiation of the rostral duodenum." *Development* no. 122 (3):983–95.

Olsson, M., B. Olsson, P. Jacobson et al. 2011. "Expression of the selenoprotein S (SELS) gene in subcutaneous adipose tissue and SELS genotype are associated with metabolic risk factors." *Metabolism* no. 60 (1):114–20.

Ozdemir, S., M. Ayaz, B. Can, and B. Turan. 2005. "Effect of selenite treatment on ultrastructural changes in experimental diabetic rat bones." *Biol Trace Elem Res* no. 107 (2):167–79.

Park, K., E. B. Rimm, D. S. Siscovick et al. 2012. "Toenail selenium and incidence of type 2 diabetes in US men and women." *Diabetes Care* no. 35 (7):1544–51.

Pepper, M. P., M. Z. Vatamaniuk, X. Yan, C. A. Roneker, and X. G. Lei. 2011. "Impacts of dietary selenium deficiency on metabolic phenotypes of diet-restricted GPX1-overexpressing mice." *Antioxid Redox Signal* no. 14 (3):383–90.

Pi, J., Y. Bai, Q. Zhang et al. 2007. "Reactive oxygen species as a signal in glucose-stimulated insulin secretion." *Diabetes* no. 56 (7):1783–91.

Pi, J., Q. Zhang, J. Fu et al. 2010. "ROS signaling, oxidative stress and Nrf2 in pancreatic beta-cell function." *Toxicol Appl Pharmacol* no. 244 (1):77–83.

Pillay, T. S., and M. W. Makgoba. 1992. "Enhancement of epidermal growth factor (EGF) and insulin-stimulated tyrosine phosphorylation of endogenous substrates by sodium selenate." *FEBS Lett* no. 308 (1):38–42.

Pinto, A., B. Speckmann, M. Heisler, H. Sies, and H. Steinbrenner. 2011. "Delaying of insulin signal transduction in skeletal muscle cells by selenium compounds." *J Inorg Biochem* no. 105 (6):812–20.

Pinto, A., D. T. Juniper, M. Sanil et al. 2012. "Supranutritional selenium induces alterations in molecular targets related to energy metabolism in skeletal muscle and visceral adipose tissue of pigs." *J Inorg Biochem* no. 114:47–54.

Prevost, G., A. Arabo, L. Jian et al. 2013. "The PACAP-regulated gene selenoprotein T is abundantly expressed in mouse and human beta-cells and its targeted inactivation impairs glucose tolerance." *Endocrinology* no. 154 (10):3796–806.

Puigserver, P., J. Rhee, J. Donovan et al. 2003. "Insulin-regulated hepatic gluconeogenesis through FOXO1-PGC-1alpha interaction." *Nature* no. 423 (6939):550–5.

Rains, J. L., and S. K. Jain. 2011. "Oxidative stress, insulin signaling, and diabetes." *Free Radic Biol Med* no. 50 (5):567–75.

Rajpathak, S., E. Rimm, J. S. Morris, and F. Hu. 2005. "Toenail selenium and cardiovascular disease in men with diabetes." *J Am Coll Nutr* no. 24 (4):250–6.

Ray, P. D., B. W. Huang, and Y. Tsuji. 2012. "Reactive oxygen species (ROS) homeostasis and redox regulation in cellular signaling." *Cell Signal* no. 24 (5):981–90.

Rayman, M. P. 2012. "Selenium and human health." *Lancet* no. 379 (9822):1256–68.

Rayman, M. P., and S. Stranges. 2013. "Epidemiology of selenium and type 2 diabetes: Can we make sense of it?" *Free Radic Biol Med* no. 65:1557–64.

Rayman, M. P., H. G. Infante, and M. Sargent. 2008. "Food-chain selenium and human health: Spotlight on speciation." *Br J Nutr* no. 100 (2):238–53.

Rayman, M. P., E. Searle, L. Kelly et al. 2014. "Effect of selenium on markers of risk of pre-eclampsia in UK pregnant women: A randomised, controlled pilot trial." *Br J Nutr* no. 112 (1):99–111.

Roberts, C. K., and K. K. Sindhu. 2009. "Oxidative stress and metabolic syndrome." *Life Sci* no. 84 (21–22):705–12.

Rochon, C., I. Tauveron, C. Dejax et al. 2003. "Response of glucose disposal to hyperinsulinaemia in human hypothyroidism and hyperthyroidism." *Clin Sci (Lond)* no. 104 (1):7–15.

Rock, C., and P. J. Moos. 2009. "Selenoprotein P regulation by the glucocorticoid receptor." *Biometals* no. 22 (6):995–1009.

Rosen, P., P. P. Nawroth, G. King, W. Moller, H. J. Tritschler, and L. Packer. 2001. "The role of oxidative stress in the onset and progression of diabetes and its complications: A summary of a Congress Series sponsored by UNESCO-MCBN, the American Diabetes Association and the German Diabetes Society." *Diabetes Metab Res Rev* no. 17 (3):189–212.

Saito, Y., N. Sato, M. Hirashima, G. Takebe, S. Nagasawa, and K. Takahashi. 2004. "Domain structure of bi-functional selenoprotein P." *Biochem J* no. 381 (Pt 3):841–6.

Salonen, J. T., K. Nyyssonen, T. P. Tuomainen et al. 1995. "Increased risk of non-insulin dependent diabetes mellitus at low plasma vitamin E concentrations: A four year follow up study in men." *BMJ* no. 311 (7013):1124–7.

Saltiel, A. R., and C. R. Kahn. 2001. "Insulin signalling and the regulation of glucose and lipid metabolism." *Nature* no. 414 (6865):799–806.

Schoenmakers, E., M. Agostini, C. Mitchell et al. 2010. "Mutations in the selenocysteine insertion sequence-binding protein 2 gene lead to a multisystem selenoprotein deficiency disorder in humans." *J Clin Invest* no. 120 (12):4220–35.

Scott, K. E., F. B. Wheeler, A. L. Davis et al. 2012. "Metabolic regulation of invadopodia and invasion by acetyl-CoA carboxylase 1 and de novo lipogenesis." *PLoS One* no. 7 (1):e29761.

Sheng, X. Q., K. X. Huang, and H. B. Xu. 2004. "New experimental observation on the relationship of selenium and diabetes mellitus." *Biol Trace Elem Res* no. 99 (1–3):241–53.

Skaff, O., D. I. Pattison, P. E. Morgan et al. 2012. "Selenium-containing amino acids are targets for myeloperoxidase-derived hypothiocyanous acid: Determination of absolute rate constants and implications for biological damage." *Biochem J* no. 441 (1): 305–16.

Southgate, R. J., C. R. Bruce, A. L. Carey et al. 2005. "PGC-1alpha gene expression is down-regulated by Akt- mediated phosphorylation and nuclear exclusion of FoxO1 in insulin-stimulated skeletal muscle." *FASEB J* no. 19 (14):2072–4.

Speckmann, B., P. L. Walter, L. Alili et al. 2008. "Selenoprotein P expression is controlled through interaction of the coactivator PGC-1alpha with FoxO1a and hepatocyte nuclear factor 4alpha transcription factors." *Hepatology* no. 48 (6):1998–2006.

Speckmann, B., H. Sies, and H. Steinbrenner. 2009. "Attenuation of hepatic expression and secretion of selenoprotein P by metformin." *Biochem Biophys Res Commun* no. 387 (1):158–63.

Squires, J. E., and M. J. Berry. 2008. "Eukaryotic selenoprotein synthesis: Mechanistic insight incorporating new factors and new functions for old factors." *IUBMB Life* no. 60 (4):232–5.

Stapleton, S. R., G. L. Garlock, L. Foellmi-Adams, and R. F. Kletzien. 1997. "Selenium: potent stimulator of tyrosyl phosphorylation and activator of MAP kinase." *Biochim Biophys Acta* no. 1355 (3):259–69.

Steinbrenner, H. 2013. "Interference of selenium and selenoproteins with the insulin-regulated carbohydrate and lipid metabolism." *Free Radic Biol Med* no. 65:1538–47.

Steinbrenner, H., and H. Sies. 2009. "Protection against reactive oxygen species by seleno-proteins." *Biochim Biophys Acta* no. 1790 (11):1478–85.

Steinbrenner, H., B. Speckmann, A. Pinto, and H. Sies. 2011. "High selenium intake and increased diabetes risk: Experimental evidence for interplay between selenium and carbohydrate metabolism." *J Clin Biochem Nutr* no. 48 (1):40–5.

Steinbrenner, H., A. L. Hotze, B. Speckmann et al. 2013. "Localization and regulation of pancreatic selenoprotein P." *J Mol Endocrinol* no. 50 (1):31–42.

Stoffers, D. A., V. Stanojevic, and J. F. Habener. 1998. "Insulin promoter factor-1 gene mutation linked to early-onset type 2 diabetes mellitus directs expression of a dominant negative isoprotein." *J Clin Invest* no. 102 (1):232–41.

Stoytcheva, Z., R. M. Tujebajeva, J. W. Harney, and M. J. Berry. 2006. "Efficient incorporation of multiple selenocysteines involves an inefficient decoding step serving as a potential translational checkpoint and ribosome bottleneck." *Mol Cell Biol* no. 26 (24):9177–84.

Stranges, S., J. R. Marshall, R. Natarajan et al. 2007. "Effects of long-term selenium supplementation on the incidence of type 2 diabetes: A randomized trial." *Ann Intern Med* no. 147 (4):217–23.

Sundaram, R. K., A. Bhaskar, S. Vijayalingam, M. Viswanathan, R. Mohan, and K. R. Shanmugasundaram. 1996. "Antioxidant status and lipid peroxidation in type II diabetes mellitus with and without complications." *Clin Sci (Lond)* no. 90 (4):255–60.

Sunde, R. A., and A. M. Raines. 2011. "Selenium regulation of the selenoprotein and nonselenoprotein transcriptomes in rodents." *Adv Nutr* no. 2 (2):138–50.

Tan, M., L. Sheng, Y. Qian et al. 2001. "Changes of serum selenium in pregnant women with gestational diabetes mellitus." *Biol Trace Elem Res* no. 83 (3):231–7.

Tanaka, Y., P. O. Tran, J. Harmon, and R. P. Robertson. 2002. "A role for glutathione peroxidase in protecting pancreatic beta cells against oxidative stress in a model of glucose toxicity." *Proc Natl Acad Sci U S A* no. 99 (19):12363–8.

Taniguchi, C. M., B. Emanuelli, and C. R. Kahn. 2006. "Critical nodes in signalling pathways: Insights into insulin action." *Nat Rev Mol Cell Biol* no. 7 (2):85–96.

Thannickal, V. J., and B. L. Fanburg. 2000. "Reactive oxygen species in cell signaling." *Am J Physiol Lung Cell Mol Physiol* no. 279 (6):L1005–28.

Thompson, J. N., and M. L. Scott. 1970. "Impaired lipid and vitamin E absorption related to atrophy of the pancreas in selenium-deficient chicks." *J Nutr* no. 100 (7):797–809.

Tiganis, T. 2011. "Reactive oxygen species and insulin resistance: The good, the bad and the ugly." *Trends Pharmacol Sci* no. 32 (2):82–9.

Trachootham, D., W. Lu, M. A. Ogasawara, R. D. Nilsa, and P. Huang. 2008. "Redox regulation of cell survival." *Antioxid Redox Signal* no. 10 (8):1343–74.

Turanov, A. A., X. M. Xu, B. A. Carlson, M. H. Yoo, V. N. Gladyshev, and D. L. Hatfield. 2011. "Biosynthesis of selenocysteine, the 21st amino acid in the genetic code, and a novel pathway for cysteine biosynthesis." *Adv Nutr* no. 2 (2):122–8.

Turner, N., and L. K. Heilbronn. 2008. "Is mitochondrial dysfunction a cause of insulin resistance?" *Trends Endocrinol Metab* no. 19 (9):324–30.

Uchizono, Y., R. Takeya, M. Iwase et al. 2006. "Expression of isoforms of NADPH oxidase components in rat pancreatic islets." *Life Sci* no. 80 (2):133–9.

Ulusu, N. N., and B. Turan. 2005. "Beneficial effects of selenium on some enzymes of diabetic rat heart." *Biol Trace Elem Res* no. 103 (3):207–16.

Veal, E. A., A. M. Day, and B. A. Morgan. 2007. "Hydrogen peroxide sensing and signaling." *Mol Cell* no. 26 (1):1–14.

Vijayalingam, S., A. Parthiban, K. R. Shanmugasundaram, and V. Mohan. 1996. "Abnormal antioxidant status in impaired glucose tolerance and non-insulin-dependent diabetes mellitus." *Diabet Med* no. 13 (8):715–9.

Walder, K., L. Kantham, J. S. McMillan et al. 2002. "Tanis: A link between type 2 diabetes and inflammation?" *Diabetes* no. 51 (6):1859–66.

Walter, P. L., H. Steinbrenner, A. Barthel, and L. O. Klotz. 2008. "Stimulation of selenoprotein P promoter activity in hepatoma cells by FoxO1a transcription factor." *Biochem Biophys Res Commun* no. 365 (2):316–21.

Wang, L., Z. Jiang, and X. G. Lei. 2012. "Knockout of SOD1 alters murine hepatic glycolysis, gluconeogenesis, and lipogenesis." *Free Radic Biol Med* no. 53 (9):1689–96.

Wang, W. C., A. L. Makela, V. Nanto, and P. Makela. 1995. "Serum selenium levels in diabetic children. A followup study during selenium-enriched agricultural fertilization in Finland." *Biol Trace Elem Res* no. 47 (1–3):355–64.

Wang, X. D., M. Z. Vatamaniuk, S. K. Wang, C. A. Roneker, R. A. Simmons, and X. G. Lei. 2008. "Molecular mechanisms for hyperinsulinaemia induced by overproduction of selenium-dependent glutathione peroxidase-1 in mice." *Diabetologia* no. 51 (8): 1515–24.

Wang, X., M. Z. Vatamaniuk, C. A. Roneker et al. 2011. "Knockouts of SOD1 and GPX1 exert different impacts on murine islet function and pancreatic integrity." *Antioxid Redox Signal* no. 14 (3):391–401.

Wang, X., W. Zhang, H. Chen et al. 2014. "High selenium impairs hepatic insulin sensitivity through opposite regulation of ROS." *Toxicol Lett* no. 224 (1):16–23.

Watson, J. D. 2014. "Type 2 diabetes as a redox disease." *Lancet* no. 383 (9919):841–3.

Whiting, P. H., A. Kalansooriya, I. Holbrook, F. Haddad, and P. E. Jennings. 2008. "The relationship between chronic glycaemic control and oxidative stress in type 2 diabetes mellitus." *Br J Biomed Sci* no. 65 (2):71–4.

Winzell, M. S., M. Coghlan, B. Leighton et al. 2011. "Chronic glucokinase activation reduces glycaemia and improves glucose tolerance in high-fat diet fed mice." *Eur J Pharmacol* no. 663 (1–3):80–6.

Yan, X., M. P. Pepper, M. Z. Vatamaniuk, C. A. Roneker, L. Li, and X. G. Lei. 2012. "Dietary selenium deficiency partially rescues type 2 diabetes-like phenotypes of glutathione peroxidase-1-overexpressing male mice." *J Nutr* no. 142 (11):1975–82.

Yang, K. C., L. T. Lee, Y. S. Lee, H. Y. Huang, C. Y. Chen, and K. C. Huang. 2010. "Serum selenium concentration is associated with metabolic factors in the elderly: A cross-sectional study." *Nutr Metab (Lond)* no. 7:38.

Yang, S. J., S. Y. Hwang, H. Y. Choi et al. 2011. "Serum selenoprotein P levels in patients with type 2 diabetes and prediabetes: Implications for insulin resistance, inflammation, and atherosclerosis." *J Clin Endocrinol Metab* no. 96 (8):E1325–9.

Yasui, S., K. Mawatari, R. Morizumi et al. 2012. "Hydrogen peroxide inhibits insulin-induced ATP-sensitive potassium channel activation independent of insulin signaling pathway in cultured vascular smooth muscle cells." *J Med Invest* no. 59 (1–2):36–44.

Ye, Y., Y. Shibata, C. Yun, D. Ron, and T. A. Rapoport. 2004. "A membrane protein complex mediates retro-translocation from the ER lumen into the cytosol." *Nature* no. 429 (6994):841–7.

Yoon, J. C., P. Puigserver, G. Chen et al. 2001. "Control of hepatic gluconeogenesis through the transcriptional coactivator PGC-1." *Nature* no. 413 (6852):131–8.

Zeng, M. S., X. Li, Y. Liu et al. 2012. "A high-selenium diet induces insulin resistance in gestating rats and their offspring." *Free Radic Biol Med* no. 52 (8):1335–42.

Zhang, C. Y., G. Baffy, P. Perret et al. 2001. "Uncoupling protein-2 negatively regulates insulin secretion and is a major link between obesity, beta cell dysfunction, and type 2 diabetes." *Cell* no. 105 (6):745–55.

Zhang, N., W. Jing, J. Cheng et al. 2011. "Molecular characterization and NF-kappaB-regulated transcription of selenoprotein S from the Bama mini-pig." *Mol Biol Rep* no. 38 (7):4281–6.

Zhang, Y., and X. Chen. 2011. "Reducing selenoprotein P expression suppresses adipocyte differentiation as a result of increased preadipocyte inflammation." *Am J Physiol Endocrinol Metab* no. 300 (1):E77–85.

Zhou, J., K. Huang, and X. G. Lei. 2013. "Selenium and diabetes—Evidence from animal studies." *Free Radic Biol Med* no. 65:1548–56.

Zmijewski, J. W., S. Banerjee, H. Bae, A. Friggeri, E. R. Lazarowski, and E. Abraham. 2010. "Exposure to hydrogen peroxide induces oxidation and activation of AMP-activated protein kinase." *J Biol Chem* no. 285 (43):33154–64.

12 Selenoproteins and the Metabolic Syndrome

Lucia A. Seale, Ann Marie Zavacki,
and Marla J. Berry

CONTENTS

12.1 Introduction 248
12.2 Insulin Resistance as a Hallmark of Metabolic Syndrome 248
12.3 Selenium and Metabolic Syndrome...249
    12.3.1 Effects of Selenium in Energy Metabolism:
        Epidemiological Studies 249
    12.3.2 Effects of Selenium in Energy Metabolism: Role
        of Selenoprotein Synthesis Factors...250
        12.3.2.1 tRNA$^{[Ser]Sec}$ 250
        12.3.2.2 Selenocysteine Lyase 251
12.4 Specific Selenoproteins and Energy Metabolism 252
    12.4.1 Iodothyronine Deiodinases...255
        12.4.1.1 Role of Deiodinases in Energy Dissipation via Brown
            Adipose Tissue..257
        12.4.1.2 Role of Local Thyroid Hormone Activation
            and Inactivation on Food Intake 258
        12.4.1.3 Role of Deiodinases in Insulin Sensitivity and Glucose
            Utilization 258
    12.4.2 Glutathione Peroxidases 259
        12.4.2.1 Glutathione Peroxidase 1 (GPx1).................................260
        12.4.2.2 Glutathione Peroxidase 3 (GPx3).................................260
        12.4.2.3 Glutathione Peroxidase 4 (GPx4).................................261
    12.4.3 Selenoprotein M (SelM) 262
    12.4.4 Selenoprotein N (SelN)...262
    12.4.5 Selenoprotein P (Sepp1) 262
    12.4.6 Selenoprotein S (SelS) 263
    12.4.7 Selenoprotein T (SelT)..263
    12.4.8 Thioredoxin Reductases (TrxR) 264
12.5 Closing Remarks..264
References..264

12.1 INTRODUCTION

We are witnessing a global epidemic of the metabolic syndrome, a health condition defined primarily by the presence of central and peripheral insulin resistance that leads to the concomitant occurrence of several risk factors for cardiovascular diseases and type 2 diabetes, such as obesity, dyslipidemia, hypertension, and hyperinsulinemia (i.e., excess insulin in circulation) (Huang, 2009). Our dietary habits and diminished physical activity are considered the main triggers of metabolic syndrome development although the contribution of a genetic component to this pathogenesis cannot be disregarded. In this chapter, we will examine how an essential dietary component, selenium, contributes to the development and manifestations of metabolic syndrome through a family of proteins that contain the rare amino acid selenocysteine (Sec).

12.2 INSULIN RESISTANCE AS A HALLMARK OF METABOLIC SYNDROME

Insulin is a central player in the metabolic syndrome due to its role as the main coordinator of glucose metabolism, and thus understanding the regulation of insulin production, secretion, and mechanism of action are fundamental to appropriate analysis of disease progression. Insulin is an anabolic peptide hormone produced by the β-cells in the islets of Langerhans of the pancreas and secreted upon elevation of glucose levels in the bloodstream. This hormone is produced as a pro-hormone, requiring intracellular processing to become an active hormone. Insulin processing pathways have been discussed in detail elsewhere (Davidson, 2004), and the importance of the oxidative environment of the β-cell for proper insulin production and secretion is well established (Cerf, 2007; Graciano et al., 2011). Once secreted into the bloodstream, insulin acts on target cells, such as hepatocytes, adipocytes, and skeletal muscle cells, by binding to its cell surface receptor and activating a tyrosine phosphorylation signaling cascade. This allows glucose to be taken up and used as fuel for the maintenance of cellular homeostasis by target cells, such as myocytes, hepatocytes, and adipocytes. Disruption of this cascade is a hallmark of insulin resistance and ultimately leads to glucose failing to enter the cell, thus affecting energy dynamics. A primary trigger for insulin resistance is hyperinsulinemia, which occurs as a direct result of increased circulating levels of glucose. During normal metabolism, a rise in glucose levels triggers an increase in the number and activity of pancreatic β-cells along with the stimulation of insulin production and secretion. However, continual hyperglycemia leads to chronic stimulation of target cells by insulin, resulting in a weakening of the insulin signal and, ultimately, failure of the downstream tyrosine phosphorylation cascade that impairs energy metabolism (Kahn and Goldfine, 1993).

Despite its classic regulatory role in carbohydrate metabolism, the role of insulin in regulating other major components of energy metabolism, such as lipids, should also be carefully analyzed because these also contribute to obesity and cardiovascular disease. Among all occurrences that characterize metabolic syndrome, obesity is pinpointed as the major independent risk factor for its pathogenesis, triggering impairment of lipid and carbohydrate handling and promoting insulin resistance and hypertension. In our bodies, excess lipid is stored in adipose tissue; however, caloric excess may promote ectopic fat storage in tissues such as pancreatic β-cells, hepatocytes, and cardiac muscle, thus impairing their normal functions. Excess lipids can also lead to its accumulation in the visceral area, or abdominal obesity, and a large waistline is a strong predictor of metabolic syndrome development in humans (Lusis et al., 2008). Moreover, obesity increases the release of free fatty acids and cytokines, such as tumor necrosis factor alpha (TNF-α) into the bloodstream, and these molecules can antagonize insulin action while decreasing the release of the adipocyte hormone adiponectin, an insulin sensitizer.

12.3 SELENIUM AND METABOLIC SYNDROME

Numerous macro- and micronutrients are essential for carbohydrate and lipid metabolism. Among them, it is well established that selenium regulates specific aspects of energy metabolism, including gluconeogenesis (Kiersztan et al., 2007). However, the mechanisms connecting selenium and energy balance with insulin pathways and lipid handling are not fully understood. Selenite supplementation has been demonstrated to stimulate insulin production in a pancreatic β-cell line via upregulation of insulin promoter factor 1 (Ipf1), a major regulator of insulin mRNA expression (Campbell et al., 2008). On the other hand, in the *db/db* type 2 diabetes mouse model, treatment with selenate (Se [VI]) protected against weight gain with decreases in plasma cholesterol and triglycerides and improved insulin sensitivity (Mueller and Pallauf, 2006). However, in the same study, treatment with supranutritional amounts of selenite (Se [IV]) did not yield the same results. More studies aimed at elucidating the mechanisms behind these contradictory findings are needed.

Here we examine the role of selenium acting through the mediators of selenium action, selenoproteins, in metabolic disorders with a focus on the specific selenoproteins that have been demonstrated to participate, directly or indirectly, in the regulation of lipid and carbohydrate handling and how changes in these will affect metabolic outcomes. Nevertheless, it should be emphasized that these elements all work together in concert to achieve physiological responses.

12.3.1 EFFECTS OF SELENIUM IN ENERGY METABOLISM: EPIDEMIOLOGICAL STUDIES

The effects of selenium treatment or supplementation on energy metabolism and human health are an unresolved issue. Despite the antidiabetic properties described for a few selenium compounds (Mueller and Pallauf, 2006), epidemiological studies have indicated that this relationship is not straightforward. Due to the complex nature of the various pathways involved in energy balance, it is unlikely that changes in selenium alone will lead to metabolic disorders, and longitudinal studies in humans have not connected selenium to metabolic outcomes (Hughes et al., 1997; Laclaustra et al., 2009; Stranges et al., 2011). However, one cannot rule out selenium levels as a potential contributor to carbohydrate and lipid metabolic pathways on a molecular level.

In humans, the selenium effects on lipid metabolism seem to vary according to selenium baseline status of the population studied. For example, the results of the

Third National Health and Nutrition Examination Survey (NHANES III) demonstrated higher selenium levels to be associated with dyslipidemia, including high total cholesterol and triglycerides (Laclaustra et al., 2010). However, in the randomized, placebo-controlled, double-blind, parallel-group United Kingdom Prevention of Cancer by Intervention of Selenium (UK PRECISE) study, selenium supplementation had a modest beneficial effect on plasma lipid levels, that is, total cholesterol and non-high-density lipoprotein (HDL) cholesterol; it should be noted that the population assessed had low selenium levels in the serum (88.8 ng/g equivalent to 91.2 µg/L) prior to supplementation (Rayman et al., 2011). Additionally, obese females who underwent bariatric surgery had significantly reduced serum selenium levels (Alasfar et al., 2011). In a study with 15 normolipidemic participants, consumption of 45 g/day of Brazil nuts, a rich dietary source of selenium, had minimal effects on plasma lipid profile after a 15-day period (Strunz et al., 2008). The short-term nature of these latter studies plus a possible contribution of gender to the effects of selenium on lipids, however, prevents further conclusions.

With regard to carbohydrate metabolism, studies of an elderly population subset of the same UK PRECISE study demonstrated that selenium supplementation had no effect on plasma adiponectin levels, a known predictor of type 2 diabetes risk (Rayman et al., 2012). A cross-sectional analysis of the NHANES III found a positive association between serum selenium and prevalence of type 2 diabetes in American adults (Bleys et al., 2007). On the other hand, three additional case-control studies found reduced incidence of type 2 diabetes with high selenium status (Rayman, 2012). However, longitudinal studies such as the Selenium and Vitamin E Cancer Prevention Trial (SELECT) do not support the link of selenium levels with increased risk of type 2 diabetes development (Rayman and Stranges, 2013). A detailed analysis of epidemiological studies targeting selenium supplementation and type 2 diabetes risk and outcomes is discussed in Chapter 11.

12.3.2 EFFECTS OF SELENIUM IN ENERGY METABOLISM: ROLE OF SELENOPROTEIN SYNTHESIS FACTORS

Genetic manipulation by which a gene encoding a protein or an RNA factor is removed or "knocked out" in the whole animal by targeted recombination or in a tissue-specific fashion by using Cre-lox systems is a powerful way to uncover protein and/or RNA function. Substantial insights into the role of selenium in energy metabolism and metabolic disorders were gained by these means in rodents in which key elements of the selenoprotein synthesis machinery were targeted. These insights will be described in the sections below. For details of the selenoprotein biosynthesis pathway, see Chapter 3.

12.3.2.1 tRNA[Ser]Sec

Selenocysteine (Sec) is a rare amino acid, incorporated into selenoproteins cotranslationally and delivered to the translating ribosome by a specific Sec containing tRNA, designated tRNA[Ser]Sec (see Chapter 4). Targeted deletion of the tRNA[Ser]Sec gene, *Trsp*, in mice resulted in complete inhibition of the synthesis of all selenoproteins and led to embryonic lethality (Bosl et al., 1997). To circumvent this fatal outcome, tissue-specific knockouts of *Trsp* were developed, allowing for survival and elucidation of the roles of selenoproteins in physiology of different tissues and organs in rodents (Carlson et al., 2009).

For example, hepatic disruption of *Trsp* and thus selenoprotein synthesis in the liver decreased selenium levels in this organ and led to severe liver necrosis (Carlson et al., 2004). Additionally, mice with hepatocyte-specific disruption of *Trsp* combined with expression of a mutant tRNA[Ser]Sec favoring synthesis of only housekeeping selenoproteins, such as GPx4, and thioredoxin reductases 1, 2, and 3 (TrxR1, TrxR2, and TrxR3 or TGR, respectively; see also Chapter 13) but not permissive for synthesis of stress-related selenoproteins, such as selenoprotein W (SelW) and glutathione peroxidase 1 (GPx1), also helped shed light on the influence of selenium on lipid metabolism. These mice had increased plasma cholesterol levels. Microarray analysis of the livers revealed that disruption of *Trsp* induced upregulation of genes from the cholesterol biosynthesis pathway, including cytochrome b5 reductase 3 (*Cyb5r3*) and 24-dehydrocholesterol reductase (*Dhcr24*), and downregulation of genes involved in cholesterol metabolism and transport, such as phosphatidylcholine transfer protein (*Pctp*) and low-density lipoprotein receptor (*Ldlr*) (Sengupta et al., 2008). These results underscore the potential for differential regulation of selenoprotein synthesis at the tRNA level even though the specific mechanism driving these changes in lipid metabolism remains to be clarified.

Nevertheless, one hypothesis involves a tRNA[Ser]Sec modification that is required for selenoprotein synthesis, the isopentenylation (i^{6}A) at adenosine 37, which is located immediately 3′ of the anticodon. This isopentenylation reaction is catalyzed by tRNA isopentenyl transferase, which uses isopentenyl pyrophosphate as a substrate to be attached to the tRNA (see Chapter 4). Deficiency in this modification reduces Sec insertion efficiency and leads to premature termination of translation (Moustafa et al., 2001). Interestingly, isopentenyl pyrophosphate is a downstream product derived from the mevalonate pathway, which ultimately synthesizes cholesterol. Statins, a class of drugs clinically used to reduce cholesterol biosynthesis, act by diminishing the activity of the enzyme 3-hydroxy-3-methyl-glutaryl-CoA reductase (HMG-CoA reductase). This enzyme specifically diminishes the production of mevalonic acid, the precursor of isoprenoids and cholesterol. As a consequence of statin treatment, the mevalonate pathway is downregulated, which decreases cholesterol levels and also diminishes isopentenyl pyrophosphate production (Beltowski et al., 2009); thus, treatment with statins might compromise selenoprotein synthesis (Kromer and Moosmann, 2009). It has been demonstrated that treatment with statins in selenium-saturated cells leads to decreases in general selenoprotein synthesis, including GPx1 and selenoprotein N (SelN) in muscle cells (Fuhrmeister et al., 2012) although in mesothelioma cells endogenous type 2 deiodinase (D2) mRNA expression and activity are increased upon the same treatment (Miller et al., 2012).

12.3.2.2 Selenocysteine Lyase

In our bodies, selenium is absorbed in the intestines and carried to the liver where most of it is utilized for selenoprotein production. Of particular note is the synthesis of Sepp1 in the liver, a glycoprotein containing multiple Sec residues, which is then released into the bloodstream, where it works as a selenium carrier to other tissues (Burk and Hill, 2009). Excess selenium is then excreted as metabolites through the kidneys (see Chapter 2).

In order to maintain homeostasis of selenium levels, that is, stabilized selenoprotein production, organisms have evolved mechanisms to recycle selenium when its levels are limiting. This mechanism is primarily coordinated by the enzyme selenocysteine lyase (Sec lyase or Scly). Sec lyase specifically decomposes Sec, releasing alanine and selenide, the latter thought to be captured by the enzyme selenophosphate synthethase (SPS) to reenter the selenoprotein synthesis pathway (Kurokawa et al., 2011; Mihara and Esaki, 2012). Interaction of Sec lyase with SPS enzymes was previously demonstrated *in vitro* (Tobe et al., 2009). Activity of Sec lyase depends on availability of Sec and of the cofactor pyridoxal 5′-phosphate (Esaki et al., 1982, 1985; Mihara et al., 2000).

Interestingly, development of knockout mice lacking the Sec lyase gene, *Scly*, (Raman et al., 2012) revealed that this enzyme plays an unexpected role in energy homeostasis. The *Scly* knockout mouse displayed several characteristics of metabolic syndrome (Seale et al., 2012), such as hyperinsulinemia, hyperleptinemia, obesity, and glucose intolerance, without significant changes in body selenium metabolism except in the liver, a key organ coordinating energy metabolism, and, coincidentally, the organ with the highest expression of Sec lyase (Esaki et al., 1982). *Scly* knockout mice had lower selenium content in the liver with an upregulation of selenoprotein mRNAs and unchanged selenoprotein levels except for selenoprotein S (SelS), which was downregulated. Upon dietary selenium deficiency, metabolic syndrome in *Scly* knockout mice worsened, specifically including downregulation of hepatic GPx1 levels, a further decrease in SelS expression in the liver, and diminishing Sepp1 in the serum. Hepatic steatosis, hyperinsulinemia, and hypercholesterolemia, all characteristics displayed by *Scly* knockout mice upon selenium deficiency, have been linked at the molecular level to increased transcription of acetyl-CoA carboxylase-1 (ACC1), which is a critical enzyme for de novo lipogenesis regulated through phosphorylation by AMPKα. Under selenium deficiency, these animals displayed lower levels of phosphorylation of key molecular switch AMPKα, a major coordinator of hepatic carbohydrate and lipid metabolism. It remains unclear whether the effect on AMPKα is a downstream consequence of other metabolic effects of *Scly* disruption or if it is being caused directly by the absence of Sec lyase through reactions not yet resolved. Thus, although the specific mechanism still needs to be defined, the disruption of Sec lyase demonstrated once again that selenium utilization is connected with energy metabolism.

12.4   SPECIFIC SELENOPROTEINS AND ENERGY METABOLISM

Various selenoproteins have been established to be closely involved in carbohydrate and lipid metabolism and regulation of energy balance. Moreover, crucial metabolic pathways involve selenium via the action of selenoproteins. For example, insulin production and secretion is highly dependent on the redox status of the pancreatic β-cell (Graciano et al., 2011), and such status is dependent on the enzyme superoxide dismutase (SOD) and the selenoenzyme GPx1, responsible for the detoxification of superoxide or hydroperoxides, respectively (Lei and Vatamaniuk, 2011). Additional studies have suggested that independent selenoproteins, such as type 2 deiodinase (D2), type 3 deiodinase (D3), glutathione peroxidase 3 (GPx3), glutathione peroxidase 4 (GPx4), selenoprotein M (SelM), selenoprotein N (SelN), selenoprotein P (Sepp1), selenoprotein S (SelS), selenoprotein T (SelT), and TrxRs are responsible for linking selenium with specific metabolic pathways. The specific role of these selenoproteins in energy homeostasis will be discussed below and summarized in Table 12.1.

TABLE 12.1
Selenoproteins and Selenoprotein Synthesis Factors Involved in Energy Metabolism

Protein/RNA	Acronym	Metabolic Effects	Reference
		Selenoprotein Synthesis	
tRNA[Ser[Sec]]	Trsp, tRNA[Ser[Sec]]	– Loss in liver leads to severe necrosis	Carlson et al., 2009
		– Expression of form that only allows expression of housekeeping selenoproteins leads increased cholesterol synthesis	Sengupta et al., 2008; Chapter 4
Selenocysteine lyase	Sec lyase, Scly	– Loss leads to hyperinsulinemia, hyperleptinemia, obesity, and glucose intolerance, which was even greater with selenium deficiency	Seale et al., 2012
		Selenoproteins	
Type 2 deiodinase	Dio2, D2	– Important for BAT activation and energy expenditure/heat production	Mullur et al., 2014; de Jesus et al., 2001; Rothwell and Stock, 1979
		– KO leads to increased weight gain and liver steatosis upon feeding of a HFD	Marsili et al., 2011; Castillo et al., 2011
		– Regulates feeding via changes in hypothalamus	Kong et al., 2004; Revel et al., 2006; Barrett et al., 2007; Yoshimura et al., 2003; Hanon et al, 2010; Yasuo et al., 2005; Coppola et al., 2007
		– KO leads to increased gluconeogenesis and insulin resistance	Marsili et al., 2011

(Continued)

TABLE 12.1 (CONTINUED)
Selenoproteins and Selenoprotein Synthesis Factors Involved in Energy Metabolism

Protein/RNA	Acronym	Metabolic Effects	Reference
		– Necessary for the amelioration of insulin resistance by compounds that relieve ER-stress	da-Silva et al., 2011
		– Polymorphisms linked to impaired glucose handling in humans	Mentuccia et al., 2002; Canani et al., 2005
		– Overexpression leads to increased glucose usage in heart	Hong et al., 2013
Type 3 deiodinase	*Dio3*, D3	– Important for normal BAT development and maturation	Hernandez et al., 2007; Charalambous et al., 2012
		– Regulates feeding via changes in hypothalamus	Barrett et al., 2007; Yasuo et al., 2005
		– Highly expressed in β-cells, loss leads to impaired insulin production	Aguayo-Mazzucato et al., 2013; Medina et al., 2011
Glutathione peroxidase 1	GPx1	– Loss leads to increased insulin signaling and glucose uptake	Loh et al., 2009
		– Decreases in humans associated with increased insulin sensitivity	Schoenmakers et al., 2010
		– Gpx1 overexpression in mice leads to obesity, hyperglycemia, hyperinsulinemia, insulin resistance	McClung et al., 2004; Wang et al., 2008
		– Alters redox state and PTP1B activity to regulate hepatic lipogenesis	Mueller et al., 2008
Glutathione peroxidase 3	GPx3	– Highly expressed in thyroid, may mediate effects indirectly by regulating thyroid hormone production	Schmutzler et al., 2007
		– Increased serum levels in humans associated with metabolic syndrome and obesity	Baez-Duarte et al., 2012, 2014
Glutathione peroxidase 4	GPx4, PHGPx	– Required for survival	Imai et al., 2003; Yoo et al., 2012
		– Overexpression in mice protects from liver damage, β-cells dysfunction from ROS injury	Ran et al., 2004; Koulajian et al., 2013
		– Decreases in fat tissue of obese mice	Long et al., 2013
Selenoprotein M	SelM	– Loss leads to decreased leptin sensitivity in the arcuate nucleus of the hypothalamus and obesity	Pitts et al., 2013

(Continued)

TABLE 12.1 (CONTINUED)
Selenoproteins and Selenoprotein Synthesis Factors Involved in Energy Metabolism

Protein/RNA	Acronym	Metabolic Effects	Reference
Selenoprotein N	SelN, SepN1	– Mutation in humans leads to hypotrophy of slow twitch muscle fibers and abnormal glucose tolerance	Clarke et al., 2006
		– Loss in mice has no impact on body weight	Rederstoff et al., 2011
Selenoprotein P	Sepp1, SelP	– Elevated serum levels associated with increased visceral obesity and fasting glucose, and insulin resistance in humans	Choi et al., 2013; Misu et al., 2010
		– Serum levels negatively correlated with adiponectin levels in humans	Misu et al., 2012
		– Loss in mice leads to decreased plasma nonesterified fatty acids and increased glucose tolerance	Schweizer et al., 2005
		– Loss leads to increased energy expenditure and resistance to diet induced obesity	Misu et al., 2010; Towler and Hardie, 2007
Selenoprotein S	SelS, Seps1	– Increases in fat of diabetic patients, and associated with increased risk of cardiovascular disease in females	Du et al., 2008; Gao et al., 2007; Alanne et al., 2007
		– Possible receptor for serum amyloid A apolipoprotein	Karlsson et al., 2004
		– Hepatic expression increased by TNFα	Gao et al., 2006
		– Increases in liver, muscle, and fat of diabetic mice during fasting	Walder et al., 2002
Selenoprotein T	SelT	– Loss in β-cells leads to impaired insulin production and glucose intolerance	Prevost et al., 2013
Thioredoxin reductases	TrxR1, TrxR2, and TrxR3	– Indirectly through reduction of thioredoxin, a necessary cofactor for many metabolic processes	Chapters 13 and 14

12.4.1 IODOTHYRONINE DEIODINASES

Thyroid hormone regulates the expression of numerous genes involved in both energy storage and expenditure. Hyperthyroid patients with elevated levels of circulating thyroid hormones have increased resting energy expenditure; weight loss despite increased food intake; lowered cholesterol levels; and increases in lipogenesis, lipolysis, and gluconeogenesis. In contrast, hypothyroid patients with low levels of thyroid hormone exhibit the opposite range of symptoms, including lower resting energy expenditure; weight gain; increased cholesterol levels; and decreased lipogenesis, lipolysis, and glucose production (reviewed in Mullur et al., 2014). The iodothyronine deiodinase family of selenoenzymes both activate and inactivate thyroid hormone, and thus changes in these enzymes can lead to changes in energy balance that could contribute to the metabolic syndrome.

In humans, a majority of the daily hormone produced by the thyroid (~80%) is in the form of the pro-hormone thyroxine (T4), which contains four iodine molecules (Bianco et al., 2002). T4 can be activated extrathyroidally by removal of an outer ring iodine by the type 1 and 2 iodothyronine deiodinases (D1 and D2) to form the biologically active thyroid hormone, T3. Conversely, D3 can turn off the thyroid hormone signal by inactivating T4 and T3 via inner-ring deiodination to generate 3, 3′, 5′ triiodothyronine (rT3) and 3,3′ diiodothyronine (3,3′ T2), respectively (Figure 12.1).

Both D1 and D2 contribute to circulating levels of T3; however, notably, increased D2 expression can also augment T3 content locally within a specific cell type or tissue, thus rendering them "hyperthyroid" while serum thyroid hormone levels remain

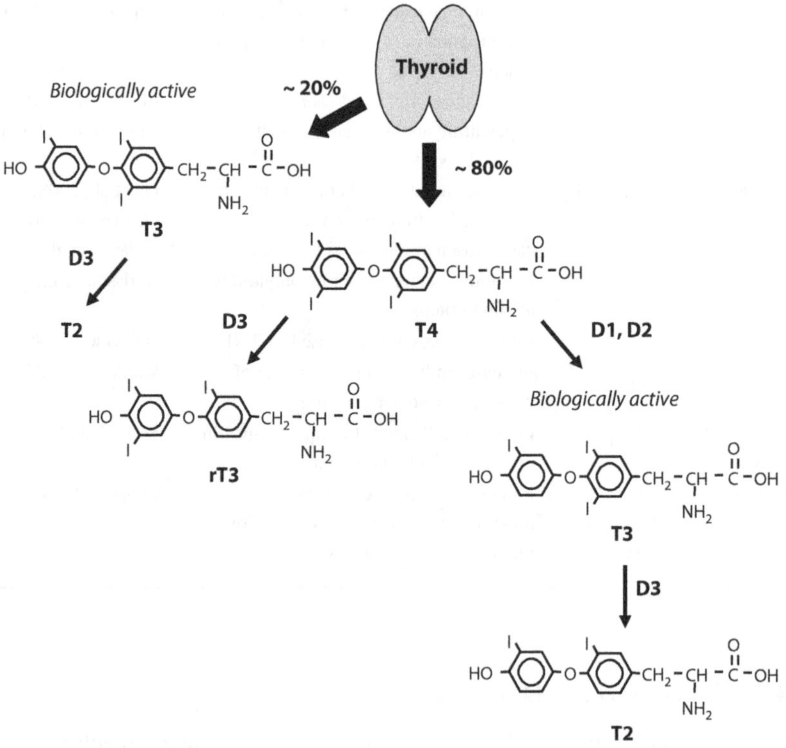

FIGURE 12.1 Extrathyroidal metabolism of thyroid hormones. Approximately 80% of the thyroid hormone produced by the thyroid is in the form of the precursor thyroxine (T4), and approximately 20% is in the form of the biologically active thyroid hormone, 3,5, 3′-triiodothyronine (T3). T4 can either be activated by the type 1 and 2 iodothyronine deiodinases (D1 and D2) to form T3 or inactivated by the type 3 iodothyronine deiodinase (D3) to form 3,5′,3′-triiodothyronine (reverse T3 or rT3). T3 can also be inactivated by D3 to form 3,3′-diiodothyronine (T2).

unchanged. A classical example of this is found in rodent brown adipose tissue (BAT). Upon cold exposure, increased sympathetic stimulation leads to a rapid increase in BAT D2 and increased intratissue T4 to T3 conversion. This results in thyroid hormone receptor occupancy going from ~50% to greater than 95%, leading to the expression of a program of genes such as uncoupling protein-1 (UCP-1), which is necessary for generation of heat and survival (Silva and Larsen, 1983, 1985; Bianco and Silva, 1988; Silva, 1988). On the other hand, expression of D3 and the resultant thyroid hormone inactivation can lead to a specific lowering of thyroid hormone levels within cells and tissues, resulting in a compartmentalized "hypothyroidism" (Gereben et al., 2008). A striking pathological example of this can be found in basal cell carcinomas (Dentice et al., 2007). This tumor expresses high amounts of D3, and the lowered thyroid hormone levels lead to increased cellular proliferation that is abolished when D3 is either knocked down by RNAi or when cells are treated with high amounts of T3. Thus, alterations in the expression and activity of both D2 and D3 are a powerful way for dynamic local changes in T3 to mediate a vast array of specific downstream effects despite circulating levels of thyroid hormone remaining constant.

12.4.1.1 Role of Deiodinases in Energy Dissipation via Brown Adipose Tissue

Increased adiposity is one of the major risk factors in developing the metabolic syndrome. In rodents, brown adipose tissue is a major tissue involved in energy balance, expending energy in the form of heat during cold exposure or diet-induced thermogenesis. Numerous studies in rodent models have found that alterations in energy expended by this tissue can lead to either weight loss or gain (Xu et al., 2011; Gilliam and Neufer, 2012). Although brown adipose tissue was primarily thought to be present only in human infants, recent studies have identified significant deposits in adults, suggesting that this tissue may also be a potential therapeutic target (Cypess et al., 2009; van Marken Lichtenbelt et al., 2009; Virtanen et al., 2009).

As mentioned above, the local T3 supplied by D2 in BAT is critical for normal function during cold exposure. *Dio2* knockout mice have impaired activation and heat generation from this tissue and only survive cold exposure by increased shivering (de Jesus et al., 2001). Under conditions of excess caloric consumption, BAT can also be activated to dissipate excess calories during diet-induced thermogenesis (Rothwell and Stock, 1979). However, this process is defective in *Dio2* knockout mice, leading to increased weight gain and liver steatosis when mice are fed a high fat diet (HFD) (Castillo et al., 2011; Marsili et al., 2011). Surprisingly, studies with tissue-specific knockouts of *Dio2* either in fat or astrocytes indicate that the loss of D2 in the brain is the primary factor that drives the phenotype of the global *Dio2* knockout mouse via alterations in the sympathetic control of BAT with changes in BAT itself being secondary (Fonseca et al., 2014). Changes in D2 in BAT are also important in mediating the protective effects of bile acids to prevent diet-induced obesity and the resulting insulin resistance. Mechanistically, bile acids have been found to activate BAT and increase D2 expression in this tissue via cAMP generated by the G-protein coupled receptor TGR5. Notably, this protective effect is lost in *Dio2* knockout mice (Watanabe et al., 2006). Changes in D2 in the pituitary can also indirectly lead to changes in BAT activation and protection from diet-induced obesity. HFD feeding increases the activation of the cJun N-terminal kinase (JNK) pathway and leads to an increase in D2 in the pituitary. However, in mice with a JNK deletion in the anterior pituitary, D2 was not increased upon HFD feeding, leading to lower T3 levels in this tissue and an increase in expression of TSH, a gene negatively regulated by T3. This elevation in TSH stimulates the thyroid to produce more T3 and T4, leading to increased oxygen consumption and resistance to diet induced obesity (Vernia et al., 2013). In agreement with this, mice that have a tissue-specific loss of D2 in their pituitary have slightly less fat in their body composition (Fonseca et al., 2013).

The thyroid hormone inactivating deiodinase, D3, is also an important contributor to BAT function via its role in the development of this tissue. D3 activity and mRNA expression correlate with the rate of proliferation of undifferentiated BAT precursor cells, and differentiation to mature brown adipocytes is associated with decreased D3 expression (Hernandez et al., 2007). Disruption of the *Dio3* gene in mice results in an impairment of BAT maturation, leading to significant lethality in the third postnatal week due to the inability of BAT to maintain body temperature and allow for the transition to independent living away from the neonatal nest (Charalambous et al., 2012).

12.4.1.2   Role of Local Thyroid Hormone Activation and Inactivation on Food Intake

Another factor that can contribute to adiposity is caloric intake, and the hypothalamic region of the brain is a major control center that regulates food consumption (Morton et al., 2014). Previous studies in rodents have shown that when T3 levels are artificially increased in the hypothalamus, either by direct T3 injection or by the implantation of T3 releasing pellets, this leads to an increase in food ingestion (Kong et al., 2004; Lopez et al., 2010; Murphy et al., 2012). Local changes in T3 in the hypothalamus mediated by changes in D2 and D3 also can lead to changes in food intake. For example, in seasonal breeders, such as quail or hamsters, increases in D2 and decreases in D3 in the medial basal hypothalamus triggered by the increase in light during the long day period leads to local increases in T3 in this region and increased feeding and weight gain associated with reproductive competency (Yoshimura et al., 2003; Revel et al., 2006; Barrett et al., 2007; Hanon et al., 2010). During short day periods that are associated with a decrease in feeding, D3 expression predominates, and D2 is low, ensuring low T3 levels in the hypothalamus (Yasuo et al., 2005; Barrett et al., 2007). D2 also increases in the arcuate nucleus of the hypothalamus during fasting, and this rise is needed for normal refeeding after fasting (Kong et al., 2004; Coppola et al., 2007). Thus changes in local thyroid hormone metabolism by the deiodinase selenoproteins in the hypothalamus mediate changes in T3 that can control food intake and adiposity.

12.4.1.3   Role of Deiodinases in Insulin Sensitivity and Glucose Utilization

Dio2 knockout mice have increased gluconeogenesis and are insulin resistant even when on a chow diet prior to increased weight gain (Marsili et al., 2011). There is also some indication that the muscle of *Dio2* knockout mice may have an impaired ability to respond to insulin with primary myoblast cultures derived from *Dio2* knockout mice generating less pAKT in response to insulin stimulation (Grozovsky et al., 2009). However, mice with a muscle-specific knockout of *Dio2* have normal metabolism with no change in insulin sensitivity (Fonseca et al., 2014).

Endoplasmic reticulum (ER) stress can lead to insulin resistance, and treating mice with compounds such as tauroursodeoxycholic acid (TUDCA) to prevent this restores insulin sensitivity (Ozcan et al., 2006). D2 is an ER-resident protein whose translation is decreased during ER-stress, and treatment of isolated brown adipocytes with TUDCA increases D2 activity and stimulates T3-regulated genes and oxygen consumption (Arrojo et al., 2011; da-Silva et al., 2011). *In vivo*, treatment of HFD-fed mice with TUDCA doubled BAT activity and normalized glucose intolerance; however, *Dio2* knockout mice were resistant to this effect, suggesting that *Dio2* is important in the activation of BAT and subsequent glucose disposal by this tissue (da-Silva et al., 2011). D2 expression also increases glucose usage in cardiac tissue with transgenic mice that overexpress *Dio2* under the control of the α-myosin heavy chain promoter having twice the amount of glucose metabolism in this tissue despite insulin-dependent glucose uptake being unchanged (Hong et al., 2013).

In humans, the *Dio2* gene has also been linked to alterations in glucose metabolism. A common polymorphism in the human *Dio2* of Thr92Ala (homozygous in ~19% of the population) was associated with a lower glucose disposal rate in a group of 972 obese women with an increased homeostasis model assessment (HOMA) index in type 2 diabetic patients (Mentuccia et al., 2002; Canani et al., 2005). Surprisingly, this amino acid substitution is not in a critical catalytic region of the D2 protein, and this change did not lead to kinetic differences in enzyme activity when studied *in vitro* (Canani et al., 2005). However, D2 activity was significantly reduced in specimens of rectus abdominis or sternocleidomastoid muscle and thyroid tissue from individuals homozygous for this polymorphism, despite no difference in D2 mRNA expression levels, leading to speculation that the *Dio2* Thr92Ala allele could be in linkage disequilibrium with another nearby gene that affected D2 function or that this polymorphism might reduce D2 protein stability or mRNA translation (Canani et al., 2005; Dora et al., 2010).

Changes in D2 and D3 have also been found during the functional maturation of the insulin secreting β-cell in the pancreas. In neonatal rats, D2 and D3 expression are high postnatal and then decrease prior to a peak in insulin production at P11. *In vivo*, neonatal T3 supplementation accelerated metabolic development due to the direct T3-dependent induction of MAFA, a transcription factor that leads to glucose responsive insulin secretion in the β-cell (Aguayo-Mazzucato et al., 2013). D3 is also highly expressed in both embryonic and adult β-cells in both humans and mice (Medina et al., 2011). *Dio3* knockout mouse neonatal and adult pancreases have less β-cell mass, insulin content, and impaired expression of key β-cell genes. *Dio3* knockout animals were found to be glucose intolerant *in vitro* and *in vivo* as a result of impaired glucose-stimulated insulin secretion while peripheral sensitivity to insulin was unchanged (Medina et al., 2011).

12.4.2 GLUTATHIONE PEROXIDASES

Glutathione is one of the most abundant thiols in cells and the primary reducing agent of the glutathione peroxidase enzymes. Glutathione peroxidases are selenoenzymes that degrade lipid and hydrogen peroxides using glutathione as substrate (Brigelius-Flohé and Maiorino, 2013). Rats with chemically induced short-term depletion of glutathione had lower hepatic lipogenesis and triglycerides in the liver, and their blood lipid profile was maintained. This effect on lipids occurred due to changes in the redox and glutathionylation states of protein tyrosine phosphatase 1B (PTP1B), leading to improvement of the insulin-signaling cascade (Brandsch et al., 2010).

12.4.2.1   Glutathione Peroxidase 1 (GPx1)

The development of GPx1 knockout mice shed light onto the effect of glutathione on insulin signaling via activation of PTP1B. Skeletal muscle cells from GPx1 knockout mice also displayed enhanced protection against HFD-induced insulin resistance (Loh et al., 2009). This effect was attributed to increased generation of reactive oxygen species (ROS) induced by insulin, which led to oxidation of inhibitory phosphatases, and a consequent increase in AKT phosphorylation and glucose uptake in target tissues. Phosphatase PTP1B activity is affected by selenium supplementation via alterations in its redox regulation, which is promoted by GPx1 activity; once PTP1B activity is increased, the insulin signaling cascade is attenuated and hepatic lipogenesis is induced downstream (Mueller et al., 2008). A detailed review of the regulatory mechanism of PTP1B by GPx1 and its effects on insulin sensitivity can be found in Chapter 11. Interestingly, human subjects with mutations in the SECIS element binding protein, SBP2, required for selenoprotein translation, also presented increased cellular ROS and GPx1 deficiencies with enhanced systemic and cellular insulin sensitivity (Schoenmakers et al., 2010), see also Chapter 16. These results are similar to what is observed in GPx1 knockout mice that also have improved insulin sensitivity, strengthening the evidence for a role of GPx1 in regulation of the insulin-signaling cascade.

GPx1 is involved in diverse aspects of lipid metabolism. Mice overexpressing GPx1 are obese, hyperglycemic, hyperinsulinemic, and insulin-resistant, effects that are consistent with the hydroperoxide removing action exerted by GPx1 (McClung et al., 2004). It was pinpointed that GPx1 overexpression dysregulates pancreatic insulin production and secretion, which ultimately would lead to chronic hyperinsulinemia (Wang et al., 2008). Thus, the effects on glucose metabolism are mediated through GPx1 acting directly on pancreatic β-cells and not only by modulation of insulin sensitivity and hepatic lipogenesis.

12.4.2.2   Glutathione Peroxidase 3 (GPx3)

Glutathione peroxidase 3 (GPx3) is a selenoprotein mostly produced by the kidneys and secreted into the bloodstream. Although serving as a local pool for selenium in the kidneys (Malinouski et al., 2012), GPx3 is mainly found in circulation, and this pool encompasses approximately 20% of circulating selenium (Koyama et al., 1999; Burk and Hill, 2005). Besides its plasma function, GPx3 is one of the most highly expressed selenoproteins in the thyroid gland (Schmutzler et al., 2007), in which selenium availability is prioritized even during selenium deficiency. The thyroid gland can regulate body energetics through its production of thyroid hormone as previously explained in Section 12.4.1 of this chapter. Because GPx3 catalyzes the reduction of lipid and hydrogen peroxides, it is possible that high levels of GPx3 confer protection against peroxidative damage in the thyrocytes during thyroid hormone synthesis because hydrogen peroxide is required for this synthesis to occur (Beckett and Arthur, 2005). It is also suggested that GPx3 might be involved in polymerization of the colloidal thyroglobulin (Schmutzler et al., 2007); however, the specific mechanism by which GPx3 would exert this function remains to be further investigated.

Interestingly, an association study found that serum GPx3 levels were increased in overweight or obese Mexican patients (Baez-Duarte et al., 2012). Additional analysis conducted as part of the Mexican Diabetes Prevention Study verified the association of different single nucleotide polymorphisms (SNP) in the GPx3 gene with adult metabolic syndrome patients without type 2 diabetes. Serum GPx3 levels were elevated in these metabolic syndrome patients, and correlated with triglycerides/high-density lipoprotein-cholesterol levels and higher risk for cardiovascular problems and insulin resistance with three sites of polymorphism identified as potential genetic markers for this specific population (Baez-Duarte et al., 2014).

12.4.2.3 Glutathione Peroxidase 4 (GPx4)

GPx4 is a Sec-containing phospholipid hydroperoxide glutathione peroxidase that reduces lipid hydroperoxides, thus inhibiting lipid peroxidation in membrane systems and regulating the activity of lipoxygenases and cyclooxygenases (reviewed in Brigelius-Flohé and Maiorino, 2013; Conrad, 2009; Conrad et al., 2007). Whole body disruption of the GPx4 gene in mice leads to embryonic lethality due to arrest during early gastrulation (Imai et al., 2003), and tamoxifen-induced ablation in adult mice also led to death due to severe mitochondrial damage with reduced ATP production in the liver (Yoo et al., 2012). Thus, it is clear that GPx4 is required for survival regardless of age with its function not being compensated by any other protein. Interestingly, mice overexpressing GPx4 improved their oxidative stress response and were protected from liver damage via reduction of apoptosis (Ran et al., 2004). Upon infusion of oleate, pancreatic β-cells of these GPx4 overexpressing mice were protected from ROS injury and did not exhibit the expected dysfunctions. Despite oleate overload, GPx4-overexpressing animals also maintained their plasma glucose and insulin and pancreatic insulin sensitivity (Koulajian et al., 2013). GPx4 is also one of the genes downregulated in the epididymal adipose tissue of obese mice, either due to *ob/ob* gene mutation or via feeding of a high fat diet (Long et al., 2013). This confers increased oxidative stress to the tissue and diminished response to lipid peroxidation and protein carbonylation derived from lipid peroxidation, a mechanism that might contribute to metabolic syndrome pathogenesis. Additionally, mice with mitochondria-specific overexpression of GPx4 were protected from cardiac dysfunction evoked by ischemia/reperfusion (Dabkowski et al., 2008), and when induced to develop type 1 diabetes, their cardiac mitochondria was protected from protein loss, guaranteeing improved heart function (Baseler et al., 2013). These results suggest that GPx4 indirectly influences outcomes from metabolic disorders, such as cardiovascular diseases, by its action in redox regulation and in apoptosis of key hormone-producing areas of the body. It would be interesting, for example, to see if GPx4 overexpression could also influence metabolism in other cell types involved in energy metabolism homeostasis, such as thyrocytes or adipocytes.

12.4.3 Selenoprotein M (SelM)

The 16-kDa thioredoxin-like ER-resident protein SelM is found to be highly expressed in neuronal tissues (Korotkov et al., 2002; Zhang et al., 2008). Development of a whole-body SelM knockout mouse led to the discovery that this selenoprotein does not contribute to cognitive function, motor coordination, or anxiety-like behavior; however, surprisingly, a role in metabolic regulation was uncovered as these mice all developed obesity due to hyperleptinemia with diminished sensitivity to the adipocyte hormone leptin in the arcuate hypothalamus, a region in the central nervous system, which coordinates energy expenditure metabolism. Baseline corticosterone was also altered in these mice. Interestingly, despite the disruption of leptin signaling and the corticosterone alteration, glycemic control was normal as was insulin signaling in the liver (Pitts et al., 2013). These data suggest that SelM might play a role in leptin signaling in the hypothalamus-pituitary axis and affect food intake.

12.4.4 Selenoprotein N (SelN)

Mutations in the SelN gene, *Sepn1*, were correlated to several muscular disorders. A fraction of patients with a G->A mutation leading to congenital hypotrophy of slow-twitch muscle fibers also had an abnormal oral glucose tolerance test (Clarke et al., 2006), suggesting a potential role of SelN in insulin resistance. The development of a *Sepn1* knockout mouse demonstrated this protein to be involved in muscle growth as these mice developed muscle weakness and spinal rigidity, serving thus as a model for SelN-dependent myopathies (Rederstorff et al., 2011). However, the animals presented normal body weight, and glucose tolerance was not investigated. Whether this model can also enlighten our understanding of the role of selenium in skeletal muscle metabolism in the context of fuel choices and energy balance is unclear.

12.4.5 Selenoprotein P (Sepp1)

Visceral obesity, elevated fasting glucose, and insulin resistance, hallmarks of metabolic syndrome and commonly leading to type 2 diabetes, have been associated with elevated serum levels of Sepp1 in humans (Misu et al., 2010; Choi et al., 2013). Moreover, serum Sepp1 levels negatively correlated with the levels of insulin-sensitizer hormone adiponectin in type 2 diabetes patients (Misu et al., 2012). Due to all these results, Sepp1 has been suggested as a biomarker for metabolic disturbances. Nevertheless, a cross-sectional study in Korean children up to 9 years of age found circulating Sepp1 levels to be independently associated with reduced risk of metabolic syndrome development (Ko et al., 2014). This finding could be a result of a lower baseline selenium in these children as Korean dietary selenium intake averages 57.5 μg/day, significantly lower than in Japan or in the United States, where averages are between 114 and 129 μg/day (Choi et al., 2009). Thus, the specific influence of Sepp1 levels in metabolic disorders is unresolved.

Elevated hepatic selenium levels were found in the Sepp1 knockout mouse model (Schweizer et al., 2005). Interestingly, the animal presented decreases in plasma non-esterified fatty acids and displayed reduced postprandial plasma insulin with improved glucose tolerance. After challenge with a HFD, Sepp1 knockout mice increased their food intake without changes in body weight, a probable consequence of their reported increase in basal energy expenditure (Misu et al., 2010). The suggested mechanism for such effect involves inactivation of AMPKα (Towler and Hardie, 2007). In general, AMPKα is inhibited by high glucose levels (Salt et al., 1998; da Silva Xavier et al., 2000), thus functioning as a sensor to allow insulin secretion. In the liver, AMPKα is regulated by several factors, including the adipocyte hormones leptin and adiponectin and by insulin via AKT signaling (Towler and Hardie, 2007).

Interestingly, most of the hepatic genes involved in gluconeogenesis and de novo lipogenesis share similar mechanisms controlling their expression. Nuclear transcription factors, such as the PPAR family and the PPARγ coactivator 1 alpha (PGC1α), have been shown to participate in the direct activation of genes involved in both pathways (Sugden and Holness, 2008). Sepp1 gene expression in HepG2 cells is enhanced by PGC1α interacting with two other gluconeogenic factors, forkhead box protein class O1a (FoxO1a) and hepatocyte nuclear factor 4 alpha (HNF-4α) (Speckmann et al., 2008). Sepp1 is, thus, under similar transcriptional control as the gluconeogenic enzymes glucose-6-phosphatase and phosphoenolpyruvate carboxykinase, suggesting Sepp1 plays a role in gluconeogenesis, possibly contributing to the interconnection of selenium and carbohydrate metabolic pathways; see also Chapter 11.

Sepp1 is also expressed in the endocrine pancreas. Although its expression is increased upon selenate treatment, it is downregulated by high glucose, which corroborates an antioxidant function of Sepp1 to cope with glucose toxicity that leads to the loss of β-cells (Steinbrenner et al., 2013).

12.4.6 SELENOPROTEIN S (SELS)

SelS is present in human plasma, and its mRNA levels were found to be higher in omental adipose tissue of a Chinese population with type 2 diabetes (Du et al., 2008; Gao et al., 2007). The SelS gene, *Seps1*, has also been associated with increased risk of cardiovascular diseases in female participants of two independent cohorts (Alanne et al., 2007). Hepatic SelS expression increases after activation with TNFα (Gao et al., 2006). *Seps1* was also suggested to be a receptor for serum amyloid A (Karlsson et al., 2004), which is an apolipoprotein associated with HDL. Serum amyloid A is secreted by liver and adipose tissue during inflammatory processes; thus, it is possible that upon chronic inflammation as in insulin-resistant states, this protein plays a role in lipid metabolism. Interestingly, hepatoma cells can also secrete SelS, despite its localization in the ER membrane. SelS was also found to increase in fasting liver, skeletal muscle, and white adipose tissue of diabetic mice (Walder et al., 2002), suggesting it is affected by glucose-handling dysfunctions as well. However, the mechanism for this effect on metabolism is still unknown, in part due to the unavailability of rodent SelS knockout models for study.

12.4.7 SELENOPROTEIN T (SELT)

With highest expression in endocrine tissues, such as thyroid, pituitary, testis, and pancreas (Tanguy et al., 2011; Prevost et al., 2013), SelT is an ER-resident selenoprotein regulated by Ca^{2+} and by pituitary adenylate cyclase-activating peptide (PACAP) (Grumolato et al., 2008). Specifically in the pancreas, SelT is expressed in β- and δ-cells, which produce insulin and somatostatin, respectively, and are major regulators of energy metabolism (Prevost et al., 2013). In β-cells, regulation of SelT by Ca^{2+} and PACAP coincides with regulation of insulin secretion. Conditional knockout of the SelT gene in mouse β-cells led to glucose intolerance due to deficiency in insulin production as a consequence of changes in islet morphology with the animal presenting more islets with reduced size (Prevost et al., 2013). These characteristics closely resemble a type 1 diabetic pancreas, and further studies of the role of SelT in insulin production may further enlighten our mechanistic understanding of this disease.

12.4.8 THIOREDOXIN REDUCTASES (TRXR)

The TrxR family of selenoenzymes has been implicated as an important coordinator of metabolic activities, and Chapter 13 is dedicated to discussing these implications.

12.5 CLOSING REMARKS

As presented in this chapter, the utility of selenium supplementation in the prevention and treatment of metabolic syndrome is questionable at this point and might be relevant only to improve specific aspects of the pathogenesis. The role played by selenium metabolism in the metabolic syndrome still lacks mechanistic evidence; however, recently, we have begun to obtain significant insight into this process. It is clear from mouse knockout models of factors necessary for selenoprotein synthesis including the tRNA[Ser]Sec and Sec lyase, and from numerous models in which specific selenoproteins have been deleted or overexpressed, that selenoproteins can have profound effects on many factors that are either associated with or contribute to the metabolic syndrome. Overall, these mouse models exhibit changes in many processes, such as cholesterol production and metabolism, glucose homeostasis, insulin production and sensitivity, and energy expenditure/food intake, and subsequent weight gain, making them attractive targets for potential modulation to prevent and treat this disease. However, our understanding of the molecular mechanisms that lead to these effects is far from being complete. Given the societal burden of the obesity epidemic and its resulting health care complications, additional research into how selenoproteins can modulate metabolism and energy balance will be an important area for future studies with potentially important implications for public health.

REFERENCES

Aguayo-Mazzucato, C., A. M. Zavacki, A. Marinelarena et al. 2013. "Thyroid hormone promotes postnatal rat pancreatic beta-cell development and glucose-responsive insulin secretion through MAFA." *Diabetes* no. 62 (5):1569–80. doi: 10.2337/db12-0849.
Alanne, M., K. Kristiansson, K. Auro et al. 2007. "Variation in the selenoprotein S gene locus is associated with coronary heart disease and ischemic stroke in two independent Finnish cohorts." *Hum Genet* no. 122 (3–4):355–65.

Alasfar, F., M. Ben-Nakhi, M. Khoursheed, E. O. Kehinde, and M. Alsaleh. 2011. "Selenium is significantly depleted among morbidly obese female patients seeking bariatric surgery." *Obes Surg* no. 21 (11):1710–3.

Arrojo e Drigo, R., T. L. Fonseca, M. Castillo et al. 2011. "Endoplasmic reticulum stress decreases intracellular thyroid hormone activation via an eIF2a-mediated decrease in type 2 deiodinase synthesis." *Mol Endocrinol* no. 25 (12):2065–75. doi: 10.1210 /me.2011-1061.

Baez-Duarte, B. G., I. Zamora-Ginez, F. Mendoza-Carrera et al. 2012. "Serum levels of glutathione peroxidase 3 in overweight and obese subjects from central Mexico." *Arch Med Res* no. 43 (7):541–7. doi: 10.1016/j.arcmed.2012.09.001.

Baez-Duarte, B. G., F. Mendoza-Carrera, A. Garcia-Zapien et al. 2014. "Glutathione peroxidase 3 serum levels and GPX3 gene polymorphisms in subjects with metabolic syndrome." *Arch Med Res.* doi: 10.1016/j.arcmed.2014.05.001.

Barrett, P., F. J. Ebling, S. Schuhler et al. 2007. "Hypothalamic thyroid hormone catabolism acts as a gatekeeper for the seasonal control of body weight and reproduction." *Endocrinology* no. 148 (8):3608–17. doi: 10.1210/en.2007-0316.

Baseler, W. A., E. R. Dabkowski, R. Jagannathan et al. 2013. "Reversal of mitochondrial proteomic loss in Type 1 diabetic heart with overexpression of phospholipid hydroperoxide glutathione peroxidase." *Am J Physiol Regul Integr Comp Physiol* no. 304 (7):R553–65. doi: 10.1152/ajpregu.00249.2012.

Beckett, G. J., and J. R. Arthur. 2005. "Selenium and endocrine systems." *J Endocrinol* no. 184 (3):455–65.

Beltowski, J., G. Wojcicka, and A. Jamroz-Wisniewska. 2009. "Adverse effects of statins—Mechanisms and consequences." *Curr Drug Saf* no. 4 (3):209–28.

Bianco, A. C., and J. E. Silva. 1988. "Cold exposure rapidly induces virtual saturation of brown adipose tissue nuclear T3 receptors." *Am J Physiol* no. 255 (4 Pt 1):E496–503.

Bianco, A. C., D. Salvatore, B. Gereben, M. J. Berry, and P. R. Larsen. 2002. "Biochemistry, cellular and molecular biology, and physiological roles of the iodothyronine selenodeiodinases." *Endocr Rev* no. 23 (1):38–89.

Bleys, J., A. Navas-Acien, and E. Guallar. 2007. "Serum selenium and diabetes in US adults." *Diabetes Care* no. 30 (4):829–34.

Bosl, M. R., K. Takaku, M. Oshima, S. Nishimura, and M. M. Taketo. 1997. "Early embryonic lethality caused by targeted disruption of the mouse selenocysteine tRNA gene (Trsp)." *Proc Natl Acad Sci U S A* no. 94 (11):5531–4.

Brandsch, C., T. Schmidt, D. Behn et al. 2010. "Glutathione deficiency down-regulates hepatic lipogenesis in rats." *Lipids Health Dis* no. 9:50.

Brigelius-Flohé, R., and M. Maiorino. 2013. "Glutathione peroxidases." *Biochim Biophys Acta* no. 1830 (5):3289–303. doi: 10.1016/j.bbagen.2012.11.020.

Burk, R. F., and K. E. Hill. 2005. "Selenoprotein P: An extracellular protein with unique physical characteristics and a role in selenium homeostasis." *Annu Rev Nutr* no. 25:215–35.

Burk, R. F., and K. E. Hill. 2009. "Selenoprotein P-expression, functions, and roles in mammals." *Biochim Biophys Acta* no. 1790 (11):1441–7.

Campbell, S. C., A. Aldibbiat, C. E. Marriott et al. 2008. "Selenium stimulates pancreatic beta-cell gene expression and enhances islet function." *FEBS Lett* no. 582 (15):2333–7. doi: 10.1016/j.febslet.2008.05.038.

Canani, L. H., C. Capp, J. M. Dora et al. 2005. "The type 2 deiodinase A/G (Thr92Ala) polymorphism is associated with decreased enzyme velocity and increased insulin resistance in patients with type 2 diabetes mellitus." *J Clin Endocrinol Metab* no. 90 (6):3472–8. doi: 10.1210/jc.2004-1977.

Carlson, B. A., S. V. Novoselov, E. Kumaraswamy et al. 2004. "Specific excision of the selenocysteine tRNA[Ser]Sec (Trsp) gene in mouse liver demonstrates an essential role of selenoproteins in liver function." *J Biol Chem* no. 279 (9):8011–7.

Carlson, B. A., M. H. Yoo, P. A. Tsuji, V. N. Gladyshev, and D. L. Hatfield. 2009. "Mouse models targeting selenocysteine tRNA expression for elucidating the role of selenoproteins in health and development." *Molecules* no. 14 (9):3509–27.

Castillo, M., J. A. Hall, M. Correa-Medina et al. 2011. "Disruption of thyroid hormone activation in type 2 deiodinase knockout mice causes obesity with glucose intolerance and liver steatosis only at thermoneutrality." *Diabetes* no. 60 (4):1082–9.

Cerf, M. E. 2007. "High fat diet modulation of glucose sensing in the beta-cell." *Med Sci Monit* no. 13 (1):RA12–7.

Charalambous, M., S. R. Ferron, S. T. da Rocha et al. 2012. "Imprinted gene dosage is critical for the transition to independent life." *Cell Metab* no. 15 (2):209–21. doi: 10.1016/j.cmet.2012.01.006.

Choi, H. Y., S. Y. Hwang, C. H. Lee et al. 2013. "Increased selenoprotein P levels in subjects with visceral obesity and nonalcoholic fatty liver disease." *Diabetes Metab J* no. 37 (1):63–71. doi: 10.4093/dmj.2013.37.1.63.

Choi, Y., J. Kim, H. S. Lee et al. 2009. "Selenium content in representative Korean foods." *Journal of Food Composition and Analysis* no. 22 (2):117–122. doi: 10.1016/J.Jfca.2008.11.009.

Clarke, N. F., W. Kidson, S. Quijano-Roy et al. 2006. "SEPN1: Associated with congenital fiber-type disproportion and insulin resistance." *Ann Neurol* no. 59 (3):546–52. doi: 10.1002/ana.20761.

Conrad, M. 2009. "Transgenic mouse models for the vital selenoenzymes cytosolic thioredoxin reductase, mitochondrial thioredoxin reductase and glutathione peroxidase 4." *Biochim Biophys Acta* no. 1790 (11):1575–85.

Conrad, M., M. Schneider, A. Seiler, and G. W. Bornkamm. 2007. "Physiological role of phospholipid hydroperoxide glutathione peroxidase in mammals." *Biol Chem* no. 388 (10):1019–25. doi: 10.1515/BC.2007.130.

Coppola, A., Z. W. Liu, Z. B. Andrews et al. 2007. "A central thermogenic-like mechanism in feeding regulation: An interplay between arcuate nucleus T3 and UCP2." *Cell Metab* no. 5 (1):21–33. doi: 10.1016/j.cmet.2006.12.002.

Cypess, A. M., S. Lehman, G. Williams et al. 2009. "Identification and importance of brown adipose tissue in adult humans." *N Engl J Med* no. 360 (15):1509–17. doi: 10.1056/NEJMoa0810780.

Dabkowski, E. R., C. L. Williamson, and J. M. Hollander. 2008. "Mitochondria-specific transgenic overexpression of phospholipid hydroperoxide glutathione peroxidase (GPx4) attenuates ischemia/reperfusion-associated cardiac dysfunction." *Free Radic Biol Med* no. 45 (6):855–65. doi: 10.1016/j.freeradbiomed.2008.06.021.

da-Silva, W. S., S. Ribich, R. Arrojo e Drigo et al. 2011. "The chemical chaperones tauroursodeoxycholic and 4-phenylbutyric acid accelerate thyroid hormone activation and energy expenditure." *FEBS Lett* no. 585 (3):539–44. doi: 10.1016/j.febslet.2010.12.044.

da Silva Xavier, G., I. Leclerc, I. P. Salt et al. 2000. "Role of AMP-activated protein kinase in the regulation by glucose of islet beta cell gene expression." *Proc Natl Acad Sci U S A* no. 97 (8):4023–8.

Davidson, H. W. 2004. "(Pro)insulin processing: A historical perspective." *Cell Biochem Biophys* no. 40 (3 Suppl):143–58.

de Jesus, L. A., S. D. Carvalho, M. O. Ribeiro et al. 2001. "The type 2 iodothyronine deiodinase is essential for adaptive thermogenesis in brown adipose tissue." *J Clin Invest* no. 108 (9):1379–85.

Dentice, M., C. Luongo, S. Huang et al. 2007. "Sonic hedgehog-induced type 3 deiodinase blocks thyroid hormone action enhancing proliferation of normal and malignant keratinocytes." *Proc Natl Acad Sci U S A* no. 104 (36):14466–71. doi: 10.1073/pnas.0706754104.

Dora, J. M., W. E. Machado, J. Rheinheimer, D. Crispim, and A. L. Maia. 2010. "Association of the type 2 deiodinase Thr92Ala polymorphism with type 2 diabetes: Case-control study and meta-analysis." *Eur J Endocrinol* no. 163 (3):427–34. doi: 10.1530/EJE -10-0419.

Du, J. L., C. K. Sun, B. Lu et al. 2008. "Association of SelS mRNA expression in omental adipose tissue with Homa-IR and serum amyloid A in patients with type 2 diabetes mellitus." *Chin Med J (Engl)* no. 121 (13):1165–8.

Esaki, N., T. Nakamura, H. Tanaka, and K. Soda. 1982. "Selenocysteine lyase, a novel enzyme that specifically acts on selenocysteine. Mammalian distribution and purification and properties of pig liver enzyme." *J Biol Chem* no. 257 (8):4386–91.

Esaki, N., N. Karai, T. Nakamura, H. Tanaka, and K. Soda. 1985. "Mechanism of reactions catalyzed by selenocysteine beta-lyase." *Arch Biochem Biophys* no. 238 (2):418–23.

Fonseca, T. L., M. Correa-Medina, M. P. Campos et al. 2013. "Coordination of hypothalamic and pituitary T3 production regulates TSH expression." *J Clin Invest* no. 123 (4):1492–500. doi: 10.1172/JCI61231.

Fonseca, T. L., J. P. Werneck-De-Castro, M. Castillo et al. 2014. "Tissue-specific inactivation of type 2 deiodinase reveals multilevel control of fatty acid oxidation by thyroid hormone in the mouse." *Diabetes* no. 63 (5):1594–604. doi: 10.2337/db13-1768.

Fuhrmeister, J., M. Tews, A. Kromer, and B. Moosmann. 2012. "Prooxidative toxicity and selenoprotein suppression by cerivastatin in muscle cells." *Toxicol Lett* no. 215 (3):219–27. doi: 10.1016/j.toxlet.2012.10.010.

Gao, Y., N. R. Hannan, S. Wanyonyi et al. 2006. "Activation of the selenoprotein SEPS1 gene expression by pro-inflammatory cytokines in HepG2 cells." *Cytokine* no. 33 (5):246–51. doi: 10.1016/j.cyto.2006.02.005.

Gao, Y., J. Pagnon, H. C. Feng et al. 2007. "Secretion of the glucose-regulated selenoprotein SEPS1 from hepatoma cells." *Biochem Biophys Res Commun* no. 356 (3):636–41.

Gereben, B., A. M. Zavacki, S. Ribich et al. 2008. "Cellular and molecular basis of deiodinase-regulated thyroid hormone signaling." *Endocr Rev* no. 29 (7):898–938. doi: 10.1210 /er.2008-0019.

Gilliam, L. A., and P. D. Neufer. 2012. "Transgenic mouse models resistant to diet-induced metabolic disease: Is energy balance the key?" *J Pharmacol Exp Ther* no. 342 (3):631–6. doi: 10.1124/jpet.112.192146.

Graciano, M. F., M. M. Valle, A. Kowluru, R. Curi, and A. R. Carpinelli. 2011. "Regulation of insulin secretion and reactive oxygen species production by free fatty acids in pancreatic islets." *Islets* no. 3 (5):213–23.

Grozovsky, R., S. Ribich, M. L. Rosene et al. 2009. "Type 2 deiodinase expression is induced by peroxisomal proliferator-activated receptor-gamma agonists in skeletal myocytes." *Endocrinology* no. 150 (4):1976–83. doi: 10.1210/en.2008-0938.

Grumolato, L., H. Ghzili, M. Montero-Hadjadje et al. 2008. "Selenoprotein T is a PACAP-regulated gene involved in intracellular Ca2+ mobilization and neuroendocrine secretion." *Faseb J* no. 22 (6):1756–68.

Hanon, E. A., K. Routledge, H. Dardente et al. 2010. "Effect of photoperiod on the thyroid-stimulating hormone neuroendocrine system in the European hamster (Cricetus cricetus)." *J Neuroendocrinol* no. 22 (1):51–5. doi: 10.1111/j.1365-2826.2009.01937.x.

Hernandez, A., B. Garcia, and M. J. Obregon. 2007. "Gene expression from the imprinted Dio3 locus is associated with cell proliferation of cultured brown adipocytes." *Endocrinology* no. 148 (8):3968–76. doi: 10.1210/en.2007-0029.

Hong, E. G., B. W. Kim, D. Y. Jung et al. 2013. "Cardiac expression of human type 2 iodothyronine deiodinase increases glucose metabolism and protects against doxorubicin-induced cardiac dysfunction in male mice." *Endocrinology* no. 154 (10):3937–46. doi: 10.1210/en.2012-2261.

Huang, P. L. 2009. "A comprehensive definition for metabolic syndrome." *Dis Model Mech* no. 2 (5–6):231–7. doi: 10.1242/dmm.001180.

Hughes, K., T. C. Aw, P. Kuperan, and M. Choo. 1997. "Central obesity, insulin resistance, syndrome X, lipoprotein(a), and cardiovascular risk in Indians, Malays, and Chinese in Singapore." *J Epidemiol Community Health* no. 51 (4):394–9.

Imai, H., F. Hirao, T. Sakamoto et al. 2003. "Early embryonic lethality caused by targeted disruption of the mouse PHGPx gene." *Biochem Biophys Res Commun* no. 305 (2):278–86.

Kahn, C. R., and A. B. Goldfine. 1993. "Molecular determinants of insulin action." *J Diabetes Complications* no. 7 (2):92–105.

Karlsson, H. K., H. Tsuchida, S. Lake, H. A. Koistinen, and A. Krook. 2004. "Relationship between serum amyloid A level and Tanis/SelS mRNA expression in skeletal muscle and adipose tissue from healthy and type 2 diabetic subjects." *Diabetes* no. 53 (6):1424–8.

Kiersztan, A., I. Lukasinska, A. Baranska et al. 2007. "Differential effects of selenium compounds on glucose synthesis in rabbit kidney-cortex tubules and hepatocytes. in vitro and in vivo studies." *J Inorg Biochem* no. 101 (3):493–505. doi: 10.1016/j.jinorgbio.2006.11.012.

Ko, B. J., S. M. Kim, K. H. Park, H. S. Park, and C. S. Mantzoros. 2014. "Levels of circulating selenoprotein P, fibroblast growth factor (FGF) 21 and FGF23 in relation to the metabolic syndrome in young children." *Int J Obes (Lond)*. doi: 10.1038/ijo.2014.45.

Kong, W. M., N. M. Martin, K. L. Smith et al. 2004. "Triiodothyronine stimulates food intake via the hypothalamic ventromedial nucleus independent of changes in energy expenditure." *Endocrinology* no. 145 (11):5252–8. doi: 10.1210/en.2004-0545.

Korotkov, K. V., S. V. Novoselov, D. L. Hatfield, and V. N. Gladyshev. 2002. "Mammalian selenoprotein in which selenocysteine (Sec) incorporation is supported by a new form of Sec insertion sequence element." *Mol Cell Biol* no. 22 (5):1402–11.

Koulajian, K., A. Ivovic, K. Ye et al. 2013. "Overexpression of glutathione peroxidase 4 prevents beta-cell dysfunction induced by prolonged elevation of lipids in vivo." *Am J Physiol Endocrinol Metab* no. 305 (2):E254–62. doi: 10.1152/ajpendo.00481.2012.

Koyama, H., K. Omura, A. Ejima et al. 1999. "Separation of selenium-containing proteins in human and mouse plasma using tandem high-performance liquid chromatography columns coupled with inductively coupled plasma-mass spectrometry." *Anal Biochem* no. 267 (1):84–91.

Kromer, A., and B. Moosmann. 2009. "Statin-induced liver injury involves cross-talk between cholesterol and selenoprotein biosynthetic pathways." *Mol Pharmacol* no. 75 (6):1421–9.

Kurokawa, S., M. Takehashi, H. Tanaka et al. 2011. "Mammalian selenocysteine lyase is involved in selenoprotein biosynthesis." *J Nutr Sci Vitaminol (Tokyo)* no. 57 (4):298–305.

Laclaustra, M., A. Navas-Acien, S. Stranges, J. M. Ordovas, and E. Guallar. 2009. "Serum selenium concentrations and diabetes in US adults: National Health and Nutrition Examination Survey (NHANES) 2003–2004." *Environ Health Perspect* no. 117 (9):1409–13.

Laclaustra, M., S. Stranges, A. Navas-Acien, J. M. Ordovas, and E. Guallar. 2010. "Serum selenium and serum lipids in US adults: National Health and Nutrition Examination Survey (NHANES) 2003–2004." *Atherosclerosis* no. 210 (2):643–8. doi: 10.1016/j.atherosclerosis.2010.01.005.

Lei, X. G., and M. Z. Vatamaniuk. 2011. "Two tales of antioxidant enzymes on beta cells and diabetes." *Antioxid Redox Signal* no. 14 (3):489–503. doi: 10.1089/ars.2010.3416.

Loh, K., H. Deng, A. Fukushima et al. 2009. "Reactive oxygen species enhance insulin sensitivity." *Cell Metab* no. 10 (4):260–72.

Long, E. K., D. M. Olson, and D. A. Bernlohr. 2013. "High-fat diet induces changes in adipose tissue trans-4-oxo-2-nonenal and trans-4-hydroxy-2-nonenal levels in a depot-specific manner." *Free Radic Biol Med* no. 63:390–8. doi: 10.1016/j.freeradbiomed.2013.05.030.

Lopez, M., L. Varela, M. J. Vazquez et al. 2010. "Hypothalamic AMPK and fatty acid metabolism mediate thyroid regulation of energy balance." *Nat Med* no. 16 (9):1001–8. doi: 10.1038/nm.2207.

Lusis, A. J., A. D. Attie, and K. Reue. 2008. "Metabolic syndrome: From epidemiology to systems biology." *Nat Rev Genet* no. 9 (11):819–30. doi: 10.1038/nrg2468.

Malinouski, M., S. Kehr, L. Finney et al. 2012. "High-resolution imaging of selenium in kidneys: A localized selenium pool associated with glutathione peroxidase 3." *Antioxid Redox Signal* no. 16 (3):185–92. doi: 10.1089/ars.2011.3997.

Marsili, A., C. Aguayo-Mazzucato, T. Chen et al. 2011. "Mice with a targeted deletion of the type 2 deiodinase are insulin resistant and susceptible to diet induced obesity." *PLoS One* no. 6 (6):e20832. doi: 10.1371/journal.pone.0020832.

McClung, J. P., C. A. Roneker, W. Mu et al. 2004. "Development of insulin resistance and obesity in mice overexpressing cellular glutathione peroxidase." *Proc Natl Acad Sci U S A* no. 101 (24):8852–7.

Medina, M. C., J. Molina, Y. Gadea et al. 2011. "The thyroid hormone-inactivating type III deiodinase is expressed in mouse and human {beta}-cells and its targeted inactivation impairs insulin secretion." *Endocrinology* no. 152 (10):3717–27. doi:10.1210/en.2011–1210.

Mentuccia, D., L. Proietti-Pannunzi, K. Tanner et al. 2002. "Association between a novel variant of the human type 2 deiodinase gene Thr92Ala and insulin resistance: Evidence of interaction with the Trp64Arg variant of the beta-3-adrenergic receptor." *Diabetes* no. 51 (3):880–3.

Mihara, H., and N. Esaki. 2012. "Selenocysteine lyase: Mechanism, structure, and biological role." In *Selenium—Its Molecular Biology and Role in Human Health*, edited by D. L. Hatfield, V. N. Gladyshev and M. J. Berry, 95–105. New York, NY: Springer.

Mihara, H., T. Kurihara, T. Watanabe, T. Yoshimura, and N. Esaki. 2000. "cDNA cloning, purification, and characterization of mouse liver selenocysteine lyase. Candidate for selenium delivery protein in selenoprotein synthesis." *J Biol Chem* no. 275 (9):6195–200.

Miller, B. T., C. B. Ueta, V. Lau et al. 2012. "Statins and downstream inhibitors of the isoprenylation pathway increase type 2 iodothyronine deiodinase activity." *Endocrinology* no. 153 (8):4039–48. doi: 10.1210/en.2012-1117.

Misu, H., T. Takamura, H. Takayama et al. 2010. "A liver-derived secretory protein, selenoprotein P, causes insulin resistance." *Cell Metab* no. 12 (5):483–95. doi: 10.1016/j.cmet.2010.09.015.

Misu, H., K. Ishikura, S. Kurita et al. 2012. "Inverse correlation between serum levels of selenoprotein P and adiponectin in patients with type 2 diabetes." *PLoS One* no. 7 (4):e34952. doi: 10.1371/journal.pone.0034952.

Morton, G. J., T. H. Meek, and M. W. Schwartz. 2014. "Neurobiology of food intake in health and disease." *Nat Rev Neurosci* no. 15 (6):367–78. doi: 10.1038/nrn3745.

Moustafa, M. E., B. A. Carlson, M. A. El-Saadani et al. 2001. "Selective inhibition of selenocysteine tRNA maturation and selenoprotein synthesis in transgenic mice expressing isopentenyladenosine-deficient selenocysteine tRNA." *Mol Cell Biol* no. 21 (11): 3840–52.

Mueller, A. S., and J. Pallauf. 2006. "Compendium of the antidiabetic effects of supranutritional selenate doses. *in vivo* and *in vitro* investigations with type II diabetic db/db mice." *J Nutr Biochem* no. 17 (8):548–60. doi: 10.1016/j.jnutbio.2005.10.006.

Mueller, A. S., S. D. Klomann, N. M. Wolf et al. 2008. "Redox regulation of protein tyrosine phosphatase 1B by manipulation of dietary selenium affects the triglyceride concentration in rat liver." *J Nutr* no. 138 (12):2328–36.

Mullur, R., Y. Y. Liu, and G. A. Brent. 2014. "Thyroid hormone regulation of metabolism." *Physiol Rev* no. 94 (2):355–82. doi: 10.1152/physrev.00030.2013.

Murphy, M., P. H. Jethwa, A. Warner et al. 2012. "Effects of manipulating hypothalamic tri-iodothyronine concentrations on seasonal body weight and torpor cycles in Siberian hamsters." *Endocrinology* no. 153 (1):101–12. doi: 10.1210/en.2011-1249.

Ozcan, U., E. Yilmaz, L. Ozcan et al. 2006. "Chemical chaperones reduce ER stress and restore glucose homeostasis in a mouse model of type 2 diabetes." *Science* no. 313 (5790):1137–40. doi: 10.1126/science.1128294.

Pitts, M. W., M. A. Reeves, A. C. Hashimoto et al. 2013. "Deletion of selenoprotein M leads to obesity without cognitive deficits." *J Biol Chem* no. 288 (36):26121–34. doi: 10.1074 /jbc.M113.471235.

Prevost, G., A. Arabo, L. Jian et al. 2013. "The PACAP-regulated gene selenoprotein T is abundantly expressed in mouse and human beta-cells and its targeted inactivation impairs glucose tolerance." *Endocrinology* no. 154 (10):3796–806. doi: 10.1210/en.2013-1167.

Raman, A. V., M. W. Pitts, A. Seyedali et al. 2012. "Absence of selenoprotein P but not selenocysteine lyase results in severe neurological dysfunction." *Genes Brain Behav* no. 11 (5):601–13. doi: 10.1111/j.1601-183X.2012.00794.x.

Ran, Q., H. Liang, M. Gu et al. 2004. "Transgenic mice overexpressing glutathione peroxidase 4 are protected against oxidative stress-induced apoptosis." *J Biol Chem* no. 279 (53):55137–46. doi: 10.1074/jbc.M410387200.

Rayman, M. P. 2012. "Selenium and human health." *Lancet* no. 379 (9822):1256–68. doi: 10.1016/S0140-6736(11)61452-9.

Rayman, M. P., and S. Stranges. 2013. "Epidemiology of selenium and type 2 diabetes: Can we make sense of it?" *Free Radic Biol Med.* doi: 10.1016/j.freeradbiomed.2013.04.003.

Rayman, M. P., S. Stranges, B. A. Griffin, R. Pastor-Barriuso, and E. Guallar. 2011. "Effect of supplementation with high-selenium yeast on plasma lipids: A randomized trial." *Ann Intern Med* no. 154 (10):656–65. doi: 10.7326/0003-4819-154-10-201105170-00005.

Rayman, M. P., G. Blundell-Pound, R. Pastor-Barriuso et al. 2012. "A randomized trial of selenium supplementation and risk of type-2 diabetes, as assessed by plasma adiponectin." *PLoS One* no. 7 (9):e45269. doi: 10.1371/journal.pone.0045269.

Rederstorff, M., P. Castets, S. Arbogast et al. 2011. "Increased muscle stress-sensitivity induced by selenoprotein N inactivation in mouse: A mammalian model for SEPN1-related myopathy." *PLoS One* no. 6 (8):e23094. doi: 10.1371/journal.pone.0023094.

Revel, F. G., M. Saboureau, P. Pevet, J. D. Mikkelsen, and V. Simonneaux. 2006. "Melatonin regulates type 2 deiodinase gene expression in the Syrian hamster." *Endocrinology* no. 147 (10):4680–7. doi: 10.1210/en.2006-0606.

Rothwell, N. J., and M. J. Stock. 1979. "A role for brown adipose tissue in diet-induced thermogenesis." *Nature* no. 281 (5726):31–5.

Salt, I. P., G. Johnson, S. J. Ashcroft, and D. G. Hardie. 1998. "AMP-activated protein kinase is activated by low glucose in cell lines derived from pancreatic beta cells, and may regulate insulin release." *Biochem J* no. 335 (Pt 3):533–9.

Schmutzler, C., B. Mentrup, L. Schomburg et al. 2007. "Selenoproteins of the thyroid gland: Expression, localization and possible function of glutathione peroxidase 3." *Biol Chem* no. 388 (10):1053–9.

Schoenmakers, E., M. Agostini, C. Mitchell et al. 2010. "Mutations in the selenocysteine insertion sequence-binding protein 2 gene lead to a multisystem selenoprotein deficiency disorder in humans." *J Clin Invest* no. 120 (12):4220–35.

Schweizer, U., F. Streckfuss, P. Pelt et al. 2005. "Hepatically derived selenoprotein P is a key factor for kidney but not for brain selenium supply." *Biochem J* no. 386 (Pt 2):221–6.

Seale, L. A., A. C. Hashimoto, S. Kurokawa et al. 2012. "Disruption of the selenocysteine lyase-mediated selenium recycling pathway leads to metabolic syndrome in mice." *Mol Cell Biol* no. 32 (20):4141–54. doi: 10.1128/MCB.00293-12.

Sengupta, A., B. A. Carlson, V. J. Hoffmann, V. N. Gladyshev, and D. L. Hatfield. 2008. "Loss of housekeeping selenoprotein expression in mouse liver modulates lipoprotein metabolism." *Biochem Biophys Res Commun* no. 365 (3):446–52.

Silva, J. E. 1988. "Full expression of uncoupling protein gene requires the concurrence of norepinephrine and triiodothyronine." *Mol Endocrinol* no. 2 (8):706–13. doi: 10.1210 /mend-2-8-706.

Silva, J. E., and P. R. Larsen. 1983. "Adrenergic activation of triiodothyronine production in brown adipose tissue." *Nature* no. 305 (5936):712–3.

Silva, J. E., and P. R. Larsen. 1985. "Potential of brown adipose tissue type II thyroxine 5'-deiodinase as a local and systemic source of triiodothyronine in rats." *J Clin Invest* no. 76 (6):2296–305. doi: 10.1172/JCI112239.

Speckmann, B., P. L. Walter, L. Alili et al. 2008. "Selenoprotein P expression is controlled through interaction of the coactivator PGC-1alpha with FoxO1a and hepatocyte nuclear factor 4alpha transcription factors." *Hepatology* no. 48 (6):1998–2006.

Steinbrenner, H., A. L. Hotze, B. Speckmann et al. 2013. "Localization and regulation of pancreatic selenoprotein P." *J Mol Endocrinol* no. 50 (1):31–42. doi: 10.1530/JME-12-0105.

Stranges, S., F. Galletti, E. Farinaro et al. 2011. "Associations of selenium status with cardiometabolic risk factors: An 8-year follow-up analysis of the Olivetti Heart study." *Atherosclerosis* no. 217 (1):274–8. doi: 10.1016/j.atherosclerosis.2011.03.027.

Strunz, C. C., T. V. Oliveira, J. C. Vinagre et al. 2008. "Brazil nut ingestion increased plasma selenium but had minimal effects on lipids, apolipoproteins, and high-density lipoprotein function in human subjects." *Nutr Res* no. 28 (3):151–5. doi: 10.1016/j .nutres.2008.01.004.

Sugden, M. C., and M. J. Holness. 2008. "Role of nuclear receptors in the modulation of insulin secretion in lipid-induced insulin resistance." *Biochem Soc Trans* no. 36 (Pt 5):891–900.

Tanguy, Y., A. Falluel-Morel, S. Arthaud et al. 2011. "The PACAP-regulated gene selenoprotein T is highly induced in nervous, endocrine, and metabolic tissues during ontogenetic and regenerative processes." *Endocrinology* no. 152 (11):4322–35. doi: 10.1210 /en.2011-1246.

Tobe, R., H. Mihara, T. Kurihara, and N. Esaki. 2009. "Identification of proteins interacting with selenocysteine lyase." *Biosci Biotechnol Biochem* no. 73 (5):1230–2.

Towler, M. C., and D. G. Hardie. 2007. "AMP-activated protein kinase in metabolic control and insulin signaling." *Circ Res* no. 100 (3):328–41. doi: 10.1161/01.RES .0000256090.42690.05.

van Marken Lichtenbelt, W. D., J. W. Vanhommerig, N. M. Smulders et al. 2009. "Cold-activated brown adipose tissue in healthy men." *N Engl J Med* no. 360 (15):1500–8. doi: 10.1056/NEJMoa0808718.

Vernia, S., J. Cavanagh-Kyros, T. Barrett et al. 2013. "Diet-induced obesity mediated by the JNK/DIO2 signal transduction pathway." *Genes Dev* no. 27 (21):2345–55. doi: 10.1101 /gad.223800.113.

Virtanen, K. A., M. E. Lidell, J. Orava et al. 2009. "Functional brown adipose tissue in healthy adults." *N Engl J Med* no. 360 (15):1518–25. doi: 10.1056/NEJMoa0808949.

Walder, K., L. Kantham, J. S. McMillan et al. 2002. "Tanis: A link between type 2 diabetes and inflammation?" *Diabetes* no. 51 (6):1859–66.

Wang, X. D., M. Z. Vatamaniuk, S. K. Wang et al. 2008. "Molecular mechanisms for hyperinsulinaemia induced by overproduction of selenium-dependent glutathione peroxidase-1 in mice." *Diabetologia* no. 51 (8):1515–24. doi: 10.1007/s00125-008-1055-3.

Watanabe, M., S. M. Houten, C. Mataki et al. 2006. "Bile acids induce energy expenditure by promoting intracellular thyroid hormone activation." *Nature* no. 439 (7075):484–9. doi: 10.1038/nature04330.

Xu, Y., J. K. Elmquist, and M. Fukuda. 2011. "Central nervous control of energy and glucose balance: Focus on the central melanocortin system." *Ann N Y Acad Sci* no. 1243:1–14. doi: 10.1111/j.1749-6632.2011.06248.x.

Yasuo, S., M. Watanabe, N. Nakao et al. 2005. "The reciprocal switching of two thyroid hormone-activating and -inactivating enzyme genes is involved in the photoperiodic gonadal response of Japanese quail." *Endocrinology* no. 146 (6):2551–4. doi: 10.1210/en.2005-0057.

Yoo, S. E., L. Chen, R. Na et al. 2012. "Gpx4 ablation in adult mice results in a lethal phenotype accompanied by neuronal loss in brain." *Free Radic Biol Med* no. 52 (9):1820–7. doi: 10.1016/j.freeradbiomed.2012.02.043.

Yoshimura, T., S. Yasuo, M. Watanabe et al. 2003. "Light-induced hormone conversion of T4 to T3 regulates photoperiodic response of gonads in birds." *Nature* no. 426 (6963):178–81. doi: 10.1038/nature02117.

Zhang, Y., Y. Zhou, U. Schweizer et al. 2008. "Comparative analysis of selenocysteine machinery and selenoproteome gene expression in mouse brain identifies neurons as key functional sites of selenium in mammals." *J Biol Chem* no. 283 (4):2427–38.

13 Thioredoxin Reductase
A Coordinator in Metabolic Activities

Sofi Eriksson and Edward E. Schmidt

CONTENTS

13.1 Introduction 273
13.2 Redox Metabolism 274
    13.2.1 Universal Need for Reductive Systems... 274
    13.2.2 GSH System in Cellular Redox Homeostasis 275
    13.2.3 Trx System in Cellular Redox Homeostasis 275
    13.2.4 Why Do We Have Both a GSH and a Trx System?.......................... 276
13.3 Coordinated Activities of Metabolic Systems 277
    13.3.1 Evidence That TrxR1 Coordinates Other Metabolic Pathways........ 277
13.4 Potential TrxR1-Interacting Regulators of Metabolic Processes.................. 280
    13.4.1 Thioredoxin-Interacting Protein (Txnip)... 280
    13.4.2 Nrf2/Keap1 Stress Response 281
    13.4.3 Protein Tyrosine Phosphatase 1B 282
    13.4.4 5′-AMP-Activated Protein Kinase (AMPK)...................................... 283
    13.4.5 p53.. 283
13.5 Integration of Metabolic Systems into a "Metabolic Ecosystem".............. 284
Acknowledgments.. 285
References... 286

13.1 INTRODUCTION

Although it is well appreciated that metabolic pathways in mammalian cells often use products of other pathways and provide their metabolites to yet different pathways, we are only beginning to appreciate how metabolic systems might directly and indirectly cross-regulate each other. Thus, all metabolic activities in a cell or organism are coordinated into a highly interactive system. In recent years, it has become apparent that cytosolic thioredoxin reductase-1, a flavin-containing NADPH-dependent selenoprotein, exerts considerable influence over other redox pathways, drug-metabolism pathways, and energy metabolism pathways. Thus, within the integrated system, thioredoxin reductase-1 behaves as a "keystone component" to coordinate the activities of other pathways. Coincidentally, it has been independently coming to light that the thioredoxin pathway is, itself, regulated by each of these other pathways and, indeed, that most if not all of these pathways might in some way influence the activities of all of the other pathways.

Thioredoxin (Trx) was discovered as a thiol-based provider of reducing power to ribonucleotide reductase (RNR) in *E. coli*, which led to unveiling of the Trx pathway (Holmgren, 1985; Laurent et al., 1964). When this system was found to be nonessential in *E. coli*, glutaredoxin, glutathione (GSH), and the GSH pathway were discovered, which form an at least partially redundant mechanism of providing reducing power to RNR (Holmgren, 1976; Holmgren and Aslund, 1995). Subsequently, the Trx pathway, the GSH pathway, or both were found to exist in all phyla (Holmgren, 1989). These pathways were soon discovered to play important roles in other reductive processes, in particular, for combatting oxidative stress and repairing oxidative damage (reviewed in Holmgren, 2000). More recently, the selenoprotein Trx-reductase-1 (TrxR1) in mammals has been shown to have very dramatic influences on other metabolic pathways, including drug metabolism pathways and bioenergetic pathways.

Recent advances in studies on seemingly distinct pathways or processes are leading to recognition of the complex crosstalk between metabolic activities within cells and organisms. Redox metabolism, energy metabolism, and drug metabolism are not separate processes, but rather, they are highly integrated. Thus metaphorically, the metabolic activities in cells appear to be akin to an ecosystem, wherein perturbation of any component can lead to widespread changes in activities throughout the system. In such an integrated system, there are likely few or no "master regulators" because all components will be checked and balanced by other components within the system. Nevertheless, just as an ecosystem can have "keystone species" whose influence on the system is perhaps more pronounced than average, cells likely have "keystone components" with more pronounced impacts across metabolic pathways. Here we review studies suggesting that TrxR1 is one such keystone component in coordinating diverse metabolic activities in cells.

13.2   REDOX METABOLISM

13.2.1   Universal Need for Reductive Systems

In all living systems, DNA replication requires a source of deoxyribonucleotides, which universally are synthesized from ribonucleotides by RNR (Holmgren, 1981; Thelander and Reichard, 1979). In addition, reductive pathways play critical roles in repair or detoxification processes and therefore function as endogenous antioxidant systems (Arnér and Holmgren, 2000; Holmgren, 2000). The importance of this is increased in aerobic organisms, wherein respiration and associated metabolic activities expose cellular macromolecules to oxidative damage. Reductive damage-prevention pathways eliminate reactive oxygen species (ROS) (Arnér and Holmgren, 2000; Holmgren, 2000). Repair mechanisms reduce oxidatively damaged cellular macromolecules, including protein disulfides, methionine sulfoxide, lipid peroxides, and others (Arnér and Holmgren, 2006).

The reducing power that cells use to sustain a reduced intracellular environment and drive critical reductive reactions comes from the cells' own energy metabolism. Most importantly, glucose oxidation via the pentose phosphate pathway reduces oxidized nicotinamide adenine dinucleotide phosphate ($NADP^+$) to the reduced form ($NADPH + H^+$) (Stanton, 2012). NADPH is the universal "currency" used within all cells to transfer reducing power from energy metabolism pathways to the cellular thiol-based GSH and Trx antioxidant systems (Berndt et al., 2007; Fernandes and Holmgren, 1989, 2004; Holmgren and Sengupta, 2010).

13.2.2 GSH System in Cellular Redox Homeostasis

GSH is a 307 Da tripeptide (L-gamma-glutamyl-L-cysteinylglycine) that carries reducing power as a transferrable electron on the thiol (–SH) of the Cys side chain. Typically, two GSH molecules will form an oxidized glutathione-disulfide (GSSG) in a reaction that transfers two electrons of reducing power to an acceptor. GSSG can be reductively recycled by glutathione reductase (GR), which oxidizes one molecule of NADPH to $NADP^+$ to obtain two electrons and uses these electrons to reduce the disulfide bond in GSSG and regain two molecules of GSH (Lu, 2013).

GSH is short-lived and must therefore be continually synthesized even in the presence of sustained GR activity (Griffith, 1999; Lu, 2013). In the first step of GSH synthesis, the gamma-glutamyl linkage between Glu and Cys is catalyzed by Glu-Cys ligase (Gcl), yielding gamma-glutamylcysteine. Gcl is a heterodimeric enzyme with a catalytic subunit, Gclc, and a modifier subunit, Gclm (Franklin et al., 2009). In a second step, GSH-synthase (Gss) forms the peptide bond with Gly (Forman et al., 2009; Lu, 2013). Mice lacking GR are robust (Rogers et al., 2004, 2006). Moreover, although mice lacking Gclc are embryonic lethal (Nakamura et al., 2011; Shi et al., 2000), the reasons for lethality are unclear because cells lacking Gclc are viable in culture (Shi et al., 2000). Also, cells and organisms of many phyla, including mice, tolerate systemic depletion of GSH following pharmacologic inhibition of Gcl (Prigge et al., 2012; Reichheld et al., 2007; Spyrou and Holmgren, 1996; Williamson et al., 1982). The absence of overt phenotypes following disruption of GSH functionality is generally explained by the redundancies between the GSH system and the Trx system.

13.2.3 Trx System in Cellular Redox Homeostasis

Trxs are small proteins (~12 kDa) that can carry two transferrable electrons on an active site Cys-pair via a reversible disulfide/dithiol motif. The disulfide in the active site of oxidized Trx can be recycled to a dithiol motif by TrxR, thereby consuming one molecule of NADPH (Arnér and Holmgren, 2000). TrxR exist in cells primarily as a homodimer, and its selenocysteine (Sec) is located within the carboxyl-(C)-terminal active site, Gly-Cys-Sec-Gly. In the process of reducing oxidized Trx, the Sec in TrxR forms a reversible selenenylsulfide. The selenenylsulfide is subsequently reduced by the N-terminal redox active dithiol of the other subunit in the homodimer (Arnér, 2009; Zhong et al., 2000).

Separate Trx systems exist in the cytosolic/nuclear compartments (Trx1 and TrxR1) and in mitochondria (Trx2 and TrxR2), encoded by the *Txn1*, *Txnrd1*, *Txn2*, and *Txnrd2* genes, respectively, in mice. Mice lacking both copies of any one of these four genes are embryonic lethal although the mechanisms of lethality are unclear (Arnér, 2009). For example, *Txnrd1*[null/null] or *Txn1*[null/null] embryos proliferate to contain several thousand cells (Jakupoglu et al., 2005; Matsui et al., 1996), but they are disorganized and fail to gastrulate (Bondareva et al., 2007; Matsui et al., 1996). Moreover, using *conditional-null* alleles of the *Txnrd1* gene (*Txnrd1*[cond]), it has been shown that many diverse *Txnrd1*[null/null] cell types and tissues are robust (Bondareva et al., 2007; Mandal et al., 2010; Soerensen et al., 2008; Suvorova et al., 2009). Also, systemic pharmacologic inhibition of TrxR activity in mice or humans is well tolerated (Arnér, 2009). *Txnrd1*[null/null] mouse livers require GSH synthesis to sustain hepatic GSH levels and replicate DNA, suggesting that, in the absence of TrxR1, only GSH can provide electrons, via glutaredoxin, to RNR for synthesis of DNA precursors (Holmgren, 1989; Prigge et al., 2012). This again demonstrates a functional redundancy between the Trx- and GSH-systems that allows survival when either one of the systems is compromised.

13.2.4 WHY DO WE HAVE BOTH A GSH AND A TRX SYSTEM?

A question thus arises: Why do mammalian cells have two distinct pathways that each depend on the same input currency, NADPH, and have apparently redundant output activities? Historically, it has been suggested that certain reductive reactions might require the GSH system, and others would require the Trx system (Fernandes and Holmgren, 2004). However, the robust nature of completely GR-null mice and of Gcl-null cells, for example, disfavors the possibility of there being essential activities in mammals that require reducing potential passed from NADPH to GSH by GR (Rogers et al., 2004; Shi et al., 2000). Moreover, there is little evidence to suggest that the Trx system plays essential redox roles that cannot be fully complemented by the GSH system. Thus, the early embryonic lethality of TrxR1-null or Trx1-null embryos is not associated with either failed cell proliferation (i.e., disrupted RNR activity) or measurable oxidative stress (Bondareva et al., 2007; Matsui et al., 1996). Instead, the phenotype of the embryos suggests they have a signaling defect that influences cell fate specification and early embryonic patterning (Bondareva et al., 2007; Matsui et al., 1996). Indeed, having now worked with the conditional-null allele of *Txnrd1* for more than a decade, we remain still unaware of evidence that any mammalian cell types are dependent on TrxR1 (unpublished observations). A more compelling argument might be made that the presence of two redundant systems simply provides robustness during severe oxidative challenges.

In recent years, it has become increasingly revealed that the Trx system plays important roles in coordinating the activities of other systems, including the GSH system, drug metabolism pathways, energy metabolism pathways, and others (Lu and Holmgren, 2014). Thus it is interesting to consider the possibility that the robustly redundant activities of the Trx and GSH systems has allowed the mammalian Trx system to evolve additional roles in the context of mechanisms that sense intracellular physiological status and thereupon coordinate diverse metabolic activities.

13.3 COORDINATED ACTIVITIES OF METABOLIC SYSTEMS

13.3.1 EVIDENCE THAT TRxR1 COORDINATES OTHER METABOLIC PATHWAYS

By using Cre-dependent *conditional-null* alleles of the *Txnrd1* gene to bypass early embryonic lethality and then inducing conversion to a TrxR1-null state at later times, it was discovered that mouse cells and tissues are robustly viable in the absence of TrxR1 (Mandal et al., 2010; Soerensen et al., 2008; Suvorova et al., 2009). Studies specifically on liver, which provides a useful model for biochemical, genetic, and histologic investigations, revealed that disruption of TrxR1 in the whole liver had no substantial effect on liver growth, proliferation, regeneration, or longevity and did not cause measurable hepatic oxidative stress (Prigge et al., 2012; Rollins et al., 2010; Suvorova et al., 2009). Transcriptome analyses on these livers revealed that the predominant response was chronically activated expression of diverse cytoprotective enzymes, notably drug metabolism phase II conjugases and phase III exporters, with a more subtle repression of lipogenic enzymes (Iverson et al., 2013; Suvorova et al., 2009). Interestingly, the TrxR1-null livers exhibited no evidence of oxidative stress, including normal GSH:GSSG ratios, no measurable accumulation of lipid hydroperoxides, and no measurable accumulation of protein carbonyls (Iverson et al., 2013; Suvorova et al., 2009).

The liver plays a predominant role in intermediary metabolism, nearly all of which is mediated by hepatocytes. Thus, plasma levels of sugars, lipids, amino acids, nucleosides, vitamins, steroid hormones, and other molecules are regulated by the liver (Dudrick and Kavic, 2002; Tappy and Minehira, 2001). In addition, hepatocytes synthesize and secrete most of the proteins in plasma, including serum albumin, alpha-fetoprotein, complement proteins, blood-clotting factors, apolipoproteins, and others (Petkov et al., 2004). However, surprisingly, the transcriptome of *TrxR1-null* livers revealed little at the mRNA level to suggest that loss of TrxR1 might have a substantial impact on these processes. Of roughly 2×10^4 probe sets on the arrays that showed above-threshold hybridization signals, suggesting these mRNAs were expressed in hepatocytes, only ~1% showed substantial mRNA level differences between wild-type and TrxR1-null livers (Suvorova et al., 2009). The apparent take-home message from the transcriptome analysis appeared to be that, overall, the status of liver in the complete absence of hepatocytic TrxR1 was "business as usual" (Suvorova et al., 2009). Moreover, there was little about the overt phenotype of the mice or their livers to suggest that hepatic metabolism had substantially shifted as a result of disrupting TrxR1. However, these implications might have been misleading. Thus, it was unexpected when ultrastructural studies on TrxR1-null livers (Figure 13.1) revealed that these livers were dramatically engorged with glycogen (Iverson et al., 2013). This was subsequently confirmed by periodic Schiff's staining and direct measurements of glycogen content (Iverson et al., 2013) (Figure 13.1).

The observation that disruption of TrxR1 in liver induced strong expression of mRNAs encoding cytoprotective enzymes, in particular enzymes related to drug or xenobiotic metabolism (Suvorova et al., 2009), prompted tests of whether this realigned metabolic profile, in the presence of a disrupted cytosolic thioredoxin system, provided substantial protection to xenobiotics. Mice were challenged with the

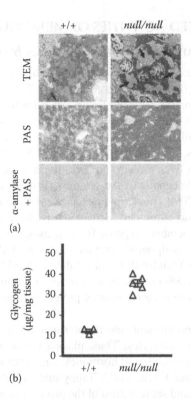

FIGURE 13.1 Disruption of TrxR1 switches hepatic energy storage from a more lipogenic to a more glycogenic balance. (a) Transmission electron micrographs (TEM, top panels) reveal predominantly lipid vesicles in wild-type liver (white arrows). Few lipid vesicles are found in TrxR1-null hepatocytes, and instead, vast cytoplasmic tracts of glycogenic hyaloplasm (black arrows) are observed. Periodic Acid Schiff's (PAS) staining, which reacts with glycogen to give a strong magenta color, reveal increased staining in TrxR1-null livers (middle panels, right) than in wild-type livers (left). Pretreatment of sections with α-amylase prior to PAS staining (lower panels) eliminates nearly all PAS-reactivity, verifying that staining arises from glycogen. (b) Quantification of glycogen content. Liver was homogenized, boiled in alkali to release glycogen and hydrolyze most other macromolecules, and then glycogen was precipitated with ethanol. Glycogen was digested with α-amylase and liberated glucose was measured. Triangles represent values from individual animals, either wild-type (+/+) or TrxR1-null (*null/null*). Bars represent mean and SEM. For all analyses in this figure, animals were harvested between 8 and 10 am, which is roughly the diurnal peak of glycogen accumulation. See Iverson et al. (2013) for details.

classical pre-hepatotoxin acetaminophen (paracetamol, APAP). APAP itself is not cytotoxic; however, it is processed by hepatocytes into an active quinone, N-acetyl-*p*-benzoquinone imine (NAPQI), which is cytotoxic (Dong et al., 2000; James et al., 2003; Thummel et al., 1993). Mice with TrxR1-null livers were remarkably resistant to APAP challenge even at three times the LD_{50} of the drug (Iverson et al., 2013).

In hepatocytes, defenses to APAP occur at several levels. APAP, itself, is conjugated to glucuronate by UDP-glucuronyl transferase and subsequently excreted by Abcc-family transporters, thereby eliminating a portion of the APAP before it is activated into the cytotoxic NAPQI form (Homolya et al., 2003; Zamek-Gliszczynski et al., 2006). Following oxidative activation, the NAPQI is conjugated to GSH by glutathione S-transferases (GSTs), also promoting excretion by Abcc transporters (Slitt et al., 2005). The GSH-based excretion pathway is thought to be most important as pathology does not occur until hepatic GSH is depleted (Hinson et al., 2010). Also, administration of N-acetylcysteine (NAC), which augments hepatic GSH biosynthesis under conditions of GSH depletion, can effectively attenuate pathological responses (Corcoran et al., 1985; Hinson et al., 2010; James et al., 2003).

Examination of the mechanisms by which TrxR1-null hepatocytes escaped APAP toxicity revealed some expected mechanisms and some surprises. Thus, due to increased expression of GST-A and -M class enzymes and components of the glutathione biosynthetic pathway as well as increased expression of Abcc exporters (Suvorova et al., 2009), the GSH-based defense pathway was more robust in TrxR1-null hepatocytes (Iverson et al., 2013). However, in addition, it was discovered that the TrxR1-null livers under nonchallenged conditions had much larger reserves of glycogen (Figure 13.1), they overexpressed the pathway for converting glycogen into UDP-glucuronate, and they overexpressed the UDP-glucuronyl transferases that conjugate glucuronate to APAP (Iverson et al., 2013). Following APAP challenge, glycogen stores in either wild-type or TrxR1-null livers were rapidly depleted. However, in the TrxR1-null livers, this amounted to approximately threefold greater glycogen consumption than in wild-type livers, corresponding to a molar amount of hexose consumption approximately equivalent to the molar amount of APAP in the challenge dose (Iverson et al., 2013). This suggests that protection of the TrxR1-null hepatocytes from APAP challenge results primarily from rapid effective glucuronidation and export of the APAP prior to it being activated into the cytotoxic NAPQI (Iverson et al., 2013). Importantly, this study also revealed that patterns of carbohydrate metabolism and storage are substantially realigned in TrxR1-null hepatocytes (Iverson et al., 2013) (Figure 13.1).

Although the TrxR1-null liver transcriptome did not reveal increased abundance of mRNAs encoding glycogenic genes (Iverson et al., 2013; Suvorova et al., 2009), the recognition of dramatic glycogen engorgement in these livers focused more attention on the repression of lipogenic genes in the TrxR1-null livers. On one hand, repression of lipogenic activities on a transcriptional level alone might be sufficient to rechannel glucose assimilation into dramatically increased glycogenesis. In addition, however, one should consider that there might be substantial impacts on pathway activities that occur at a post-transcriptional level, which would not be revealed in the transcriptome data, but could play dramatic roles in the glycogenic response to TrxR1 disruption. For example, TrxR1-dependent redox regulation of the activities of growth factor receptors via certain protein tyrosine phosphatases (PTPs), which are active in regulating carbohydrate metabolism and storage, might participate in the glycogen engorgement seen in TrxR1-null livers (Figure 13.1).

13.4 POTENTIAL TrxR1-INTERACTING REGULATORS OF METABOLIC PROCESSES

13.4.1 THIOREDOXIN-INTERACTING PROTEIN (TXNIP)

The Trx-interacting protein, Txnip, also known as Trx-binding protein-2 (TBP2) or vitamin D3 upregulated protein (Vdup1), is an alpha-arrestin family protein that binds to and antagonizes the activity of Trx1 in a redox-dependent fashion (Parikh et al., 2007). Thus, on the most fundamental level, Txnip is a direct antagonist of the Trx1 system. More recently, however, it has also been revealed to have large impacts on cellular and systemic metabolism (Figure 13.2a). Txnip protein inhibits adipogenesis (Chutkow and Lee, 2011), and Txnip-null mice exhibit increased muscle glycogen levels under fasting (Andres et al., 2011) and increased adipogenesis (Chutkow

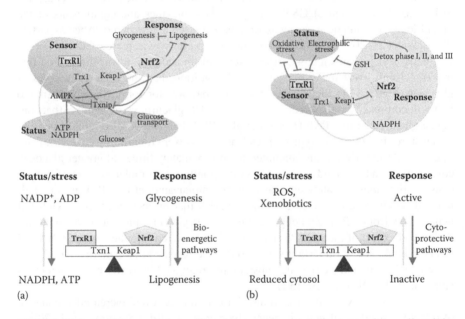

FIGURE 13.2 TrxR1 as an intermediary between the metabolic status of the cell and the output of metabolic pathways. (a) Energy metabolism pathways. TrxR1 has been shown to regulate the balance between lipid versus glycogen storage (see text). Model shown is based on activities known to participate in this process and their known or inferred interactions with TrxR1 and other components shown. At top, TrxR1 and some other regulators serve as an intermediary "sensor" system that recognizes the energy status of the cell and correspondingly affects output metabolic pathways, many of which are transcriptionally regulated by Nrf2 (top). Below, the balance shows how metabolic inputs, via TrxR1, Trx1, and Keap1, modulate Nrf2 activity to influence metabolic output. (b) Drug/xenobiotic metabolism pathways. At top, as in panel A, TrxR1 is at the core of a sensor system that recognized oxidative or electrophilic challenges to the cells and influences the output of cytoprotective pathways, many of which are transcriptionally regulated by Nrf2. Below, TrxR1, Trx1, and Keap1 transmit information about the stress status of the cell to Nrf2, thereby modulating the output of cytoprotective pathways.

et al., 2010). Inhibition of glycolysis is correlated to Txnip upregulation in tumors (Chen et al., 2010); Txnip has been proposed to function as a sensor of glycolytic flux (Yu et al., 2010); and human Txnip is upregulated in diabetic conditions (Parikh et al., 2007).

Txnip expression is induced by high glucose levels via the carbohydrate-response element located in the Txnip promoter, which interacts with the transcription factor MondoA (Parikh et al., 2007; Spindel et al., 2012; Yu et al., 2010). In heart, it has been shown that both glucose-induced expression of Txnip via MondoA and basal expression of Txnip are antagonized by the transcription factor nuclear factor, erythroid derived 2, like 2 (Nrf2, Nfe2l2) (He and Ma, 2012). Thus, under normal conditions, high glucose induces Txnip, which inhibits Trx1 activity. However, under conditions of oxidative stress, when Nrf2 is activated, high glucose does not induce expression of Txnip, allowing cells to sustain Trx1-dependent antioxidant pathways under these conditions (He and Ma, 2012). Recently, Txnip was also shown to prevent the degradation of the tumor suppressor p53, increasing p53 transcriptional activity (Jung et al., 2013; Suh et al., 2013).

13.4.2 Nrf2/Keap1 Stress Response

The Nrf2/Kelch-like ECH-associated protein 1 (Keap1) pathway normally serves to activate a rapid transcriptional response to oxidative or electrophilic challenges (Hayes and Dinkova-Kostova, 2014). Nrf2 is a member of the "cap-'n'-collar" bZIP transcription factor family of proteins that is ubiquitously and constitutively expressed (Moi et al., 1994). Under unstressed conditions, Nrf2 protein interacts with Keap1, a Cys-rich protein that acts as an adapter to facilitate ubiquitylation and rapid degradation of Nrf2 by the Cullin-3/RBX1 ubiquitin E3 ligase complex (Itoh et al., 1999, 2003; McMahon et al., 2003). However, under conditions of oxidative or electrophilic stress, specific Cys residues in Keap1 become oxidized (Brigelius Flohé and Kipp, 2013). This prevents ubiquitylation of Nrf2, essentially blocking Keap1 and allowing newly synthesized Nrf2, instead, to interact with the ubiquitous Maf proteins, which are small bZIP proteins that heterodimerize with Nrf2 to form an active transcription factor (Marini et al., 1997). The Nrf2/Maf heterodimer binds to DNA motifs entitled "electrophilic response elements" (EpRE) or "antioxidant response elements" (ARE), which are found in the promoters or regulatory regions of genes encoding a variety of cytoprotective enzymes (Hayes and Dinkova-Kostova, 2014).

The livers of mice with TrxR1-null hepatocytes show increased expression of mRNAs for many cytoprotective genes that are known to be upregulated by the Nrf2 pathway (Figure 13.2b) (Suvorova et al., 2009). In mouse embryo fibroblast (MEF) cell cultures as well as in liver, TrxR1-deficiency causes increased levels of Nrf2 protein, increased nuclear localization of Nrf2 protein in cells, and increased occupancy of Nrf2 protein on the regulatory EpREs in the promoters of established Nrf2 target genes as measured by chromatin immunoprecipitation (ChIP) (Suvorova et al., 2009). In HeLa cells, knockdown of TrxR1 levels results in formation of an intermolecular disulfide bond between Cys^{226} and Cys^{613} in Keap1 and stabilization of Nrf2 protein, and this is further augmented by simultaneous disruption of GSH biosynthesis (Fourquet et al., 2010). In addition, however, mRNA levels for several genes associated with fatty acid metabolism are downregulated in TrxR1-null liver, including, for example, ATP-citrate lyase (Acly), fatty acid desaturase 2 (Fads2), fatty acid-binding protein 5 (Fabp5), stearoyl-CoA desaturase 1 (Scd1), and elongation of very long-chain fatty acids 6 (Elovl6) (Iverson et al., 2013; Suvorova et al., 2009). Interestingly, whereas Nrf2 is most well known for its ability to induce expression of genes encoding cytoprotective functions, the lipogenic genes whose expression is repressed in TrxR1-null livers are known to be negatively regulated by Nrf2 (Hayes and Dinkova-Kostova, 2014). Work in recent years by several laboratories has been increasingly showing that Nrf2 plays an important role in lipid metabolism (Hayes and Dinkova-Kostova, 2014). For example, enhanced Nrf2 activity following knockdown of Keap1 mRNA causes decreased lipogenesis (Xu et al., 2013) whereas the lack of Nrf2 in Nrf2-null mice causes increased hepatic lipid accumulation in response to a high-fat diet (Tanaka et al., 2008).

13.4.3 PROTEIN TYROSINE PHOSPHATASE 1B

Protein tyrosine phosphatases (PTPs) cause dephosphorylation-dependent inactivation of active (ligand-bound tyrosine-phosphorylated) growth factor receptors (Bae et al., 2011). Growth factors and hormones, such as insulin and insulin-like growth factor-1 (IGF-1), upon binding to their tyrosine kinase receptor, induce superoxide generation via receptor-associated NADPH-oxidases. The resultant locally increased ROS levels can cause inhibition of PTPs through oxidation of cysteine residues (Dagnell et al., 2013; Goldstein et al., 2005). In this context, PTPs are negative regulators of receptor signaling, and inhibition of PTP activity potentiates the effects of tyrosine phosphorylation and phosphoinositide-3 kinase (PI3K)/Akt and *Ras*-Raf-MEK-*ERK* signaling. The Trx system participates in regulating this signaling by (1) reactivating oxidized Phosphatase and Tensin (PTEN) homolog protein, which suppresses PI3K/Akt signaling (Lee et al., 2002; Meuillet et al., 2004; Schwertassek et al., 2014); (2) fueling peroxiredoxins, which participate in eliminating the ROS that is generated at the receptors (Bae et al., 2011; Rhee et al., 2005, 2012); and (3) by reactivating oxidized PTP1B (Dagnell et al., 2013). Cells genetically deficient in TrxR1 also display increased levels of oxidized PTP1B and increased phosphorylation of the platelet-derived growth factor tyrosine kinase receptor (PDGF-β) (Dagnell et al., 2013).

PTP1B deficiency in hepatocytes has been shown to protect against APAP-induced cell death (Mobasher et al., 2013). Compared to wild-type mice, PTP1B-null mice showed delayed nuclear export, ubiquitylation, and degradation of Nrf2. The prolonged nuclear Nrf2 activity in PTP1B-deficient mice was suggested to be due to glycogen synthase kinase-3β (GSK3β) inhibition. Hence, cellular Nrf2 activity is also regulated via phosphorylation. In the cytosol, Nrf2 can be phosphorylated at Ser_{40} by protein kinase C (PKC) or Ras-regulated kinases (Niture et al., 2014). Phosphorylation of Ser_{40}, together with Keap1 oxidation, promotes Nrf2 stabilization and nuclear translocation. Within the nucleus, Nrf2 can be phosphorylated at Tyr_{568}, facilitated by the GSK3β and its downstream targets, such as Src family tyrosine kinases (Hayes and Dinkova-Kostova, 2014; Niture et al., 2014). Interestingly, GSK3β is also known to negatively regulate glycogen synthase (UDP-glucose-glycogen glucosyltransferase) (Kaidanovich-Beilin and Woodgett, 2011). GSK3β inhibition can, for instance, be achieved by enhanced PI3K/Akt signaling. Because GSK3β antagonizes both Nrf2 activity and glycogen synthesis, the increased Nrf2 activity and glycogen accumulation described in TrxR1-deficient livers (Iverson et al., 2013) might partly result from increased PTP1B oxidation and subsequent enhanced PI3K/Akt-mediated GSK3β inhibition.

13.4.4 5′-AMP-Activated Protein Kinase (AMPK)

Trx1 was recently shown to prevent oxidation of 5′-AMP-activated protein kinase (AMPK) (Shao et al., 2014). AMPK is a low-energy sensor, which is activated by AMP and inhibited by glucose. AMPK activity is also induced during conditions of oxidative stress (Wu et al., 2014; Zhang et al., 2008). In general, AMPK activation switches "on" catabolic pathways, thereby causing increased glucose availability, glycolysis, and production of ATP and NADPH. Fatty acid synthesis, which consumes ATP, is negatively regulated by AMPK. AMPK is activated by phosphorylation; however, if AMPK gets oxidized, it cannot be phosphorylated. Balancing this, Trx1 can reduce the redox-active disulfides in AMPK, thereby making the enzyme available for phosphorylation again (Shao et al., 2014). Glucose deprivation in mice caused increased Trx1 expression and increased AMPK activity (Shao et al., 2014). Conversely, mice feed a high fat diet showed lower levels of Trx1 and less phosphorylation of AMPK. Keap1 knockout mice, which have increased Nrf2 activity, also show increased AMPK activity (Kulkarni et al., 2013b; More et al., 2013b), which might be explained by Nrf2-dependent induced expression of Trx1 and TrxR1. Furthermore, AMPK has been suggested to be a negative regulator of TXNIP (Shaked et al., 2011; Wu et al., 2013). During low-energy conditions, TXNIP can be phosphorylated, most likely by AMPK, consequently, targeting TXNIP for degradation.

13.4.5 p53

The tumor-suppressor protein p53 is most well known for its activities in modulating DNA repair and apoptosis (Polager and Ginsberg, 2009; Zuckerman et al., 2009). However, it has additionally been implicated in playing roles, both regulatory and responsive, in metabolism and oxidative stress (Arnér and Holmgren, 2006; Burhans and Heintz, 2009; Menendez et al., 2009; Vogelstein and Kinzler, 2004; Wang et al., 2014). For example, it has been reported that Trx can regulate p53 redox status and DNA-binding indirectly by reducing redox effector factor 1 (Ref-1, also named apurinic/apyrmidinic endonuclease 1, APE1) (Lu and Holmgren, 2014; Seemann and Hainaut, 2005; Ueno et al., 1999). APE1/Ref-1 is a multifunctional enzyme reducing and activating a number of transcription factors. Two examples of transcription factors involved in energy metabolism regulation are p53 and hypoxia inducible factor 1α (HIF 1α) (Hafsi and Hainaut, 2011; Thakur et al., 2014).

p53-dependent regulation of energy metabolism is probably important for p53-dependent suppression of tumor development. p53-dependent regulation occurs through several mechanisms. p53 can decrease NADPH synthesis by transcriptional repression of malic enzyme genes (ME1 and ME2) or by direct interaction and inhibition of glucose-6-phosphate dehydrogenase (G6PDH). Interestingly, there is also a p53-regulated protein promoting NADPH synthesis. TP53-induced glycolysis and apoptosis regulator (TIGAR), is an enzyme reducing cellular fructose-2,6-bisphosphate levels, subsequently lowering glycolytic rate while increasing NADPH production (Gorrini et al., 2013; Zhuang et al., 2012).

13.5   INTEGRATION OF METABOLIC SYSTEMS INTO A "METABOLIC ECOSYSTEM"

Historically, redox metabolism, drug/xenobiotic metabolism, and energy metabolism have been viewed as fairly independent processes, and each has been considered an independent field of study. However, in recent years, studies on these apparently unrelated metabolic systems have revealed unexpected interconnections (Figure 13.3). Work on redox metabolism systems, especially the selenoprotein TrxR1, revealed unexpectedly that perturbations in TrxR1 impacted expression of (i) other redox metabolism enzymes, in particular induction of mRNAs involved in the Trx1 pathway, GSH biosynthesis or use, and some NADPH-dependent oxidoreductases; (ii) cytoprotective enzymes, in particular induction of drug metabolism enzymes; and

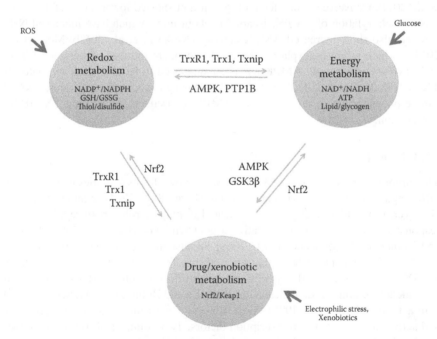

FIGURE 13.3   The metabolic ecosystem. The redox metabolism system, energy metabolism system, and drug/xenobiotic metabolism system (ovals) are cross-coordinated via a small number of "keystone components" (vectors) that mediate a global metabolic response to diverse perturbations of any of the systems (dark gray vectors).

(iii) bioenergetic enzymes, in particular repression of mRNAs for enzymes involved in lipogenesis (Iverson et al., 2013; Suvorova et al., 2009). Coincidentally, work on drug/xenobiotic metabolism systems have revealed that Nrf2 induction (i) impacts redox metabolism pathways, specifically inducing expression of mRNAs for TrxR1, mRNAs involved in GSH biosynthesis or use, and some NADPH-dependent oxido-reductases; (ii) induces cytoprotective enzymes, in particular induction drug metabolism enzymes; and (iii) affects expression of bioenergetic enzymes, in particular repressing mRNAs for enzymes involved in lipogenesis (Aleksunes et al., 2006; Cheng et al., 2011; Hayes and Dinkova-Kostova, 2014; Itoh et al., 2010; More et al., 2013b; Singh et al., 2010; Xu et al., 2012a, 2013). And finally, work on bioenergetic pathways and metabolic disorders revealed that conditions such as obesity or fasting impacted expression of mRNAs encoding (i) redox metabolism enzymes, including components of the Trx1 and the GSH system, and some NADPH-dependent oxido-reductases; (ii) cytoprotective enzymes, including drug metabolism enzymes; and (iii) bioenergetic pathways (Kulkarni et al., 2013a,b, 2014; More et al., 2013a; Wen et al., 2013; Xu et al., 2012b; Yalcin et al., 2013). Thus, in each case, perturbation of one "metabolic system" had impacts within that system as well as on the other two "metabolic systems."

The coordinate regulation of redox-, bioenergetic-, and drug/xenobiotic-metabolism pathways requires mechanisms for each pathway to sense and respond to the activities of each of the other pathways. Studies to date suggest that some proteins within each pathway play dominant roles in this, making them "keystone components" in integrating the individual metabolic systems into the global "metabolic ecosystem." Within the redox systems, the selenoprotein, TrxR1 has predominant influences over bioenergetic- and drug/xenobiotic-metabolism systems. Within the drug/xenobiotic-metabolism system, Nrf2 has predominant influences over redox- and bioenergetic-metabolism systems; and within the bioenergetic metabolism system, AMPK has predominant influences on the redox- and drug/xenobiotic-metabolism systems. Thus, these components likely qualify as "keystone" components for system coordination. In addition, other proteins also participate directly in signaling between the systems. Keap1 relays signals to Nrf2. Trx1 and Txnip likely relay signals to and from the redox metabolism system (Figure 13.3). Regulatory proteins including p53, GSK3β, APE1/Ref-1, MondoA, TIGAR, Hif1-alpha, and others play roles in coordinating these metabolic systems and others into an integrated metabolic eco-system that provides effective global responses to the diverse gamut of conditions that cells and organisms encounter routinely and stresses or perturbations that can occur occasionally.

ACKNOWLEDGMENTS

The authors thank their present and past colleagues for the contributions each has made to our understanding of thioredoxin reductase in cellular homeostasis. For our research on the thioredoxin system, SE was supported by a fellowship from the Swedish Foundation Blanceflor Boncompagni Ludovisi, née Bildt. EES was supported by National Institutes of Health grants from the National Cancer Institute and the National Institutes on Aging, Allergy and Infectious Diseases, Child Health and Human Development, and the Office of the Director as well as by grants from the National Science Foundation, the March of Dimes Foundation, and the United States Department of Agriculture. Additional support has been provided by the Montana Agricultural Experiment Station, Montana State University, the College of Agriculture, the Department of Immunology and Infectious Disease. An NIH IDeA grant to Montana State University provided infrastructure support.

REFERENCES

Aleksunes, L. M., A. L. Slitt, J. M. Maher et al. 2006. "Nuclear factor-E2-related factor 2 expression in liver is critical for induction of NAD(P)H:Quinone oxidoreductase 1 during cholestasis." *Cell Stress Chaperones* no. 11 (4):356–63.

Andres, A. M., E. P. Ratliff, S. Sachithanantham, and S. T. Hui. 2011. "Diminished AMPK signaling response to fasting in thioredoxin-interacting protein knockout mice." *FEBS Lett* no. 585 (8):1223–30. doi: 10.1016/j.febslet.2011.03.042.

Arnér, E. S. J. 2009. "Focus on mammalian thioredoxin reductases—Important selenoproteins with versatile functions." *Biochim Biophys Acta* no. 1790 (6):495–526.

Arnér, E. S. J., and A. Holmgren. 2000. "Physiological functions of thioredoxin and thioredoxin reductase." *Eur J Biochem* no. 267 (20):6102–9.

Arnér, E. S. J., and A. Holmgren. 2006. "The thioredoxin system in cancer." *Semin Cancer Biol* no. 16 (6):420–6.

Bae, Y. S., H. Oh, S. G. Rhee, and Y. D. Yoo. 2011. "Regulation of reactive oxygen species generation in cell signaling." *Mol Cells* no. 32 (6):491–509. doi: 10.1007/s10059-011-0276-3.

Berndt, C., C. H. Lillig, and A. Holmgren. 2007. "Thiol-based mechanisms of the thioredoxin and glutaredoxin systems: Implications for diseases in the cardiovascular system." *Am J Physiol Heart Circ Physiol* no. 292 (3):H1227–36.

Bondareva, A. A., M. R. Capecchi, S. V. Iverson et al. 2007. "Effects of thioredoxin reductase-1 deletion on embryogenesis and transcriptome." *Free Radic Biol Med* no. 43 (6):911–23.

Brigelius-Flohé, R., and A. P. Kipp. 2013. "Selenium in the redox regulation of the Nrf2 and the Wnt pathway." *Methods Enzymol* no. 527:65–86. doi: 10.1016/B978-0-12-405882-8 .00004-0.

Burhans, W. C., and N. H. Heintz. 2009. "The cell cycle is a redox cycle: Linking phase-specific targets to cell fate." *Free Radic Biol Med* no. 47 (9):1282–93. doi: 10.1016/j .freeradbiomed.2009.05.026.

Chen, J. L., D. Merl, C. W. Peterson et al. 2010. "Lactic acidosis triggers starvation response with paradoxical induction of TXNIP through MondoA." *PLoS Genet* no. 6 (9). doi: 10.1371 /journal.pgen.1001093.

Cheng, Q., K. Taguchi, L. M. Aleksunes et al. 2011. "Constitutive activation of nuclear factor-E2-related factor 2 induces biotransformation enzyme and transporter expression in livers of mice with hepatocyte-specific deletion of Kelch-like ECH-associated protein 1." *J Biochem Mol Toxicol* no. 25 (5):320–9. doi: 10.1002/jbt.20392.

Chutkow, W. A., and R. T. Lee. 2011. "Thioredoxin regulates adipogenesis through thioredoxin-interacting protein (Txnip) protein stability." *J Biol Chem* no. 286 (33):29139–45. doi: 10.1074/jbc.M111.267666.

Chutkow, W. A., A. L. Birkenfeld, J. D. Brown et al. 2010. "Deletion of the alpha-arrestin protein Txnip in mice promotes adiposity and adipogenesis while preserving insulin sensitivity." *Diabetes* no. 59 (6):1424–34. doi: 10.2337/db09-1212.

Corcoran, G. B., E. L. Todd, W. J. Racz, H. Hughes, C. V. Smith, and J. R. Mitchell. 1985. "Effects of N-acetylcysteine on the disposition and metabolism of acetaminophen in mice." *J Pharmacol Exp Ther* no. 232 (3):857–63.

Dagnell, M., J. Frijhoff, I. Pader et al. 2013. "Selective activation of oxidized PTP1B by the thioredoxin system modulates PDGF-beta receptor tyrosine kinase signaling." *Proc Natl Acad Sci U S A* no. 110 (33):13398–403. doi: 10.1073/pnas.1302891110.

Dong, H., R. L. Haining, K. E. Thummel, A. E. Rettie, and S. D. Nelson. 2000. "Involvement of human cytochrome P450 2D6 in the bioactivation of acetaminophen." *Drug Metab Dispos* no. 28 (12):1397–400.

Dudrick, S. J., and S. M. Kavic. 2002. "Hepatobiliary nutrition: History and future." *J Hepatobiliary Pancreat Surg* no. 9 (4):459–68. doi: 10.1007/s005340200057.

Fernandes, A. P., and A. Holmgren. 2004. "Glutaredoxins: Glutathione-dependent redox enzymes with functions far beyond a simple thioredoxin backup system." *Antioxid Redox Signal* no. 6 (1):63–74.

Forman, H. J., H. Zhang, and A. Rinna. 2009. "Glutathione: Overview of its protective roles, measurement, and biosynthesis." *Mol Aspects Med* no. 30 (1–2):1–12. doi: 10.1016/j.mam.2008.08.006.

Fourquet, S., R. Guerois, D. Biard, and M. B. Toledano. 2010. "Activation of NRF2 by nitrosative agents and H2O2 involves KEAP1 disulfide formation." *J Biol Chem* no. 285 (11):8463–71. doi: 10.1074/jbc.M109.051714.

Franklin, C. C., D. S. Backos, I. Mohar, C. C. White, H. J. Forman, and T. J. Kavanagh. 2009. "Structure, function, and post-translational regulation of the catalytic and modifier subunits of glutamate cysteine ligase." *Mol Aspects Med* no. 30 (1–2):86–98. doi: 10.1016/j.mam.2008.08.009.

Goldstein, B. J., K. Mahadev, X. Wu, L. Zhu, and H. Motoshima. 2005. "Role of insulin-induced reactive oxygen species in the insulin signaling pathway." *Antioxid Redox Signal* no. 7 (7–8):1021–31. doi: 10.1089/ars.2005.7.1021.

Gorrini, C., I. S. Harris, and T. W. Mak. 2013. "Modulation of oxidative stress as an anticancer strategy." *Nat Rev Drug Discov* no. 12 (12):931–47. doi: 10.1038/nrd4002.

Griffith, O. W. 1999. "Biologic and pharmacologic regulation of mammalian glutathione synthesis." *Free Radic Biol Med* no. 27 (9–10):922–35.

Hafsi, H., and P. Hainaut. 2011. "Redox control and interplay between p53 isoforms: Roles in the regulation of basal p53 levels, cell fate, and senescence." *Antioxid Redox Signal* no. 15 (6):1655–67. doi: 10.1089/ars.2010.3771.

Hayes, J. D., and A. T. Dinkova-Kostova. 2014. "The Nrf2 regulatory network provides an interface between redox and intermediary metabolism." *Trends Biochem Sci* no. 39 (4):199–218. doi: 10.1016/j.tibs.2014.02.002.

He, X., and Q. Ma. 2012. "Redox regulation by nuclear factor erythroid 2-related factor 2: Gatekeeping for the basal and diabetes-induced expression of thioredoxin-interacting protein." *Mol Pharmacol* no. 82 (5):887–97. doi: 10.1124/mol.112.081133.

Hinson, J. A., D. W. Roberts, and L. P. James. 2010. "Mechanisms of acetaminophen-induced liver necrosis." *Handb Exp Pharmacol* (196):369–405.

Holmgren, A. 1976. "Hydrogen donor system for Escherichia coli ribonucleoside-diphosphate reductase dependent upon glutathione." *Proc Natl Acad Sci U S A* no. 73 (7):2275–9.

Holmgren, A. 1981. "Regulation of ribonucleotide reductase." *Curr Top Cell Regul* no. 19:47–76.

Holmgren, A. 1985. "Thioredoxin." *Annu Rev Biochem* no. 54:237–71.

Holmgren, A. 1989. "Thioredoxin and glutaredoxin systems." *J Biol Chem* no. 264 (24):13963–6.

Holmgren, A. 2000. "Antioxidant function of thioredoxin and glutaredoxin systems." *Antioxid Redox Signal* no. 2 (4):811–20.

Holmgren, A., and F. Aslund. 1995. "Glutaredoxin." *Methods Enzymol* no. 252:283–92.

Holmgren, A., and R. Sengupta. 2010. "The use of thiols by ribonucleotide reductase." *Free Radic Biol Med* no. 49 (11):1617–28.

Homolya, L., A. Varadi, and B. Sarkadi. 2003. "Multidrug resistance-associated proteins: Export pumps for conjugates with glutathione, glucuronate or sulfate." *Biofactors* no. 17 (1–4):103–14.

Itoh, K., N. Wakabayashi, Y. Katoh et al. 1999. "Keap1 represses nuclear activation of antioxidant responsive elements by Nrf2 through binding to the amino-terminal Neh2 domain." *Genes Dev* no. 13 (1):76–86.

Itoh, K., N. Wakabayashi, Y. Katoh, T. Ishii, T. O'Connor, and M. Yamamoto. 2003. "Keap1 regulates both cytoplasmic-nuclear shuttling and degradation of Nrf2 in response to electrophiles." *Genes Cells* no. 8 (4):379–91.

Itoh, K., J. Mimura, and M. Yamamoto. 2010. "Discovery of the negative regulator of Nrf2, Keap1: A historical overview." *Antioxid Redox Signal* no. 13 (11):1665–78. doi: 10.1089/ars.2010.3222.

Iverson, S. V., S. Eriksson, J. Xu et al. 2013. "A Txnrd1-dependent metabolic switch alters hepatic lipogenesis, glycogen storage, and detoxification." *Free Radic Biol Med* no. 63:369–80. doi: 10.1016/j.freeradbiomed.2013.05.028.

Jakupoglu, C., G. K. Przemeck, M. Schneider et al. 2005. "Cytoplasmic thioredoxin reductase is essential for embryogenesis but dispensable for cardiac development." *Mol Cell Biol* no. 25 (5):1980–8.

James, L. P., P. R. Mayeux, and J. A. Hinson. 2003. "Acetaminophen-induced hepatotoxicity." *Drug Metab Dispos* no. 31 (12):1499–506.

Jung, H., M. J. Kim, D. O. Kim et al. 2013. "TXNIP maintains the hematopoietic cell pool by switching the function of p53 under oxidative stress." *Cell Metab* no. 18 (1):75–85. doi: 10.1016/j.cmet.2013.06.002.

Kaidanovich-Beilin, O., and J. R. Woodgett. 2011. "GSK-3: Functional insights from cell biology and animal models." *Front Mol Neurosci* no. 4:40. doi: 10.3389/fnmol.2011.00040.

Kulkarni, S. R., L. E. Armstrong, and A. L. Slitt. 2013a. "Caloric restriction-mediated induction of lipid metabolism gene expression in liver is enhanced by Keap1-knockdown." *Pharm Res* no. 30 (9):2221–31. doi: 10.1007/s11095-013-1138-9.

Kulkarni, S. R., J. Xu, A. C. Donepudi, W. Wei, and A. L. Slitt. 2013b. "Effect of caloric restriction and AMPK activation on hepatic nuclear receptor, biotransformation enzyme, and transporter expression in lean and obese mice." *Pharm Res* no. 30 (9):2232–47. doi: 10.1007/s11095-013-1140-2.

Kulkarni, S. R., A. C. Donepudi, J. Xu et al. 2014. "Fasting induces nuclear factor E2-related factor 2 and ATP-binding cassette transporters via protein kinase A and Sirtuin-1 in mouse and human." *Antioxid Redox Signal* no. 20 (1):15–30. doi: 10.1089/ars.2012.5082.

Laurent, T. C., E. C. Moore, and P. Reichard. 1964. "Enzymatic synthesis of deoxyribonucleotides. IV. Isolation and characterization of thioredoxin, the hydrogen donor from Escherichia Coli B." *J Biol Chem* no. 239:3436–44.

Lee, S. R., K. S. Yang, J. Kwon, C. Lee, W. Jeong, and S. G. Rhee. 2002. "Reversible inactivation of the tumor suppressor PTEN by H2O2." *J Biol Chem* no. 277 (23):20336–42. doi: 10.1074/jbc.M111899200.

Lu, J., and A. Holmgren. 2014. "The thioredoxin antioxidant system." *Free Radic Biol Med* no. 66:75–87. doi: 10.1016/j.freeradbiomed.2013.07.036.

Lu, S. C. 2013. "Glutathione synthesis." *Biochim Biophys Acta* no. 1830 (5):3143–53. doi: 10.1016/j.bbagen.2012.09.008.

Mandal, P. K., M. Schneider, P. Kolle et al. 2010. "Loss of thioredoxin reductase 1 renders tumors highly susceptible to pharmacologic glutathione deprivation." *Cancer Res* no. 70 (22):9505–14.

Marini, M. G., K. Chan, L. Casula, Y. W. Kan, A. Cao, and P. Moi. 1997. "hMAF, a small human transcription factor that heterodimerizes specifically with Nrf1 and Nrf2." *J Biol Chem* no. 272 (26):16490–7.

Matsui, M., M. Oshima, H. Oshima et al. 1996. "Early embryonic lethality caused by targeted disruption of the mouse thioredoxin gene." *Dev Biol* no. 178 (1):179–85.

McMahon, M., K. Itoh, M. Yamamoto, and J. D. Hayes. 2003. "Keap1-dependent proteasomal degradation of transcription factor Nrf2 contributes to the negative regulation of antioxidant response element-driven gene expression." *J Biol Chem* no. 278 (24):21592–600. doi: 10.1074/jbc.M300931200.

Menendez, D., A. Inga, and M. A. Resnick. 2009. "The expanding universe of p53 targets." *Nat Rev Cancer* no. 9 (10):724–37.

Meuillet, E. J., D. Mahadevan, M. Berggren, A. Coon, and G. Powis. 2004. "Thioredoxin-1 binds to the C2 domain of PTEN inhibiting PTEN's lipid phosphatase activity and membrane binding: A mechanism for the functional loss of PTEN's tumor suppressor activity." *Arch Biochem Biophys* no. 429 (2):123–33. doi: 10.1016/j.abb.2004.04.020.

Mobasher, M. A., A. Gonzalez-Rodriguez, B. Santamaria et al. 2013. "Protein tyrosine phosphatase 1B modulates GSK3beta/Nrf2 and IGFIR signaling pathways in acetaminophen-induced hepatotoxicity." *Cell Death Dis* no. 4:e626. doi: 10.1038/cddis.2013.150.

Moi, P., K. Chan, I. Asunis, A. Cao, and Y. W. Kan. 1994. "Isolation of NF-E2-related factor 2 (Nrf2), a NF-E2-like basic leucine zipper transcriptional activator that binds to the tandem NF-E2/AP1 repeat of the beta-globin locus control region." *Proc Natl Acad Sci U S A* no. 91 (21):9926–30.

More, V. R., Q. Cheng, A. C. Donepudi et al. 2013a. "Alcohol cirrhosis alters nuclear receptor and drug transporter expression in human liver." *Drug Metab Dispos* no. 41 (5):1148–55. doi: 10.1124/dmd.112.049676.

More, V. R., J. Xu, P. C. Shimpi et al. 2013b. "Keap1 knockdown increases markers of metabolic syndrome after long-term high fat diet feeding." *Free Radic Biol Med* no. 61C:85–94. doi: 10.1016/j.freeradbiomed.2013.03.007.

Nakamura, B. N., T. J. Fielder, Y. D. Hoang et al. 2011. "Lack of maternal glutamate cysteine ligase modifier subunit (Gclm) decreases oocyte glutathione concentrations and disrupts preimplantation development in mice." *Endocrinology* no. 152 (7):2806–15. doi: 10.1210/en.2011-0207.

Niture, S. K., R. Khatri, and A. K. Jaiswal. 2014. "Regulation of Nrf2-an update." *Free Radic Biol Med* no. 66:36–44. doi: 10.1016/j.freeradbiomed.2013.02.008.

Parikh, H., E. Carlsson, W. A. Chutkow et al. 2007. "TXNIP regulates peripheral glucose metabolism in humans." *PLoS Med* no. 4 (5):e158. doi: 10.1371/journal.pmed.0040158.

Petkov, P. M., J. Zavadil, D. Goetz et al. 2004. "Gene expression pattern in hepatic stem/progenitor cells during rat fetal development using complementary DNA microarrays." *Hepatology* no. 39 (3):617–27.

Polager, S., and D. Ginsberg. 2009. "p53 and E2f: Partners in life and death." *Nat Rev Cancer* no. 9 (10):738–48.

Prigge, J. R., S. Eriksson, S. V. Iverson et al. 2012. "Hepatocyte DNA replication in growing liver requires either glutathione or a single allele of txnrd1." *Free Radic Biol Med* no. 52 (4):803–10. doi: 10.1016/j.freeradbiomed.2011.11.025.

Reichheld, J. P., M. Khafif, C. Riondet, M. Droux, G. Bonnard, and Y. Meyer. 2007. "Inactivation of thioredoxin reductases reveals a complex interplay between thioredoxin and glutathione pathways in Arabidopsis development." *Plant Cell* no. 19 (6):1851–65.

Rhee, S. G., S. W. Kang, W. Jeong, T. S. Chang, K. S. Yang, and H. A. Woo. 2005. "Intracellular messenger function of hydrogen peroxide and its regulation by peroxiredoxins." *Curr Opin Cell Biol* no. 17 (2):183–9.

Rhee, S. G., H. A. Woo, I. S. Kil, and S. H. Bae. 2012. "Peroxiredoxin functions as a peroxidase and a regulator and sensor of local peroxides." *J Biol Chem* no. 287 (7):4403–10. doi: 10.1074/jbc.R111.283432.

Rogers, L. K., T. Tamura, B. J. Rogers, S. E. Welty, T. N. Hansen, and C. V. Smith. 2004. "Analyses of glutathione reductase hypomorphic mice indicate a genetic knockout." *Toxicol Sci* no. 82 (2):367–73.

Rogers, L. K., C. M. Bates, S. E. Welty, and C. V. Smith. 2006. "Diquat induces renal proximal tubule injury in glutathione reductase-deficient mice." *Toxicol Appl Pharmacol* no. 217 (3):289–98. doi: 10.1016/j.taap.2006.08.012.

Rollins, M. F., D. M. van der Heide, C. M. Weisend et al. 2010. "Hepatocytes lacking thioredoxin reductase 1 have normal replicative potential during development and regeneration." *J Cell Sci* no. 123 (Pt 14):2402–12.

Schwertassek, U., A. Haque, N. Krishnan et al. 2014. "Reactivation of oxidized PTP1B and PTEN by thioredoxin 1." *Febs J* no. 281 (16):3545–58. doi: 10.1111/febs.12898.

Seemann, S., and P. Hainaut. 2005. "Roles of thioredoxin reductase 1 and APE/Ref-1 in the control of basal p53 stability and activity." *Oncogene* no. 24 (24):3853–63. doi: 10.1038/sj.onc.1208549.

Shaked, M., M. Ketzinel-Gilad, E. Cerasi, N. Kaiser, and G. Leibowitz. 2011. "AMP-activated protein kinase (AMPK) mediates nutrient regulation of thioredoxin-interacting protein (TXNIP) in pancreatic beta-cells." *PLoS ONE* no. 6 (12):e28804. doi: 10.1371/journal.pone.0028804.

Shao, D., S. Oka, T. Liu et al. 2014. "A redox-dependent mechanism for regulation of AMPK activation by Thioredoxin1 during energy starvation." *Cell Metab* no. 19 (2):232–45. doi: 10.1016/j.cmet.2013.12.013.

Shi, Z. Z., J. Osei-Frimpong, G. Kala et al. 2000. "Glutathione synthesis is essential for mouse development but not for cell growth in culture." *Proc Natl Acad Sci U S A* no. 97 (10):5101–6.

Singh, S., S. Vrishni, B. K. Singh, I. Rahman, and P. Kakkar. 2010. "Nrf2-ARE stress response mechanism: A control point in oxidative stress-mediated dysfunctions and chronic inflammatory diseases." *Free Radic Res* no. 44 (11):1267–88. doi: 10.3109/10715762.2010.507670.

Slitt, A. M., P. K. Dominick, J. C. Roberts, and S. D. Cohen. 2005. "Effect of ribose cysteine pretreatment on hepatic and renal acetaminophen metabolite formation and glutathione depletion." *Basic Clin Pharmacol Toxicol* no. 96 (6):487–94. doi: 10.1111/j.1742-7843.2005.pto_13.x.

Soerensen, J., C. Jakupoglu, H. Beck et al. 2008. "The role of thioredoxin reductases in brain development." *PLoS ONE* no. 3 (3):e1813.

Spindel, O. N., C. World, and B. C. Berk. 2012. "Thioredoxin interacting protein: Redox dependent and independent regulatory mechanisms." *Antioxid Redox Signal* no. 16 (6):587–96. doi: 10.1089/ars.2011.4137.

Spyrou, G., and A. Holmgren. 1996. "Deoxyribonucleoside triphosphate pools and growth of glutathione-depleted 3T6 mouse fibroblasts." *Biochem Biophys Res Commun* no. 220 (1):42–6.

Stanton, R. C. 2012. "Glucose-6-phosphate dehydrogenase, NADPH, and cell survival." *IUBMB Life* no. 64 (5):362–9. doi: 10.1002/iub.1017.

Suh, H. W., S. Yun, H. Song et al. 2013. "TXNIP interacts with hEcd to increase p53 stability and activity." *Biochem Biophys Res Commun* no. 438 (2):264–9. doi: 10.1016/j.bbrc.2013.07.036.

Suvorova, E. S., O. Lucas, C. M. Weisend et al. 2009. "Cytoprotective Nrf2 pathway is induced in chronically txnrd 1-deficient hepatocytes." *PLoS One* no. 4 (7):e6158.

Tanaka, Y., L. M. Aleksunes, R. L. Yeager et al. 2008. "NF-E2-related factor 2 inhibits lipid accumulation and oxidative stress in mice fed a high-fat diet." *J Pharmacol Exp Ther* no. 325 (2):655–64. doi: 10.1124/jpet.107.135822.

Tappy, L., and K. Minehira. 2001. "New data and new concepts on the role of the liver in glucose homeostasis." *Curr Opin Clin Nutr Metab Care* no. 4 (4):273–7.

Thakur, S., B. Sarkar, R. P. Cholia, N. Gautam, M. Dhiman, and A. K. Mantha. 2014. "APE1/Ref-1 as an emerging therapeutic target for various human diseases: Phytochemical modulation of its functions." *Exp Mol Med* no. 46:e106. doi: 10.1038/emm.2014.42.

Thelander, L., and P. Reichard. 1979. "Reduction of ribonucleotides." *Annu Rev Biochem* no. 48:133–58.

Thummel, K. E., C. A. Lee, K. L. Kunze, S. D. Nelson, and J. T. Slattery. 1993. "Oxidation of acetaminophen to N-acetyl-p-aminobenzoquinone imine by human CYP3A4." *Biochem Pharmacol* no. 45 (8):1563–9.

Ueno, M., H. Masutani, R. J. Arai et al. 1999. "Thioredoxin-dependent redox regulation of p53-mediated p21 activation." *J Biol Chem* no. 274 (50):35809–15.

Vogelstein, B., and K. W. Kinzler. 2004. "Cancer genes and the pathways they control." *Nat Med* no. 10 (8):789–99.

Wang, D. B., C. Kinoshita, Y. Kinoshita, and R. S. Morrison. 2014. "p53 and mitochondrial function in neurons." *Biochim Biophys Acta.* no. 1842 (8):1186–97. doi: 10.1016/j.bbadis.2013.12.015.

Wen, X., A. C. Donepudi, P. E. Thomas, A. L. Slitt, R. S. King, and L. M. Aleksunes. 2013. "Regulation of hepatic phase II metabolism in pregnant mice." *J Pharmacol Exp Ther* no. 344 (1):244–52. doi: 10.1124/jpet.112.199034.

Williamson, J. M., B. Boettcher, and A. Meister. 1982. "Intracellular cysteine delivery system that protects against toxicity by promoting glutathione synthesis." *Proc Natl Acad Sci U S A* no. 79 (20):6246–9.

Wu, N., B. Zheng, A. Shaywitz et al. 2013. "AMPK-dependent degradation of TXNIP upon energy stress leads to enhanced glucose uptake via GLUT1." *Mol Cell* no. 49 (6):1167–75. doi: 10.1016/j.molcel.2013.01.035.

Wu, S. B., Y. T. Wu, T. P. Wu, and Y. H. Wei. 2014. "Role of AMPK-mediated adaptive responses in human cells with mitochondrial dysfunction to oxidative stress." *Biochim Biophys Acta* no. 1840 (4):1331–44. doi: 10.1016/j.bbagen.2013.10.034.

Xu, J., S. R. Kulkarni, A. C. Donepudi, V. R. More, and A. L. Slitt. 2012a. "Enhanced Nrf2 activity worsens insulin resistance, impairs lipid accumulation in adipose tissue, and increases hepatic steatosis in leptin-deficient mice." *Diabetes* no. 61 (12):3208–18. doi: 10.2337/db11-1716.

Xu, J., S. R. Kulkarni, L. Li, and A. L. Slitt. 2012b. "UDP-glucuronosyltransferase expression in mouse liver is increased in obesity- and fasting-induced steatosis." *Drug Metab Dispos* no. 40 (2):259–66. doi: 10.1124/dmd.111.039925.

Xu, J., A. C. Donepudi, J. E. Moscovitz, and A. L. Slitt. 2013. "Keap1-knockdown decreases fasting-induced fatty liver via altered lipid metabolism and decreased fatty acid mobilization from adipose tissue." *PLoS ONE* no. 8 (11):e79841. doi: 10.1371/journal.pone.0079841.

Yalcin, E. B., V. More, K. L. Neira et al. 2013. "Downregulation of sulfotransferase expression and activity in diseased human livers." *Drug Metab Dispos* no. 41 (9):1642–50. doi: 10.1124/dmd.113.050930.

Yu, F. X., T. F. Chai, H. He, T. Hagen, and Y. Luo. 2010. "Thioredoxin-interacting protein (Txnip) gene expression: Sensing oxidative phosphorylation status and glycolytic rate." *J Biol Chem* no. 285 (33):25822–30. doi: 10.1074/jbc.M110.108290.

Zamek-Gliszczynski, M. J., K. A. Hoffmaster, K. Nezasa, M. N. Tallman, and K. L. Brouwer. 2006. "Integration of hepatic drug transporters and phase II metabolizing enzymes: Mechanisms of hepatic excretion of sulfate, glucuronide, and glutathione metabolites." *Eur J Pharm Sci* no. 27 (5):447–86.

Zhang, M., Y. Dong, J. Xu et al. 2008. "Thromboxane receptor activates the AMP-activated protein kinase in vascular smooth muscle cells via hydrogen peroxide." *Circ Res* no. 102 (3):328–37. doi: 10.1161/CIRCRESAHA.107.163253.

Zhong, L., E. S. J. Arnér, and A. Holmgren. 2000. "Structure and mechanism of mammalian thioredoxin reductase: The active site is a redox-active selenolthiol/selenenylsulfide formed from the conserved cysteine-selenocysteine sequence." *Proc Natl Acad Sci U S A* no. 97 (11):5854–9.

Zhuang, J., W. Ma, C. U. Lago, and P. M. Hwang. 2012. "Metabolic regulation of oxygen and redox homeostasis by p53: Lessons from evolutionary biology?" *Free Radic Biol Med* no. 53 (6):1279–85. doi: 10.1016/j.freeradbiomed.2012.07.026.
Zuckerman, V., K. Wolyniec, R. V. Sionov, S. Haupt, and Y. Haupt. 2009. "Tumour suppression by p53: The importance of apoptosis and cellular senescence." *J Pathol* no. 219 (1):3–15.

14 Selenium Mediates a Switch in Macrophage Polarization

K. Sandeep Prabhu, Avinash K. Kudva, and Shakira M. Nelson

CONTENTS

14.1 Introduction 293
14.2 Overview of the Inflammatory Response...294
14.3 Resolution of Inflammation 295
14.4 Macrophages in Inflammation and Anti-Inflammation.........................296
14.5 Macrophage Phenotypes..298
14.6 Selenium and Its Role in Asthma and Helminth Infections 301
14.7 Conclusions..302
Acknowledgments..302
References...303

14.1 INTRODUCTION

Selenium (Se) is important for optimal functioning of the immune system (Ebert-Dumig et al., 1999; Kiremidjian-Schumacher and Roy, 1998; Kubena and McMurray, 1996; McKenzie et al., 1998). Several studies demonstrate Se deficiency to be associated with impaired immune responsiveness, and Se-supplementation results in increased immunocompetence (McKenzie et al., 1998). For example, Se supplementation significantly increased tumor cytotoxicity of macrophages and natural killer cells (Kubena and McMurray, 1996). Selenium deficiency has been associated with several pathological conditions, including cardiovascular diseases (Azoicai et al., 1997), rheumatoid arthritis (Knekt et al., 2000), AIDS (Hori et al., 1997), systemic inflammatory response syndrome (SIRS) (Kocyigit et al., 2004; Qujeq et al., 2003; Sakr et al., 2007), and most notably cancer (Combs, 1999; Ganther, 1999), for which oxidative stress and inflammation appear to be the common denominators. It has been proposed that Se, in the form of selenoproteins, mitigates inflammation in addition to preventing malignant transformation of cells (Ganther, 1999). In fact, an investigation of the causal relationship between Se in forage crops and county levels of cancer mortality in the United States and cancer mortality rates for the major cancer sites were found to be higher in counties with low Se (Clark et al., 1991; Rayman, 2000, 2002). Selenium intake may be suboptimal with respect to disease risk, notably in the population of adults in the United Kingdom, Europe, China, and New Zealand. Moreover, with the changes in lifestyle and food habits, plasma Se levels appear to be on the downward trend, tilting more toward deficiency. Particularly in cigarette smokers, HIV, and breast cancer patients, the plasma Se levels were significantly reduced (Cole et al., 2005; Fleming, 1989; Preston, 1991). Heyland et al. (2005) conducted a meta-analysis demonstrating the correlation between plasma Se levels and tissue inflammation, infection, and organ failure. This study also showed that Se supplementation was associated with lowered mortality rate. Studies by Sakr et al. (2007) suggested a correlation between decreased plasma Se levels and multi-organ dysfunction with SIRS or septic shock. Angstwurm et al. (2007) reported the ability of selenite (1000 µg IV) to improve survival of patients with SIRS, sepsis, and septic shock.

Although there is a growing body of evidence suggesting that intake of Se above the normal nutritional range could confer benefit in treating an inflammation (Rayman, 2000, 2002), it is not clear how exactly Se or selenoproteins mitigate inflammation. In this review, we examine the process of inflammation in detail and address the possible mechanisms by which Se promotes a switch in the inflammatory response with an emphasis on the differential regulation of eicosanoids.

14.2 OVERVIEW OF THE INFLAMMATORY RESPONSE

Inflammation is a broad term used to depict an adaptive response elicited upon exposure to diverse stimuli to restore homeostasis (Barton, 2008). Although classical inflammation, usually referred to as pathological inflammation, is a result of infection and/or tissue injury that is well understood as the basis of acute inflammation, the systemic response to chronic inflammation is less well understood (Medzhitov, 2008). Chronic inflammation as in type 2 diabetes, obesity, and cardiovascular disease are thought to originate from imbalances in physiological homeostasis, including changes in cellular redox homeostasis that result from Se deficiency. Tissue stress and malfunction rely on innate immune cells, such as macrophages, that are one of the key cells responsible for chronic inflammatory diseases prevalent in the modern world.

It is generally assumed that controlled inflammation is beneficial to the host, and on the other hand, uncontrolled or dysregulated inflammation can be detrimental. The physiological counterpart of dysregulated inflammation may be afforded by blood components (plasma and leukocytes), in which an inflammatory trigger is associated with diverse changes in innate and adaptive cells in the form of cytokines (for example, TNFα), chemokines, enzymes, and small molecule mediators. It is these early events that are thought to be important in setting the tone for a transient inflammatory response. It is important to note that efficient expression of selenoproteins is likely to impact this seminal step. Earlier studies in our laboratory have shown the benefit of Se supplementation of mice at levels higher than those required for nutritional adequacy in effectively mitigating the expression of many proinflammatory cytokines (TNFα, IL-1β, and IFNγ) upon challenge with bacterial endotoxin lipopolysaccharide (LPS) (Vunta et al., 2008). However, it has also been shown that some of the early events during the inflammatory phase are essential to trigger a switch in the response to initiate anti-inflammatory or resolution pathways that ultimately restore homeostasis. Thus, for the successful treatment of inflammation, it is not only important to decrease the expression of proinflammatory mediators, but also to simultaneously increase the expression of anti-inflammatory mediators to initiate timely and effective return to homeostasis.

14.3   RESOLUTION OF INFLAMMATION

The struggle to combat inflammation has shifted to looking outside of conventional anti-inflammatory treatments that utilize nonsteroidal anti-inflammatory drugs (NSAIDs) and steroids. Specifically, regulation of inflammation has gravitated toward enhancing proresolution mediators. Seminal studies from the Serhan laboratory at Harvard Medical School have demonstrated an important role for lipid mediator families (lipoxins, resolvins, and protectins) as proresolving metabolites, which are part of an active biochemical and metabolic resolution process (Serhan, 2009, 2010). Studies in our laboratory have focused on the enhanced production of the endogenous cyclopentenone derivatives of prostaglandin D_2 (PGD_2) in macrophages supplemented with Se (Gandhi et al., 2011; Vunta et al., 2007). These cyclopentenone PGs (CyPGs), most notably Δ^{12}-PGJ_2 and 15-deoxy-$\Delta^{12,14}$-PGJ_2, are derived from arachidonic acid metabolism by cyclooxygenases (COX-1 and COX-2) to PGH_2 that is further transferred to PGD_2 synthases (hematopoietic H-PGDS and lipocalin-type L-PGDS) to form PGD_2. PGD_2 undergoes two dehydration reactions to form Δ^{12}-PGJ_2 and 15-deoxy-$\Delta^{12,14}$-PGJ_2. Production of these CyPGs, as a result of eicosanoid class switching, is stimulated by selenoprotein expression particularly in macrophages (Gandhi et al., 2011). Studies from many laboratories, including ours, have shown that CyPGs impact numerous cellular processes, of which some are redox-dependent, leading to the downregulation of the expression of proinflammatory genes (Gandhi et al., 2011). On the other hand, PGH_2 is also metabolized to PGE_2 and thromboxane (TXA_2) that are reported to have proinflammatory functions. However, recent studies have shown that PGE_2 is also essential to negatively regulate the production of inflammatory cytokines/chemokines and of IL-17 in fatal visceral leishmaniasis (Saha et al., 2014). In this study, macrophages were shown to be the key for the production of PGE_2 upon infection that led to the inhibition of anti-leishmanial activity of IL-17. Interestingly, PGE_2 is further metabolized by 15-hydroxy PG dehydrogenase (15-PGDH) to 15-keto-PGE_2 and 13, 14-dihydro-15-keto-PGE_2. These metabolites may have anti-inflammatory activities through a feed-forward activation of the transcription factor, peroxisome proliferator activator receptor (PPAR)-γ, for which these PGE_2 metabolites potentially serve as endogenous ligands (Lu et al., 2013). In fact, using a metabolomics approach, we have recently demonstrated a causal relationship between plasma and urine 15-keto-PGE_2 metabolite and the Se status during inflammation *in vivo* and *in vitro* (Kaushal et al., 2014). In close agreement with the metabolomics results, the expression of 15-PGDH was also found to be upregulated by a supranutritional Se status in inflamed tissues and macrophages, suggesting an important role for Se in the metabolism of PGE_2 to bioactive metabolites with proresolution properties. Additional mechanisms utilized by these lipid mediators include the activation of the redox-sensitive transcription factor Nrf-2 that drives the expression of enzymes involved in phase-2 metabolism as part of a stress-adaptive response system (Kawamoto et al., 2000; Levonen et al., 2001). Furthermore, ligand-dependent activation of PPARγ suppresses nuclear factor-κB (NF-κB) that primarily drives the expression of proinflammatory genes (Harmon et al., 2011). This process exemplifies a complex feedback regulation by an inflammatory molecule (such as PGE_2) that initiates pathways of resolution.

Studies from our laboratory have shown that Se supplementation decreases the activation of NF-κB in macrophages treated with LPS while also activating PPARγ (Vunta et al., 2007, 2008). Further studies have confirmed that Se-dependent changes in the metabolism of arachidonic acid oxidation in macrophages were critical in the differential regulation of NF-κB and PPARγ. In collaboration with the Hatfield Laboratory, we have recently reported the central role of macrophage selenoproteins in not only mitigating inflammation, but also increasing the resolution from chemical injury of the colonic epithelium (Kaushal et al., 2014). $Trsp^{fl/fl}LysM^{Cre}$ mice that lack selenocysteinyl tRNA (see Chapter 4) in macrophages were susceptible to dextran sodium sulfate (DSS)-induced colitis associated with high inflammation and poor resolution (Kaushal et al., 2014). Interestingly, analysis of the colonic extracts of DSS-treated $Trsp^{fl/fl}LysM^{Cre}$ mice suggested that selenoproteins were essential for the upregulation of 15-PGDH. It should be noted that even though 15-PGDH preferentially oxidizes PGE_2, 15-PGDH can also oxidize other eicosanoids (Ruckrich et al., 1976; Sun et al., 1976). However, given the spatiotemporal control of eicosanoid biogenesis, it is possible that 15-PGDH activity may be tightly regulated to preferentially metabolize PGE_2 on demand.

14.4 MACROPHAGES IN INFLAMMATION AND ANTI-INFLAMMATION

Macrophages are part of the monocyte phagocyte system comprised of a heterogeneous population of innate immune cells. Originally identified by Metchnikoff at the beginning of the 20th century as phagocytic cells responsible for eliminating pathogens (Tauber, 2003), macrophages are one of the most important immune effector cells that play a key role in tissue homeostasis, antigen presentation, immunity, cancer, metabolism, and wound healing (Wynn et al., 2013). Found essentially in all tissues, adult tissue macrophages are terminally differentiated cells of the mononuclear phagocytic lineage derived from circulating monocytes that originate from the hematopoietic stem cells in the bone marrow. However, macrophages that originate during ontogeny persist into adulthood thus lending credence to the idea of self-renewal (Wynn et al., 2013). Regardless of the multiple routes of origin, macrophages have been classified to functionally pertain to inflammatory states. The importance of macrophages has been recognized in many chronic diseases, such as atherosclerosis, multiple sclerosis, sepsis, inflammatory bowel disease (IBD), and rheumatoid arthritis. In these diseases, macrophage activation unleashes multiple key proinflammatory cytokines, such as TNFα, IL-6, IL-12, IL-18, and IL-23 that drive autoimmune inflammation (Wynn et al., 2013).

Their ability to recognize "danger" signals through detection of necrotic debris or pathogen-associated molecular patterns (PAMPs) is effected through toll-like receptors (TLRs), triggering cell activation within tissues (Mosser, 2003; Mosser and Edwards, 2008). In addition to endogenous signal detection, macrophages are also activated in response to cellular microenvironment changes (Edwards et al., 2006; Lawrence and Natoli, 2011), encompassing incredible plasticity to alter phenotypes depending on the signals received (Edwards et al., 2006; Mosser, 2003; Mosser and Edwards, 2008). The conventional macrophage activation phenotypes are broadly classified into classically activated (also referred to as CAM or M1) or alternatively activated (also referred to as AAM or M2) (Figure 14.1) macrophages, with many newly discovered intermediary phenotypes (Laskin, 2009).

The classical M1 macrophages upon activation express proinflammatory mediators, such as TNFα, IL-1β, NO, and PGE$_2$. However, under supplemented condition with selenium, these macrophages are skewed toward alternatively activated macrophages (M2), leading to resolution of inflammation. These M2 macrophages express arginase-1 (Arg-1) and other M2-specific markers, such as Fizz1, YM1, and IL-10, in addition to promoting release of anti-inflammatory cytokines, such as IL-4, by Th2 cells. The M2 macrophages also favor cyclooxygenase (COX)-mediated production of cyclopentenone prostaglandins (CyPG), Δ^{12}-prostaglandin J$_2$ (Δ^{12}-PGJ$_2$)

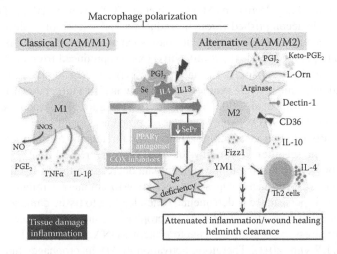

FIGURE 14.1 Role of selenium and selenoproteins in the polarization of macrophages. Schematic illustration of the key role of Se and selenoproteins in the polarization of macrophages from a classically activated (M1) phenotype to an alternatively activated (M2) phenotype that represents a continuum of proinflammatory and anti-inflammatory macrophages, respectively. Se-dependent metabolism of arachidonic acid to 15d-PGJ$_2$ along with T-cell (Th2)-derived cytokines are critical for the polarization toward an M2 phenotype. Although this polarization is aided by selenoprotein expression, inhibitors of cyclooxygenase (COX) enzymes or antagonists of the peroxisome proliferator activated receptor (PPAR)-γ block the Se-dependent skewing of macrophage phenotypes. L-Orn, L-ornithine. Fizz1, resistin-like secreted protein found in inflammatory zone 1. YM1, chitinase-like lectin. For further details, see text.

and its dehydration product 15-deoxy-$\Delta^{12,14}$-PGJ$_2$, and an oxidation product of PGE$_2$, 15-keto-PGE$_2$, and 13,14-dihydro-15-keto-PGE$_2$ that can all function as PPARγ ligands. Also accumulation of these mediators exerts a positive feedback loop by increasing the polarization into alternatively activated macrophages. On the contrary, Se deficiency and inhibition of COX by nonsteroidal anti-inflammatory drugs (NSAIDs) and PPARγ by antagonists can block the polarization to favor accumulation of M1 macrophages, thus maintaining inflammation. Therefore, it would be interesting to explore if NSAIDs serve as potential confounders in clinical trials to limit the spectrum of biological activities of Se.

14.5   MACROPHAGE PHENOTYPES

The nomenclature of M1 macrophages is derived from cytokines responsible for their activation, namely IFNγ and TNFα, which are associated as Th1-type cytokines (Lawrence and Natoli, 2011). Intriguingly, both signals have been shown to only prime the macrophage for activation but not actually activate it (Mosser, 2003). Activation occurs in response to exposure to a microbial product, such as LPS (Gordon, 2007; Gordon and Taylor, 2005; Louis et al., 1999; Mosser, 2003; Olefsky and Glass, 2010). Once activated, M1 macrophages are responsible for microbicidal activity, that is, migrating to areas of inflammation, where encountered pathogens are phagocytosed and destroyed (Mantovani et al., 2002; Mosser, 2003). The killing of pathogens is accomplished through production and release of proinflammatory molecules, such as nitric oxide and cytokines, such as interleukin (IL)-1β and IL-12 (Ginderachter, 2008; Vats et al., 2006). Nitric oxide is produced from the L-arginine (L-Arg), an amino acid found in macrophage cells. L-Arg is acted upon by the enzyme inducible nitric oxide synthase (iNOS) that is also a target downstream gene of NF-κB (Mantovani et al., 2002; Vats et al., 2006) (Figure 14.2). Under Se-deficient conditions, increased iNOS expression was observed as a result of enhanced activation of NF-κB (Prabhu et al., 2002). In contrast, treatment of macrophages with Se (as inorganic selenite) at supraphysiological doses decreased the LPS-dependent activation of NF-κB, leading to the concomitant decrease in the production of nitric oxide. Although M1 macrophages are necessary to counteract inflammation and microbes, their persistence is detrimental, often leading to tissue damage and tumor development (Ginderachter, 2008). M1 macrophages contribute to increased levels of oxidative stress, prompting damage to proteins, DNA, and tissue (Laskin, 2009; Murray and Wynn, 2011). Therefore, activation of M1 macrophages must be controlled. Thus far, all our studies clearly indicate that Se and selenoproteins negatively impact the expression of characteristic M1 markers, such as iNOS, IL-1β, and PGE$_2$.

Similar to M1 macrophages, M2 macrophages develop in response to immune signals, the most well known being Th2-produced cytokines IL-4 and IL-13 (Chawla, 2010; Gordon, 2003, 2007; Levings and Schrader, 1999). A vital function of M2 macrophages is the downregulation of the production of cytotoxic inflammatory mediators (Laskin, 2009). A second, more intriguing, function of M2 macrophages is the stimulation of tissue repair leading to wound healing (Gordon, 2003; Laskin, 2009; Mosser, 2003; Mosser and Edwards, 2008). Various markers are used to

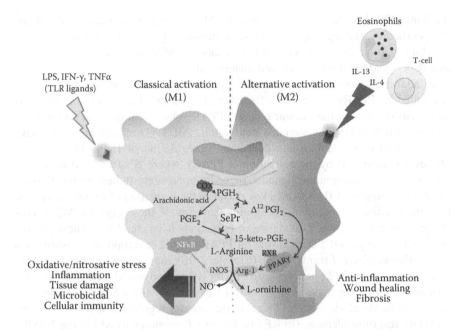

FIGURE 14.2 Classical versus alternatively activated macrophages. Classical (M1) macrophages are activated by proinflammatory stimulants, such as LPS, and Th1 cytokines, such as IFN-γ, TNFα, and IL-1β. Products such as PGE_2 produced by M1 macrophage induce inflammation in the body, leading to microbial killing and tissue damage. Conversely, alternative (M2) macrophage polarization is aided by Th2 cytokines IL-4 and IL-13 through a STAT6-specific pathway. Arginase-1 (Arg-1) is a key enzyme that catalyzes the production of L-ornithine from L-Arg that is subsequently converted to proline and polyamines, leading to wound healing and fibrosis. Selenoprotein (SePr)-dependent shunting of the arachidonic acid metabolism toward Δ^{12}-PGJ_2 and 15d-PGJ_2, along with PGE_2 metabolites, also contributes to the activation of anti-inflammatory pathways via the differential regulation of NF-κB and PPARγ. For further details, see text.

identify M2 macrophages, such as arginase-1 (Arg-1), Ym1, Fizz1, and the mannose receptor (Mrc-1) (Fairweather and Cihakova, 2009; Menzies et al., 2010), which are expressed upon treatment with IL-4 (Louis et al., 1999; Morris et al., 1998; Raes et al., 2002, 2005). Arg-1 is often used as a prototypic marker of alternative macrophage activation. In a move that illustrates M2 macrophage regulation of M1 macrophages, Arg-1 uses L-Arg as its substrate, and thus competes with iNOS for the intracellular L-Arg pool (Chawla, 2010; Martinez et al., 2009) (Figure 14.2). In M2 macrophages, L-Arg is converted to urea and L-ornithine by Arg-1. L-ornithine is converted into proline and polyamines, which contribute to wound healing and fibrosis (Gordon, 2003; Laskin, 2009; Vats et al., 2006). A fascinating discovery of M2 macrophages are the multiple subtypes that are characterized on the basis of their mechanism of induction (Luzina et al., 2012): M2a, M2b, and M2c. M2a macrophages are specifically activated by IL-4 and IL-13 and are the subtype most usually studied (Laskin, 2009). The M2b subtype is induced by immune complexes in combination with LPS

(Laskin, 2009; Luzina et al., 2012), and the M2c subtype is induced by IL-10 and glucocorticoids, playing a significant role in immunosuppression and tissue remodeling (Luzina et al., 2012). Recently a fourth subtype, M2d, has been discovered, but its inducers and function are not well understood.

In addition to cytokines as immune regulators, M2 macrophage activation is controlled by PPARγ and STAT6 (Bouhlel et al., 2007; Luzina et al., 2012). As previously discussed, the nuclear receptor PPARγ plays an important role in macrophage differentiation (Lawrence and Natoli, 2011). Gandhi et al. (2011) have demonstrated that selenoprotein-dependent upregulation of H-PGDS in macrophages was mediated by the activation of PPARγ, where Δ^{12}-PGJ_2 and its metabolite served as endogenous ligands. In addition, recent studies by the Chawla Laboratory suggested IL-4 to be a key player in the activation of PPARγ in macrophages, coordinating metabolic programs that provide the energy for M2 macrophage activation, such as β-oxidation of fatty acids (Chawla, 2010). Subsequently, studies have suggested a critical role for PPARγ in M2 macrophage activation via the regulation of *Arg-1* transcription (Chawla, 2010), thus supporting the long-term maintenance of the M2 phenotype. In PPARγ KO mouse models, IL-4 stimulation resulted in a 40% decrease in *Arg-1* mRNA expression and a 50% decrease in its activity (Chawla et al., 2001). Direct transcriptional regulation was identified by a PPARγ response element (PPRE) in the *Arg-1* promoter to which the PPARγ/RXR heterodimer binds (Chawla, 2010). Selenite enhanced the PPARγ-mediated upregulation of Arg-1 and the other M2 markers, *Mrc-1* and *Fizz-1* (Nelson et al., 2011). This indicates cooperation between Se and IL-4 because IL-4 and PPARγ were both required for the expression of M2 activation markers (Chawla, 2010). In macrophages cultured in the presence of Se (adequate to supranutritional levels) a PPARγ antagonist or a chemical inhibitor of H-PGDS decreased the M2 markers to levels seen under Se-deficient conditions suggesting the importance of Se in the polarization of macrophages. Intriguingly, incubation of cells with 1,4-phenylenebis (methylene)selenocyanate (p-XSC), an organic form of Se that does not release Se for incorporation into selenoproteins or the use of macrophages isolated from the bone marrow of $Trsp^{fl/fl}LysM^{Cre}$ that do not express selenoproteins (Kaushal et al., 2014) led to decreased expression of prototypical M2 marker genes, suggesting a relationship between selenoprotein expression and macrophage polarization. Similarly, inhibition of the transcription factor STAT6 (signal transducer and activator of transcription) that is required for the induction of PPARγ prevented the IL-4-mediated expression of M2 genes in macrophages in the presence of Se (Nelson et al., 2011). To explore the effects of different forms of Se on the macrophage phenotype shunting, two organic forms of Se, selenomethionine (SeMet) and methylseleninic acid (MSA), with prior history of chemoprevention were used. Surprisingly, MSA increased Arg-1 activity in a fashion similar to sodium selenite while SeMet failed (Nelson et al., 2011). Thus, the biochemical form of Se and its efficiency by which it is incorporated into selenoproteins cannot be ignored. Taken together, it appears that supplementation with bioavailable Se (as characterized by selenoprotein synthesis) effects macrophage polarization from an M1- to an M2-like phenotype. Such a polarization is thought to support wound healing, granuloma formation, or attenuation of inflammation as in the case of infection by pathogens. Changes in the cellular redox status might be one of the key signals that affect such a polarization upon activation of the IL4 receptor (IL-4R), particularly during immune activation pathways involving T-cells as in asthma. Thus, M2 macrophages function to suppress immune activation, leading to recovery from cell injury. Current studies in our laboratory are underway to validate these findings in an *in vivo* system using models that trigger polarized immune responses, as in helminth infection models, to examine if Se supplementation could promote expulsion of pathogen followed by enhanced wound healing.

14.6 SELENIUM AND ITS ROLE IN ASTHMA AND HELMINTH INFECTIONS

In chronic inflammatory diseases, such as asthma, studies utilizing Se supplementation have demonstrated varying data. Typically, asthma is characterized by airway hyper-responsiveness, alveolar macrophage presence that depends on high expression of Th2 cytokines (IL-4 and IL-13), airway inflammation, and mucous secretions (Erle and Sheppard, 2014; Hoffmann, 2012; Hoffmann and Berry, 2008; Norton and Hoffmann, 2012). Direct effects of IL-13 expression include increased production and storage of mucin within airway epithelial cells in both mouse and human models (Erle and Sheppard, 2014). Other factors that play a role in asthma development include JAK kinase and its phosphorylation of STAT6. STAT6 activation is required for the IL-13–induced increase in mucus production (Kuperman et al., 2002). Moreover, oxidative stress has been shown to significantly contribute to the pathogenesis of asthma, specifically detected in the lower airways of asthmatics (Norton and Hoffmann, 2012). Selenium, through redox-active selenoproteins, is believed to rescue redox homeostasis thus reducing the incidence of asthma. However, further research is required to consistently prove the role of Se in asthma.

IL-13 and IL-4 cytokines have also been demonstrated to play a positive role in the clearance of parasites. Some parasites utilize these cytokines to clear themselves from their hosts, avoiding the classical inflammatory pathway response (Maizels and Balic, 2004; Maizels et al., 2004). With different life cycle characteristics, different species of parasites induce chronic infection, such as the trematode *Schistosoma mansoni* (*S. mansoni*), in which eggs laid by adult female worms become lodged in the liver. Such an event leads to the activation of granulocytes (that produce IL-4 and IL-13) and T-cells leading to granulomatous lesions that ultimately develop into liver fibrosis with increased portal blood pressure and bleeding (Kreider et al., 2007; MacDonald et al., 2002). A trichostrongyle nematode roundworm gastrointestinal parasite, *Heligmosomoides bakeri* (*H. bakeri*) is commonly used to study chronic intestinal helminth infections due to its skewed STAT6-dependent Th2 immune response and its ability to elevate the expression of IL-13, IL-4, and IL-10 without changes in expression of Th1 cytokines, such as IFNγ (Maizels and Balic, 2004; Smith et al., 2013). Although effective *H. bakeri* clearance is dependent on Th2 cytokine production (Au Yeung et al., 2005), this response was decreased in Se-deficient mice compared to Se-adequate mice, providing additional protective abilities of selenium in a parasite infection.

Similar to asthma, gastrointestinal helminth infections are affected by the Se status. However, in contrast to some asthma studies, Se has consistently been shown to play a positive role in parasite clearance. In mice deficient in Se, a delayed expulsion of adult *H. bakeri* worms from a challenge infection has been observed (Smith et al., 2005, 2013). Other studies have shown that the absence of both Se and vitamin E delays adult worm expulsion when compared to mice supplemented with Se alone, vitamin E alone, or both (Smith et al., 2005). Furthermore, it has been demonstrated that the delayed expulsion of adult worms in Se-deficient mice is an effect of the impaired Th2-dependent gene expression within the small intestine (Smith et al., 2005, 2013). We feel that a possible mechanism behind this impaired expression may be afforded by alternatively activated macrophages that may be a protective response against *H. bakeri* infections; however, this warrants further studies. Taken together, these studies demonstrate that Se plays an integral role in resolving parasite infections by strongly biasing the response toward Th2 immunity.

14.7 CONCLUSIONS

In this review, we describe the ability of selenium, in the form of selenoproteins, to affect multiple metabolic pathways within macrophages to polarize them more toward an M2-like phenotype. M2 macrophages are a staple in Th2-type responses, infiltrating different tissues in response to different infections, such as helminthic parasites, or as a means to promote wound healing during inflammatory responses. Recent studies have identified novel subsets of M2 macrophages (M2a, M2b, M2c, and M2d). Each subset has been identified by its mechanism of induction, but further studies are required to fully elucidate if the transition from M1 toward M2 and the various subtypes within are regulated by the selenoproteome in general or by the expression of specific selenoproteins. More importantly, specific selenoproteins that are key to macrophage polarization need to be identified because our studies, thus far, have only utilized macrophages that lack tRNA[Ser]Sec, leading to the deletion of the entire selenoproteome. This is important because our studies have demonstrated the ability of selenoproteins to skew eicosanoid metabolism toward anti-inflammatory mediators that not only inhibit NF-κB, but also activate PPARγ in a ligand-dependent manner. Given that endogenous CyPGs are key for Se-dependent macrophage polarization, it would be prudent to explore if nonsteroidal anti-inflammatory drugs (NSAIDs) serve as potential confounders in clinical trials to limit the spectrum of biological activities of Se and selenoproteins.

ACKNOWLEDGMENTS

This review was made possible by Public Health Service grants awarded to KSP from the National Institutes of Health and those from Penn State University. We gratefully acknowledge all current and past members of the Prabhu Laboratory and many collaborators and colleagues for reagents, timely help, and advice.

REFERENCES

Angstwurm, M. W., L. Engelmann, T. Zimmermann et al. 2007. "Selenium in intensive care (SIC): Results of a prospective randomized, placebo-controlled, multiple-center study in patients with severe systemic inflammatory response syndrome, sepsis, and septic shock." *Crit Care Med* no. 35 (1):118–26.

Au Yeung, K. J., A. Smith, A. Zhao et al. 2005. "Impact of vitamin E or selenium deficiency on nematode-induced alterations in murine intestinal function." *Exp Parasitol* no. 109 (4):201–8.

Azoicai, D., A. Ivan, M. Bradatean et al. 1997. [The importance of the use of selenium in the role of an antioxidant in preventing cardiovascular diseases]. *Rev Med Chir Soc Med Nat Iasi* no. 101 (3–4):109–15.

Barton, G. M. 2008. "A calculated response: Control of inflammation by the innate immune system." *J Clin Invest* no. 118 (2):413–20.

Bouhlel, M. A., B. Derudas, E. Rigamonti et al. 2007. "PPARgamma activation primes human monocytes into alternative M2 macrophages with anti-inflammatory properties." *Cell Metab* no. 6 (2):137–43.

Chawla, A. 2010. "Control of macrophage activation and function by PPARs." *Circ Res* no. 106 (10):1559–69.

Chawla, A., Y. Barak, L. Nagy, D. Liao, P. Tontonoz, and R. M. Evans. 2001. "PPAR-gamma dependent and independent effects on macrophage-gene expression in lipid metabolism and inflammation." *Nat Med* no. 7 (1):48–52.

Clark, L. C., K. P. Cantor, and W. H. Allaway. 1991. "Selenium in forage crops and cancer mortality in US counties." *Arch Environ Health* no. 46 (1):37–42.

Cole, S. B., B. Langkamp-Henken, B. S. Bender, K. Findley, K. A. Herrlinger-Garcia, and C. R. Uphold. 2005. "Oxidative stress and antioxidant capacity in smoking and nonsmoking men with HIV/acquired immunodeficiency syndrome." *Nutr Clin Pract* no. 20 (6):662–7.

Combs, G. F., Jr. 1999. "Chemopreventive mechanisms of selenium." *Med Klin (Munich)* no. 94 Suppl 3:18–24.

Ebert-Dumig, R., J. Seufert, D. Schneider, J. Köhrle, N. Schutze, and F. Jakob. 1999. [Expression of selenoproteins in monocytes and macrophages—Implications for the immune system]. *Med Klin (Munich)* no. 94 Suppl 3:29–34.

Edwards, J. P., X. Zhang, K. A. Frauwirth, and D. M. Mosser. 2006. "Biochemical and functional characterization of three activated macrophage populations." *J Leukoc Biol* no. 80 (6):1298–307.

Erle, D. J., and D. Sheppard. 2014. "The cell biology of asthma." *J Cell Biol* no. 205 (5):621–31.

Fairweather, D., and D. Cihakova. 2009. "Alternatively activated macrophages in infection and autoimmunity." *J Autoimmun* no. 33 (3–4):222–30.

Fleming, C. R. 1989. "Trace element metabolism in adult patients requiring total parenteral nutrition." *Am J Clin Nutr* no. 49 (3):573–9.

Gandhi, U. H., N. Kaushal, K. C. Ravindra et al. 2011. "Selenoprotein-dependent up-regulation of hematopoietic prostaglandin D2 synthase in macrophages is mediated through the activation of peroxisome proliferator-activated receptor (PPAR) gamma." *J Biol Chem* no. 286 (31):27471–82.

Ganther, H. E. 1999. "Selenium metabolism, selenoproteins and mechanisms of cancer prevention: Complexities with thioredoxin reductase." *Carcinogenesis* no. 20 (9):1657–66.

Ginderachter, J. V. 2008. "Classical and alternative activation of macrophages: Different pathways of macrophage-mediated tumor promotion." In *Selected aspects of cancer progression: Metastasis, apoptosis, and immune response*, edited by H. E. Nasir. Dordrecht, Netherlands: Springer Science + Business Media B.V.

Gordon, S. 2003. "Alternative activation of macrophages." *Nat Rev Immunol* no. 3 (1):23–35.

Gordon, S. 2007. "The macrophage: Past, present and future." *Eur J Immunol* no. 37 Suppl 1:S9–17.

Gordon, S., and P. R. Taylor. 2005. "Monocyte and macrophage heterogeneity." *Nat Rev Immunol* no. 5 (12):953–64.

Harmon, G. S., M. T. Lam, and C. K. Glass. 2011. "PPARs and lipid ligands in inflammation and metabolism." *Chem Rev* no. 111 (10):6321–40.

Heyland, D. K., R. Dhaliwal, U. Suchner, and M. M. Berger. 2005. "Antioxidant nutrients: A systematic review of trace elements and vitamins in the critically ill patient." *Intensive Care Med* no. 31 (3):327–37.

Hoffmann, P. R. 2012. "Asthma in children and nutritional selenium get another look." *Clin Exp Allergy* no. 42 (4):488–9.

Hoffmann, P. R., and M. J. Berry. 2008. "The influence of selenium on immune responses." *Mol Nutr Food Res* no. 52 (11):1273–80.

Hori, K., D. Hatfield, F. Maldarelli, B. J. Lee, and K. A. Clouse. 1997. "Selenium supplementation suppresses tumor necrosis factor alpha-induced human immunodeficiency virus type 1 replication *in vitro*." *AIDS Res Hum Retroviruses* no. 13 (15):1325–32.

Kaushal, N., A. K. Kudva, A. D. Patterson et al. 2014. "Crucial role of macrophage selenoproteins in experimental colitis." *J Immunol* no. 193 (7):3683–92.

Kawamoto, Y., Y. Nakamura, Y. Naito et al. 2000. "Cyclopentenone prostaglandins as potential inducers of phase II detoxification enzymes. 15-deoxy-delta(12,14)-prostaglandin j2-induced expression of glutathione S-transferases." *J Biol Chem* no. 275 (15):1 1291–9.

Kiremidjian-Schumacher, L., and M. Roy. 1998. "Selenium and immune function." *Z Ernahrungswiss* no. 37 Suppl 1:50–6.

Knekt, P., M. Heliovaara, K. Aho, G. Alfthan, J. Marniemi, and A. Aromaa. 2000. "Serum selenium, serum alpha-tocopherol, and the risk of rheumatoid arthritis." *Epidemiology* no. 11 (4):402–5.

Kocyigit, A., F. Armutcu, A. Gurel, and B. Ermis. 2004. "Alterations in plasma essential trace elements selenium, manganese, zinc, copper, and iron concentrations and the possible role of these elements on oxidative status in patients with childhood asthma." *Biol Trace Elem Res* no. 97 (1):31–41.

Kreider, T., R. M. Anthony, J. F. Urban, Jr., and W. C. Gause. 2007. "Alternatively activated macrophages in helminth infections." *Curr Opin Immunol* no. 19 (4):448–53.

Kubena, K. S., and D. N. McMurray. 1996. "Nutrition and the immune system: A review of nutrient-nutrient interactions." *J Am Diet Assoc* no. 96 (11):1156–64; 1165–6.

Kuperman, D. A., X. Huang, L. L. Koth et al. 2002. "Direct effects of interleukin-13 on epithelial cells cause airway hyperreactivity and mucus overproduction in asthma." *Nat Med* no. 8 (8):885–9.

Laskin, D. L. 2009. "Macrophages and inflammatory mediators in chemical toxicity: A battle of forces." *Chem Res Toxicol* no. 22 (8):1376–85.

Lawrence, T., and G. Natoli. 2011. "Transcriptional regulation of macrophage polarization: Enabling diversity with identity." *Nat Rev Immunol* no. 11 (11):750–61.

Levings, M. K., and J. W. Schrader. 1999. "IL-4 inhibits the production of TNF-alpha and IL-12 by STAT6-dependent and -independent mechanisms." *J Immunol* no. 162 (9):5224–9.

Levonen, A. L., D. A. Dickinson, D. R. Moellering, R. T. Mulcahy, H. J. Forman, and V. M. Darley-Usmar. 2001. "Biphasic effects of 15-deoxy-delta(12,14)-prostaglandin J(2) on glutathione induction and apoptosis in human endothelial cells." *Arterioscler Thromb Vasc Biol* no. 21 (11):1846–51.

Louis, C. A., V. Mody, W. L. Henry, Jr., J. S. Reichner, and J. E. Albina. 1999. "Regulation of arginase isoforms I and II by IL-4 in cultured murine peritoneal macrophages." *Am J Physiol* no. 276 (1 Pt 2):R237–42.

Lu, D., C. Han, and T. Wu. 2013. "15-hydroxyprostaglandin dehydrogenase-derived 15-keto-prostaglandin E2 inhibits cholangiocarcinoma cell growth through interaction with peroxisome proliferator-activated receptor-gamma, SMAD2/3, and TAP63 proteins." *J Biol Chem* no. 288 (27):19484–502.

Luzina, I. G., A. D. Keegan, N. M. Heller, G. A. Rook, T. Shea-Donohue, and S. P. Atamas. 2012. "Regulation of inflammation by interleukin-4: A review of "alternatives." *J Leukoc Biol* no. 92 (4):753–64.

MacDonald, A. S., M. I. Araujo, and E. J. Pearce. 2002. "Immunology of parasitic helminth infections." *Infect Immun* no. 70 (2):427–33.

Maizels, R. M., and A. Balic. 2004. "Resistance to helminth infection: The case for interleukin-5-dependent mechanisms." *J Infect Dis* no. 190 (3):427–9.

Maizels, R. M., A. Balic, N. Gomez-Escobar, M. Nair, M. D. Taylor, and J. E. Allen. 2004. "Helminth parasites—Masters of regulation." *Immunol Rev* no. 201:89–116.

Mantovani, A., S. Sozzani, M. Locati, P. Allavena, and A. Sica. 2002. "Macrophage polarization: Tumor-associated macrophages as a paradigm for polarized M2 mononuclear phagocytes." *Trends Immunol* no. 23 (11):549–55.

Martinez, F. O., L. Helming, and S. Gordon. 2009. "Alternative activation of macrophages: An immunologic functional perspective." *Annu Rev Immunol* no. 27:451–83.

McKenzie, R. C., T. S. Rafferty, and G. J. Beckett. 1998. "Selenium: An essential element for immune function." *Immunol Today* no. 19 (8):342–5.

Medzhitov, R. 2008. "Origin and physiological roles of inflammation." *Nature* no. 454 (7203):428–35.

Menzies, F. M., F. L. Henriquez, J. Alexander, and C. W. Roberts. 2010. "Sequential expression of macrophage anti-microbial/inflammatory and wound healing markers following innate, alternative and classical activation." *Clin Exp Immunol* no. 160 (3):369–79.

Morris, S. M., Jr., D. Kepka-Lenhart, and L. C. Chen. 1998. "Differential regulation of arginases and inducible nitric oxide synthase in murine macrophage cells." *Am J Physiol* no. 275 (5, Pt. 1):E740–7.

Mosser, D. M. 2003. "The many faces of macrophage activation." *J Leukoc Biol* no. 73 (2):209–12.

Mosser, D. M., and J. P. Edwards. 2008. "Exploring the full spectrum of macrophage activation." *Nat Rev Immunol* no. 8 (12):958–69.

Murray, P. J., and T. A. Wynn. 2011. "Protective and pathogenic functions of macrophage subsets." *Nat Rev Immunol* no. 11 (11):723–37.

Nelson, S. M., X. Lei, and K. S. Prabhu. 2011. "Selenium levels affect the IL-4-induced expression of alternative activation markers in murine macrophages." *J Nutr* no. 141 (9):1754–61.

Norton, R. L., and P. R. Hoffmann. 2012. "Selenium and asthma." *Mol Aspects Med* no. 33 (1):98–106.

Olefsky, J. M., and C. K. Glass. 2010. "Macrophages, inflammation, and insulin resistance." *Annu Rev Physiol* no. 72:219–46.

Prabhu, K. S., F. Zamamiri-Davis, J. B. Stewart, J. T. Thompson, L. M. Sordillo, and C. C. Reddy. 2002. "Selenium deficiency increases the expression of inducible nitric oxide synthase in RAW 264.7 macrophages: Role of nuclear factor-kappaB in up-regulation." *Biochem J* no. 366 (Pt. 1):203–9.

Preston, A. M. 1991. "Cigarette smoking-nutritional implications." *Prog Food Nutr Sci* no. 15 (4):183–217.

Qujeq, D., B. Hidari, K. Bijani, and H. Shirdel. 2003. "Glutathione peroxidase activity and serum selenium concentration in intrinsic asthmatic patients." *Clin Chem Lab Med* no. 41 (2):200–2.

Raes, G., P. De Baetselier, W. Noel, A. Beschin, F. Brombacher, and G. Hassanzadeh Gh. 2002. "Differential expression of FIZZ1 and Ym1 in alternatively versus classically activated macrophages." *J Leukoc Biol* no. 71 (4):597–602.

Raes, G., R. Van den Bergh, P. De Baetselier et al. 2005. "Arginase-1 and Ym1 are markers for murine, but not human, alternatively activated myeloid cells." *J Immunol* no. 174 (11):6561; author reply 6561–2.

Rayman, M. P. 2000. "The importance of selenium to human health." *Lancet* no. 356 (9225):233–41.

Rayman, M. P. 2002. "The argument for increasing selenium intake." *Proc Nutr Soc* no. 61 (2):203–15.

Ruckrich, M. F., W. Schlegel, and A. Jung. 1976. "Prostaglandin endoperoxide analogues and prostaglandid D2 as substrates of human placental 15-hydroxy prostaglandin dehydrogenase." *FEBS Lett* no. 68 (1):59–62.

Saha, A., A. Biswas, S. Srivastav, M. Mukherjee, P. K. Das, and A. Ukil. 2014. "Prostaglandin E2 negatively regulates the production of inflammatory cytokines/chemokines and IL-17 in visceral leishmaniasis." *J Immunol* no. 193 (5):2330–9.

Sakr, Y., K. Reinhart, F. Bloos et al. 2007. "Time course and relationship between plasma selenium concentrations, systemic inflammatory response, sepsis, and multiorgan failure." *Br J Anaesth* no. 98 (6):775–84.

Serhan, C. N. 2009. "Systems approach to inflammation resolution: Identification of novel anti-inflammatory and pro-resolving mediators." *J Thromb Haemost* no. 7 Suppl 1:44–8.

Serhan, C. N. 2010. "Novel lipid mediators and resolution mechanisms in acute inflammation: To resolve or not?" *Am J Pathol* no. 177 (4):1576–91.

Smith, A., K. B. Madden, K. J. Yeung et al. 2005. "Deficiencies in selenium and/or vitamin E lower the resistance of mice to Heligmosomoides polygyrus infections." *J Nutr* no. 135 (4):830–6.

Smith, A. D., L. Cheung, E. Beshah, T. Shea-Donohue, and J. F. Urban, Jr. 2013. "Selenium status alters the immune response and expulsion of adult *Heligmosomoides bakeri* worms in mice." *Infect Immun* no. 81 (7):2546–53.

Sun, F. F., S. B. Armour, V. R. Bockstanz, and J. C. McGuire. 1976. "Studies on 15-hydroxy-prostaglandin dehydrogenase from monkey lung." *Adv Prostaglandin Thromboxane Res* no. 1:163–9.

Tauber, A. I. 2003. "Metchnikoff and the phagocytosis theory." *Nat Rev Mol Cell Biol* no. 4 (11):897–901.

Vats, D., L. Mukundan, J. I. Odegaard et al. 2006. "Oxidative metabolism and PGC-1beta attenuate macrophage-mediated inflammation." *Cell Metab* no. 4 (1):13–24.

Vunta, H., F. Davis, U. D. Palempalli et al. 2007. "The anti-inflammatory effects of selenium are mediated through 15-deoxy-Delta12,14-prostaglandin J2 in macrophages." *J Biol Chem* no. 282 (25):17964–73.

Vunta, H., B. J. Belda, R. J. Arner, C. Channa Reddy, J. P. Vanden Heuvel, and K. Sandeep Prabhu. 2008. "Selenium attenuates pro-inflammatory gene expression in macrophages." *Mol Nutr Food Res* no. 52 (11):1316–23.

Wynn, T. A., A. Chawla, and J. W. Pollard. 2013. "Macrophage biology in development, homeostasis and disease." *Nature* no. 496 (7446):445–55.

Section VI

Selenoprotein Polymorphisms and Mutations

15 Genetic Polymorphisms in Selenoprotein Genes

Functionality and Disease Risk

Catherine Méplan, Janaina Lombello Santos Donadio, and John Hesketh

CONTENTS

15.1 Introduction 309
15.2 Origin of SNPs 311
15.3 SNPs in Selenoprotein Genes 312
15.4 SNPs in Selenoprotein Genes and Diseases 315
    15.4.1 Breast Cancer 332
    15.4.2 Prostate Cancer 333
    15.4.3 Colorectal Cancer 333
    15.4.4 Pathway Approach to SNP in Selenoprotein Genes 335
15.5 Factors Influencing the Significance of SNPs 336
    15.5.1 Physiological Factors 336
    15.5.2 Genetic Interactions 336
    15.5.3 Biomarkers of Se Status in Association Studies 336
15.6 Conclusions and Perspectives 337
Acknowledgment 338
References 338

15.1 INTRODUCTION

The relationship between diet and disease has long been known to human medicine. A great effort has been made over the years to understand the consequences of nutrient deficiencies or excess intake on clinical conditions. Recently, nutrition research has incorporated genomics technology to study the reciprocal interactions between diet and genes, particularly in relation to the modulation of risk of chronic diseases. These technological advances led to the development of nutritional genomics, which comprises the study of both how genetic makeup affects response to nutrients (nutrigenetics) and how nutritional compounds modulate gene expression patterns and molecular pathways (nutrigenomics). Nutritional genomics has not only contributed to a change in our perception of disease prevention at the individual and population levels by high-lighting the needs to evaluate disease risk in the context of genetic background and exposures to causative agents and potential preventive dietary compounds, but as well it has provided a unique opportunity to use the wealth of data produced by genomic projects to understand the dynamic aspects of the relationship between genes and nutrients, in particular when individuals are exposed to a change of dietary resources.

One of the dietary factors that particularly benefited from such approaches to further understand its role in chronic diseases is selenium (Se). Se is unique as a micronutrient as its biological activity results from its incorporation into the amino acid selenocysteine (Sec), a cysteine analogue containing a selenol group in place of the sulfur-containing thiol. Sec is present in the structure of a family of proteins called selenoproteins and is required for selenoproteins to be synthetised and active (see Chapter 3 for details). Sec incorporation is complex and involves two major regulatory elements present in selenoprotein mRNAs: a stem-loop structure called the SECIS element (Sec-insertion sequence) present in the 3′ untranslated region of the mRNA and the recoding of the UGA codon from a premature termination codon to a Sec (see Chapter 3 for details) present within the coding sequence. The biologi-cal activity of selenoproteins depends on the presence of Sec in their structure, and

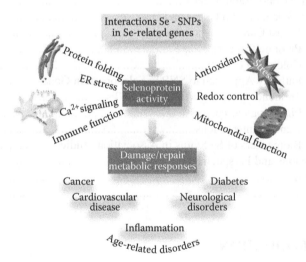

FIGURE 15.1 Schematic representation of the downstream consequences of the interactions between Se intake and SNPs in selenoprotein and Se-related genes. Selenoproteins exert their biological functions in several metabolic pathways involved in the maintenance of homeosta-sis and response to stress. In particular, they prevent oxidative damage and control cellular redox status; they contribute to mitochondrial function and immune response; and they are involved in endoplasmic reticulum (ER) stress response, calcium signaling and protein folding. Alteration of these processes contributes to multifactorial age-related diseases. Combination of suboptimal Se status with SNPs in Se metabolism have the potential to reduce selenoprotein synthesis or activity, contributing to the accumulation of damage, the reduction of the repair capacity and the alterations metabolic responses, thus impacting disease risk.

because of the strong redox properties of the selenol group most of them are involved in redox-active defenses, redox control, and stress response biochemical pathways, such as the endoplasmic reticulum unfolded protein response (see Bellinger et al., 2009, for more details and Figure 15.1). Alteration of these biochemical pathways has been proposed to be associated with the development of chronic diseases, such as cancer, diabetes, and dementia, and there is both epidemiological and mechanistic evidence supporting the view that suboptimal Se intake influences the risk of developing chronic diseases, such as cancer, as a consequence of reduced expression of selenoproteins (Bellinger et al., 2009; Méplan, 2011; Méplan and Hesketh, 2014). In addition, several genetic factors influencing selenoprotein synthesis or activity have now been associated with increased risk for multifactorial diseases, including, in particular, cancer. In this chapter, we first introduce the origin of genetic variations within the human population and the consequences of these genetic variations on selenoprotein activity and synthesis and then discuss the effects of single nucleotide polymorphisms (SNPs) in the Se metabolic pathway on risk for several diseases. In addition, we discuss how biomarkers of Se status influence disease risk by interacting with SNPs in selenoprotein genes. Finally, this leads us to propose a new perspective on the potential use of genomics approaches combined with nutrient status to understand the role of selenium in disease prevention in the human population.

15.2 ORIGIN OF SNPs

Exploring the role of human genetic variations in the etiology of disease risk in the context of our dietary exposure brings us back to the origin of interindividual genetic variations. SNPs or single nucleotide polymorphisms represent more than 90% of the genetic variations between individuals; they correspond to a change of a single base in the genomic DNA, and their frequencies vary greatly depending on the study population, reflecting the fact that they originate from mutations that occurred during human evolution and were fixed in the populations. A catalogue of common genetic variants that occur in human beings is provided by the International HapMap Project along with the frequency of these SNPs in different populations and the linkage disequilibrium between these SNPs (http://hapmap.ncbi.nlm.nih.gov/what ishapmap.html.en). As in many other simpler organisms, the fixation of polymorphisms in the human population has been influenced by changes in dietary exposure that affected the individual fitness and adaptation to a given environment (Nielsen, 2005). Probably one of the best examples is the fixation, during the transition from foraging to farming economy, of a polymorphism in the *APOE* gene determined by three common alleles: *E2*, *E3*, and *E4*. Each variant confers a different affinity for binding to ApoE and low-density lipoprotein (LDL) receptors and its affinity for lipoprotein particles. The ancestral allele *E4*, which is associated with high blood cholesterol and confers an advantage to populations dependent upon food collection and for whom food supply is scarce, was progressively substituted, in all human groups, by the *E3* allele, which confers an advantage to populations relying on food production (Corbo and Scacchi, 1999). However, in populations exposed to a high-fat Western diet and who have an extended lifespan, *E4* variant has been now shown to contribute to the risk for age-related complex disorders, such as cardiovascular and

Alzheimer's diseases. This example illustrates that the same polymorphic variant can confer a risk or an advantage depending on an individual's dietary exposure, and corollary dietary factors can modulate the genetic risk carried by an individual (Corbo and Scacchi, 1999).

Although the quantification and evaluation of the effect of diet on genetic diversity in the human population remains to be fully assessed, the availability of large-scale SNP data sets from diverse populations will offer the possibility to map the fixations of some genetic variations with human migration patterns to geographical areas with different distributions of micronutrients. In that respect, Se represents a strong candidate for such studies as the soil content in Se exhibits large variations between different geographical areas around the world (White et al. 2015) and expression of selenoproteins and associated metabolic pathways are affected by mild deficiency and suboptimal intake of Se with potential effects on the individual's fitness.

Moreover the application of genomics approaches offers a novel opportunity to identify, at a population level, genetic components underlying the susceptibility to complex, multifactorial diseases (e.g., cancer, diabetes, and dementia) and to understand the impact of interactions between genetic and environmental factors on disease risk. One major issue associated with this effort has been the difficulty in reproducing some of the associations between a particular SNP and a disease trait in different populations. This partially reflects the biases introduced by small or middle-size homogenous populations compared with large-scale, heterogeneous populations, such as the ones used for Genome Wide Association Studies (GWAS) and the complexity of studying multifactorial disorders determined by the interactions of genetic factors with environmental factors, such as diet. Integrating individuals' nutritional status into analysis of such associations is required to fully understand the effect of genetic components.

15.3 SNPs IN SELENOPROTEIN GENES

The case of SNPs in selenoprotein genes has not been an exception in regard to the frustrating exercise of identifying genetic associations with disease risk or susceptibility. Mechanistic studies have identified a number of polymorphisms in selenoprotein genes with functional consequences affecting either selenoprotein activity or selenoprotein synthesis (Table 15.1). Several SNPs in the 3′ untranslated region have been shown to affect selenoprotein synthesis by affecting UGA-readthrough efficiency: rs5859 in the 15 kDa selenoprotein (*SEP15*) gene (see also Chapter 10) and rs713041 in the glutathione peroxidase 4 (*GPX4*) gene (Bermano et al., 1996; Hu et al., 2001; Korotkov et al., 2001; Méplan et al., 2008). In addition, rs1050450 (a T/C variant in the *GPX1* gene) induces a Proline to Leucine amino acid change at position 198 of the protein sequence (Pro198Leu) that has been shown to reduce cytosolic GPx1 enzyme activity (Hu and Diamond, 2003). In addition, evidence suggests that some functional SNPs affect not only the corresponding protein, but also influence the competition between selenoprotein mRNAs for Sec-tRNA[Ser]Sec and thus the synthesis of other selenoproteins, highlighting the complex interactions between selenoproteins (Méplan et al., 2007, 2008, 2009). Selenoproteins exert different functions in different cellular and extracellular compartments (see Figure 15.1), yet

TABLE 15.1

Functional SNPs in Selenoprotein Genes

Gene Symbol	Chromosomal Location	SNPs	Location/ Modification	Base Change	Functional Consequences	Associated Diseases/ Biomarker	References
SEP15	1p22.3	rs5845	3'UTR	G > A	Sec insertion	Colorectal, prostate cancer	Jablonska et al., 2008;
		rs5859	3'UTR	G > A or C > T	Sec insertion	Colorectal, prostate, lung cancer	Karunasinghe et al., 2012a; Méplan et al., 2010; Steinbrecher et al., 2010; Sutherland et al., 2010
		rs540049	3'UTR	C > T	Enzyme activity (GPX3)	Prostate cancer	
GPX1	3p21.3	rs1050450	Pro198Leu	C > T	Enzyme activity	Breast, lung, prostate, and bladder cancer	Méplan et al., 2013; Pellatt et al., 2013; Peters and Takata, 2008;
		rs1800668	Promoter	C > T	Gene expression	Breast cancer	Ravn-Haren et al., 2006;
		rs3811699	Promoter	G > A	Gene expression	Diabetes	Steinbrecher et al., 2010; Zhao et al., 2005
GPX2	14q23.3	rs4902347	tagSNP	C > T	?	Rectal cancer	Haug et al., 2012
GPX3	5q33.1	rs8177447	tagSNP	C > T	?	Rectal cancer	Haug et al., 2012; Peters et al., 2008
		rs3828599	tagSNP	C > T	?	Rectal cancer	
		rs736775	tagSNP	C > T	?	Rectal cancer	
GPX4	19p13.3	rs713041	3'UTR	C > T	Sec insertion	Colorectal, breast, and lung cancer	Jaworska et al., 2013; Méplan et al., 2008, 2010, 2013; Villette et al., 2002
SELK	3p21.31	rs9880056	Tagging SNP/ promoter	T > C	?	Prostate cancer	Méplan et al., 2012

(Continued)

TABLE 15.1 (CONTINUED)
Functional SNPs in Selenoprotein Genes

Gene Symbol	Chromosomal Location	SNPs	Location/ Modification	Base Change	Functional Consequences	Associated Diseases/ Biomaker	References
SEPP1	5p12	rs3877899	Ala234Thr	G > A	Enzyme activity	Breast cancer	Méplan et al., 2007, 2010, 2013; Peters et al., 2008; Steinbrecher et al., 2010; Takata et al., 2012
		rs7579	3'UTR	G > A	Sec insertion?	Colorectal and prostate cancer	
		rs12055266	tagSNP	A > G	?	Colorectal cancer	
		rs3797310	tagSNP	G > A	?	Colorectal cancer	
		rs2972994	tagSNP	C > T	?	Colorectal cancer	
		rs230813	tagSNP (3'UTR)	C > G	?	Malondialdehyde levels	
		rs230819	Intron	C > A	?	Malondialdehyde levels	
SELS	15q26.3	rs34713741	Promoter	C > T	Gene expression?	Colorectal cancer	Curran et al., 2005; Méplan et al., 2010; Pellatt et al., 2013; Shibata et al., 2009; Sutherland et al., 2010
		rs28665122	Promoter	C > T	Gene expression	Gastric cancer	
TXNRD1	12q23.3	rs7310505	Intron	C > A	?	Prostate cancer	Méplan et al., 2012
TXNRD2	22q11.21	rs9605030	Intron	C > T	?	Prostate cancer	Méplan et al., 2010, 2012
		rs9605031	Intron	C > T		Colorectal cancer	

Note: SNPs for which there is evidence of functionality are listed by publication.

their synthesis depends upon a common, complex mechanism (see Chapter 3) and the bioavailability of Se to synthesize Sec. In conditions in which Se intake is limited, a hierarchy of distribution of Sec between the selenoproteins to support their synthesis takes place with selenoproteins high in the hierarchy, having their expression maintained to the detriment of proteins low in the hierarchy (Bermano et al., 1995). Importantly, several SNPS have been shown to affect this hierarchy. In the *GPX4* gene, rs713041, which affects selenocysteine insertion, was shown to affect the synthesis of other selenoproteins in plasma and blood cells of healthy individuals and the capacity of these individuals to respond to Se supplementation, suggesting that the distribution of Sec between selenoproteins could be affected by a change in one selenoprotein (Méplan et al., 2008), and this was supported by cell culture studies (Gautrey et al., 2011). In the *SEPP1* gene, which codes for the Se transporter selenoprotein P (SePP), rs7579 (in the 3′UTR) and rs3877899 (a coding SNP resulting in an Alanine to Threonine amino acid change in the coding region) have been shown to affect distribution of the two plasma isoforms of SePP and to affect the synthesis of other selenoproteins and response to supplementation in a healthy population with a relatively low Se intake. This suggests that these polymorphisms affect Se bioavailability for the synthesis of other selenoproteins by modulating the SePP isoform pattern (Méplan et al., 2007, 2009).

Figure 15.1 illustrates how functional genetic variants in Se-related genes can affect selenoprotein synthesis or activity. As selenoproteins play a key role in many stress response mechanisms, functional genetic variants in Se-related genes that affect selenoprotein synthesis or activity have the potential, in conjunction with low Se intake, to contribute to a reduction of selenoprotein activity, thereby reducing an individual's capacity to respond to additional stress. The direct consequence will be that some functional SNPs could predispose an individual to the accumulation of damage and to development of age-related disease, and thus genotype for these SNPs will be associated with increased disease risk (Méplan, 2011; Méplan and Hesketh, 2012).

15.4 SNPs IN SELENOPROTEIN GENES AND DISEASES

Epidemiological studies have explored the role of Se in cancer prevention, and mechanistic studies have highlighted that selenoproteins mediate cancer-preventive effects of selenium. In addition, several association studies have investigated the association of genetic variants in selenoprotein and pathways of Se metabolism with disease risk (Tables 15.2–15.4). However, the disparity of Se intake between populations often caused inconsistency between genetic association studies investigating the effect of SNP on disease risk without considering Se biomarkers and/or levels of Se intake. This is illustrated in Tables 15.2 through 15.4. Se status measurements were not always feasible, depending on the source of samples used or the size of the population studied.

In addition to association of single SNPs with disease risk, it became apparent that the effect of one SNP could be modified by (i) the Se status or biomarker thereof, and (ii) genetic interaction between SNPs in other selenoprotein genes and/or with other redox-active enzymes. The interaction with environmental factors (as quantified by plasma Se, SePP levels, erythrocyte GPx1 activity, and plasma GPx3 activity) is

TABLE 15.2
SNPs in Selenoprotein Genes and Se Biomarkers in Healthy Populations

Reference	Country/ Population	n	CC (%)	CT (%)	TT (%)	C (%)	T (%)	Biomarkers of Se Status	Values	Main Results
						GPX1: rs1050450				
Bastaki et al., 2006	USA/Asians	115	86	11	3	0.92	0.08	eGPx activity	13.2 U/g Hb	No effect of ethnicity on GPX1 activity.
	USA/Caucasians	63	46	46	8	0.69	0.31			
	USA/Hispanics	20	60	35	5	0.78	0.22			
	USA/others	19	58	21	21	0.68	0.32			
Combs et al., 2012	USA/adults	261	46	43	11	0.67	0.33	Plasma Se SEPP GPx3 activity	142 ng/mL 3.43 µg/L 3.64 nmol/m/mg	Individuals with TT genotype had lower plasma Se status than CC at baseline. GPx3 and SEPP did not increase in response to SeMet supplementation.
Jablonska et al., 2009	Poland/adults	405	48	42	10	0.69	0.31	eGPx activity Plasma Se	15.1 U/g Hb 54.4 µg/L	Weaker correlation between eGPx activity and plasma Se in TT compared with CC.
Suzen et al., 2010	Turkey/adults	250	42	44	14	0.64	0.36	–	–	Second highest T allele frequency in Turkish population.
Karunasinghe et al., 2012b	New Zealand/ men	503	48	44	8	0.70	0.30	eGPx activity Serum Se TR activity	14.1 U/g Hb 111.6 µg/L 1.02 U	Strong correlation in CC and CT men between serum Se and GPx1 activity, no correlation in TT men. Reduction of DNA damage in as plasma Se increases in CC.
Hiragi et al., 2011	Brazil/FD	172	47	40	13	0.67	0.47	–	–	Differences in allele frequencies between the three ethnic groups.
	Brazil/Africans	72	71	22	7	0.82	0.18			
	Brazil/Indians	60	95	3	2	0.97	0.03			

(Continued)

TABLE 15.2 (CONTINUED)

SNPs in Selenoprotein Genes and Se Biomarkers in Healthy Populations

Reference	Country/ Population	n	CC (%)	CT (%)	TT (%)	C (%)	T (%)	Biomarkers of Se Status	Values	Main Results
						GPX4: rs713041				
Combs et al., 2012	USA/adults	261	28	52	20	0.54	0.46	Plasma Se	142 ng/mL	No effect of genotype on biomarkers.
								SEPP	3.43 µg/L	
								GPx3 activity	3.64 nmol/m/mg	
Takata et al., 2012	USA/adults	195	27	48	24	0.51	0.49	eGPx activity	43.1U/g prot	No effect of this SNP on biomarkers; however, two SNPs in GPX1 were associated with GPX1 activity (rs1987628 and rs8179164).
								Serum Se	136 µg/L	
								SEPP	5.8 µg/L	
								GPX3 activity	727 U/L	
Villette et al., 2002	Scotland/adults	66	33	42	24	0.55	0.45	Plasma Se	67.9 µg/L	CC genotype associated with higher 5′-lipoxygenase concentration.
								GPX4 activity	10.5 U/g prot	
Karunasinghe et al., 2012b	New Zealand/ men	503	31	48	21	0.55	0.45	eGPx activity	14.5 U/g Hb	TT genotype associated with a reduction in DNA damage with increasing serum Se.
								Serum Se	111.8 µg/L	
								TR activity	1.05 U	

Reference	Country Population	n	GG (%)	GA (%)	AA (%)	G (%)	A (%)	Biomarkers of Se Status	Values	Main Results
						SEPP1: rs3877899				
Combs et al., 2012	USA/adults	261	58	38	4	0.77	0.23	Plasma Se	142 ng/mL	No effect of genotype on biomarkers.
								SEPP	3.43 µg/L	
								GPx3 activity	3.64 nmol/m/mg	

(Continued)

TABLE 15.2 (CONTINUED)
SNPs in Selenoprotein Genes and Se Biomarkers in Healthy Populations

Reference	Country Population	n	GG (%)	GA (%)	AA (%)	G (%)	A (%)	Biomarkers of Se Status	Values	Main Results
Takata et al., 2012	USA/adults	189	62	32	5	0.79	0.21	eGPx activity Serum Se SEPP GPX3 activity	43.1 U/g prot 136 µg/L 5.8 µg/L 727 U/L	No effect of this SNP on biomarkers; however, other SNPs in SEPP1 gene (rs230813 and rs230819) were associated with malondialdehyde levels.
Méplan et al., 2007	UK/Caucasians	98	46	47	7	0.69	0.31			
	UK/South Asians	49	65	29	6	0.80	0.20			
	UK/Chinese	49	98	2	0	0.99	0.01			
Karunasinghe et al., 2012b	New Zealand/ men	503	57	38	5	0.75	0.25	eGPx activity Serum Se TR activity	14.0 U/g Hb 109.9 µg/L 1.08 U	Stronger correlation in GG between serum Se and Thioredoxin reductase activity compared with AA and GA.
SEPP1: rs7579										
Combs et al., 2012	USA/adults	261	44	44	12	0.66	0.34	Plasma Se SEPP GPx3 activity	142 ng/mL 3.43 µg/L 3.64 nmol/m/mg	No effect of genotype on biomarkers.
Méplan et al., 2007	UK/Caucasians	46	48	46	7	0.71	0.29			
	UK/South Asians	47	62	34	4	0.79	0.21			
	UK/Chinese	49	51	43	6	0.72	0.28			

Note: Allele distribution in healthy populations is indicated, and where measured, the relationship between genotype and biomarkers of Se status is indicated.

TABLE 15.3
Association between Pro198Leu (rs1050450) in *GPX1* and Disease Risk

Reference	Country/ Population	Case/ Control (*n*)	CC (%)	CT (%)	TT (%)	C (%)	T (%)	Biomarkers of Se Status	Values	Main Results
Breast Cancer						*GPX1:* rs1050450				
Ahn et al., 2005	USA	1038/1088	45/48	44/42	11/10	0.67/0.69	0.33/0.31	–	–	No association
Cox et al., 2004	USA	1229/1629	47/48	42/43	11/10	0.68/0.69	0.32/0.31	–	–	No association
Hu and Diamond, 2003	USA	79/517	46/47	32/40	23/12	0.61/0.67	0.39/0.33	–	–	T allele more frequent in breast cancer cells
Knight et al., 2004	Canada	399/372	48/45	43/44	9/10	0.70/0.67	0.30/0.33	–	–	No association
Cebrian et al., 2006	UK	2293/2278	48/47	42/44	10/10	0.69/0.69	0.31/0.31	–	–	No association
Udler et al., 2007	UK	4371	48	45	7	0.71	0.29	–	–	No association
Ravn-Haren et al., 2006	Denmark	377/377	47/54	45/36	9/10	0.69/0.72	0.31/0.28	eGPx activity	87/89 (U/g Hb)	T allele increases risk T allele decreases GPx activity

(*Continued*)

TABLE 15.3 (CONTINUED)

Association between Pro198Leu (rs1050450) in *GPX1* and Disease Risk

Reference	Country/ Population	Case/ Control (n)	CC (%)	CT (%)	TT (%)	C (%)	T (%)	Biomarkers of Se Status	Values	Main Results
Méplan et al., 2013	Denmark	933/959	50/52	42/39	8/9	0.71/0.72	0.29/0.28	—	—	T allele increases risk of nonductal tumors; interaction with rs3877899 (*SEPP1*): heterozygotes for both had higher risk of nonductal tumors
Prostate Cancer										
Abe et al., 2011	USA	745	46	42	12	0.67	0.33	Plasma Se	121.4 (µg/L)	No association
Choi et al., 2007	USA	500/1391	50/51	43/42	8/8	0.71/0.71	0.29/0.29	—	—	No association
Steinbrecher et al., 2010	Germany	247/487	50/54	44/37	6/9	0.72/0.73	0.28/0.27	GPx3 activity Serum Se SEPP	647/656 (U/L) 86.2/87.7 (µg/L) 2.88/2.92 (mg/L)	T allele decreases risk with increasing serum Se levels

(Continued)

TABLE 15.3 (CONTINUED)

Association between Pro198Leu (rs1050450) in *GPX1* and Disease Risk

Reference	Country/Population	Case/Control (n)	CC (%)	CT (%)	TT (%)	C (%)	T (%)	Biomarkers of Se Status	Values	Main Results
Arsova-Sarafinovska et al., 2009	Macedonia	82/123	66/46	21/38	13/15	0.76/0.65	0.24/0.34	eGPx activity	6.57/8.13 (U/mL)	T allele decreases risk
Karunasinghe et al., 2012a	New Zealand	262/435	47/50	42/43	11/8	0.68/0.71	0.32/0.29	eGPx activity Serum Se	13.6/15 101.2/112.9	T allele increase risk
Colorectal Cancer										
Peters et al., 2008	USA/adults	656/743	49/48	40/45	11/8	0.69/0.70	0.31/0.30	Serum Se	135.2/137.2 (ng/mL)	No association
Hansen et al., 2005	Norway/adults	981/397	51/49	41/41	9/10	0.71/0.70	0.29/0.30	–	–	No association
Hansen et al., 2009	Denmark/adults	375/779	46/44	44/45	10/11	0.68/0.66	0.32/0.34	–	–	No association
Méplan et al., 2010	Czech Republic/adults	681/637	52/56	45/41	3/4	0.74/0.76	0.26/0.24	–	–	No association alone, but an interaction with rs37413471 in *SELS* increases risk
Sutherland et al., 2010	Korea/adults	752/668	85/82	15/18*	0	0.92/0.91	0.08/0.09	–	–	No association

(Continued)

TABLE 15.3 (CONTINUED)

Association between Pro198Leu (rs1050450) in *GPX1* and Disease Risk

Reference	Country/Population	Case/Control (n)	CC (%)	CT (%)	TT (%)	C (%)	T (%)	Biomarkers of Se Status	Values	Main Results
Lung Cancer										
Yang et al., 2004	USA/young and old adults	162/165	47/52	46/35	7/13	0.70/0.69	0.30/0.31	—	—	Among old smokers; CC genotype was associated with lung cancer
		67/63	52/46	34/43	13/11	0.69/0.67	0.31/0.33			
Ratnasinghe et al., 2000	Finland/men	315/313	29/42	50/43	21/15	0.54/0.64	0.46/0.36	—	—	T allele increases risk
Jaworska et al., 2013	Poland/adults	95/176	56/45	35/47	9/8	0.73/0.68	0.27/0.32	Plasma Se	63.2/74.6 (µg/L)	T allele decreases risk
Raaschou-Nielsen et al., 2007	Denmark/adults	432/798	48/43	43/45	9/11	0.70/0.66	0.30/0.34	—	—	T allele decreases risk
Rosenberger et al., 2008	Germany/adults	186/207	61/47	34/43	5/10	0.78/0.68	0.22/0.32	—	—	T allele decreases risk
Laryngeal Cancer										
Jaworska et al., 2013	Poland/adults	111/213	59/46	33/45	8/10	0.75/0.68	0.25/0.32	Plasma Se	64.8/77.1 (µg/L)	T allele decreases risk
Bladder Cancer										
Zhao et al., 2005	USA/adults	224	52	46	2	0.75	0.25	—	—	T allele increases risk
Ichimura et al., 2004	Japan/adults	213/209	78/89	22/11	0	0.89/0.95	0.11/0.05	—	—	T allele increases risk

(Continued)

TABLE 15.3 (CONTINUED)
Association between Pro198Leu (rs10500450) in *GPX1* and Disease Risk

Reference	Country/Population	Case/Control (n)	CC (%)	CT (%)	TT (%)	C (%)	T (%)	Biomarkers of Se Status	Values	Main Results
Cardiovascular Disease										
Hamanishi et al., 2004	Japan/diabetic adults	184	82	18	0	0.91	0.18	—	—	T allele increases intima-media thickness and risk of cardiovascular disease in diabetic patients
Kashin-Beck										
Huang et al., 2013	China/adults	638/324	88/84	12/16	0	0.94/0.92	0.06/0.08	—	—	No association with Kashin-Beck disease
Metabolic Syndrome and Obesity										
Kuzuya et al., 2008	Japan/adults	2183	84	15	0.3	0.92	0.08	—	—	T allele increases risk of Metabolic Syndrome
Cominetti et al., 2011	Brazil/obese women	37	49	38	13	0.68	0.32	eGPx activity Plasma Se Erythrocyte Se	36.6 (U/g Hb) 55.7 (µg/L) 60.5 (µg/L)	T allele was associated with higher DNA damage compared to C allele
Stroke										
Forsberg et al., 2000	Swedish/adults	101/214	55/53	38/40	7/7	0.74/0.73	0.26/0.27	GPx1 activity	1.28 (µkat/g Hb)	No effect of genotype in GPx1 activity; No association with stroke

Note: The table summarizes data from various association studies of the allele frequency for rs10450450 and risk of diseases indicated.

TABLE 15.4

Association between SNPs in *GPX4*, *SEPP1*, and *SEP15* Genes and Disease Risk

Reference	Country/ Population	Cases/ Controls (*n*)	CC (%)	CT (%)	TT (%)	C (%)	T (%)	Biomarkers of Se Status	Values (unit) Cases/ Controls	Results
					GPX4: rs713041					
Breast Cancer										
Cebrian et al., 2006	UK	2182/ 2264	31/30	51/48	18/22	0.56/0.54	0.44/0.46	–	–	No association
Udler et al., 2007	UK	4356	30	50	19	0.55	0.45	–	–	SNP associated with increased risk of mortality by breast cancer
Méplan et al., 2013	Denmark	939/960	34/35	47/45	19/20	0.57/0.57	0.43/0.43	–	–	T allele associated with decreased GPX1 activity
Colorectal Cancer										
Peters et al., 2008	USA/adults	745/758	31/35	49/47	20/18	0.56/0.59	0.44/0.41	Serum Se	135.2/ 137.2 (ng/mL)	No association
Bermano et al., 2007	UK/adults	252/187	30/20	52/60	17/20	0.56/0.50	0.44/0.50	–	–	TT genotype associated with decreased risk

(Continued)

TABLE 15.4 (CONTINUED)
Association between SNPs in *GPX4*, *SEPP1*, and *SEP15* Genes and Disease Risk

Reference	Country/ Population	Cases/ Controls (n)	CC (%)	CT (%)	TT (%)	C (%)	T (%)	Biomarkers of Se Status	Values (unit) Cases/ Controls	Results
Méplan et al., 2010	Czech Republic/ adults	729/664	31/ 38	50/ 45	19/ 17	0.56/0.60	0.44/0.40	–	–	CT genotype associated with increased risk of CRC; interaction with rs4880 (*SOD2*) and rs9605031 (*TXNRD2*) associated with increased risk, but with rs3877899 (*SEPP1*) associated with reduced risk
Sutherland et al., 2010	Korea/ adults	796/697	43/39	45/47	13/14	0.65/0.63	0.35/0.37	–	–	No association
Prostate Cancer										
Abe et al., 2011	USA	739	29	49	22	0.53	0.47	Plasma Se	120 (µg/L)	No association
Steinbrecher et al., 2010	Germany	245/490	31/32	47/47	22/21	0.55/055	0.45/0.45	GPx3 act Serum Se SEPP	647/656 (U/L) 86.2/87.7 (µg/L) 2.88/2.92 (mg/L)	No association
Karunasinghe et al., 2012a	New Zealand	260/439	32/33	50/48	18/19	0.57/0.57	0.43/0.43	eGPx activity Serum Se	13.6/15 (U/g Hb) 101.2/112.9 (µg/L)	No association

(Continued)

TABLE 15.4 (CONTINUED)
Association between SNPs in *GPX4*, *SEPP1*, and *SEP15* Genes and Disease Risk

Reference	Country/ Population	Cases/ Controls (n)	CC (%)	CT (%)	TT (%)	C (%)	T (%)	Biomarkers of Se Status	Values (unit) Cases/ Controls	Results
Lung Cancer										
Jaworska et al., 2013	Poland/ adults	95/176	41/31	49/55	9/14	0.66/0.58	0.34/0.42	Plasma Se	63.2/ 74.6 (µg/L)	T allele associated with reduced risk
Laryngeal Cancer										
Jaworska et al., 2013	Poland/ adults	111/213	38/32	41/51	21/17	0.59/0.57	0.41/0.43	Plasma Se	64.8/ 77.1 (µg/L)	T allele associated with reduced risk
Kashin-Beck Disease										
Du et al., 2012	China/ adults	219/194	31/35	57/53	11/13	0.60/0.61	0.40/0.39	–	–	No association; GPX4 mRNA expression was lower in individuals with Kashin-Beck disease than controls

(Continued)

TABLE 15.4 (CONTINUED)
Association between SNPs in *GPX4*, *SEPP1*, and *SEP15* Genes and Disease Risk

Reference	Country/Population	Cases/Controls (n)	GG (%)	GA (%)	AA (%)	G (%)	A (%)	Biomarkers of Se Status	Values (unit) Cases/Controls	Results
							SEPP1: rs3877899			
Breast Cancer										
Méplan et al., 2013	Denmark	937/959	63/62	34/33	3/5	0.80/0.78	0.20/0.22	–	–	A allele associated with reduced risk; interaction with rs1050450 (*GPX1*): heterozygotes for both associated with increased risk of nonductal cancers
Colorectal Cancer										
Peters et al., 2008	USA/adults	752/766	59/60	35/35	6/5	0.77/0.77	0.23/0.23	Serum Se	135.2/137.2 (ng/mL)	No association
Al-Taie et al., 2004	Germany/adults	193/127	63/63	33/33	4/4	0.79/0.79	0.21/0.21	–	–	No association
Méplan et al., 2010	Czech Republic/adults	732/657	58/62	35/31	6/7	0.76/0.78	0.24/0.22	–	–	No association alone, but interaction with rs5859 in *SEP15* associated with increased risk and with rs713041 in *GPX4* decreased risk
Sutherland et al., 2010	Korea/adults	802/711	99/100	1/0	0	1/1	0	–	–	No association

(Continued)

TABLE 15.4 (CONTINUED)
Association between SNPs in *GPX4*, *SEPP1*, and *SEP15* Genes and Disease Risk

Reference	Country/ Population	Cases/ Controls (n)	GG (%)	GA (%)	AA (%)	G (%)	A (%)	Biomarkers of Se Status	Values (unit) Cases/ Controls	Results
Prostate Cancer										
Cooper et al., 2008	Sweden	2643/1570	58/56	36/38	7/6	0.76/0.75	0.24/0.25	Plasma Se	76.0 (µg/L)	No association
Steinbrecher et al., 2010	Germany	248/492	61/55	35/39	4/5	0.79/0.75	0.21/0.25	Serum GPx Serum Se SEPP	656 (U/L) 87.7 (µg/L) 2.92 (mg/L)	No association
Geybels et al., 2014	Netherlands	951	59	34	6	0.77	0.23	Toenail Se	0.54 (µg/g)	Low Se status is associated with increased risk for prostate cancer independent of genotype
Karunasinghe et al., 2012a	New Zealand	259/436	59/58	34/37	7/4	0.76/0.77	0.24/0.23	eGPx act Serum Se	13.6/15 (U/g Hb) 101.2/112.9 (µg/L)	No association
Kashin–Beck Disease						*SEPP1*: rs7579				
Sun et al., 2010	China/adults	167/166	51/59	40/32	9/9	0.71/0.75	0.29/0.25	–	–	No association
Breast Cancer										
Méplan et al., 2013	Denmark	937/957	49/46	42/44	9/11	0.70/0.68	0.30/0.32	–	–	No association

(*Continued*)

TABLE 15.4 (CONTINUED)
Association between SNPs in *GPX4*, *SEPP1*, and *SEP15* Genes and Disease Risk

Reference	Country/Population	Cases/Controls (n)	GG (%)	GA (%)	AA (%)	G (%)	A (%)	Biomarkers of Se Status	Values (unit) Cases/Controls	Results
Colorectal Cancer										
Méplan et al., 2010	Czech Republic/adults	690/629	38/43	53/51	9/6	0.64/0.68	0.36/0.32	–	–	AA genotype associated with increased risk, but interaction with rs5859 (SEP15) associated with decreased risk
Sutherland et al., 2010	Korea/adults	745/669	55/54	36/36	9/10	0.73/0.72	0.27/0.28	–	–	No association
Prostate Cancer										
Steinbrecher et al., 2010	Germany	248/492	47/51	42/42	11/7	0.68/0.72	0.32/0.28	Serum GPx Serum Se SEPP	656 (U/L) 87.7 (µg/L) 2.92 (mg/L)	AA genotype associated with increased risk; effect on SePP concentrations with GA + AA having higher plasma SePP concentrations
Geybels et al., 2014	Netherlands	951	48	42	10	0.69	0.31	Toenail Se	0.54 (µg/g)	Low Se status is associated with increased risk for prostate cancer independent of genotype

(Continued)

TABLE 15.4 (CONTINUED)

Association between SNPs in *GPX4*, *SEPP1*, and *SEP15* Genes and Disease Risk

Reference	Country/ Population	Cases/ Controls (n)	GG (%)	GA (%)	AA (%)	G (%)	A (%)	Biomarkers of Se Status	Values (unit) Cases/ Controls	Results
						SEP15: rs5859				
Lung Cancer										
Jablonska et al., 2008	Poland/ adults	325/287	58/56	36/38	6/6	0.76/0.75	0.24/0.25	Plasma Se	49.9/ 53.3 (ng/mL)	A allele associated with increased risk in individuals with low Se status
Prostate Cancer										
Penney et al., 2010	USA	1195/1186	63/62	32/33	5/5	0.79/0.79	0.21/0.21	–	–	No association
Steinbrecher et al., 2010	Germany	248/492	67/63	31/32	2/5	0.82/0.79	0.18/0.21	Serum GPx Serum Se SEPP	656 (U/L) 87.7 (µg/L) 2.92 (mg/L)	Effect on GPX3 activity: AA associated with a decrease in GPX3 activity
Colorectal Cancer										
Méplan et al., 2010	Czech Republic/ adults	682/626	62/62	34/34	4/4	0.79/0.79	0.21/0.21	–	–	No association alone, however, interaction with rs3877899 (*SEPP1*) associated with increased risk and with three other SNPs in *SEPP1* decreases *(Continued)*

TABLE 15.4 (CONTINUED)
Association between SNPs in *GPX4*, *SEPP1*, and *SEP15* Genes and Disease Risk

Reference	Country/Population	Cases/Controls (n)	GG (%)	GA (%)	AA (%)	G (%)	A (%)	Biomarkers of Se Status	Values (unit) Cases/Controls	Results
Sutherland et al., 2010	Korea/adults	824/730	96/97	4/3	0/0	0.98/0.98	0.02/0.02	–	–	A allele associated with increased risk
SEP15: rs5845										
Colorectal Cancer										
Sutherland et al., 2010	Korea/adults	827/732	96/97	4/3	0/0	0.98/0.98	0.02/0.02	–	–	No association
Prostate Cancer										
Karunasinghe et al., 2012a	New Zealand	258/430	62/63	33/34	5/3	0.78/0.80	0.22/0.20	eGPx activity Serum Se	13.6/15 (U/g Hb) 101.2/112.9 (µg/L)	AA genotype associated with increased risk
Lung Cancer										
Jaworska et al., 2013	Poland/adults	95/176	57/57	37/39	6/5	0.75/0.76	0.25/0.24	Plasma Se	63.2/74.6 (µg/L)	No association
Laryngeal Cancer										
Jaworska et al., 2013	Poland/adults	111/213	55/60	42/34	3/6	0.76/0.77	0.24/0.23	Plasma Se	64.8/77.1 (µg/L)	No association

Note: The table summarizes data from various association studies of the allele frequency for SNPs in *GPX4*, *SEPP1*, and *SEP15* genes and risk of diseases indicated.

consistent with the notion that genes and nutrients interact in the determination of disease risk and highlight the need of including biomarkers of Se status in association studies. However, as we will discuss later, there is still a level of uncertainty as to which biomarker would be the best candidate to include. The observation that there can be a polygenic contribution to multifactorial diseases is significant when considering the functional links between selenoproteins and between selenoproteins and other redox-active enzymes. For example the genetic interaction (i.e., combination of genotype effects) between *SOD2* and *GPX4*, mirrors the functional interactions observed in the mitochondria between the two proteins to maintain redox status (Cooper et al., 2008; Méplan et al., 2010).

15.4.1 BREAST CANCER

A number of studies have established a link between genotype for the rs1050450 SNP in the *GPX1* gene and breast cancer risk (Table 15.3). Initially, an association between rs1050450 and breast cancer risk was observed in a US population, and in addition, loss of heterozygosity at the *GPX1* locus was observed in about 36% of breast tumors analyzed (Hu and Diamond, 2003). This association was also observed in a Danish cohort (Ravn-Haren et al., 2006) and subsequently confirmed in an expanded analysis with 975 cases and controls matched for age and hormone replacement therapy (Méplan et al., 2013). In addition, in the expanded study, rs3877899, a SNP in *SEPP1* gene causing an Ala to Thr amino acid change at position 234 was also found to be associated with breast cancer risk (Table 15.4). However, other studies of US and Canadian populations have failed to find an association of rs1050450 with breast cancer (Ahn et al., 2005; Cox et al., 2004; Knight et al., 2004), and a meta-analysis suggests an overall lack of association despite an increased risk among African women (Hu et al., 2010). The results from these studies, similar to findings from genotyping studies in US populations in relation to colorectal cancer, may reflect differences in Se status between study populations, especially between North American and European studies, and may explain the inconsistency of observations relating a possible association between Se status and genotype for rs1050450. It appears that there may be an association between breast cancer risk and genotype for rs1050450 in populations with low Se intake, such as a European (e.g., Danish) population, but not in female populations with high Se intake (such as US and Canadian populations), suggesting that when Se is not limiting, increased production of selenoproteins could potentially compensate for the reduced GPx1 activity observed in Leu carrier for rs1050450 (Hu and Diamond, 2003). In addition, evidence from the Danish study population revealed that carriers of the Leu variant for rs1050450 have an increased risk of having a high-grade ductal breast tumor or nonductal breast tumors, suggesting that GPx1 activity is crucial to prevent oxidative damage occurring during the initiation or progression of breast tumors (Méplan et al., 2013).

Supporting the role of GPx1 activity in breast function or breast cancer etiology, a GCG repeat polymorphism in the *GPX1* gene, coding for a variable number of alanines (five to seven) in the protein sequence, was associated with an increased risk of breast cancer in premenopausal women (Knight et al., 2004).

Finally, an association between rs5859 in *SEP15* and breast cancer was observed in African Americans (Table 15.3) but not in Caucasians (Hu et al., 2001). In addition, loss of heterozygosity was observed at the *SEP15* locus in tumor tissue (Hu et al., 2001). In a second study, *SEP15* allelic loss has as well been implicated with the development of breast cancer among African American women (Nasr et al., 2003), highlighting differences between ethnic groups, including differences in genetic background (including the potential polygenic contribution to disease risk) or differences in exposure to dietary/environmental factors interacting with genetic background. The observed loss of heterozygosity at both *GPX1* and *SEP15* loci suggests a potential tumor-suppressor role of the two selenoproteins in breast cancer etiology.

15.4.2 PROSTATE CANCER

Most of the studies carried out to examine the influence of genetic variants in selenoprotein genes on the risk of prostate cancer have focused on SNPs, which are known to be functionally significant (Tables 15.3 and 15.4); they include variants in *SEPP1*, *GPX1*, and *SEP15*, and the studies have been carried out in both US and European populations (Geybels et al., 2014; Penney et al., 2010, 2013; Steinbrecher et al., 2010). Analysis of *SEP15* genotype in a nested case-control study within Physicians Health Study in the United States suggested that prostate cancer risk is modified by a combination of low Se status and genetic variation in *SEP15*. Subsequently, further analysis focused on variants in *SEPP1* has suggested that SNPs in *SEPP1* also influence prostate incidence (Penney et al., 2013). This latter observation is compatible with earlier findings from two studies of European populations that indicated prostate cancer risk is modified by variants in *SEPP1*; in one study, rs7579 was shown to modulate the risk for prostate cancer (Cooper et al., 2008; Steinbrecher et al., 2010), and in a second, disease risk was modified by low Se status and a genetic interaction between rs4880 in *SOD2* and rs3877899 in *SEPP1* (Cooper et al., 2008). Recently, the association of genotype for rs7579 with advanced prostate cancer risk has been confirmed in a second European population (Geybels et al., 2014). In addition, two studies of European populations have suggested that genetic variation in *GPX1* modifies prostate cancer risk. Initially, rs1050450 in *GPX1* was found to modify the relationship between serum Se concentration and disease risk (Steinbrecher et al., 2010), and recently two other variants in *GPX1* (rs17650792 and rs1800668) have been associated with advanced prostate cancer risk (Geybels et al., 2014).

15.4.3 COLORECTAL CANCER

Several studies have identified the genetic association of known functional SNPs in selenoprotein genes with colorectal cancer (CRC) risk (Tables 15.3 and 15.4). Analysis of frequencies of SNPs in selenoprotein genes in a Korean population (827 cases, 727 controls; Sutherland et al., 2010) and in a Czech population (832 cases, 705 controls; Méplan et al., 2010) showed that in both populations a SNP (rs34713741) in the promoter region of *SELS* (coding for the endoplasmic reticulum protein selenoprotein S) was found to modulate cancer risk. The T allele of rs34713741 was associated with greater CRC risk (odds ratio of 1.68) in the Czech population, and a second variant in close proximity led to increased risk in females in the Korean population (odds ratio 2.25). The replication of this association in two distinct populations strongly indicates that, independently of lifestyle and dietary factors, SNPs in *SELS* promoter influence CRC risk. Interestingly, rs34713741 has also been linked to gastric cancer risk (Shibata et al., 2009). In addition, genotypes for SNPs in the 3'UTR of the *SEPP1* (rs7579) and *GPX4* (rs713041) altered risk of CRC in the Czech cohort (Méplan et al., 2010). In addition, in this study population, significant genetic interactions between SNPs in the manganese superoxide dismutase *SOD2* (rs4880), *GPX4* (rs713041), and *TXNRD2* (rs960531) were found to further modulate CRC risk. Such interactions mirror the metabolic links between GPx4, TrxR2, and MnSOD in the response to oxidative stress in the mitochondria, suggesting that these interactions could exert their functional effects by modulating the capacity of an individual to respond to additional stress, thus contributing to cancer development. In addition, genetic interactions between SNPS in *SEPP1* (rs7579, rs3877899, rs3797310, and rs12055266) and either *SEP15* (rs5859) or *GPX4* (rs713041) were found to affect CRC risk (Méplan et al., 2010) and could reflect that selenoprotein synthesis could be affected in individuals carrying a combination of SNPs in *SEPP1* that modulate Se delivery and SNPs in *GPX4* or *SEP15* that modulate selenocysteine incorporation (Méplan et al., 2010). These studies highlight potential mechanistic links that could contribute to the understanding of Se anticarcinogenic properties (Figure 15.1).

In addition to the candidate SNP approach, a hypothesis-generating approach has been developed to study the effects of multiple genetic variants across one gene. This approach is based on the genotyping of *tagSNP*, which represents a SNP in high linkage disequilibrium with other SNPs grouped in a haplotype, to tag, that is, to genotype, all genetic variants that belong to the same haplotype group. However, association studies carried out in US populations and using tagSNPs to genotype all SNPs in specific selenoprotein genes provided contradictory findings. Genotyping of subjects from the Prostate, Lung, Colorectal, and Ovarian Cancer Screening Trial identified four variants in *SEPP1* (in the 5' gene region of *SEPP1* at −4166; three in the 3' gene region of *SEPP1*) and one in the *TXNRD1* (in the 5' gene region at position −181) significantly associated with risk of advanced distal colorectal adenoma (Peters et al., 2008). Applying a similar tagSNP approach to a combination of three US case-control studies, Haug et al. (2012) investigated the association of variants in *GPX1-4* with colorectal cancer risk. Three SNPs in *GPX3* (rs8177447, rs3828599, and rs736775) and one in *GPX2* (rs4902347) were significantly associated with rectal cancer risk but not with risk of either colon cancer or adenoma (Haug et al., 2012). On the contrary, genotyping of 804 cases and 805 matched controls from the Women's Health Initiative Study for tagSNPs covering genetic variations in *GPX1-4* and *SEPP1* showed no significant association of these variants with CRC risk in a population with high Se levels (Takata et al., 2011). In this study, only one tagSNP (rs8178974 in *GPX4*) was associated with colorectal cancer risk, but overall genetic variations in *GPX4* showed no association. More recently, another US–based study (Slattery et al., 2012) has indicated that the risk for rectal cancer was modified by tagSNPs in *TXRND3* and in selenoprotein N (*SEPN*) genes. In addition, interaction between tagSNPs in several selenoprotein genes and lifestyle factors were shown to contribute to both colon and rectal cancer risk. Indeed tagSNPs in *TXNRD1*, and *TXNRD2* genes were shown to interact with aspirin/NSAID use to modify risk of colon cancer whereas the interaction between tagSNPs in *TXNRD3* and aspirin/NSAID use altered risk for rectal cancer. Finally, tagSNPs in *TXNRD2*, *SELS*, *SEP15*, and *SEPW1* were shown to interact with estrogen status to modify colon cancer risk (Slattery et al., 2012). Discrepancies between the results of these studies could reflect differences in Se status between the studied populations. In this regard, it is important to note that in the study of Haug et al. (2012), the Se status was high, and this could have obscured effects of genotype on disease risk. It is critical in future work to combine genotyping for selenoprotein SNPs with measures of Se status and carry out such studies in populations in which the Se status is relatively low, and indeed such a study using samples from the EPIC study in Europe is at present underway. In addition, the use of tagSNPs is limited to the degree of genetic linkage between the "tagging SNP" and "tagged SNPs," thus, like in GWAS, important functional SNP can be missed due to the study design.

15.4.4 PATHWAY APPROACH TO SNP IN SELENOPROTEIN GENES

Several GWAS have now identified some selenoprotein genes as strong candidates associated with disease traits, in particular in relation to inflammatory-related conditions (Franke et al., 2010; Jostins et al., 2012; Porcu et al., 2013; Vithana et al., 2012; Wallace et al., 2010). However, one limitation of GWAS is that they do not take into account the interactions between different genetic variants and between these variants and biomarkers of Se status. On the other hand, the analysis of a restricted number of functional SNPs, as described above, revealed key genetic interactions between SNPs in different selenoprotein genes, stressing the importance of considering the effects of multiple genetic factors to further understand the complexity underlying multifactorial diseases. A novel, alternative approach, using an Illumina™ GoldenGate assay, was designed to simultaneously genotype 384 tag-SNPs that capture genotype for all SNPs in 72 genes, including all selenoprotein genes and related genes involved in Se metabolism (Méplan et al., 2012). Such a broad approach provides opportunities (i) to identify novel SNPs associated with a disease trait, (ii) to identify the importance of a particular selenoprotein in a tissue function/disease development, and (iii) to integrate the effect of multiple SNPs across a whole metabolic pathway on disease risk. This assay was used to determine if SNPs across the Se metabolic pathway were associated with prostate cancer risk in an EPIC-Heidelberg population (Méplan et al., 2012). The study revealed that the risk of high-grade or advanced stage prostate cancer was modified by interactions between serum markers of Se status (namely plasma SePP, Se, and GPx3 activity) with SNPS in *SELK* (rs9880056), *TXNRD2* (rs9605030 and rs9605031), and *TXNRD1* (rs7310505) genes (Méplan et al., 2012), identifying potential novel functional SNPs or functional SNPs in strong linkage disequilibrium with these SNPs and further highlighting the interaction between genetic and dietary factors to modulate disease risk.

15.5   FACTORS INFLUENCING THE SIGNIFICANCE OF SNPs

15.5.1   PHYSIOLOGICAL FACTORS

Physiological factors have been found to modulate the influence of SNPs on seleno-protein metabolism. The effect of rs3877899 and rs7579 in *SEPP1* on plasma Se are modulated by body mass index, and the effect of rs713041 in *GPX4* on lymphocyte GPx4 levels following withdrawal of Se supplementation was observed in females but not males (Méplan et al., 2007, 2008, 2009). Similarly, the association of a SNP in *SELS* with colorectal cancer risk differs in males and females (Méplan et al., 2010; Sutherland et al., 2010). In addition, several SNPs associated with breast cancer were shown to affect the risk for another sex hormone–dependent cancer, prostate cancer (e.g., rs3877899 in *SEPP1* gene; Cooper et al., 2008; Méplan et al., 2013) suggest-ing that interaction between sex hormones and genetic factors could play an impor-tant role in the etiology of these cancers. In particular, evidence from both animal and cell culture models suggest a dual effect of estrogen in relation to breast cancer with (i) an increased production of reactive oxygen species as a result of the redox cycling of the catechol estrogens, suggesting that they contribute to carcinogene-sis by increasing DNA damage (Yang et al., 2013), and (ii) estrogens upregulating redox-active enzymes (MnSOD and GPx) in animal models (Roy and Liehr, 1989; Yager, 2000; Yager and Liehr, 1996).

15.5.2   GENETIC INTERACTIONS

As mentioned above, there is also evidence that the physiological impact of one SNP may be affected by another genetic variant, leading to a SNP–SNP interaction. For example, risk for colorectal cancer was not only increased by genotypes for SNPs in the *SEPP1*, *GPX4*, and *SELS* genes, but was further modulated by SNP–SNP inter-actions with polymorphisms in other selenoprotein genes, and these interactions reflect, in particular, the function of SePP1 in delivery of Se for synthesis of the other selenoproteins (Méplan et al., 2010). Genetic interactions with SNPs in genes coding for redox-active enzymes have been identified. In particular, rs4880 in the *SOD2* gene coding for MnSOD has been shown to interact with rs713041 (*GPX4*) and rs960531 (*TXNRD2*) to determine colorectal cancer risk (Méplan et al., 2010) or with rs3877899 in *SEPP1* gene in relation to prostate cancer risk (Cooper et al., 2008). The biological mechanisms underlying the interactions observed in these and other studies could reflect either the known selenoprotein hierarchy and selenocys-teine incorporation mechanism or complementary roles of various selenoproteins and redox active enzymes in functions, such as antioxidant protection or redox control.

15.5.3   BIOMARKERS OF SE STATUS IN ASSOCIATION STUDIES

As highlighted by several studies, disease risk is modified by the interactions between Se status, or biomarkers of Se status, and the genetic component carried by SNPs in selenoprotein genes. Because all selenoproteins require Se for their synthesis, they all have the potential to act as biomarkers of Se status. Lessons from the SelGen study showed that supplementation of healthy volunteers with physiological levels of Se was associated with a loss of differences between genotype for SNPs in *SEPP1* (rs3877899 and rs7579) or in *GPX4* (rs713041) in the supplemented population (Méplan et al., 2007, 2008, 2009). This observation is compatible with the observation that several associations between genetic factors and disease traits (Table 15.2) could not be reproduced in populations with different Se status. In particular, associations in populations with low Se intake, such as Europe, were lost in populations with high Se intake, such as in the United States (see Tables 15.2–15.4 for examples). So far, the observation of an interaction between a particular Se biomarker and a genetic factor that modulates disease risk has been regarded as a suggestion for a role of a specific selenoprotein in the etiology of the disease or the function of the normal tissue. However, the key question is to identify biomarkers that could predict the disease risk and interactions between SNP and biomarkers that are relevant to a specific disease trait or tissue. In that respect, the report that GPx1 activity was altered in erythrocytes (eGPx1) of women who later in life developed breast cancer (Méplan et al., 2013) suggested that eGPx1 could become a biomarker for breast disease risk. However, further functional studies are required to understand the relationship of causality.

So far, interactions between biomarkers of Se and genetic variations in selenoprotein genes have been reported to influence (i) the disease risk and (ii) the severity of the disease (grade/stage/mortality). For example, a significant interaction between Se status and mortality associated with prostate cancer was observed with carriers of the minor G allele for rs561104 (*SEP15*) having a lower mortality risk compared with AA individuals; in addition, the mortality risk was further reduced in GG/GA individuals with higher selenium status (Penney et al., 2010). Similarly, interactions between serum markers of Se status and genotypes for rs9880056 in *SELK*, rs9605030, and rs9605031 in *TXNRD2* and rs7310505 in *TXNRD1* modified the risk of high-grade or advanced stage prostate cancer (Méplan et al., 2012).

In addition, selenoproteins exert specific roles in some tissues (e.g., deiodinases play a major role in thyroid but are less important in some other tissues) or play a key role in response to specific stress (e.g., oxidative stress vs. misfolded proteins), and there is a tissue-specific hierarchy of selenoprotein synthesis (Bermano et al., 1995). It is therefore predictable that the significance of Se biomarkers will vary between tissues and diseases types.

15.6 CONCLUSIONS AND PERSPECTIVES

Animal studies, ex vivo cell experiments, and epidemiological data all support a key role for Se and selenoproteins in the maintenance of biological pathways affected in age-related disease with, in particular, correct protein folding, maintenance of redox status, and response against oxidative stress. This concept is supported by genetic data suggesting that some genetic variants are modulating the risk of developing chronic diseases by potentially altering selenoprotein synthesis, the distribution of Sec between selenoproteins, and the enzymatic activity of the corresponding proteins. An extra level of complexity is added by hierarchic distribution of Sec between selenoproteins in conditions of low Se supply, such as the ones encountered in European countries. As result, the interactions between the Se status or biomarkers of the Se status with variants that alter selenoprotein synthesis and the allocation of Sec between the different selenoproteins act as modulators of the genetic risk carried by the SNPs. To strengthen the evidence for association of SNP in selenoprotein genes with disease risk requires analysis of larger cohorts, replication in multiple cohorts, and the analysis of the genetic data together with measures of Se status. The understanding of these interactions has the potential, in the future, to directly impact human health and disease prevention by adapting a *personalized dietary recommendation* of Se to individuals depending on genetic makeup, biomarkers of Se status, and additional risk factors. From an evolutionary perspective, it would be interesting to understand if SNPs in selenoprotein genes associated with disease risks have been selected based on Se bioavailability in the environment or because they conferred a specific advantage to some subgroup populations to cope with environmental stresses. This could contribute to further understanding the importance of some selenoproteins in specific tissues or chronic diseases.

ACKNOWLEDGMENT

JLSD is supported by FAPESP (São Paulo Research Foundation, process 2013/03224-7).

REFERENCES

Abe, M., W. Xie, M. M. Regan et al. 2011. "Single-nucleotide polymorphisms within the antioxidant defence system and associations with aggressive prostate cancer." *BJU Int* no. 107 (1):126–34.

Ahn, J., M. D. Gammon, R. M. Santella et al. 2005. "No association between glutathione peroxidase Pro198Leu polymorphism and breast cancer risk." *Cancer Epidemiol Biomarkers Prev* no. 14 (10):2459–61.

Al-Taie, O. H., N. Uceyler, U. Eubner et al. 2004. "Expression profiling and genetic alterations of the selenoproteins GI-GPx and SePP in colorectal carcinogenesis." *Nutr Cancerl* no. 48 (1):6–14.

Arsova-Sarafinovska, Z., N. Matevska, A. Eken et al. 2009. "Glutathione peroxidase 1 (GPX1) genetic polymorphism, erythrocyte GPX activity, and prostate cancer risk." *Int Urol Nephrol* no. 41 (1):63–70.

Bastaki, M., K. Huen, P. Manzanillo et al. 2006. "Genotype-activity relationship for Mn-superoxide dismutase, glutathione peroxidase 1 and catalase in humans." *Pharmacogenet Genomics* no. 16 (4):279–86.

Bellinger, F. P., A. V. Raman, M. A. Reeves, and M. J. Berry. 2009. "Regulation and function of selenoproteins in human disease." *Biochem J* no. 422 (1):11–22.

Bermano, G., F. Nicol, J. A. Dyer et al. 1995. "Tissue-specific regulation of selenoenzyme gene expression during selenium deficiency in rats." *Biochem J* no. 311 (Pt 2):425–30.

Bermano, G., J. R. Arthur, and J. E. Hesketh. 1996. "Role of the 3′ untranslated region in the regulation of cytosolic glutathione peroxidase and phospholipid-hydroperoxide glutathione peroxidase gene expression by selenium supply." *Biochem J* no. 320 (Pt 3):891–5.

Bermano, G., V. Pagmantidis, N. Holloway et al. 2007. "Evidence that a polymorphism within the 3′UTR of glutathione peroxidase 4 is functional and is associated with susceptibility to colorectal cancer." *Genes Nutr* no. 2 (2):225–32.

Cebrian, A., P. D. Pharoah, S. Ahmed et al. 2006. "Tagging single-nucleotide polymorphisms in antioxidant defense enzymes and susceptibility to breast cancer." *Cancer Res* no. 66 (2):1225–33.

Choi, J. Y., M. L. Neuhouser, M. Barnett et al. 2007. "Polymorphisms in oxidative stress-related genes are not associated with prostate cancer risk in heavy smokers." *Cancer Epidemiol Biomarkers Prev* no. 16 (6):1115–20.

Combs, G. F., Jr., M. I. Jackson, J. C. Watts et al. 2012. "Differential responses to selenome-thionine supplementation by sex and genotype in healthy adults." *Br J Nutr* no. 107 (10):1514–25.

Cominetti, C., M. C. de Bortoli, E. Purgatto et al. 2011. "Associations between glutathione peroxidase-1 Pro198Leu polymorphism, selenium status, and DNA damage levels in obese women after consumption of Brazil nuts." *Nutrition* no. 27 (9):891–6.

Cooper, M. L., H. O. Adami, H. Gronberg, F. Wiklund, F. R. Green, and M. P. Rayman. 2008. "Interaction between single nucleotide polymorphisms in selenoprotein P and mito-chondrial superoxide dismutase determines prostate cancer risk." *Cancer Res* no. 68 (24):10171–7.

Corbo, R. M., and R. Scacchi. 1999. "Apolipoprotein E (APOE) allele distribution in the world. Is APOE*4 a 'thrifty' allele?" *Ann Hum Genet* no. 63 (Pt 4):301–10.

Cox, D. G., S. E. Hankinson, P. Kraft, and D. J. Hunter. 2004. "No association between GPX1 Pro198Leu and breast cancer risk." *Cancer Epidemiol Biomarkers Prev* no. 13 (11 Pt 1): 1821–2.

Curran, J. E., J. B. Jowett, K. S. Elliott et al. 2005. "Genetic variation in selenoprotein S influences inflammatory response." *Nat Genet* no. 37 (11):1234–41.

Du, X. H., X. X. Dai, R. Xia Song et al. 2012. "SNP and mRNA expression for glutathione peroxidase 4 in Kashin-Beck disease." *Br J Nutr* no. 107 (2):164–9.

Forsberg, L., U. de Faire, S. L. Marklund, P. M. Andersson, B. Stegmayr, and R. Morgenstern. 2000. "Phenotype determination of a common Pro-Leu polymorphism in human gluta-thione peroxidase 1." *Blood Cells Mol Dis* no. 26 (5):423–6.

Franke, A., D. P. B. McGovern, J. C. Barrett et al. 2010. "Genome-wide meta-analysis increases to 71 the number of confirmed Crohn's disease susceptibility loci." *Nature Genetics* no. 42 (12):1118+.

Gautrey, H., F. Nicol, A. A. Sneddon, J. Hall, and J. Hesketh. 2011. "A T/C polymorphism in the GPX4 3'UTR affects the selenoprotein expression pattern and cell viability in trans-fected Caco-2 cells." *Biochim Biophys Acta* no. 1810 (6):284–91.

Geybels, M. S., P. A. van den Brandt, L. J. Schouten et al. 2014. "Selenoprotein gene vari-ants, toenail selenium levels, and risk for advanced prostate cancer." *J Natl Cancer Inst* no. 106 (3):dju003.

Hamanishi, T., H. Furuta, H. Kato et al. 2004. "Functional variants in the glutathione peroxi-dase-1 (GPx-1) gene are associated with increased intima-media thickness of carotid arter-ies and risk of macrovascular diseases in japanese type 2 diabetic patients." *Diabetes* no. 53 (9):2455–60.

Hansen, R., M. Saebo, C. F. Skjelbred et al. 2005. "GPX Pro198Leu and OGG1 Ser326Cys polymorphisms and risk of development of colorectal adenomas and colorectal cancer." *Cancer Lett* no. 229 (1):85–91.

Hansen, R. D., B. N. Krath, K. Frederiksen et al. 2009. "GPX1 Pro(198)Leu polymorphism, erythrocyte GPX activity, interaction with alcohol consumption and smoking, and risk of colorectal cancer." *Mutat Res* no. 664 (1–2):13–9.

Haug, U., E. M. Poole, L. Xiao et al. 2012. "Glutathione peroxidase tagSNPs: Associations with rectal cancer but not with colon cancer." *Genes Chromosomes Cancer* no. 51 (6):598–605.

Hiragi, C. D., A. L. Miranda-Vilela, D. M. S. Rocha, S. F. de Oliveira, A. Hatagima, and M. D. Klautau-Guimaraes. 2011. "Superoxide dismutase, catalase, glutathione peroxidase and gluthatione S-transferases M1 and T1 gene polymorphisms in three Brazilian population groups." *Genetics and Molecular Biology* no. 34 (1):11–18.

Hu, J., G. W. Zhou, N. Wang, and Y. J. Wang. 2010. "GPX1 Pro198Leu polymorphism and breast cancer risk: A meta-analysis." *Breast Cancer Res Treat* no. 124 (2):425–31.

Hu, Y. J., and A. M. Diamond. 2003. "Role of glutathione peroxidase 1 in breast cancer: Loss of heterozygosity and allelic differences in the response to selenium." *Cancer Res* no. 63 (12):3347–51.

Hu, Y. J., K. V. Korotkov, R. Mehta et al. 2001. "Distribution and functional consequences of nucleotide polymorphisms in the 3'-untranslated region of the human Sep15 gene." *Cancer Res* no. 61 (5):2307–10.

Huang, L., Y. Shi, F. Lu et al. 2013. "Association study of polymorphisms in selenoprotein genes and Kashin-Beck disease and serum selenium/iodine concentration in a Tibetan population." *PLoS One* no. 8 (8):e71411.

Ichimura, Y., T. Habuchi, N. Tsuchiya et al. 2004. "Increased risk of bladder cancer associated with a glutathione peroxidase 1 codon 198 variant." *J Urol* no. 172 (2):728–32.

Jablonska, E., J. Gromadzinska, W. Sobala, E. Reszka, and W. Wasowicz. 2008. "Lung cancer risk associated with selenium status is modified in smoking individuals by Sep15 polymorphism." *Eur J Nutr* no. 47 (1):47–54.

Jablonska, E., J. Gromadzinska, E. Reszka et al. 2009. "Association between GPx1 Pro198 Leu polymorphism, GPx1 activity and plasma selenium concentration in humans." *Eur J Nutr* no. 48 (6):383–6.

Jaworska, K., S. Gupta, K. Durda et al. 2013. "A low selenium level is associated with lung and laryngeal cancers." *PLoS One* no. 8 (3):e59051.

Jostins, L., S. Ripke, R. K. Weersma et al. 2012. "Host-microbe interactions have shaped the genetic architecture of inflammatory bowel disease." *Nature* no. 491 (7422):119–124.

Karunasinghe, N., D. Y. Han, M. Goudie et al. 2012a. "Prostate disease risk factors among a New Zealand cohort." *J Nutrigenet Nutrigenomics* no. 5 (6):339–351.

Karunasinghe, N., D. Y. Han, S. Zhu et al. 2012b. "Serum selenium and single-nucleotide polymorphisms in genes for selenoproteins: Relationship to markers of oxidative stress in men from Auckland, New Zealand." *Genes Nutr* no. 7 (2):179–90.

Knight, J. A., U. V. Onay, S. Wells et al. 2004. "Genetic variants of GPX1 and SOD2 and breast cancer risk at the Ontario site of the Breast Cancer Family Registry." *Cancer Epidemiol Biomarkers Prev* no. 13 (1):146–9.

Korotkov, K. V., E. Kumaraswamy, Y. Zhou, D. L. Hatfield, and V. N. Gladyshev. 2001. "Association between the 15-kDa selenoprotein and UDP-glucose:glycoprotein glucosyltransferase in the endoplasmic reticulum of mammalian cells." *J Biol Chem* no. 276 (18):15330–6.

Kuzuya, M., F. Ando, A. Iguchi, and H. Shimokata. 2008. "Glutathione peroxidase 1 Pro198Leu variant contributes to the metabolic syndrome in men in a large Japanese cohort." *Am J Clin Nutr* no. (6):1939–44.

Méplan, C. 2011. "Trace elements and ageing, a genomic perspective using selenium as an example." *J Trace Elem Med Biol* no. 25 Suppl 1:S11–6.

Méplan, C., and J. Hesketh. 2012. "The influence of selenium and selenoprotein gene variants on colorectal cancer risk." *Mutagenesis* no. 27 (2):177–86.

Méplan, C., and J. Hesketh. 2014. "Selenium and cancer: A story that should not be forgotten-insights from genomics." *Cancer Treat Res* no. 159:145–66.

Méplan, C., L. K. Crosley, F. Nicol et al. 2007. "Genetic polymorphisms in the human selenoprotein P gene determine the response of selenoprotein markers to selenium supplementation in a gender-specific manner (the SELGEN study)." *Faseb J* no. 21 (12):3063–74.

Méplan, C., L. K. Crosley, F. Nicol et al. 2008. "Functional effects of a common single-nucleotide polymorphism (GPX4c718t) in the glutathione peroxidase 4 gene: Interaction with sex." *Am J Clin Nutr* no. 87 (4):1019–27.

Méplan, C., F. Nicol, B. T. Burtle et al. 2009. "Relative abundance of selenoprotein P isoforms in human plasma depends on genotype, se intake, and cancer status." *Antioxid Redox Signal* no. 11 (11):2631–40.

Méplan, C., D. J. Hughes, B. Pardini et al. 2010. "Genetic variants in selenoprotein genes increase risk of colorectal cancer." *Carcinogenesis* no. 31 (6):1074–9.

Méplan, C., S. Rohrmann, A. Steinbrecher et al. 2012. "Polymorphisms in thioredoxin reductase and selenoprotein K genes and selenium status modulate risk of prostate cancer." *PLoS ONE* no. 7 (11):e48709.

Méplan, C., L. O. Dragsted, G. Ravn-Haren, A. Tjonneland, U. Vogel, and J. Hesketh. 2013. "Association between polymorphisms in glutathione peroxidase and selenoprotein P genes, glutathione peroxidase activity, HRT use and breast cancer risk." *PLoS One* no. 8 (9):e73316.

Nasr, M. A., Y. J. Hu, and A. M. Diamond. 2003. "Allelic loss at the SEP15 locus in breast cancer." *Cancer Therapy* no. 1:293–8.

Nielsen, R. 2005. "Molecular signatures of natural selection." *Annu Rev Genet* no. 39:197–218.

Pellatt, A. J., R. K. Wolff, E. M. John et al. 2013. "SEPP1 influences breast cancer risk among women with greater Native American ancestry: The Breast Cancer Health Disparities Study." *Plos One* no. 8 (11):e80554.

Penney, K. L., F. R. Schumacher, H. Li et al. 2010. "A large prospective study of SEP15 genetic variation, interaction with plasma selenium levels, and prostate cancer risk and survival." *Cancer Prev Res (Phila)* no. 3 (5):604–10.

Penney, K. L., H. Li, L. A. Mucci et al. 2013. "Selenoprotein P genetic variants and mRNA expression, circulating selenium, and prostate cancer risk and survival." *Prostate* no. 73 (7):700–5.

Peters, U., and Y. Takata. 2008. Selenium and the prevention of prostate and colorectal cancer. *Mol Nutr Food Res* no. 52 (11):1261–72.

Peters, U., N. Chatterjee, R. B. Hayes et al. 2008. "Variation in the selenoenzyme genes and risk of advanced distal colorectal adenoma." *Cancer Epidemiol Biomarkers Prev* no. 17 (5):1144–54.

Porcu, E., M. Medici, G. Pistis et al. 2013. "A meta-analysis of thyroid-related traits reveals novel loci and gender-specific differences in the regulation of thyroid function." *Plos Genetics* no. 9 (2).

Raaschou-Nielsen, O., M. Sorensen, R. D. Hansen et al. 2007. "GPX1 Pro198Leu polymorphism, interactions with smoking and alcohol consumption, and risk for lung cancer." *Cancer Lett* no. 247 (2):293–300.

Ratnasinghe, D., J. A. Tangrea, M. R. Andersen et al. 2000. "Glutathione peroxidase codon 198 polymorphism variant increases lung cancer risk." *Cancer Res* no. 60 (22):6381–3.

Ravn-Haren, G., A. Olsen, A. Tjonneland et al. 2006. "Associations between GPX1 Pro198Leu polymorphism, erythrocyte GPX activity, alcohol consumption and breast cancer risk in a prospective cohort study." *Carcinogenesis* no. 27 (4):820–5.

Rosenberger, A., T. Illig, K. Korb et al. 2008. "Do genetic factors protect for early onset lung cancer? A case control study before the age of 50 years." *BMC Cancer* no. 8:60.

Roy, D., and J. G. Liehr. 1989. "Changes in activities of free radical detoxifying enzymes in kidneys of male Syrian hamsters treated with estradiol." *Cancer Res* no. 49 (6): 1475–80.

Shibata, T., T. Arisawa, T. Tahara et al. 2009. "Selenoprotein S (SEPS1) gene -105G>A promoter polymorphism influences the susceptibility to gastric cancer in the Japanese population." *BMC Gastroenterol* no. 9:2.

Slattery, M. L., A. Lundgreen, B. Welbourn, C. Corcoran, and R. K. Wolff. 2012. "Genetic variation in selenoprotein genes, lifestyle, and risk of colon and rectal cancer." *PLoS ONE* no. 7 (5):e37312.

Steinbrecher, A., C. Méplan, J. Hesketh et al. 2010. "Effects of selenium status and polymorphisms in selenoprotein genes on prostate cancer risk in a prospective study of European men." *Cancer Epidemiol Biomarkers Prev* no. 19 (11):2958–68.

Sun, W., X. Wang, X. Zou et al. 2010. "Selenoprotein P gene r25191g/a polymorphism and quantification of selenoprotein P mRNA level in patients with Kashin-Beck disease." *Br J Nutr* no. 104 (9):1283–7.

Sutherland, A., D. H. Kim, C. Relton, Y. O. Ahn, and J. Hesketh. 2010. "Polymorphisms in the selenoprotein S and 15-kDa selenoprotein genes are associated with altered susceptibility to colorectal cancer." *Genes Nutr* no. 5 (3):215–23.

Suzen, H. S., E. Gucyener, O. Sakalli et al. 2010. "CAT C-262T and GPX1 Pro198Leu polymorphisms in a Turkish population." *Mol Biol Rep* no. 37 (1):87–92.

Takata, Y., A. R. Kristal, I. B. King et al. 2011. "Serum selenium, genetic variation in selenoenzymes, and risk of colorectal cancer: Primary analysis from the Women's Health Initiative observational study and meta-analysis." *Cancer Epidemiol Biomarkers Prev* no. 20 (9):1822–30.

Takata, Y., I. B. King, J. W. Lampe et al. 2012. "Genetic variation in GPX1 is associated with GPX1 activity in a comprehensive analysis of genetic variations in selenoenzyme genes and their activity and oxidative stress in humans." *J Nutr* no. 142 (3):419–26.

Udler, M., A. T. Maia, A. Cebrian et al. 2007. "Common germline genetic variation in antioxidant defense genes and survival after diagnosis of breast cancer." *J Clin Oncol* no. 25 (21):3015–23.

Villette, S., J. A. Kyle, K. M. Brown et al. 2002. "A novel single nucleotide polymorphism in the 3' untranslated region of human glutathione peroxidase 4 influences lipoxygenase metabolism." *Blood Cells Mol Dis* no. 29 (2):174–8.

Vithana, E. N., C. C. Khor, C. Y. Qiao et al. 2012. "Genome-wide association analyses identify three new susceptibility loci for primary angle closure glaucoma." *Nature Genetics* no. 44 (10):1142–6.

Wallace, C., D. J. Smyth, M. Maisuria-Armer, N. M. Walker, J. A. Todd, and D. G. Clayton. 2010. "The imprinted DLK1-MEG3 gene region on chromosome 14q32.2 alters susceptibility to type 1 diabetes." *Nature Genetics* no. 42 (1):68–U85.

White, L., Romagne, F., Muller et al. 2015. "Genetic Adaptation to Levels of Dietary Selenium in Recent Human History." *Molec Biol Evol*. [Epub ahead of print]

Yager, J. D. 2000. "Endogenous estrogens as carcinogens through metabolic activation." *J Natl Cancer Inst Monogr* no. (27):67–73.

Yager, J. D., and J. G. Liehr. 1996. "Molecular mechanisms of estrogen carcinogenesis." *Annu Rev Pharmacol Toxicol* no. 36:203–32.

Yang, L., M. Zahid, Y. Liao et al. 2013. "Reduced formation of depurinating estrogen-DNA adducts by sulforaphane or KEAP1 disruption in human mammary epithelial MCF-10A cells." *Carcinogenesis* no. 34 (11):2587–92.

Yang, P., W. R. Bamlet, J. O. Ebbert, W. R. Taylor, and M. de Andrade. 2004. "Glutathione pathway genes and lung cancer risk in young and old populations." *Carcinogenesis* no. 25 (10):1935–44.

Zhao, H., D. Liang, H. B. Grossman, and X. Wu. 2005. "Glutathione peroxidase 1 gene polymorphism and risk of recurrence in patients with superficial bladder cancer." *Urology* no. 66 (4):769–74.

16 Mutations in *SECISBP2*

Erik Schoenmakers, Carla Moran,
Nadia Schoenmakers, and Krishna Chatterjee

CONTENTS

16.1 Introduction 343
16.2 Structure and Function of SBP2 344
16.3 Function of Mammalian Selenoproteins 347
    16.3.1 Deletion of *Secisbp2* 350
16.4 Disorder Due to Human *SECISBP2* Mutations 351
    16.4.1 Human SBP2 Mutations 351
    16.4.2 Biochemistry 359
    16.4.3 Growth and Skeletal Phenotype 359
    16.4.4 Hearing Loss 360
    16.4.5 Developmental and Neurological Phenotype 360
    16.4.6 Axial Muscular Dystrophy 360
    16.4.7 Metabolic Phenotype 361
    16.4.8 Azoospermia with Spermatogenic Maturation Arrest 362
    16.4.9 Cutaneous Photosensitivity 362
    16.4.10 Impaired T-Cell Proliferation and Abnormal Cytokine Production ... 363
    16.4.11 Effect of Treatment 364
16.5 Conclusions 364
References 367

16.1 INTRODUCTION

There are 25 known genes coding for human selenoproteins, whose expression requires decoding of an UGA codon as the amino acid selenocysteine (Sec) instead of a premature stop. This process is mediated by the interaction of a Sec insertion sequence (SECIS) element in the 3′-untranslated region of selenoprotein mRNAs with SECIS binding protein 2 (SBP2) (Copeland and Driscoll, 1999) (see also Chapter 3). SBP2 is a limiting, obligate factor for selenoprotein synthesis as first shown by absence of selenoprotein synthesis in SBP2-depleted cell lysates with restoration of production by repletion with SBP2. Furthermore, SBP2 is essential for survival as murine *Secisbp2* deletion is embryonic lethal (Seeher et al., 2014a). Here, we will describe characteristics and pathogenesis of a human disorder due to mutations in *SECISBP2*, disrupting synthesis of multiple selenoproteins. The indispensable role of selenoproteins in diverse physiological processes is demonstrated by the multisystem phenotype of this disorder, encompassing defects in thyroid hormone metabolism, spermatogenesis, muscle function, and antioxidant defense.

16.2 STRUCTURE AND FUNCTION OF SBP2

Human SBP2 is a large, 854 amino acids, protein with the amino (N-) terminal first 400 residues being dispensable for its Sec incorporation function *in vitro* (Copeland et al., 2001; de Jesus et al., 2006) (see Figure 16.1). This N-terminal domain is not conserved in SBP2 from lower eukaryotes (being absent in insects and worms), contains no discernible functional motif, and has no known homologues. In contrast, the carboxy (C-) terminal region (amino acids 399–784) of SBP2 is both essential and sufficient for SECIS-binding and Sec incorporation. The Sec incorporation domain (SID, amino acids 399–517) is located centrally and dispensable for SECIS binding but required for Sec incorporation activity. The RNA binding domain (RBD, amino acids 517–784) contains an L7Ae-type RNA interaction motif (620–745) that is shared with a large family of ribosomal proteins (e.g., RPL30, SUP1, eRF-1, and 15.5-kD/Snu13p) (Allmang et al., 2002, 2006; Caban and Copeland, 2006; Copeland et al., 2001; de Jesus et al., 2006) and mediates interaction with the SECIS element (Fletcher et al., 2001; Walczak et al., 1998) and 28S ribosomal RNA (Copeland et al., 2001; Lescure et al., 2002). Additional contacts formed between RNA and a domain N-terminal to the L7Ae module, referred to as the bipartite/SID/K-rich region (amino acids 517–544) (Caban et al., 2007; Donovan et al., 2008; Takeuchi et al., 2009) are also absolutely required for specific recognition of the SECIS element. SBP2 contains several functional motifs in addition to the RBD: an N-terminal lysine-rich nuclear localization signal (NLS), two C-terminal leucine-rich nuclear export signals (NES), and a redox-sensitive cysteine-rich domain (CRD, amino acids 584–854) (Papp et al., 2006).

The major cis-acting determinant for Sec incorporation is the SECIS element, a stem-loop RNA structure invariably located within the 3′-untranslated region (UTR) of all eukaryotic selenoprotein mRNAs (Berry et al., 1991, 1993). SECIS elements share common structural features, consisting of two helices separated by an internal loop, a GA quartet (four consecutive non-Watson–Crick base pairs), and an apical loop (Chapple et al., 2009; Fagegaltier et al., 2000; Grundner-Culemann et al., 1999; Walczak et al., 1996, 1998). There are two structurally different types of SECIS elements: Type I has a longer apical loop whereas type II, which is the most common element, has an additional small stem-loop structure. Another RNA motif, named the Sec redefinition element, a few nucleotides downstream of the UGA codon, has been identified in selenoprotein N (SELN) transcript and appears to fulfill a fine-tuning role while the SECIS element is absolutely required for Sec incorporation (Howard et al., 2005, 2007). All selenoprotein mRNAs contain one SECIS element except for selenoprotein P (SEPP1), whose 3′ UTR contains two, tandemly repeated SECIS elements, acting in concert to efficiently decode the 10 Sec codons present in SEPP1.

Synthesis, delivery, and incorporation of Sec into polypeptide chains is mediated by a multiprotein complex (~500 kDa) (Small-Howard et al., 2006), including SBP2, a specialized tRNA (selenocysteyl-tRNA[Ser]Sec[Sec]), Sec-specific elongation factor

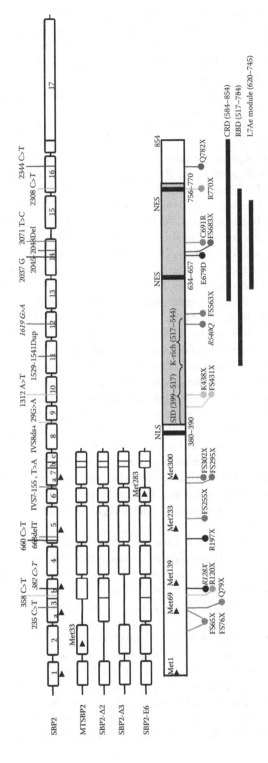

FIGURE 16.1 Genomic organization of *SECISBP2* and functional domains of SBP2, with location of human mutations. The genomic organization of *SECISBP2* (top) and functional domains of SBP2 protein (bottom) are depicted. The varying composition of natural N-terminal splice variants, each containing distal exons (8 to 17) is shown. The position of compound heterozygous mutations and homozygous mutations (italicized) is superimposed. Arrowheads denote the location of ATG codons, which could function as alternative sites for translation initiation. Functional domains identified in SBP2: 1–399: N-terminal domain; 399–784: Minimal functional protein (shaded gray); 399–517: Sec incorporation domain (SID); 517–784: minimal RNA-binding domain (RBD); 517–544: Lysine-rich domain involved in RNA specificity and ribosome binding; 620–745: L7Ae homology module; 380–390: nuclear localization signal (NLS); 584–854: redox-sensitive cysteine-rich domain (CRD) and two nuclear export signals (NES1: 634–657; NES2: 756–770).

eEFSec, ribosomal protein L30, a 43-kDa RNA binding protein (SECp43) and Sec synthase. SBP2 binds to the ribosome monomerically with the discrete ES7L-E helix of 28S rRNA interacting with the k-rich region in SBP2 (Kossinova et al., 2014). SBP2 interaction with the ribosome and SECIS element during translation is thought to be dynamic and important for recruitment of the eEFsec/tRNA[Ser]sec complex to the UGA codon. Selenoprotein biosynthesis is discussed in detail in Chapters 3 and 4.

The localization of SBP2 has been studied by several groups with compelling evidence that its steady-state subcellular localization is cytoplasmic, stably complexed with ribosomes (Copeland et al., 2001; Kinzy et al., 2005; Mehta et al., 2004). However, SBP2 is also capable of shuttling between the nucleus and cytoplasm via intrinsic NLS and NES motifs (de Jesus et al., 2006; Papp et al., 2006). One evolutionarily conserved functional NLS (amino acids 380–390) in the N-terminal region and two functional NES (NES1 amino acids 634–657; NES2 amino acids 756–770) in the C-terminal region have been identified. Additional, potential NLS motifs in the C-terminal region and NES motifs in the N-terminal region of SBP2 have been delineated but not functionally verified (de Jesus et al., 2006). Interestingly, the cellular localization of SBP2 is influenced by its redox state with the redox-sensitive CRD overlapping with the RBD and the identified NESs (Figure 16.1). Its role in regulating SBP2 localization or function, for example, protecting selenoprotein-encoding mRNAs from nonsense mediated decay (NMD), require further investigations.

Alternative splicing events in human SBP2 RNA result in a complex splicing pattern in its 5′-region with generation of at least eight splice variants, encoding five protein isoforms of varying N-terminal sequence composition (Figure 16.1) (Copeland et al., 2000; Papp et al., 2008). Some of the splicing events provoke shifts in open reading frame, leading to premature stop codons with reinitiation of translation from downstream ATG start codons. An *in vivo* splicing assay showed that at least four additional ATG start codons in exons 2, 3a, 3b, 5, and 7 can be used to generate proteins with differing N-terminal amino acid composition. One isoform, mtSBP2, has been shown to contain a targeting sequence that localizes it to mitochondria. All alternative splicing events within SBP2 are confined to its N-terminal domain, which is dispensable for its Sec incorporation function *in vitro*; nevertheless, this does not exclude the possibility that this region may yet fine-tune or regulate SBP2-dependent Sec incorporation *in vivo*.

In summary, SBP2 and the SECIS element have been shown to direct Sec incorporation at UGA codons located within mRNAs as opposed to at the end of their coding regions. Although there is a substantial body of information about how SBP2 interacts with SECIS elements, ribosome, and the Sec incorporation complex, detailed understanding of how the SECIS-bound protein complex reaches back to connect with the Sec-specifying UGA codon within the preceding ribosome are far from being elucidated. Selective association of SBP2 with particular SECIS elements in different selenoprotein mRNAs, together with variably concerted action of all the cis- and trans-acting factors described or as yet undiscovered, has been proposed to be responsible for a hierarchy of selenoprotein synthesis, in which production of some selenoproteome members is preserved at the expense of others. The importance of different splice variant isoforms, subcellular location, tissue expression levels, and redox state on SBP2 function are still open questions that need further investigation.

16.3 FUNCTION OF MAMMALIAN SELENOPROTEINS

The human selenoproteome is encoded by 25 genes with approximately half of these encoding proteins with known functions (Table 16.1). Essentially, all functionally characterized selenoproteins are oxidoreductases in which Sec is the catalytic residue. Clearly, the unique catalytic properties of Sec account for the role of selenium in these proteins although there is no consensus on their specific nature. Knowledge of mechanisms of selenoprotein action in human physiological processes is far from complete, and finding causal links between alteration in specific selenoproteins and human diseases is a major challenge (Roman et al., 2014). Some of these aspects are described in more detail elsewhere (Chapters 8–10 and Chapter 15). It is important to emphasize that the role of individual selenoproteins has to be considered in the context of a complex cellular biochemical environment in which antagonistic, additive, and synergistic effects occur.

The family of glutathione peroxidases (GPx) comprises eight proteins, of which five members contain Sec at their active site (GPx1, 2, 3, 4, 6) (Brigelius-Flohé and Maiorino, 2013). These enzymes catalyze reduction of hydrogen peroxide or other peroxides (e.g., lipid hydroperoxides) principally using glutathione but also thioredoxin or other thiol oxidoreductases as reducing factors. GPxs have a wide range of physiological functions, being involved in hydrogen peroxide (H_2O_2) signaling, detoxification of hydroperoxides, and maintaining cellular redox homeostasis (Labunskyy et al., 2014; Roman et al., 2014). Three thioredoxin reductases (TrxR1, 2, 3), which comprise the major cellular disulfide reduction system, use thioredoxin as their major substrate (reviewed in Arner, 2009; Labunskyy et al., 2014; Roman et al., 2014).

The deiodinases are a family of three membrane proteins with DIO1 and 3 situated at plasma membrane and DIO2 localizing to endoplasmic reticulum (ER). All DIOs are oxidoreductases, catalyzing reductive deiodination of thyroid hormones (TH); thus, conversion of prohormone (thyroxine, T4) to its active metabolite (triiodothyronine, T3) is mediated by DIO2 and DIO1; DIO3 catalyzes TH conversion to inactive metabolites. DIO1 is largely responsible for generating circulating T3 from T4 derived from the thyroid gland whereas DIO2 and DIO3 regulate deiodination at the local tissue level. However, their relative role in these processes is not well understood and also varies depending on selenium status, type of tissue, and stage of development (Gereben et al., 2008). Methionine-R-sulfoxide reductase 1 (MsrB1 or SELR) catalyzes repair of the R-enantiomer of oxidized methionine residues, whose presence in proteins can adversely influence their structure and function (Kryukov et al., 2002). Selenophosphate synthetase 2 (SPS2) is a selenoenzyme catalyzing ATP-dependent synthesis of selenophosphate, a selenium donor compound for Sec biosynthesis (Guimaraes et al., 1996). Human selenoprotein P (SEPP1), containing 10 Sec residues, is synthesized primarily in the liver and secreted into the bloodstream where it accounts for ~50% of total circulating selenium. The main function of SEPP1 is thought to be delivery of selenium substrate to other organs (Carlson et al., 2004; Schweizer et al., 2005).

The specific functions of several other human selenoproteins are unknown although some details of their biology have been established (Lobanov et al., 2009).

TABLE 16.1

Main Phenotypes Identified in Knockout Mouse Models of Specific Selenoproteins

Protein	Tissue Distribution	Known Physiologic Phenotype	Reference
Gpx1	Ubiquitous, high in erythrocytes, liver, kidney, lung	Complete ko: no spontaneous phenotype, cardiac and vascular dysfunction function at stress insult, protected from diet-induced obesity and improved hepatic insulin signalling; overexpression affects TNFa, insulin, EGF and insulin signalling leads to metabolic syndrome, conditional brain-ko: susceptible to ischemic and toxic at stress insult, conditional liver-ko is protected, conditional pancreatic Beta cell-ko affects insulin secretion	Fu et al., 1999; Cheng et al., 1998; Kretz-Remy et al., 1996; Lei and Wang, 2012; Handy et al., 2009; McClung et al., 2004; Wang et al., 2008; Ho et al., 1997; Pepper et al., 2011; Merry et al., 2014
Gpx2	Liver, epithelium of the gastrointestinal tract, epithilium derived tumours	Complete ko: no spontaneous phenotype, insult increases intestinal inflammation, development of cancer	Esworthy et al., 2005; Banning et al., 2012
Gpx3	Plasma and kidney	Prothrombotic state with vascular dysfunction promoting platelet-dependent arterial thrombosis	Jin et al., 2011
Gpx4	Various tissues; high in testes	Complete ko: embryonic lethal; cytosolic GPx4-ko: embryonic lethal: Neurodegeneration in hippocampus and cortex, nuclear or mitochondrial form GPx4-ko male: infertility	Liang et al., 2009; Yant et al., 2003; Imai et al., 2003; Schneider et al., 2009; Garry et al., 2008; Seiler et al., 2008; Conrad et al., 2005
Gpx6	Olfactory epithelium, embryos		
Gpx1+2		Null mice develop colitis and increased susceptibility to intestinal cancer	Esworthy et al., 2001; Chu et al., 2004
TnxR1	Ubiquitous	Complete: embryonic lethal, cerebellar hypoplasia and ataxia; conditional liver-ko: increased susceptibility to cancer with increased expression of genes involved in detoxification, repression of lipogenesis, increased glycogen storage; conditional brain-ko: cerebellar hypoplasia and motor deficits; conditional macrophage-ko: immune function impaired	Jakupoglu et al., 2005; Bondareva et al., 2007; Soerensen et al., 2008; Carlson et al., 2012; Iverson et al., 2013; Kalantari et al., 2008
TnxR2	Ubiquitous; high in the prostate, ovary, liver, testes, uterus, colon, small intestine, heart	Complete ko: embryonic lethal; conditional heart-ko: failure and postnatal death	Geisberger et al., 2007; Soerensen et al., 2008
TnxR3	Testes	Sperm maturation	Su et al., 2005

(Continued)

TABLE 16.1 (CONTINUED)
Main Phenotypes Identified in Knockout Mouse Models of Specific Selenoproteins

Protein	Tissue Distribution	Known Physiologic Phenotype	Reference
Dio1	Liver, kidney, thyroid, pituitary gland, ovary	Complete ko: increased fecal iodine excretion, increased T4 and rT3, normal levels of T3 and TSH	Schneider, 2006
Dio2	Thyroid, heart, brain, spinal cord, skeletal muscle, placenta, kidney, pancreas, pituitary gland, cochlea, hypothalamus	Complete ko: high T4 and TSH, normal T3, hearing deficits, delayed muscle repair after injury, some growth retardation, impaired thermogenesis	Dentice et al., 2010; Ng et al., 2004; Schneider et al., 2001; de Jesus et al., 2001
Dio3	Placenta, fetal tissues, skin, cochlea, cerebellum	Complete ko: partial embryonic and neonatal lethality. Surviving mice: severe growth retardation, impaired reproduction, central hypothyrodism, hearing deficit, altered cerebellar morphology and impaired locomotor function	Hernandez et al., 2006; Ng et al., 2009
Dio1+2	Various tissues; high in brain: cochlea, olfactory bulb, cerebellum, hypothalamus	Complete ko: impaired locomotion, learning and memory skills, central hypothyroidism high serum T4, TSH and rT3, normal T3	Galton et al., 2009
SelM		Complete ko: mild obesity	Pitts et al., 2013
Sel15	Various tissues; high in prostate, liver, thyroid gland, lung, brain, kidney	Complete ko: no spontaneous phenotype, increased oxidative stress markers in the liver and cataract development, protected against chemically-induced aberrant crypt formation	Kasaikina et al., 2011; Tsuji et al., 2012
SelK	Various tissues; abundant in heart	Complete ko: impaired immune response, reduction in atherosclerotic lesion formation	Verma et al., 2011; Meiler et al., 2013
SelT	Ubiquitous	Pancreatic B-cell-ko: decreased insulin secretion leading to increased plasma glucose	Prevost et al., 2013
SelP	Ubiquitous low levels; high in liver	Complete ko: impaired Se trafficking, male infertility; low Se brain/testis, high Se in liver; vascular endothelial cells affected, cognitive and motor deficit; seizures, ataxia, and neurod-generation; protected from high fat diet induced adipocyte hypertrophy, glucose intolerance, and insulin resistance	Schomburg et al., 2003; Hill et al., 2003; Misu et al., 2010; Hill et al., 2007; Mao and Teng, 2013; Hill et al., 2004; Valentine et al., 2005; Ishikura et al., 2014; Burk et al., 2006; Olson et al., 2005
MsrB1	Heart, liver, muscle, kidney	Complete ko: no spontaneous phenotype, reduced innate immunity linked with macrophage disfunction	Fomenko et al., 2009; Lee et al., 2013
SelN	Ubiquitous: abundant in skeletal muscle, brain, lung, placenta	Complete ko: no phenotype, develop severe kyphosis and hypothrophy after repeated forced exercise, muscle regeneration impaired after injury	Rederstorff et al., 2011; Castets et al., 2011

Thus, selenoproteins S and K are ER transmembrane proteins, involved in regulating ER-associated degradation of misfolded proteins (Shchedrina et al., 2011; Ye et al., 2004) and may have a role in immune function (Alanne et al., 2007; Moses et al., 2008; Nelson et al., 2011; Shibata et al., 2009; Sutherland et al., 2010; Verma et al., 2011; Vunta et al., 2007). The selenoproteins 15 (SEL15) and M (SELM), a distinct family of thiol-disulfide oxidoreductases, are also ER-localized and implicated in quality control of protein folding. Selenoprotein T, V, H, and W form a subfamily, characterized by a thioredoxin-like fold with Sec located in the N-terminal part of this domain. This contrasts with SELS, SELK, SELI, and SELO, in which Sec is C-terminally located. The biochemical and physiological role of SELN is important for skeletal muscle function with mutations in this gene being associated with specific myopathies (congenital muscular dystrophy, including multiminicore myopathy, rigid spine muscular dystrophy, and desmin-related myopathy with Mallory bodies).

The use of transgenic mouse models has been instrumental in uncovering the role of selenoproteins in tissue development and function (Table 16.1). At least four mammalian selenoproteins, TrxR1, TrxR2, Dio3, and GPx4, are essential for development with targeted murine gene disruption being embryonic lethal whereas knockout mice deficient in GPx1, GPx2, GPx3, Sel15, MsrB1, Dio1, Dio2, SelK, SepP1, SelN, SelM, and two double null mouse models GPx1/2 and Dio1/2 are viable, exhibiting only mild phenotypes in the absence of stress.

16.3.1 Deletion of *Secisbp2*

Homozygous *Secisbp2* null mice perish in utero with heterozygous animals exhibiting little phenotype when unstressed (Seeher et al., 2014a). Two conditional (liver- or brain-specific) murine *Secisbp2* knockout lines have been generated, providing models for analyzing tissue-specific SBP2 function (Seeher et al., 2014a,b). Liver-specific *Secisbp2* null mice exhibit no obvious phenotype (Seeher et al., 2014a) with normal serum hepatic transaminase levels. Interestingly, low-level synthesis of some selenoproteins (SelS, SepP1, GPx1, GPx4, Dio1) remained with SelK and TrxR1 expression being similar to wild-type littermates. Hepatic expression of nuclear factor, erythroid 2-like 2 (Nrf2)-dependent target genes was markedly upregulated, probably being induced secondary to lack of glutathione-dependent antioxidant selenoenzymes.

Neuron-specific *Secisbp2* null mice exhibited more profound abnormalities (Seeher et al., 2014b). On postnatal day 21, body weight of mutant mice was reduced by 50%, and tail length was 20% shorter with longitudinal growth also being affected. Mutant mice also exhibited a marked locomotor phenotype with impaired balance and awkward, broad-based, dystonic gait, reminiscent of that seen in *SepP1*-deficient mice (Hill et al., 2004; Renko et al., 2008; Schweizer et al., 2004). Cortical expression of many selenoproteins (TrxR1, GPx1, SelW, SelM, SelT and GPx4) was reduced but not abolished whereas SelS and SelK levels were unaffected. On the other hand, GPx1, SelW, SelM, and SelT mRNA levels were reduced whereas TrxR1 and GPx4 mRNA were unaltered. Expression of the Nrf2-dependent target gene was only moderately induced, possibly due to residual, low levels of GPx protein. Detailed analysis of neuron-specific *Secisbp2* null brain showed that parvalbumin positive (PV⁺) interneuron density was reduced in the somatosensory cortex, hippocampus, and striatum with Gad67 expression patterns confirming reduction in GABAergic interneurons. The obvious locomotor phenotype suggested basal ganglia dysfunction, and although normal tyrosine hydroxylase expression in substantia nigra neurons and their striatal terminals was detected, the densities of striatal PV⁺ and Gad67⁺ neurons were decreased as are the striatal cholinergic neurons. These findings indicate that several classes of striatal interneurons are dependent on selenoprotein expression. Limited reduction of cortical PV⁺ interneurons in younger neuron-specific *Secisbp2* null mice indicates that PV⁺ neurons are able to mature but are lost at a later stage, suggesting that PV⁺/Gad67⁺ neurons develop normally but are susceptible to later, accelerated, degenerative loss.

Several explanations have been put forward to explain residual, detectable expression of several selenoproteins in both conditional *Secisbp2* null mouse models, including the presence of nontargeted cells, incomplete Cre recombinase-mediated gene excision, expression of Secisbp2 isoforms initiated from a downstream methionine (amino acid 302), Secisbp2-independent mechanisms for selenoprotein synthesis or incorporation of alternative amino acids instead of Sec, permitting low-level synthesis of selenoprotein, albeit, probably nonfunctional.

An intriguing finding of these studies is the role that Secisbp2 plays in selenoprotein mRNA abundance as shown by the massive reduction of most selenoprotein mRNAs in the absence of Secisbp2. Clearly, it was expected that failure to translate selenoprotein mRNAs may subject these transcripts to NMD, but several transcripts do not follow the canonical rules for NMD and are more susceptible for NMD than expected, supporting the view that Secisbp2 may play a role both in translation and stability of selenoprotein mRNAs (Seeher et al., 2014a).

Several mouse models of individual selenoproteins, two Secisbp2 mouse models and various *Trsp* null mouse models (Chapter 4) have been generated and contribute to elucidating selenoprotein function. From these studies, it became clear that there is a complex interplay of the selenoproteome with correct organ function and compensatory mechanisms of non-selenoproteins. Selenium and selenoproteins have direct or indirect links to a large number of human health disorders, which are discussed in detail in other chapters. Investigation of patients with *SECISBP2* mutations identified many different phenotypes described below with several being linked to deficiency of individual selenoproteins or to increased ROS levels as result of decreased oxidoreduction capacity due to deficiency of antioxidant selenoenzymes.

16.4   DISORDER DUE TO HUMAN *SECISBP2* MUTATIONS

16.4.1   HUMAN SBP2 MUTATIONS

To date, nine families with either compound heterozygous or homozygous *SECISBP2* defects and selenoprotein deficiency have been described (Azevedo et al., 2010; Di Cosmo et al., 2009; Dumitrescu and Refetoff, 2013; Dumitrescu et al., 2005; Fu et al., 2014; Hamajima et al., 2012; Schoenmakers et al., 2010) (Table 16.2). Consistent with the recessive mode of inheritance, documented in some kindred, heterozygous individuals are unaffected. Although patients are from diverse ethnic backgrounds, they do share similar clinical phenotypes (Tables 16.3 and 16.4). Most cases presented

TABLE 16.2

Human *SECISBP2* Gene Defects

Family	Gene Mutation	Protein Mutation	Alleles Affected	Suggested Mechanism	Author
A	c.1619 G>A	R540Q	Homozygous	SID/bipartite/K-rich region affecting SECIS/ribosome binding	Dumitrescu et al., 2005
B	c.1312 A>T IVS8ds+29 G>A	K438X fs431X	Compound heterozygous	Premature stop missing full length protein	Dumitrescu et al., 2005
C	c.382 C>T	R128X	Homozygous	Premature stop absence of full length protein/splice variants affected	Di Cosmo et al., 2009
D	c.358 C>T c.2308 C>T	R120X R770X	Compound heterozygous	Premature stop absence of full length protein/splice variants affected Premature stop C-terminal end of RNA binding domain missing	Azevedo et al., 2010
E	c.668delT IVS7 -155, T>A	F223fs255X fs295X + fs302X	Compound heterozygous	Premature stop missing full length protein Splicing affected between exon 6-7 resulting in premature stop absence of full length protein	Schoenmakers et al., 2010
F	c.2017 T>C 1-5 intronic SNP's	C691R fs65X + fs76X	Compound heterozygous	Predicted to affect SECIS/ribosomebinding and increase proteasomal degradation Transcripts lacking exons 2-3-4 or 3-4 resulting in premature stop absence of full length protein	Schoenmakers et al., 2010
G	c.1529_1541dup CCAGCGCCCCACT c.235 C>T	M515fs563X Q79X	Compound heterozygous	Premature stop missing full length protein	Hamajima et al., 2012
H	c.2344 C>T c.2045-2048 delAACA	Q782X K682fs683X	Compound heterozygous	Premature stop missing full length protein Premature stop missing full length protein	Dumitrescu and Refetoff, 2013
I	660 C>T 2108 G> T or C	R197X E679D	Compound heterozygous	Premature stop missing full length protein Predicted to affect SECIS & ribosome binding	Fu et al., 2014

TABLE 16.3

Blood Biochemistry in SBP2 Deficient Patients

Family	Plasma Se	FT4	T3	rT3	TSH
A	↓	↑	↓	↑	↑
B	↓	↑	↓	↑	normal
C	↓	↑	↓	↑	normal
D	↓	↑	normal	↑	↑
E	↓	↑	normal	↑	↑
F	↓	↑	↓	↑	normal
G	↓	↑	↓	?	normal
H	?	↑	↓	↑	normal
I	?	↑	↓	↑	normal

Note: ↑ increased level; ↓ decreased level; ? unknown.

were diagnosed in childhood; only a single adult male patient has been described hitherto.

Family A (Dumitrescu et al., 2005)

Affected members ($n = 3$) of this family have a homozygous missense mutation (c.1619 G > A; p.R540Q) in exon 12 of *SECISBP2*. The R540Q mutation localizes to the RBD (Figure 16.1); consistent with this, *in vitro* analyses showed decreased R540Q mutant SBP2 protein binding to SECIS elements from GPx1 and DIO2 mRNAs, correlating with diminished GPx1 and DIO2 enzyme activity in patient-derived primary cells studied *ex vivo*. Functional studies using the homologous rodent mutant (R517Q) SBP2 protein confirmed markedly diminished binding to SECIS elements from Gpx1 and DIO2 transcripts, but interaction with elements from Gpx4 and TrxR1 was only slightly reduced (Bubenik and Driscoll, 2007). Surprisingly, in transient transfection assays, the R540Q SBP2 mutant was able to direct Sec incorporation mediated by SECIS elements from Gpx4, Gpx1, or Dio1 when they were tested individually. However, with inclusion of "decoy" SECIS element-containing constructs to mimic an intracellular environment in which multiple SECIS elements compete for SBP2 binding, wild-type and mutant SBP2 proteins did direct different patterns of synthesis. Thus, consistent with its preserved interaction with either mutant or wild-type SBP2 proteins, a decoy SECIS element from Gpx4 inhibited Sec incorporation mediated by both mutant and wild-type SBP2 equally. In contrast, decoy SECIS elements from DIO2 and Gpx1 were much less effective in blocking Sec incorporation by mutant SBP2 than the wild type, concordant with their greater binding affinity for wild-type SBP2.

The basis for differential binding of R540Q mutant SBP2 to different SECIS elements is unclear, but R540 resides in the K-rich region (amino

TABLE 16.4
Main Physiological Phenotypes in Patients with SBP2 Deficiency

Family	Age (years) Gender	Ethnicity	Hearing	Muscle and Neurological	Metabolic	Growth and Skeletal	Other	Effect of Treatment
A	14, M 7, M; 4, F	Saudi Arabian	Normal	–	–	Short stature, DBA	Normal mental development	Se: no effect
B	6, M	Irish/ Kenyan	Normal	–	–	Short stature, DBA	–	–
C	8, M	Ghanaian	–	–	–	Short stature, DBA	–	Se: no effect T3: improved growth skeletal maturation
D	12, F	Brazilian	Bilateral sensorineural loss	Peripheral sensory neuropathy, hypotonia, hip girdle weakness fatty infiltration of muscle Spirometry: reduced expiratory and inspiratory flow	Increased fat mass	Short stature, kyphoscoliosis DBA	Impaired mental development and motor coordination Bilateral clinodactyly Failure to thrive Asymmetric leg length	Se: increased plasma Se and GPx but not SEPP levels, or thyroid hormone metabolism dysfunction
E	35, M	British	Bilateral sensorineural loss, Vertigo	Lumbar spinal rigidity, reduced axial and neck strength, fatty infiltration of muscle Spirometry: reduced vital capacity, nocturnal hypoventilation	Increased fat mass Low insulin and high adiponectin levels; favourable blood lipid profile; low intrahepatic lipid	Genu valgus	Developmental delay Azoospermia Raynauds disease Photosensitivity Mild lymphopenia and reduced red cell mass	–

(Continued)

TABLE 16.4 (CONTINUED)
Main Physiological Phenotypes in Patients with SBP2 Deficiency

Family	Age (years) Gender	Ethnicity	Hearing	Muscle and Neurological	Metabolic	Growth and Skeletal	Other	Effect of Treatment
F	2, M	British	Bilateral sensorineural loss	Proximal and axial myopathy, lumbar rigidity, fatty infiltration of adductor muscle	Increased fat mass Hypoglycaemia with low fasting insulin	Short stature	Failure to thrive Eosinophilic colitis Developmental delay	T3: normalized FT3, improved growth, speech and development
G	2.6, M	Japanese	Bilateral mild conductive loss Rotatory vertigo	Fatty infiltration of muscle IQ 70	Increased fat mass	Short stature, DBA	Delayed motor and intellectual development Failure to thrive Hypoplastic thyroid gland No photosensitivity	GH alone: improved growth GH plus T3: normalized FT3, lowered FT4, improved growth and skeletal maturation
H	11, F	Turkish	–	Mental and motor retardation IQ 50	–	Short stature	–	–
I	5, F	Argentinian	–		–	DBA	Failure to thrive	–

Note: DBA, delayed bone age; F, female; IQ, intelligence quotient; Se, selenium; T3, triiodothyronine; M, male.

acids 517–544), which is involved in specific recognition of type I and type II SECIS elements (Bubenik and Driscoll, 2007). SECIS elements (Gpx4, TrxR1) with preserved binding to R540Q mutant SBP2 are type II elements (Fletcher et al., 2001; Ramos et al., 2004) whereas mutant protein interaction with type I SECIS elements (Gpx1 and Dio1) is diminished (Fletcher et al., 2001). Decreased interaction with the human DIO2 element may be an exception to this, as SECISearch predicts it is a type II variety (Driscoll et al., 2004) although this has not been functionally tested. Perhaps such subtle structural differences in elements, which do not appear to affect binding to wild-type protein, nevertheless destabilize interaction with R540Q mutant SBP2. Overall, these observations also suggest that long-lasting binding of SBP2 to a SECIS element is not a prerequisite for recoding functions with a transient interaction being sufficient to support Sec incorporation.

It has also been reported that SBP2 can distinguish between structurally similar SECIS elements (Low et al., 2000) with such differential binding determining, at least in part, the expression pattern of the selenoproteome. Compounding such subtle differences in SBP2-SECIS interaction with inherently differential recognition of structurally distinct SECIS elements by SBP2 may explain selective loss of expression of a subset of selenoproteins. With cellular SBP2 protein levels being limiting (Low et al., 2000), differential interaction of R540Q mutant SBP2 with different SECIS elements might result in different patterns of selenoproteome expression in patients. Thus, in a competitive cellular environment, weaker binding of R540Q mutant SBP2 to SECIS elements (GPx1, DIO2) in nonessential selenoproteins versus more stable interaction with elements (GPx4, TrxR1) that are essential for survival could result in an unbalanced selenoprotein expression profile in patients. Analysis of murine tissues confirms that GPx4 mRNA is highly expressed and always more abundant than Dio1, Dio2, or Gpx1 transcripts (Hoffmann et al., 2007), consistent with GPx4 and TrxR1 being essential selenoproteins with their murine knockout being embryonic lethal (Jakupoglu et al., 2005; Yant et al., 2003; Foresta et al., 2002). Similarly, due to loss of such essential housekeeping, selenoproteins (GPx4, TrxR1) being much more deleterious in humans too, mechanisms to preserve their synthesis even when fully functional SBP2 is limited might exist.

Family B (Dumitrescu et al., 2005)

Sequencing of *SECISBP2* identified a paternally inherited heterozygous, missense mutation in exon 10 (c.1312 A > T, p.K438X), resulting in a premature stop at codon 438, and a maternally transmitted nucleotide change (c.IVS8ds + 29 G > A) creates an alternative donor splice sequence TGGTATGA (consensus AGGTAAGT), generating an alternative transcript incorporating extra sequences (26 bp) into exon 8 but with a shift of reading frame leading to a premature stop. Fifty-two percent of SBP2 transcripts in patient-derived lymphocytes were aberrantly spliced with the remainder being normal. Both mutations are predicted to disrupt generation of full-length SBP2 protein but with the residual, normally spliced (48%) transcripts derived from the maternal allele being capable of directing intact SBP2 protein. With SBP2 being a limiting factor in selenoprotein synthesis, reduction of SBP2 protein levels to ~25% of normal would be predicted to enable their production in a limited amount with a mild phenotype.

Family C (Di Cosmo et al., 2009)

SECISBP2 sequencing identified a homozygous missense mutation in exon 3 (c.382 C > T, p.R128X). Although this premature stop mutation precludes production of full-length SBP2 and some N-terminal splice variants, elegant minigene experiments have shown that the use of alternative, downstream, ATG codons in exons 5 and 7 permits low-level synthesis of shorter SBP2 isoforms. Further analyses showed that three, alternatively spliced, SBP2 isoforms (an isoform containing all of exon 3; an isoform lacking 121 bp of exon 3; an isoform lacking exon 3) are present in a similar amount. Interestingly, analyses of SBP2 transcripts in heterozygous parents showed reduced (40%) levels of mutant mRNA, presumably due to NMD. On the other hand, expression of shorter SBP2 proteins was two- to three-fold higher in minigene experiments, perhaps suggesting the existence of compensatory mechanisms to counteract NMD.

Family D (Azevedo et al., 2010)

The patient is compound heterozygous for missense *SECISBP2* mutations (paternally inherited: c.358 C > T, p.R120X; maternally inherited: c.2308 C > T, p.R770X), each predicted to prematurely truncate the protein. The R770X mutation, located at the end of the RBD in the CRD, thus possibly affecting redox sensitivity and cellular localization in all predicted SBP2 splice variants. Functional studies indicate that this mutant binds SECIS elements from DIO2 and GPx4 with significantly lower affinity. The R120X mutation truncates SBP2 N-terminally and is expected to disrupt synthesis of some splice variants (e.g., mtSBP2) with translation from alternative, downstream ATG codons (e.g., M300) directing synthesis of shorter, functional SBP2 isoforms, explaining the mild phenotype.

Family E (Schoenmakers et al., 2010)

Sequencing of *SECISBP2* identified that the proband was heterozygous for a paternally inherited deletion mutation in exon 5 (c.668delT) leading to a frame shift and a premature stop (p.F223fs255X); on the other allele, a de novo deletion (IVS7-155, T > A) results in misincorporation of additional intronic sequences between exons 6 and 7, which is also predicted to terminate the SBP2 transcript prematurely. Consistent with this, full-length SBP2 protein was not detected in Western blots of patient-derived fibroblasts. Minigene experiments, using F223fs255X mutation-containing constructs, showed synthesis of shorter SBP2 proteins, likely initiated from alternative ATG codons (e.g., M300), explaining the mild phenotype.

Family F (Schoenmakers et al., 2010)

The proband was heterozygous for maternally inherited missense mutation in exon 14 (c.2071 T > C, p.C691R) in the RBD/CRD and a paternally inherited defect generating aberrantly spliced SBP2 transcripts, lacking either exons 2-3-4 or exons 3-4, leading to a frame shift/premature stop (fs65X & fs76X) with the underlying mutation causing such missplicing not identified. Usage of alternative ATG codons (e.g., M300) downstream of the frame shift premature stop is predicted to direct translation of shorter SBP2 isoforms. Interestingly, the stability of the C691R mutant SBP2 protein was also reduced, being susceptible to enhanced proteasomal degradation, possibly because the cysteine to arginine change destabilizes the hydrophobic core of the RBD and possibly also alters redox sensitivity of SBP2.

Family G (Hamajima et al., 2012)

The proband was heterozygous for a maternally inherited premature stop mutation in exon 3 (c.235 C > T, p.Q79X), and a 13 bp duplication in exon 11 (1529_1541dupCCAGCGCCCCACT) was identified on the other allele, resulting in the insertion of four amino acids (QRPT) and a frame shift/ premature stop after a further 48 residues (M515fsX563). Both mutations are predicted to truncate full-length SBP2 protein prematurely with the M515fsX563 mutation affecting all splice variants effectively resulting in a null allele; however, shorter SBP2 isoforms are expected to be generated using alternative methionine codons (e.g., M233, M300), downstream of the Q79X mutation, ameliorating the phenotype in this patient.

Family H (Dumitrescu et al., 2013)

The proband is compound heterozygous with a premature stop mutation in exon 16 (c.2344 C > T, p.Q782X) on one allele; the other allele contains a deletion in exon 14 (c.2045-2048 delAACA, p.K682fs683X), which leads to a frame shift/premature stop mutation within the RBD. Similar to the R770X SBP2 mutant (Family D), the Q782X mutation might be expected to impair SBP2 binding to the SECIS element or affect its redox sensitivity, being located within the CRD, although it has not been functionally characterized. Presumably, residual function of Q782X mutant SBP2 accounts for a mild phenotype.

Family I (Fu et al., 2014)

The proband has biallelic *SECISBP2* mutations: a maternally inherited premature stop mutation in exon 5 (c.660 C > T, p.R197X) and a paternally inherited missense mutation in exon 14 (c.2037 A > G, p.E679D). The R197X mutation truncates SBP2 N-terminally and affects several natural splice variants with translation initiation from the alternative start codons (e.g., M300), predicted to generate shorter but functional SBP2 protein. The E679D mutation, located within the L7Ae-type RNA-binding domain (620–745) is predicted to be deleterious (PolyPhen-2 algorithm score of 0.998), possibly affecting mRNA-binding activity of SBP2. It is conceivable that some residual SBP2 function derived from shorter SBP2 isoforms (R197X) and of the E679D mutant could account for a mild patient phenotype.

16.4.2 BIOCHEMISTRY

Abnormal thyroid function tests are a universal finding in patients with *SECISBP2* mutations described hitherto with typical abnormalities being raised circulating thyroxine (T4), low or normal triiodothyronine (T3), elevated reverse T3 (rT3), and normal/high plasma thyroid-stimulating hormone (TSH) levels (Table 16.3). This pattern likely reflects abnormal metabolism of thyroid hormones due to deficiencies of three Sec-containing, deiodinase enzymes (Bianco et al., 2002). Such abnormal thyroid function, together with low plasma selenium levels (reflecting deficiencies of SEPP1 and GPx3), provides a biochemical signature that facilitates identification of putative *SECISBP2* defect cases.

Dio1 null mice exhibit raised T4 and rT3 but normal T3 and TSH levels (Schneider et al., 2006). Murine Dio2 deficiency also results in high T4 but normal T3 and TSH levels (Schneider et al., 2001), together with growth retardation and defective auditory function (Ng et al., 2004). Dio3 deficiency is associated with partial embryonic and neonatal lethality with surviving mice showing severe growth retardation, impaired fertility, and central hypothyroidism (Hernandez et al., 2006). The thyroid phenotype of SBP2-deficient patients is most closely recapitulated by a murine *Dio1* and *Dio2* double knockout (Galton et al., 2009), suggesting that it is mediated by combined deiodinase deficiency. However, unlike human SBP2 deficiency, even the double *Dio1/Dio2* null mice have normal T3 levels, suggesting that diminished DIO3 activity probably also contributes to the human thyroid phenotype. The nature of the SBP2 defect and the environmental factors (e.g., iodine status) might contribute to interindividual differences in thyroid function tests in patients from different ethnic and geographical backgrounds.

16.4.3 GROWTH AND SKELETAL PHENOTYPE

Failure to thrive in infancy is seen often, and short stature is the commonest presenting symptom in most SBP2-deficient patients; associated abnormalities include delayed bone age, asymmetric leg length, and clinodactyly. The pathogenesis of such growth retardation and delayed bone development is probably multifactorial with no clear link to a single selenoprotein defect. Growth retardation is a recognized feature in *Dio2* and *Dio3* null mice (Hernandez et al., 2006; Ng et al., 2004), suggesting that the human phenotype is also mediated by abnormal thyroid hormone metabolism. Consistent with this notion, treatment of SBP2-deficient children with T3 alone (Families C and F) or in conjunction with growth hormone (Family G) resulted in catch-up growth. Growth retardation was also noted in neuron-specific *Secisbp2* null mice although its mechanistic basis is unknown (Seeher et al., 2014b). Selective depletion of multiple selenoproteins in murine osteochondroprogenitor cells results in epiphyseal abnormalities and chondronecrosis (Downey et al., 2009), suggesting an important, cell autonomous role for selenoproteins in skeletal development. However, radiological skeletal survey and bone mineral density were found to be normal in a single SBP2 case (Family E).

16.4.4 Hearing Loss

Three patients (Families D, E, F) exhibit sensorineural hearing loss with the adult male being most severely affected and requiring hearing aids (Family E), perhaps reflecting progressive, age-related dysfuntion. One child (Family G) has conductive hearing deficit associated with recurrent otitis media. Although hearing loss and retarded cochlear development are recognized features in *Dio2* null mice (Ng et al., 2004), cochlear anatomy is normal in human SBP2 cases, and DIO2 activity is only partially reduced in the disorder. An alternative hypothesis is that hearing loss reflects damage mediated by elevated cellular ROS. In that context, Gpx1 null mice are susceptible to noise-induced hearing loss (McFadden et al., 2001). ROS-mediated cochlear damage can be cumulative (Riva et al., 2006), which could explain both the progressive nature of hearing loss and the observation that auditory deficit is more severe in the older, adult, SBP2-deficient case.

16.4.5 Developmental and Neurological Phenotype

Many affected children exhibit developmental delay, which may be global, speech-related, or intellectual. The adult patient had delayed motor milestones but developed normally otherwise. IQ is variably affected, ranging from normal to severely reduced. Motor delay in these patients may be a consequence of muscular dystrophy (see below). However, intellectual impairment, seen in some cases, cannot be ascribed to deficiency of specific selenoproteins. Reduced CNS exposure to TH due to abnormal cellular thyroid hormone metabolism is one possibility, and *Dio1/Dio2* double null mice exhibit impaired learning and memory. However, several other selenoproteins have been implicated in normal brain development in mice (Table 16.1).

16.4.6 Axial Muscular Dystrophy

Fatigue and weakness, affecting proximal (leg and hip girdle), neck, and axial muscles, is a recognized feature in several patients (Families D, E, F). In childhood cases, magnetic resonance imaging typically shows selective fatty infiltration of thigh muscles (e.g., adductors, sartorius, biceps femoris; sparing of gracilis, rectus femoris) even prior to onset of clinical symptoms; more marked abnormalities in the adult patient included loss of cervical lordosis and paraspinal muscle fatty infiltration. Circulating skeletal muscle-specific creatine kinase (CK-MM) levels can be elevated. Reduced lung function (e.g., expiratory and inspiratory flow, total vital capacity) reflecting respiratory muscle weakness, progressing to nocturnal respiratory insufficiency requiring positive pressure ventilatory support, may be observed.

This musculoskeletal phenotype in SBP2-deficient patients is highly analogous to that seen in a spectrum of myopathic disorders of varying clinical severity and age of onset, including rigid spine muscular dystrophy, desmin-related myopathy with mallory body-like inclusions, and congenital fiber type disproportion, collectively termed the SEPN1-myopathies, due to defects in SELN (Ferreiro et al., 2002, 2004; Moghadaszadeh et al., 2001). Consistent with this, histological features (type 1 oxidative fiber predominance, areas of sarcomere disorganization termed minicores) in muscle from the SBP2-deficient adult, resemble those seen in SEPN1-myopathies (Schoenmakers et al., 2010). Although ubiquitously expressed, SELN may be particularly important in skeletal muscle, protecting cells from oxidative/nitrosative stress; absence of SELN is associated with abnormal nitrosylation of the ryanodine receptor and excessive leakage of calcium from sarcoplasmic reticulum to cytosol, limiting its availability for contractile processes (Arbogast and Ferreiro, 2010).

Motor incoordination in some SBP2-deficient patients could be linked to SELN myopathy and axial muscle weakness. However, neuron-specific *Secisbp2* null mice (Seeher et al., 2014b) did exhibit an obvious locomotor phenotype, together with degenerative loss of PV[+]/Gad67[+] interneurons in basal ganglia, providing a possible alternative explanation for the movement disorder and raising the possibility that this phenotype is progressive.

16.4.7 METABOLIC PHENOTYPE

The adult SBP2-deficient patient (Family E) has a high total body fat mass but paradoxically associated with favorable metabolic parameters, including low fasting insulin and raised adiponectin levels, a favorable lipid profile, and negligible intrahepatic lipid levels. Elevated fat mass index and high circulating adiponectin were also recorded in a childhood case (Family F), who also experiences recurrent, symptomatic, hypoglycemia in the presence of low circulating insulin levels, necessitating nocturnal feeds by gastrostomy tube (Schoenmakers et al., 2010). Imaging has also documented increased fat mass in other SBP2-deficient cases (Families D and F) with remaining cases not having been investigated for this phenotype. Given the established link between obesity/fat mass, insulin resistance, and metabolic syndrome, the combination of increased fat mass and enhanced insulin sensitivity in SBP2-deficient patients was surprising.

Transgenic mouse models that affect expression of selenoproteins globally show conflicting results. In a mouse model expressing a dominant negative mutant form of tRNA[Ser]Sec, selenoprotein deficiency resulted in enhanced exercise-induced muscle growth, together with increased Akt phosphorylation (Hornberger et al., 2003); in contrast, another mutant tRNA[Ser]Sec model demonstrated impaired glucose tolerance and reduced insulin sensitivity (Labunskyy et al., 2011). Three other selenoprotein knockout mouse models raise the possibility of a link between ROS and insulin sensitivity. *Sepp1* null mice are protected from high fat diet–induced adipocyte hypertrophy, glucose intolerance, and insulin resistance (Misu et al., 2010). Because Sepp1 deficiency can affect delivery of Se to tissues and the synthesis of multiple selenoproteins in different organs, the model does not necessarily indicate a specific role of Sepp1. Patients with type 2 diabetes exhibit elevated plasma SEPP1 levels, which correlate with their fasting glucose and glycated hemoglobin (Kaur et al., 2012; Misu et al., 2010; Yang et al., 2011). However, identification of FoxO1a and HNF-4-alpha binding sites in the *SEPP1* promoter indicate that its expression can be regulated by glucose concentration and thus indirectly by insulin. Accordingly, it is conceivable that low circulating SEPP1 levels simply reflect systemic insulin resistance (Steinbrenner, 2013) although its direct involvement in insulin signaling cannot be excluded. *Gpx1* null mice exhibit raised ROS levels, greater insulin-dependent Akt phosphorylation, and muscle glucose uptake and are protected from high fat diet–induced insulin resistance (Loh et al., 2009; Merry et al., 2014; Wang et al., 2011). In contrast, mice overexpressing Gpx1 develop insulin resistance (Wang et al., 2008). Elevated ROS levels enhance Erk phosphorylation in a murine Sirt3 deficiency model (Sundaresan et al., 2009), and a similar mechanism could account for increased insulin-stimulated phosphoErk levels in cells from an SBP2-deficient patient (Family E). Known cross-talk between Erk and Akt-mediated insulin signaling pathways (von Kriegsheim et al., 2009) could account for increased insulin sensitivity.

Overall, a substantial body of evidence suggests a link between selenoproteins and systemic insulin sensitivity. Experimental models suggest that elevated ROS levels could operate at multiple levels, affecting not only insulin signaling and action but perhaps also pancreatic beta-cell function and insulin secretion. However, different models have sometimes yielded conflicting results, indicating complexity and the need for further investigation in this area (Steinbrenner, 2013; Szypowska and Burgering 2011; Watson, 2014).

16.4.8 AZOOSPERMIA WITH SPERMATOGENIC MATURATION ARREST

The adult SBP2-deficient patient (Family F) is infertile with azoospermia and spermatogenic maturation arrest on testicular histology. Whether similarly compromised fertility will manifest in the other childhood cases remains unknown. The testis is highly selenium-enriched with deficiency of this trace element in rodents being associated with infertility (Flohé, 2007). SEPP1, acting as a circulating supply of selenium to meet tissue demand, has also been implicated in spermatogenesis with *SePP1*-null mice showing abnormal sperm morphology and diminished testicular selenium content (Kehr et al., 2009; Olson et al., 2005). Immunocytochemistry of the patient's testis revealed deficiency of three selenoproteins (mGPx4, TrxR3, SELV) with recognized roles in spermatogenesis. Mitochondrial GPx4 (mGPx4), which forms cross-links with other proteins, is a major structural component of the mitochondrial capsule in the midpiece of spermatozoa (Ursini et al., 1999). Murine mGpx4 inactivation causes male infertility (Schneider et al., 2009), and reduced human seminal mGPx4 activity correlates with oligospermia (Foresta et al., 2002). TrxR3, also highly enriched in spermatids, may catalyze protein disulfide bond isomerization in sperm development (Su et al., 2005). Human SELV expression is known to be testis restricted, but its function is unknown (Kryukov et al., 2003). Consistent with the known importance of mGPx4 and TrxR3 in latter spermatogenic stages, testicular histology in the adult SBP2-deficient patient showed maturation arrest with preservation of early cell types (e.g., spermatogonia and spermatocytes) but lack of mature spermatids and spermatozoa.

16.4.9 CUTANEOUS PHOTOSENSITIVITY

Abnormal cutaneous photosensitivity was first documented in the adult SBP2-deficient patient (Family E) at age 13 years, following a history of skin reddening and blistering upon sun exposure. Such symptoms persist into adulthood, necessitating constant use of high-strength sunblock. Skin biopsy showed deficiency of several selenoproteins (GPx1, TrxR1, MsrB1) with increased ROS levels in dermal fibroblasts confirming reduced cellular antioxidant capacity, leading to increased DNA damage. Markedly enhanced oxidative DNA damage and lipid peroxidation following UV exposure of his skin fibroblasts, provides a potential pathogenic basis for his photosensitivity and correlates with UV skin testing, which confirms that this patient remains very abnormally photosensitive. Abnormal photosensitivity has not been recorded in other SBP2-deficient cases, perhaps because of naturally pigmentation in non-Caucasian individuals or absence of formal testing. Alternatively, it is well recognized that cellular accumulation of ROS can itself damage enzymes (e.g., catalase) that mediate its removal (Wood et al., 2006) or pathways (e.g., MsrB1) that repair oxidized proteins (Schallreuter et al., 2008). As most other patients are still relatively young, it is conceivable that such cumulative ROS-mediated damage to antioxidant defense pathways could expose future photosensitivity. Finally, a keratinocyte-specific *Trsp* null mouse exhibits a severe cutaneous phenotype with hyperplastic epidermis aberrant hair follicle morphogenesis and progressive alopecia, highlighting the importance of selenoproteins for healthy skin (Sengupta et al., 2010).

16.4.10 IMPAIRED T-CELL PROLIFERATION AND ABNORMAL CYTOKINE PRODUCTION

In the adult SBP2-deficient patient (Family E), mildly reduced red blood cell and total lymphocyte counts were noted with normal levels of other cell lineages (e.g., platelets, neutrophils, monocytes), lymphocyte subsets and immunoglobulin levels. His T-cells showed selenoprotein deficiency and impaired proliferation following polyclonal stimulation (CD3/CD28). Peripheral blood mononuclear cells (PBMCs) showed significant telomere shortening and, when stimulated, showed markedly enhanced secretion of IL-6 and TNF-α with diminished IFN-γ production. Hematological and immune cell phenotypes of other SBP2-deficient cases have not been studied.

The defective T-cell proliferation we observed in the adult SBP2-deficient patient mirrors similar findings in a T-cell–specific *Trsp* null mouse in which excess cellular ROS generated by T-cell activation inhibits cellular proliferation (Shrimali et al., 2008). Markedly shortened telomere length seen in PBMCs from two SBP2-deficient cases (Families E and F) correlates with similar telomere length reduction seen in SBP2 knockdown human cell lines (Squires et al., 2009). The degree of telomere shortening in their PBMCs is comparable to that seen in subjects with TERT mutations, a disorder associated with aplastic anemia (Yamaguchi et al., 2005); mice engineered to have short telomeres exhibit hematologic abnormalities, including low blood counts and impaired lymphocyte proliferation (Armanios et al., 2009). Overall, these observations suggest that telomere shortening may contribute to reduced red blood cell number and lymphopenia in this disorder. Although overt immunodeficiency is not a feature in these SBP2-deficient subjects, telomere erosion is likely to limit self-renewal of highly proliferative cell compartments (e.g., bone marrow); accordingly, it is conceivable that added insults (e.g., viral infection, drug toxicity) could compromise hemopoiesis or immune function in these cases. Intriguingly, shortened telomere length in T-cells has also been specifically associated with connective tissue disorders, including lupus, rheumatoid arthritis, and scleroderma (Fujii et al., 2009; Wallace, 2010); in this context, it was noted that patient E has marked Raynaud's disease albeit without other clinical or serological features of a connective tissue disorder.

LPS-induced TNF-α secretion from macrophages cultured in selenium-depleted medium (Vunta et al., 2008) or with siRNA-mediated knockdown of SELS is markedly enhanced and a SNP in *SELS* correlates with circulating TNF-α/IL-6 levels (Curran et al., 2005). Accordingly, reduced SELS levels in PBMCs might account for the excess cellular TNF-α and IL-6 production; whether particular selenoprotein deficiencies mediate defective IFN-γ production by immune cells in SBP2-deficient patients remains to be elucidated. Finally, *SelK* null mice exhibit reduced cellular calcium flux following immune cell activation and impaired immune responses (Verma et al., 2011); Trsp^{tG37} transgenic mice show altered immunological responses after exposure to influenza virus (Sheridan et al., 2007). Accordingly, it is conceivable that other selenoprotein deficiencies contribute to additional immune phenotypes in SBP2 deficiency, which remain to be elucidated.

Patient F exhibited colitis of unknown etiology with no known gastrointestinal symptoms in other cases. Ileocolonic inflammation is a feature in *Gpx1/Gpx2* double knockout mice (Chu et al., 2004) and Trsp^{tG37} transgenic animals show aberrant crypt formation in the colon (Irons et al., 2006), suggesting involvement of selenoproteins in normal colonic function (see Chapters 4 and 9).

16.4.11 EFFECT OF TREATMENT

Although selenium supplementation in patients, with either sodium selenite or a selenomethionine-rich yeast, elevated serum Se-levels, there was no clinical (Azevedo et al., 2010; Di Cosmo et al., 2009) or biochemical (change in circulating GPx3, SEPP1, and TH level) (Schomburg et al., 2009) effect, probably because the disorder involves defective incorporation of selenium into the selenoproteins rather than deficiency of this trace element per se. On the other hand, treatment with T3 (T4 was not effective in one case) is more beneficial with an improvement in growth, development, and bone maturation seen in some patients (Di Cosmo et al., 2009; Schoenmakers et al., 2010). In the future, trials of antioxidants to counteract effects of elevated cellular ROS may be a rational therapeutic approach.

16.5 CONCLUSIONS

Compared to the lethality of murine *Secisbp2* inactivation, the phenotype of patients with *SECISBP2* defects is modest (Tables 16.3 and 16.4). This suggests that the human mutations either selectively abrogate specific functions of SBP2 or represent hypomorphic alleles, in which the altered gene product retains residual expression or functional activity.

With the exception of three missense abnormalities, all *SECISBP2* defects are predicted to prematurely truncate SBP2 protein (Table 16.2). Functional studies suggest that the missense mutants (R540Q, C691R, and E679D) retain some residual function. With premature stop mutations, initiation of translations from alternative start codons (e.g., Met300) can generate shorter but fully functional SBP2 products (Copeland et al., 2001; de Jesus et al., 2006). Premature stop mutations downstream of Met300 may eliminate synthesis of functional protein completely; however, mutations distal to the RBD but in the CRD (e.g., R770X; Azevedo et al., 2010; or Q782X; Dumitrescu et al., 2013) might generate truncated proteins retaining some function but with possibly affected RNA binding and nuclear localization. An intronic mutation IVS8ds + 29G > A (Dumitrescu et al., 2005) affecting splicing only reduces wild-type transcript levels by 50% and a similar mechanism might preserve mRNA levels with some other splice site defects.

Although a limited number of patients have been described, the *SECISBP2* mutations identified hitherto are evenly distributed with no particular "hot spot" for mutations in the gene (Figure 16.1). It is also important to note that SBP2 is rate limiting for Sec incorporation, explaining why a reduction in functional SBP2 protein levels will compromise selenoprotein synthesis. Although it has been suggested that the degree of functional SBP2 deficiency might be linked to the severity of clinical phenotype, there are however too few cases with differing environmental and other variables to make clear-cut correlations.

Several mechanisms have been put forward to explain residual selenoprotein synthesis in patients with a defective *SEBISP2* (Seeher et al., 2014b). As described above, one possibility includes retention of SBP2 synthesis from hypomorphic alleles or production of short isoforms retaining Sec incorporation function. Compensation for the defect by another SECIS-binding protein is a theoretical possibility, but no such protein has been described. Another mechanism could be incorporation of another amino acid in the position of Sec, preventing the UGA codon being read as a nonsense mutation. Such nonsense suppression is known to occur during synthesis of some selenoproteins. For example, the aminoglycoside G418 can stimulate the incorporation of Arginine into GPx1 (Handy et al., 2006), dietary selenium deficiency results in incorporation of cysteine at the UGA codon of TrxR1, generating a variant with ~10% normal activity (Lu et al., 2009; Xu et al., 2007). In Families E and F, 20% of TRXR activity was preserved (Schoenmakers et al., 2010) and 10%–25% of GPx activity is preserved in all SBP2 cases. Lastly, existence of a low-efficiency, SBP2-independent, Sec incorporation pathway analogous to pyrrolysine insertion (Srinivasan et al., 2002) has been postulated. Carefully designed studies will be required to uncover such a mechanism.

Selenoproteins mRNA contain in-frame UGA codons, such that failure to decode these appropriately as Sec could lead to the transcripts being degraded by NMD. In this context, it is interesting to note that selenoprotein mRNA levels in SBP2-deficient patients do not always follow the canonical rules for NMD (Schoenmakers et al., 2010). Tissues from liver and neuron-specific Secisbp2 null mice (Seeher et al., 2014a,b) exhibit similar selenoprotein mRNA expression profiles. These finding suggests that SBP2 deficiency might affect selenoprotein mRNA stability via mechanisms beyond susceptibility of transcripts to NMD as has been suggested previously (de Jesus et al., 2006; Papp et al., 2006). Such effects, together with differing affinity of SBP2 mutants (e.g., R540Q) for SECIS elements could contribute to varying hierarchies of selenoprotein expression in *SECISBP2* defect cases.

Increased oxidative stress and DNA damage secondary to deficiency of antioxidant selenoenzymes could heighten cancer risk in SBP2-deficient patients. However, neoplasia has not been recorded, even in the adult patient, but the possibility that it might yet develop cannot be discounted. Alternatively, it is conceivable that additional mechanisms might mitigate such risk; specifically while SBP2-deficient cells exhibit increased oxidative DNA damage and telomere shortening, cellular senescence is also enhanced with upregulation of a marker (p16INK4a), providing a potential basis for suppression of tumorigenesis (Ohtani et al., 2004).

Overall, because *SECISBP2* defects affect expression of most members of the selenoproteome, it is not surprising that the phenotype of this disorder is complex (Tables 16.3–16.5). Several features (abnormal thyroid function, azoospermia, myopathy) are attributable to deficiencies of particular selenoproteins that are expressed in specific tissues. Deficiency of antioxidant selenoenzymes mediates raised cellular ROS, which may manifest as photosensitivity, age-dependent hearing loss, and contribute to enhanced insulin sensitivity. The possibility of additional phenotypes secondary to oxidative damage (e.g., neoplasia, neurodegeneration, and premature aging) emerging with time cannot be discounted. Other clinical features (e.g., growth retardation, delayed bone maturation, delayed mental development, and motor milestones) may have a complex, multifactorial basis, due to abnormalities in multiple selenoprotein pathways. Future identification of further cases may also reveal additional phenotypes, linked to deficiencies of selenoproteins of unknown function. In addition to careful evaluation of SBP2-deficient patients, transgenic mouse models are of great utility with cellular and biochemical characterization of tissues not accessible in human patients providing mechanistic insights.

TABLE 16.5
Links between Selenoprotein Deficiencies and Clinical Phenotypes

Proteins	Role	Phenotype
DIO 1, 2, 3	Thyroid hormone metabolism	Raised FT4, normal/low FT3, normal TSH, raised reverse T3
SEPP1, GPX3	Se transport	Low plasma selenium
SEPN1	Skeletal muscle contraction	Muscular dystrophy
GPX's, TXNRDs, MSRB	Antioxidant enzymes	Photosensitivity, hearing loss, enhanced insulin sensitivity
SELV, GPX4, TXNRD3	Spermatogenesis	Azoospermia
?TXNRDs and GPXs	Blood cells	Abnormal T cell proliferation, borderline lymphopenia
		Reduced red cell mass
Unknown	Unknown	Growth retardation, delayed bone maturation
		Delayed mental and motor milestones

REFERENCES

Alanne, M., K. Kristiansson, K. Auro et al. 2007. "Variation in the selenoprotein S gene locus is associated with coronary heart disease and ischemic stroke in two independent Finnish cohorts." *Hum Genet* no. 122:355–65.

Allmang, C., P. Carbon, and A. Krol. 2002. "The SBP2 and 15.5kD/Snu13p proteins share the same RNA binding domain: Identification of SBP2 amino acids important to SECIS RNA binding." *RNA* no. 8:1308–18.

Allmang, V., P. Richard, A. Lescure et al. 2006. "A single homozygous point mutation in a 3' untranslated region motif of selN mRNA causes SEPN1-related myopathy." *EMBO Rep* no. 7:450–4.

Arbogast, S., and A. Ferreiro 2010. "Selenoproteins and protection against oxidative stress: Selenoprotein N as a novel player at the crossroads of redox signaling and calcium homeostasis." *Antioxid Redox Signal* no. 12:893–904.

Armanios, M., J. Alder, E. Parry et al. 2009. "Short telomeres are sufficient to cause the degenerative defects associated with aging." *Am J Hum Genet* no. 85:823–32.

Arner, E. 2009. "Focus on mammalian thioredoxin reductases—Important selenoproteins with versatile functions." *Biochim Biophys Acta* no. 790:495–526.

Azevedo, M., G. Barra, L. Naves et al. 2010. "Selenoprotein-related disease in a young girl caused by nonsense mutations in the SBP2 gene." *J Clin Endocrinol Metab* no. 95:4066–71.

Banning, A., A. Kipp, and R. Brigelius-Flohé. 2012. "Glutathione peroxidase 2 and its role in cancer." In: *Selenium: Its Molecular Biology and Role in Human Health* (Hatfield, D. L., Berry, M. J., and Gladyshev, V. N., eds.). New York: Springer. pp. 271–82.

Berry, M., L. Banu, Y. Chen et al. 1991. "Recognition of UGA as a selenocysteine codon in type I deiodinase requires sequences in the 3'-untranslated region." *Nature* no. 353:273–6.

Berry, M., L. Banu, J. Harney, and P. Larsen. 1993. "Functional characterization of the eukaryotic SECIS elements which direct selenocysteine insertion at UGA codons." *EMBO J* no. 12:3315–22.

Bianco, A., D. Salvatore, B. Gereben, M. Berry, and P. Larsen. 2002. "Biochemistry, cellular and molecular biology, and physiological roles of the iodothyronine selenodeiodinases." *Endocr Rev* no. 23:38–89.

Bondareva, A. A., M. R. Capecchi, and S. V. Iverson. 2007. "Effects of thioredoxin reductase-1 deletion on embryogenesis and transcriptome." *Free Radic Biol Med* no. 43:911–23.

Brigelius-Flohé, R., and M. Maiorino. 2013. "Glutathione peroxidases." *Biochim Biophys Acta* no. 1830:3289–303.

Bubenik, J., and D. Driscoll. 2007. "Altered RNA binding activity underlies abnormal thyroid hormone metabolism linked to a mutation in selenocysteine insertion sequence-binding protein 2." *J Biol Chem* no. 282:34653–62.

Burk, R., K. Hill, A. Motley, L. Austin, and B. Norsworthy. 2006. "Deletion of selP upregulates urinary selenium excretion and depresses whole-body selenium content." *Biochim Biophys Acta* no. 1760:1789–93.

Caban, K., and P. R. Copeland. 2006. "Size matters: A view of selenocysteine incorporation from the ribosome." *Cell Mol Life Sci* no. 63:73–81.

Caban, K., S. Kinzy, and P. R. Copeland. 2007. "The L7Ae RNA binding motif is a multifunctional domain required for the ribosome-dependent Sec incorporation activity of Sec insertion sequence binding protein 2." *Mol Cell Biol* no. 27:6350–60.

Carlson, B. A., S. V. Novoselov, E. Kumaraswamy et al. 2004. "Specific excision of the selenocysteine tRNASerSec (Trsp) gene in mouse liver demonstrates an essential role of selenoproteins in liver function." *J Biol Chem* no. 279:8011–7.

Carlson, B. A., M. Yoo, R. Tobe et al. 2012. "Thioredoxin reductase 1 protects against chemically induced hepatocarcinogenesis via control of cellular redox homeostasis." *Carcinogenesis* no. 33:1806–13.

Castets, P., A. T. Bertrand, M. Beuvin et al. 2011. "Satellite cell loss and impaired muscle regeneration in selenoprotein N deficiency." *Hum Mol Genet* no. 20:694–704.

Chapple, C., R. Guigo, and A. Krol. 2009. "SECISaln, a web-based tool for the creation of structure-based alignments of eukaryotic SECIS elements." *Bioinformatics* no. 25:674–5.

Cheng, W., Y. Ho, B. Valentine et al. 1998. "Cellular glutathione peroxidase is the mediator of body selenium to protect against paraquat lethality in transgenic mice." *J Nutr* no. 128:1070–6.

Chu, F. F., R. S. Esworthy, P. G. Chu et al. 2004. "Bacteria-induced intestinal cancer in mice with disrupted Gpx1 and Gpx2 genes." *Cancer Res* no. 64:962–8.

Conrad, M., S. G. Moreno, F. Sinowatz et al. 2005. "The nuclear form of phospholipid hydroperoxide glutathione peroxidase is a protein thiol peroxidase contributing to sperm chromatin stability." *Mol Cell Biol* no. 25:7637–44.

Copeland, P., and D. Driscoll 1999. "Purification, redox sensitivity, and RNA binding properties of SECIS-binding protein 2, a protein involved in selenoprotein biosynthesis." *Biol Chem* no. 274:25447–54.

Copeland, P., J. Fletcher, B. Carlson, D. L. Hatfield, and D. Driscoll. 2000. "A novel RNA binding protein, SBP2, is required for the translation of mammalian selenoprotein mRNAs." *EMBO J* no. 19:306–14.

Copeland, P., V. Stepanik, and D. Driscoll. 2001. "Insight into mammalian selenocysteine insertion: Domain structure and ribosome binding properties of Sec insertion sequence binding protein 2." *Mol Cell Biol* no. 21:1491–8.

Curran, J. E., J. B. Jowett, K. S. Elliott et al. 2005. "Genetic variation in selenoprotein S influences inflammatory response." *Nat Gene* no. 37 (11):1234–41.

de Jesus, L. A., S. D. Carvalho, M. O. Ribeiro et al. 2001. "The type 2 iodothyronine deiodinase is essential for adaptive thermogenesis in brown adipose tissue." *J Clin Invest* no. 108:1379–85.

de Jesus, L., P. Hoffmann, T. Michaud et al. 2006. "Nuclear assembly of UGA decoding complexes on selenoprotein mRNAs: A mechanism for eluding nonsense-mediated decay?" *Mol Cell Biol* no. 26:1795–805.

Dentice, M., A. Marsili, R. Ambrosio et al. 2010. "The FoxO3/type 2 deiodinase pathway is required for normal mouse myogenesis and muscle regeneration." *J Clin Invest* no. 120:4021–30.

Di Cosmo, C., N. McLellan, X. Liao et al. 2009. "Clinical and molecular characterization of a novel selenocysteine insertion SBP2 gene mutation (R128X)." *J Clin Endocrinol Metab* no. 94:4003–9.

Donovan, J., K. Caban, R. Ranaweera, J. Gonzalez-Flores, and P. Copeland. 2008. "A novel protein domain induces high affinity selenocysteine insertion sequence binding and elongation factor recruitment." *J Biol Chem* no. 283:35129–39.

Downey, C., C. Horton, B. Carlson et al. 2009. "Osteo-chondroprogenitor-specific deletion of the selenocysteine tRNA gene, Trsp, leads to chondronecrosis and abnormal skeletal development: A putative model for Kashin-Beck disease." *PLoS Genet* no. 5:e1000616.

Driscoll D., and L. Chavatte. 2004. "Finding needles in a haystack. In silico identification of eukaryotic selenoprotein genes." *EMBO Rep* no. 5:140–1.

Dumitrescu, A. M., X. H. Liao, M. S. Abdullah et al. 2005. "Mutations in SECISBP2 result in abnormal thyroid hormone metabolism." *Nat Genet* no. 37:1247–52.

Dumitrescu, A. M., and S. Refetoff. 2013. "The syndromes of reduced sensitivity to thyroid hormone." *Biochim Biophys Acta* no. 1830:3987–4003.

Esworthy, R., R. Aranda, M. Martin et al. 2001. "Mice with combined disruption of Gpx1 and Gpx2 genes have colitis." *Am J Physiol Gastro Liver Physiol* no. 281:G848–55.

Esworthy, R., L. Yang, P. Frankel, and F. F. Chu. 2005. "Epithelium specific glutathione peroxidase, Gpx2, is involved in the prevention of intestinal inflammation in selenium-deficient mice." *J Nutr* no. 135:740–5.

Fagegaltier, D., A. Lescure, R. Walczak et al. 2000. "Structural analysis of new local features in SECIS RNA hairpins." *Nucleic Acids Res* no. 28:2679–89.

Ferreiro, A., S. Quijano-Roy, C. Pichereau et al. 2002. "Mutations of the selenoprotein N gene, which is implicated in rigid spine muscular dystrophy, cause the classical phenotype of multiminicore disease: Reassessing the nosology of early-onset myopathies." *Am J Hum Genet* no. 71:739–49.

Ferreiro, A., C. Ceuterick-de Groote, J. J. Marks et al. 2004. "Desmin-related myopathy with Mallory body-like inclusions is caused by mutations of the selenoprotein N gene." *Ann Neurol* no. 55:676–86.

Fletcher, J., P. Copeland, D. Driscoll, and A. Krol. 2001. "The selenocysteine incorporation machinery: Interactions between the SECIS RNA and the SECIS-binding protein SBP2." *RNA* no. 7:1442–53.

Flohé, L. 2007. "Selenium in mammalian spermiogenesis." *Biol Chem* no. 388:987–95.

Fomenko, D., S. Novoselov, and S. Natarajan. 2009. "MsrB1 knock-out mice: Roles of MsrB1 in redox regulation and identification of a novel selenoprotein form." *J Biol Chem* no. 284:5986–93.

Foresta, C., L. Flohé, A. Garolla et al. 2002. "Male fertility is linked to the selenoprotein phospholipid hydroperoxide glutathione peroxidase" *Biol Reprod* no. 67:967–971.

Fu, J., X. Liao, M. Menucci et al. 2014. "Thyroid hormone metabolism defect caused by novel compound heterozygous mutations in the SBP2 Gene." *16th Int congress of endocrinol.* Abstract 0508.

Fu, Y., W. Cheng, J. Porres, D. Ross, and X. Lei. 1999. "Knockout of cellular glutathione peroxidase gene renders mice susceptible to diquat-induced oxidative stress." *Free Radical Biol Med* no. 27: 605–11.

Fujii, H., L. Shao, I. Colmegna, J. J. Goronzy, and C. M. Weyand. 2009. "Telomerase insufficiency in rheumatoid arthritis." *PNAS U S A* no. 106:4360–5.

Galton, V., M. Schneider, A. Clark, and D. St Germain. 2009. "Life without thyroxine to 3,5,3'-triiodothyronine conversion: Studies in mice devoid of the 5'-deiodinases." *Endocrinol* no. 150:2957–63.

Garry, M., T. Kavanagh, E. Faustman et al. 2008. "Sensitivity of mouse lung fibroblasts heterozygous for GPx4 to oxidative stress." *Free Radic Biol Med* no. 44:1075–87.

Geisberger, R., C. Kiermayer, C. Homig et al. 2007. "B- and T-cell-specific inactivation of thioredoxin reductase 2 does not impair lymphocyte development and maintenance." *Biol Chem* no. 388:1083–90.

Gereben, B., A. M. Zavacki, S. Ribich et al. 2008. "Cellular and molecular basis of deiodinase-regulated thyroid hormone signaling." *Endocr Rev* no. 29 (7):898–938.

Grundner-Culemann, E., G. Martin, J. Harney, and M. Berry. 1999. "Two distinct SECIS structures capable of directing SEC incorporation in eukaryotes." *RNA* no. 5:625–35.

Guimaraes, M. J., D. Peterson, A. Vicari et al. 1996. "Identification of a novel selD homolog from eukaryotes, bacteria, and archaea: Is there an autoregulatory mechanism in selenocysteine metabolism?" *PNAS USA* no. 93:15086–91.

Hamajima, T., Y. Mushimoto, H. Kobayashi, Y. Saito, and K. Onigata. 2012. "Novel compound heterozygous mutations in the SBP2 gene: Characteristic clinical manifestations and the implications of GH and triiodothyronine in longitudinal bone growth and maturation." *Eur J Endocrinol* no. 166:757–64.

Handy, D., G. Hang, J. Scolaro et al. 2006. "Aminoglycosides decrease Gpx1 activity by interfering with selenocysteine incorporation." *J Biol Chem* no. 281:3382–8.

Handy, D., E. Lubos, Y. Yang et al. 2009. "Gpx1 regulates mitochondrial function to modulate redox-dependent cellular responses." *J Biol Chem* no. 284:11913–21.

Hernandez, A., M. Martinez, S. Fiering et al. 2006. "Type 3 deiodinase is critical for the maturation and function of the thyroid axis." *J Clin Invest* no. 116:476–84.

Hill, K., J. Zhou, W. J. McMahan, A. K. Motley et al. 2003. "Deletion of selenoprotein P alters distribution of selenium in the mouse." *J Biol Chem* no. 278:13640–46.

Hill, K., J. Zhou, W. McMahan, A. Motley, and R. Burk. 2004. "Neurological dysfunction occurs in mice with targeted deletion of the selenoprotein p gene." *J Nutr* no. 134:157–61.

Hill, K., J. Zhou, L. Austin et al. 2007. "The selenium rich C-terminal domain of mouse selenoprotein P is necessary for the supply of selenium to brain and testis but not for the maintenance of whole body selenium." *J Biol Chem* no. 282:10972–80.

Ho, Y., J. Magnenat, R. Bronson, J. Cao et al. 1997. "Mice deficient in cellular glutathione peroxidase develop normally and show no increased sensitivity to hyperoxia." *J Biol Chem* no. 272:16644–51.

Hoffmann, P., S. Höge, P. Li et al. 2007. "The selenoproteome exhibits widely varying, tissue-specific dependence on selenoprotein P for selenium supply." *Nucleic Acids Res* no. 35 (12):3963–73.

Hornberger, T., T. McLoughlin, J. Leszczynski et al. 2003. "Selenoprotein-deficient transgenic mice exhibit enhanced exercise-induced muscle growth." *J Nutr* no. 133:3091–7.

Howard, M., G. Aggarwal, C. Anderson et al. 2005. "Recoding elements located adjacent to a subset of eukaryal selenocysteine-specifying UGA codons." *EMBO J* no. 24:1596–607.

Howard, M., M. Moyle, G. Aggarwal et al. 2007. "A recoding element that stimulates decoding of UGA codons by Sec tRNA[Ser]Sec." *RNA* no. 13:912–20.

Imai, H., F. Hirao, T. Sakamoto et al. 2003. "Early embryonic lethality caused by targeted disruption of the mouse PHGPx gene." *Biochem Biophys Res Commun* no. 305:278–86.

Irons, R., B. Carlson, D. L. Hatfield, and C. Davis. 2006. "Both selenoproteins and low molecular weight selenocompounds reduce colon cancer risk in mice with genetically impaired selenoprotein expression." *J Nutr* no. 136:1311–7.

Ishikura, K., H. Misu, M. Kumazaki et al. 2014. "Selenoprotein P as a diabetes-associated hepatokine that impairs angiogenesis by inducing VEGF resistance in vascular endothelial cells." *Diabetologia* no. 57 (9):1968–76. doi: 10.1007/s00125-014-3306-9.

Iverson, S., S. Eriksson, J. Xu et al. 2013. "A Txnrd1-dependent metabolic switch alters hepatic lipogenesis, glycogen storage, and detoxification." *Free Radical Biol Med* no. 63:369–80.

Jakupoglu, C., G. K. Przemeck, M. Schneider et al. 2005. "Cytoplasmic thioredoxin reductase is essential for embryogenesis but dispensable for cardiac development." *Mol Cell Biol* no. 25:1980–8.

Jin, R. C., C. E. Mahoney, L. Coleman Anderson et al. 2011. "Glutathione peroxidase-3 deficiency promotes platelet-dependent thrombosis *in vivo*." *Circulation* no. 123:1963–73.

Kalantari, P., V. Narayan, S. Natarajan et al. 2008. "Thioredoxin reductase-1 negatively regulates HIV-1 transactivating protein Tat-dependent transcription in human macrophages." *J Biol Chem* no. 283:33183–90.

Kasaikina, M., D. Fomenko, V. Labunskyy, S. A. Lachke, W. Qiu et al. 2011. "Roles of the 15-kDa selenoprotein (Sep15) in redox homeostasis and cataract development revealed by the analysis of Sep 15 knockout mice." *J Biol Chem* no. 286:33203–12.

Kaur, P., N. Rizk, S. Ibrahim et al. 2012. "iTRAQ-based quantitative protein expression profiling and MRM verification of markers in type 2 diabetes." *J Prot Res* no. 11:5527–39.

Kehr, S., M. Malinouski, L. Finney et al. 2009. "X-ray fluorescence microscopy reveals the role of selenium in spermatogenesis." *J Mol Biol* no. 389:808–18.

Kinzy, S., K. Caban, and P. Copeland. 2005. "Characterization of the SECIS binding protein 2 complex required for the co-translational insertion of selenocysteine in mammals." *Nucl Acids Res* no. 33:5172–80.

Kossinova, O., A. Malygin, A. Krol, and G. Karpova. 2014. "The SBP2 protein central to sele-noprotein synthesis contacts the human ribosome at expansion segment 7L of the 28S rRNA." *RNA* no. 20:1046–56.

Kretz-Remy, C., P. Mehlen, M. Mirault, and A. Arrigo. 1996. "Inhibition of I kappa B-alpha phosphorylation and degradation and subsequent NF-kappa B activation by glutathione peroxidase overexpression." *J Cell Biol* no. 133:1083–93.

Kryukov, G. V., R. A. Kumar, A. Koc, Z. Sun, and V. N. Gladyshev. 2002. "Selenoprotein R is a zinc-containing stereo-specific methionine sulfoxide reductase." *PNAS USA* no. 99: 4245–50.

Kryukov, G., S. Castellano, S. Novoselov et al. 2003. "Characterization of mammalian seleno-proteome." *Science* no. 300:1439–43.

Labunskyy, V., B. Lee, D. Handy et al. 2011. "Both maximal expression of selenoproteins and selenoprotein deficiency can promote development of type 2 diabetes-like phenotype in mice." *Antioxid Redox Signal* no. 14:2327–36.

Labunskyy, V. M., D. L. Hatfield, and V. N. Gladyshev. 2014. "Selenoproteins: Molecular pathways and physiological roles." *Physiol Rev* no. 94:739–77.

Lee, B., Z. Péterfi, F. Hoffmann et al. 2013. "MsrB1 and MICALs regulate actin assembly and macrophage function via reversible stereoselective methionine oxidation." *Mol Cell* no. 51:397–404.

Lei, X., and X. Wang. 2012. "Glutathione peroxidase 1, diabetes." In: *Selenium: Its Molecular Biology and Role in Human Health.* (Hatfield, D. L., Berry, M. J., and Gladyshev, V. N., eds.), New York: Springer. pp. 261–270.

Lescure, A., C. Allmang, K. Yamada, P. Carbon, and A. Krol. 2002. "cDNA cloning, expres-sion pattern and RNA binding analysis of human selenocysteine insertion sequence binding protein 2." *Gene* no. 291:279–85.

Liang, H., S. Yoo, R. Na et al. 2009. "Short form glutathione peroxidase 4 is the essential isoform required for survival and somatic mitochondrial functions." *J Biol Chem* no. 284:30836–44.

Lobanov, A. V., D. L. Hatfield, and V. N. Gladyshev. 2009. "Eukaryotic selenoproteins and selenoproteomes." *Biochim Biophys Acta* no. 1790:1424–8.

Loh, K., H. Deng, A. Fukushima et al. 2009. "Reactive oxygen species enhance insulin sensi-tivity." *Cell Metab* no. 10:260–272.

Low, S., E. Grundner-Culemann, J. Harney, and M. Berry. 2000. "SECIS-SBP2 interactions dictate selenocysteine incorporation efficiency and selenoprotein hierarchy." *EMBO J* no. 19:6882–90.

Lu, J., L. Zhong, M. Lonn, R. Burk, K. Hill, and A. Holmgren. 2009. "Penultimate selenocys-teine residue replaced by cysteine in thioredoxin reductase from selenium-deficient rat liver." *FASEB J.* no. 23:2394–402.

Mao, J., and W. Teng. 2013. "The relationship between selenoprotein P and glucose metabo-lism in experimental studies." *Nutrients* no. 5:1937–48.

McClung, J., C. Roneker, W. Mu et al. 2004. "Development of insulin resistance and obesity in mice overexpressing cellular glutathione peroxidase." *PNAS USA* no. 101:8852–7.

McFadden, S., K. Ohlemiller, D. Ding, M. Shero, and R. Salvi. 2001. "The influence of super-oxide dismutase and glutathione peroxidase deficiencies on noise-induced hearing loss in mice." *Noise Health* no. 3:49–64.

Mehta, A., C. M. Rebsch, S. A. Kinzy, J. E. Fletcher, and P. R. Copeland. 2004. "Efficiency of mammalian selenocysteine incorporation." *J Biol Chem* no. 279:37852–9.

Meiler, S., Y. Baumer, Z. Huang et al. 2013. "Selenoprotein K is required for palmitoylation of CD36 in macrophages: Implications in foam cell formation and atherogenesis." *J Leukoc Biol* no. 93:771–80.

Merry, T. L., M. Tran, M. Stathopoulos et al. 2014. "High-fat-fed obese glutathione peroxidase 1-deficient mice exhibit defective insulin secretion but protection from hepatic steatosis and liver damage." *Antioxid Redox Signal* no. 20:2114–29.

Misu, H., T. Takamura, H. Takayama et al. 2010. "A liver-derived secretory protein, selenoprotein P, causes insulin resistance." *Cell Metab* no. 12:483–95.

Moghadaszadeh, B., N. Petit, C. Jaillard et al. 2001. "Mutations in SEPN1 cause congenital muscular dystrophy with spinal rigidity and restrictive respiratory syndrome." *Nat Genet* no. 29:17–8.

Moses, E., M. Johnson, L. Tommerdal et al. 2008 "Genetic association of preeclampsia to the inflammatory response gene SEPS1." *Am J Obstet Gynecol* no. 198 (3):336e1–e5.

Nelson, S., X. Lei, and K. Prabhu. 2011. "Se-levels affect the IL-4-induced expression of alternative activation markers in murine macrophages." *J Nutr* no. 141:1754–61.

Ng, L., R. Goodyear, C. Woods et al. 2004. "Hearing loss and retarded cochlear development in mice lacking type 2 iodothyronine deiodinase." *PNAS USA* no. 101:3474–9.

Ng, L., A. Hernandez, W. He et al. 2009. "Protective role for type 3 deiodinase, a thyroid hormone inactivating enzyme, in cochlear development and auditory function." *Endocrinol* no. 150:1952–60.

Ohtani, N., K. Yamakoshi, A. Takahashi, and E. Hara. 2004. "The p16INK4a-RB pathway: Molecular link between cellular senescence and tumor suppression." *J Med Invest* no. 51:146–53.

Olson, G. E., V. P. Winfrey, S. K. Nagdas, K. E. Hill, and R. F. Burk. 2005. "Selenoprotein P is required for mouse sperm development." *Biol Reprod* no. 73:201–11.

Papp, L., J. Lu, F. Striebel et al. 2006. "The redox state of SBP 2 controls its localization and selenocysteine incorporation function." *Mol Cell Biol* no. 26:4895–910.

Papp, L., J. Wang, D. Kennedy et al. 2008. "Functional characterization of alternatively spliced human SECISBP2 transcript variants." *Nucleic Acids Res* no. 36:7192–206.

Pepper, M., M. Vatamaniuk, X. Yan, C. Roneker, and X. Lei. 2011. "Impacts of dietary selenium deficiency on metabolic phenotypes of diet-restricted GPX1-overexpressing mice." *Antiox Redox Sign* no. 14:383–90.

Pitts, M. W., M. A. Reeves, A. C. Hashimoto et al. 2013. "Deletion of selenoprotein M leads to obesity without cognitive deficits." *J Biol Chem* no. 288:26121–34.

Prevost, G., A. Arabo, L. Jian et al. 2013. "The PACAP-regulated gene selenoprotein T is abundantly expressed in mouse and human β-cells and its targeted inactivation impairs glucose tolerance." *Endocrinol* no. 154:3796–806.

Ramos, A., A. Lane, D. Hollingworth, and T. Fan. 2004. "Secondary structure and stability of the selenocysteine insertion sequences (SECIS) for human thioredoxin reductase and glutathione peroxidase." *Nucleic Acids Res* no. 32:1746–55.

Rederstorff, M., P. Castets, S. Arbogast et al. 2011. "Increased muscle stress-sensitivity induced by selN inactivation in mouse: A mammalian model for SEPN1-related myopathy." *PloS One* no. 6 (8):e23094.

Renko, K., M. Werner, I. Renner-Muller et al. 2008. "Hepatic selenoprotein P (SePP) expression restores selenium transport and prevents infertility and motorincoordination in Sepp-knockout mice." *Biochem J* no. 409:741–9.

Riva, C., E. Donadieu, J. Magnan, and J. Lavieille. 2006. "Age-related hearing loss in CD/1 mice is associated to ROS formation and HIF target proteins up-regulation in the cochlea." *Exp Gerontol* no. 42:327–36.

Roman, M., P. Jitarub, and C. Barbante. 2014. "Selenium biochemistry and its role for human health." *Metallomics* no. 6:25–54.

Schallreuter, K. U., K. Rübsam, N. C. Gibbons et al. 2008. "Methionine sulfoxide reductases A and B are deactivated by hydrogen peroxide (H_2O_2) in the epidermis of patients with vitiligo." *J Invest Dermatol* no. 128:808–15.

Schneider, M., S. Fiering, S. Pallud et al. 2001. "Targeted disruption of the type 2 selenodeiodinase gene (DIO2) results in a phenotype of pituitary resistance to T4." *Mol Endocrinol* no. 15:2137–48.

Schneider, M., S. Fiering, B. Thai et al. 2006. "Targeted disruption of the type 1 selenodeiodinase gene (Dio1) results in marked changes in thyroid hormone economy in mice." *Endocrinol* no. 147:580–9.

Schneider, M., H. Forster, A. Boersma et al. 2009. "Mitochondrial glutathione peroxidase 4 disruption causes male infertility." *FASEB J* no. 23:3233–42.

Schoenmakers, E., M. Agostini, C. Mitchell et al. 2010. "Mutations in the selenocysteine insertion sequence binding protein 2 gene lead to a multisystem selenoprotein deficiency disorder in humans." *J Clin Invest* no. 120:4220–35.

Schomburg, L., U. Schweizer, B. Holtmann et al. 2003. "Gene disruption discloses role of selenoprotein P in selenium delivery to target tissues." *Biochem J* no. 370:397–402.

Schomburg, L., A. M. Dumitrescu, X. H. Liao et al. 2009. "Selenium supplementation fails to correct the selenoprotein synthesis defect in subjects with SBP2 gene mutations." *Thyroid* no. 19:277–81.

Schweizer, U., M. Michaelis, J. Köhrle, and L. Schomburg. 2004. "Efficient selenium transfer from mother to offspring in selP-deficient mice enables dose dependent rescue of phenotypes associated with selenium deficiency." *Biochem J* no. 378:21–6.

Schweizer, U., F. Streckfuss, P. Pelt et al. 2005. "Hepatically derived selenoprotein P is a key factor for kidney but not for brain selenium supply." *Biochem J* no. 386:221–6.

Seeher, S., T. Atassi, Y. Mahdi et al. 2014a. "Secisbp2 is essential for embryonic development and enhances selenoprotein expression." *Antioxid Redox Signal* no. 21 (6):835–49.

Seeher, S., B. A. Carlson, A. Miniard et al. 2014b. "Impaired selenoprotein expression in brain triggers striatal neuronal loss leading to coordination defects in mice." *Biochem J* no. 462:67–75.

Seiler, A., M. Schneider, H. Forster et al. 2008. "Glutathione peroxidase 4 senses and translates oxidative stress into 12/15-lipoxygenase dependent and AIF-mediated cell death." *Cell Metab* no. 8:237–48.

Sengupta, A., U. F. Lichti, B. A. Carlson et al. 2010. "Selenoproteins are essential for proper keratinocyte function and skin development." *PLoS One* no. 18:e12249.

Shchedrina, V., R. Everley, Y. Zhang et al. 2011. "Selenoprotein K binds multiprotein complexes and is involved in the regulation of endoplasmic reticulum homeostasis." *J Biol Chem* no. 286:42937–48.

Sheridan, P., N. Zhong, B. A. Carlson et al. 2007. "Decreased selenoprotein expression alters the immune response during influenza virus infection in mice." *J Nutr* no. 137:1466–71.

Shibata, T., T. Arisawa, T. Tahara et al. 2009. "Selenoprotein S gene 105G>A promoter polymorphism influences the susceptibility to gastric cancer in the Japanese population." *BMC Gastroenterol.* no. 9:2.

Shrimali, R. K., R. D. Irons, B. A. Carlson et al. 2008. "Selenoproteins mediate T cell immunity through an antioxidant mechanism." *J Biol Chem* no. 283:20181–5.

Small-Howard, A., N. Morozova, Z. Stoytcheva et al. 2006. "Supramolecular complexes mediate selenocysteine incorporation *in vivo*." *Mol Cell Biol* no. 26:2337–46.

Soerensen, J., C. Jakupoglu, H. Beck et al. 2008. "The role of thioredoxin reductases in brain development." *PLoS ONE* no. 3:e1813.

Squires, J., P. Davy, M. Berry, and R. Allsopp. 2009. "Attenuated expression of SECIS binding protein 2 causes loss of telomeric reserve without affecting telomerase." *Exp Gerontol* no. 44:619–23.

Srinivasan, G., C. James, and J. Krzycki. 2002. "Pyrrolysine encoded by UAG in Archaea: Charging of a UAG-decoding specialized tRNA." *Science* no. 296:1459–62.

Steinbrenner, H. 2013. "Interference of selenium and selenoproteins with the insulin-regulated carbohydrate and lipid metabolism." *Free Radic Biol Med* no. 65:1538–47.

Su, D, S. V. Novoselov, Q. A. Sun et al. 2005. "Mammalian selenoprotein thioredoxin-glutathione reductase. Roles in disulfide bond formation and sperm maturation." *J Biol Chem* no. 280:26491–8.

Sundaresan, N. R., M. Gupta, G. Kim et al. 2009. "Sirt3 blocks the cardiac hypertrophic response by augmenting Foxo3a dependent antioxidant defense mechanisms in mice." *J Clin Invest* no. 119:2758–71.

Sutherland, A., D. Kim, C. Relton, Y. Ahn, and J. Hesketh. 2010. "Polymorphisms in the selS and 15-kDa selenoprotein genes are associated with altered susceptibility to colorectal cancer." *Genes Nutr* no. 5:215–23.

Szypowska, A. A., and B. M. Burgering. 2011. "The peroxide dilemma: Opposing and mediating insulin action." *Antioxid Redox Signal* no. 15:219–32.

Takeuchi, A., D. Schmitt, C. Chapple et al. 2009. "A short motif in Drosophila SECIS Binding Protein 2 provides differential binding affinity to SECIS RNA hairpins." *Nucleic Acids Res* no. 37:2126–41.

Tsuji, P. A., B. A. Carlson, S. Naranjo-Suarez et al. 2012. "Knockout of the 15 kDa selenoprotein protects against chemically-induced aberrant crypt formation in mice." *PLoS One* no. 7:e50574.

Ursini, F., S. Heim, M. Kiess et al. 1999. "Dual function of the selenoprotein PHGPx during sperm maturation." *Science* no. 285:1393–6.

Valentine, W. M., K. E. Hill, L. M. Austin et al. 2005. "Brainstem axonal degeneration in mice with deletion of selenoprotein p." *Toxicol Pathol* no. 33:570–6.

Verma, S., F. W. Hoffmann, M. Kumar et al. 2011. "Selenoprotein K knockout mice exhibit deficient calcium flux in immune cells and impaired immune responses." *J Immunol* no. 186:2127–37.

von Kriegsheim, A., D. Baiocchi, M. Birtwistle et al. 2009. "Cell fate decisions are specified by the dynamic ERK interactome." *Nat Cell Biol* no. 11:1458–64.

Vunta, H., F. Davis, U. D. Palempalli et al. 2007. "The anti-inflammatory effects of selenium are mediated through 15-deoxy-Delta12,14-prostaglandin J2 in macrophages." *J Biol Chem* no. 282:17964–73.

Vunta, H., B. J. Belda, R. J. Arner et al. 2008. "Selenium attenuates pro-inflammatory gene expression in macrophages." *Mol Nutr Food Res* no. 52:1316–23.

Walczak, R., E. Westhof, P. Carbon et al. 1996. "A novel RNA structural motif in the selenocysteine insertion element of eukaryotic selenoprotein mRNAs." *RNA* no. 2: 367–9.

Walczak, R., P. Carbon, and A. Krol. 1998. "An essential non-Watson-Crick base pair motif in 3UTR to mediate selenoprotein translation." *RNA* no. 4:74–84.

Wallace, D. J. 2010. "Telomere diseases." *N Engl J Med* no. 362:1150.

Wang, X., M. Vatamaniuk, S. Wang et al. 2008. "Molecular mechanisms for hyperinsulinaemia induced by overproduction of selenium-dependent glutathione peroxidase-1 in mice." *Diabetologia* no. 51:1515–24.

Wang, X., M. Z. Vatamaniuk, C. A. Roneker et al. 2011. "Knock outs of SOD1 and GPX1 exert different impacts on murine islet function and pancreatic integrity." *Antioxid Redox Signal* no. 14:391–401.

Watson, J. D. 2014. "Type 2 diabetes as a redox disease." *The Lancet* no. 383:841–3.

Wood, J. M., and K. U. Schallreuter. 2006. "UVA-irradiated pheomelanin alters the structure of catalase and decreases its activity in human skin." *J Invest Dermatol* no. 126:13–4.

Xu, X., B. A. Carlson, H. Mix et al. 2007. "Biosynthesis of selenocysteine on its tRNA in eukaryotes." *PLoS Biol* no. 5:e4.

Yamaguchi, H., R. T. Calado, H. Ly et al. 2005. "Mutations in TERT, the gene for telomerase reverse transcriptase, in aplastic anemia." *N Engl J Med* no. 352:1413–24.

Yang, S., S. Hwang, H. Choi et al. 2011. "Serum selenoprotein P levels in patients with type 2 diabetes and prediabetes: Implications for insulin resistance, inflammation and athero-sclerosis." *J Clin Endocrinol Metab* no. 96:E1325–9.

Yant, L., Q. Ran, L. Rao et al. 2003. "The selenoprotein GPX4 is essential for mouse develop-ment and protects from radiation and oxidative damage insults." *Free Radic Biol Med* no. 34:496–502.

Ye, Y., Y. Shibata, C. Yun, D. Ron, and T. Rapoport. 2004. "A membrane protein complex mediates retro-translocation from the ER lumen into the cytosol." *Nature* no. 429 (6994):841–7.

Index

Page numbers followed by f and t indicate figures and tables, respectively.

A

α-1-Antitrypsin, 156
Abnormal cytokine production, 363–364
ACC1 (acetyl-CoA carboxylase-1), 225, 252
A549 cells, growth of, 176
Acetaminophen (APAP), 278–279
Acetyl-CoA carboxylase-1 (ACC1), 225, 252
Acquired immune deficiency syndrome (AIDS), 6
Action mechanism, of Se forms, 149–150
Activator protein 1 (AP-1) transcription factor, modulation by selenite, 121
Adenomatous polyposis coli (APC), 192
Adenosine monophosphate-activated protein kinase (AMPK)
    p53, 224
    production and/or function of, 228
    TrxR1-interacting regulators, 283
Adenosine 5'-phosphoselenate (APSe), formation, 23
Adenosine 5'-phosphosulfate (APS), formation, 23
AE1 (anionic exchanger 1) in RBCs, 115–116
AIDS (acquired immune deficiency syndrome), 6
Akt phosphorylation, 362
Alkylating agents, 123
Allium plants, 19
Alpha-tocopherol, beta-carotene cancer prevention study (ATBC), 151
Alternatively activated macrophage (AAM/M2), 297–301
    activated by IL-4 and IL-13, 299–300
    CAM *vs.*, 298–299, 299f
    discovery of, 299
    function of, 298, 301
    polarization, 297f
    by PPARγ, 300
Alzheimer's diseases, 311–312
Amblyomma, 39
Amblyomma maculatum, 38
AMGs (aminoglycosides), action of, 43
Aminoglycosides (AMGs), action of, 43
Animal studies
    metabolic pathway, 19, 20f, 21
    metabolites, 22–23
    selenite, chemotherapeutic agent
        in vivo studies, 113–115

supranutritional Se intake as potential risk
        factor for T2DM, 221–222, 222f
Anionic exchanger 1 (AE1) in RBCs, 115–116
Anoikis, 195
Antibiotics
    amino acids, induction, 181
    antitumor, 123
    mammalian cells exposure to, 63
    in Sec incorporation, 43
Antibodies
    B72.3, monoclonal, 98
    prostate cancer, 97–98
    TAG-72 monoclonal, 98
Anti-inflammation, macrophages in, 296–298, 297f
Antimetabolites, 123
Antioxidant responsive element (ARE)
    in GSH system, 177
    within human GPx2 promoter, 191
    Nrf2/Maf heterodimer with, 281
Antioxidant(s)
    pro-oxidant functions and, selenite, 111–112, 111f
    system in hallmarks of cancer, 179
Antitumor antibiotics, 123
AOM (azoxymethane), 148, 191, 195–196, 195f, 210
APAP (acetaminophen), 278–279
APC (adenomatous polyposis coli), 192
Apoptosis, modulation of
    GPx2 and, 194–196, 195f
Apoptosis signal-regulating kinase (ASK1), 175
APS (adenosine 5'-phosphosulfate), formation, 23
APSe (adenosine 5'-phosphoselenate), formation, 23
AP-1 (activator protein 1) transcription factor, modulation by selenite, 121
Arabidopsis, 59
Arginase-1 (Arg-1), 299, 300
Arginine, 63
Argon plasma mass spectrometry, inductively coupled, 21
ASK1 (apoptosis signal-regulating kinase), 175
Association studies, biomarkers of Se status in, 336–337
Asthma, 301–302
Astragalus bisulcatus, 85f
Astragalus genus, 84

ATBC (Alpha-Tocopherol, Beta-Carotene Cancer
    Prevention Study), 151
ATP sulfurylase enzyme, in APS formation, 23
Availability, Se, 43–44
Avastin, 102
Axial muscular dystrophy, *SECISBP2* mutation,
    360–361
Azo dye–induced liver carcinogenesis rat model,
    selenite in, 113–115
Azoospermia, with spermatogenic maturation
    arrest, 362
Azoxymethane (AOM), 148, 191, 195–196, 195f,
    210

B

B72.3, monoclonal antibody, 98
BAT (brown adipose tissue), deiodinases in
    energy dissipation, 257–258
Beckman-120 Amino Acid Analyzer, 86
Beneficiary, Se, 9–10
Benzyldiselenide, 99
Benzyl selenocyanate (BSC), 148
Berzelius, Jöns Jakob, 83–84
Biallelic *SECISBP2* mutations, 358–359
Biochemistry, human *SECISBP2* mutations, 359
Biological sources, Se, 24–25, 25f
Biomarkers
    of Se status in association studies, 336–337
    studies, with healthy individuals
    forms of Se in cancer prevention,
        156–160, 158t–159t
    SNPs in selenoprotein genes and,
        316t–318t
Biosynthesis
    GluMeSeCys, 19
    MeSeCys, 19
    Sec, 61–63, 62f
    SeCys tRNA, 24
    SeHLan, 19
    selenoproteins, 9
    Sec-tRNA in, *See* Selenocysteine tRNA
        (Sec-tRNA[Ser]Sec)
Bipartite/SID/K-rich region, 344
Bladder cancer, SNPs in selenoprotein genes
    and, 322t
Bolton-Hunter reagent, 101–102, 101f, 102f
Bombyx mori, 63–64
Boylan, Mallory, 89–91
Brassica plants, 19
Brazil nuts, Se source, 139
Breast cancer, SNPs in selenoprotein genes and,
    332–333
    Gpx1, 319t–320t, 332–333
    GPx4, 324t
    SEP15, 324t, 333
    Sepp1, 327t, 328t

Brefeldin A, 204
Brennan, A., Dr., 95
Broccoli, 140
Brown adipose tissue (BAT), deiodinases in
    energy dissipation, 257–258
BSC (benzyl selenocyanate), 148

C

Cabbage, 140
Caenorhabditis elegans, 38, 203
Cancer-associated signaling pathways
    modulation by selenite, 118–123
        AP-1 transcription factor, 121
        cell cycles, 119
        epigenetic signature, 121–123, 122f
        HIF-1α, 120
        NF-κB, 120–121
        p53, 119–120
        translational machinery, 118–119
Cancers, *See also specific* entries
    bladder, 322t
    breast, SNPs in selenoprotein genes and,
        332–333
        Gpx1, 319t–320t, 332–333
        GPx4, 324t
        SEP15, 324t, 333
        Sepp1, 327t, 328t
    colon
        B72.3, monoclonal antibody, 98
        SEP15 and, mechanistic studies, 208–210
    CRC
        colitis-associated, 194
        HT-29 cells, 195
        incidence, 153
        oncogenic Ras in, 192
        polymorphisms in *SEP15* gene, 208
        SNPs in selenoprotein genes and, 321t,
            327t, 329t, 333–335
        Wnt in, 192
    GPX2 in, *See* Glutathione peroxidase 2
        (GPx2)
    incidence, 152–153
    laryngeal, SNPs in selenoprotein genes
        Gpx1, 322t
        GPx4, 326t
        SEP15, 331t
    liver
        Se supplementation for, 152
        TrxR1 deletion in promoting/sustaining,
            176–177
    lung
        doxorubicin resistant cells, 123
        incidence, 153
        LLC1 cells, 209
        SNPs in selenoprotein genes, 321t–322t,
            326t, 330t, 331t

mouse models involving tRNA[Ser] Sec gene
  (*Trsp*) and transgene (*Trsp^t^*) elucidate
  functions of selenoproteins in, 65–71
*Trsp^Δ^* and *Trsp^c^Δ* mice, 68, 69t–70t
*Trsp^Δ^* and *Trsp^c^Δ* mice with transgenes,
  *Trsp^t^*, *Trsp^tG37^*, *Trsp^tA34^*, 68, 71
*Trsp^t,^ Trsp^tG37^*, and *Trsp^tA34^* mice, 66,
  67t–68t
nonmelanoma skin cancer, 152
Nrf2 in, 190–191
polymorphisms in *SEP15* gene
  colorectal, 208
  prostate, 207–208
prostate, *See* Prostate cancer
redox cycling Se for, 95–98
reducing risk, 7–8
Se for, 3
  forms in prevention, 137–161, *See also*
    Forms of Se
  TrxR1 and, 179–181
selenite in, 95, 109–129
  antioxidant and pro-oxidant functions,
    111–112, 111f
  human studies, 124–129
  molecular basis of cytotoxicity, 115–124,
    See also Cytotoxicity, selenite
  overview, 110
  potent chemotherapeutic agent, 112–115,
    113t
  rationale for Se use, 124–125
  ROS, generation, 117–118
  safety and maximal tolerable dose,
    determination, 129
  Se in combination with surgery,
    radiotherapy, and chemotherapy, 125,
    126t–127t, 128t
  Se use as chemotherapeutic drug, 125, 129
  transport in cells, 115–117
selenoenzyme overexpression in, 175
SEP15, *See* SEP15
SNPs in selenoprotein genes
  breast, 319t–320t, 324t, 327t, 328t,
    332–333
  CRC, 321t, 327t, 329t, 333–335
  laryngeal, 322t, 326t, 331t
  lung, 321t–322t, 326t, 330t, 331t
  prostate, 320t–321t, 325t, 328t, 329t, 330t,
    331t, 333
TrxR1 in, *See also* Thioredoxin reductase 1
    (TrxR1)
  deletion, for protection, 176
  deletion effect in others, 178
  deletion in promoting/sustaining in liver
    cancer, 176–177
  hallmarks, Trx system and, 178–179
  in prevention and progression, 182f
  Se in, 179–181

selenium-free forms, 180–181
  tumor-promoting effects, 175–178
  varies in different tissues providing
    avenues for cancer therapy, 178
  Wnt in, 192–193
Capitella teleta, 40
Carboplatin, 123
Carcinogenesis *in vitro* and *in vivo* studies,
    parameters, 142t–144t
Carcinogen-induced and xenograft models,
    147–148
Cardiomyopathy Keshan disease, *See* Keshan
    disease
Cardiovascular disease, SNPs in selenoprotein
    genes, 323t
B-catenin, 192
CDF (cerebellar deficient folia) mice, 114
Celecoxib, 148
Cell cycles, in modulation by selenite, 119
Cellular protection through H₂O₂-degrading
    activity of GPx1, 226
Cellular redox homeostasis
  GSH system in, 275
  Trx system in, 275–276
Cerebellar deficient folia (CDF) mice, 114
Chaudière, Jean, 93
Chemiluminescence (CL) assay, lucigenin
  for superoxide detection, 90–93
Chemotherapeutic agent, selenite, 112–115
  in vitro studies, 112, 113t
  in vivo studies with animal models, 113–115
Chemotherapeutic drugs
  mechanisms and evidence for potentiation
    of cytotoxic effects of selenite,
    123–124
  Se use in humans, 125, 129
Chemotherapy, Se in combination, 125,
    126t–127t, 128t
Chen, J. J., 89–91
Chen, Lugen, 98
ChIP (chromatin immunoprecipitation), 281
Chlamydomonas, 59
Chloramphenicol (Cp), 43, 63
1-chloro-2,4-dinitrobenzene (DNCB), 176
Chromatin immunoprecipitation (ChIP), 281
Cisplatin, 124
Classically activated macrophage (CAM/M1)
  AAM *vs.*, 298–299, 299f
  nomenclature of, 298
  polarization, 297, 297f
Clinical trials
  phase III
    cost, 161
    expense and lack of success, 156
    initiation of future, 149
    organoselenium testing in, 157
    selenomethionine for, 149

preclinical, Se forms in, 140–150
   chemical structures, 146f
   mechanism of action, 149–150
   metabolism and disposition, 141
   in vitro studies, 141–147, 142t–144t, 146f
   in vivo studies, 147–149
  Se forms in, 152–156, 154t–155t
Clostridia, 32
C-myc oncogene, 178
Colon cancer
  B72.3, monoclonal antibody, 98
  SEP15 and, mechanistic studies, 208–210
Colorectal cancer (CRC)
  colitis-associated, 194
  HT-29 cells, 195
  incidence, 153
  oncogenic Ras in, 192
  polymorphisms in *SEP15* gene, 208
  SNPs in selenoprotein genes and, 333–335
    Gpx1, 321t
    GPx4, 324t–325t, 334
    SEP15, 330t, 331t, 334
    Sepp1, 327t, 329t, 334
  Wnt in, 192
Combretastatin A-4 (diaryl selenide), 145
Corticosteroids, 123
COX-2 (cyclooxygenase 2), upregulation of, 196–197
Coxsackie B4, RNA virus, 6
Coxsackie B3 virus, 6
CRC, *See* Colorectal cancer (CRC)
CRD (cysteine rich domain) in SBP2, 47, 344
Cross-sectional studies, supranutritional Se
    intake as potential risk factor for
    T2DM, 222–223
CT26 cells, mouse colon carcinoma, 208–209
C-terminal portion, of SECISBP2L
    (CT-SECISBP2L), 40
CT-SECISBP2L (C-terminal portion, of
    SECISBP2L), 40
Cutaneous photosensitivity, *SECISBP2*
    mutations, 362–363
Cyb5r3 (cytochrome b5 reductase 3) gene, 251
Cyclooxygenase 2 (COX-2), upregulation of,
    196–197
Cyclopentenone PGs (CyPGs), 295
CyPGs (cyclopentenone PGs), 295
Cysteine (Cys), 63, 180–181
Cysteine rich domain (CRD) in SBP2, 47, 344
Cytochrome b5 reductase 3 (*Cyb5r3*) gene, 251
Cytogen Corporation, 97–98
Cytokine production, abnormal, 363–364
Cytotoxicity, selenite
  molecular basis in cancer, 115–124
    mechanisms and evidence for potentiation
      with chemotherapeutic drugs,
      123–124
    modulation of cancer-associated
      signaling pathways, 118–123, *See also*
      Modulation
    ROS, generation, 117–118
    selenite transport, 115–117

D

D2 (type 2 deiodinase), 253, 253t–254t
D3 (type 3 deiodinase), 253, 254t
Db/db mice
  hyperglycemia in, 224
  T2DM in, 220
Deficiency, Se, 5–6
Degrading activity of GPx1, cellular protection
    through H_2O_2, 226
24-Dehydrocholesterol reductase (*Dhcr24*) gene,
    251
Deletion
  of *SECISBP2*, 350–351
  TrxR1
    for cancer protection, 176
    effect in other cancer, 178
    in promoting/sustaining liver cancer,
      176–177
Deltaproteobacteria, 32
Dental sealants, containing redox cycling Se,
    100, 100f
Developmental delay, human *SECISBP2*
    mutations, 360
Dexamethasone, 230
Dextran sodium sulfate (DSS), 191, 196, 296
Dhcr24 (24-dehydrocholesterol reductase) gene,
    251
DHT (dihydrotestosterone), 148
Diabetes, Se in, 218–235
  attempted use of compounds for,
    220–221
  beneficial effects, 221
  GPx1 and, 223–228
    paradoxical roles, 223–226
    regulatory mechanisms on insulin-
      regulated glucose metabolism, 226–228,
      See also Regulatory mechanisms of
      GPx1
  others, 231–233
  overview, 218
  pathways and coordinated mechanisms for
    selenoproteins, 234f
  redox control of insulin secretion and
    signaling, 218–220
  Sepp1 and, 228–230
    elevation of circulating levels in T2DM
      patients, 230
    functions, 228–229
    as negative regulator of insulin signaling,
      229–230

supranutritional intake as potential risk factor
  for T2DM, 221–223
    animal studies, 221–222, 222f
    cross-sectional studies, 222–223
    RCTs, 223
Diaryl selenide (combretastatin A-4), 145
Dictyostelium, 59
Diethylnitrosamine, liver carcinogen, 176
Dihydrotestosterone (DHT), 148
7,12-Dimethylbenz[a]anthracene (DMBA), 147
Dimethylselenide (DMSe)
  animal metabolite, 22
  excreted by respiration, 141
Dio3 gene, disruption of, 258
Diphenylmethyl selenocyanate (DPMSC), 148
Diplock, Anthony (Tony), Dr., 94–95
Diseases, SNPs in selenoprotein genes and, 315–335
  biomarkers in healthy populations,
    316t–318t
  bladder cancer, 322t
  breast cancer, 332–333
    Gpx1, 319t–320t, 332–333
    GPx4, 324t
    SEP15, 324t, 333
    Sepp1, 327t, 328t
  cardiovascular disease, 323t
  CRC risk, 333–335
    Gpx1, 321t
    GPx4, 324t–325t, 334
    SEP15, 330t, 331t, 334
    Sepp1, 327t, 329t, 334
  GPX4, *SEPP1*, and *SEP15* genes and disease
    risk, 324t–331t
  Kashin-Beck disease
    GPx4, 326t
    Gpx1 gene, 323t
    Sepp1, 328t
  laryngeal cancer
    Gpx1, 322t
    GPx4, 326t
    SEP15, 331t
  lung cancer
    Gpx1, 321t–322t
    GPx4, 326t
    SEP15, 330t, 331t
  metabolic syndrome
    Gpx1 gene, 323t
  obesity, 323t
  pathway approach, 335
  Pro198Leu (rs1050450) in *GPX1* and disease
    risk, 319t–323t
  prostate cancer and
    GPx4, 325t
    Gpx1 gene, 320t–321t, 333
    SEP15, 325t, 330t, 331t, 333
    Sepp1, 328t, 329t
  stroke, 323t

Diselenides, in superoxide generation, 91, 93
Disorders, Human *SECISBP2* mutations,
    351–364
  axial muscular dystrophy, 360–361
  azoospermia with spermatogenic maturation
    arrest, 362
  biochemistry, 359
  cutaneous photosensitivity, 362–363
  developmental and neurological phenotype,
    360
  effect of treatment, 364
  growth and skeletal phenotype, 359
  hearing loss, 360
  impaired T-cell proliferation and abnormal
    cytokine production, 363–364
  metabolic phenotype, 361–362
  SBP2 mutations, 351–359, 352t
    biallelic *SECISBP2* mutations, 358–359
    blood biochemistry, 353t
    heterozygous missense, 357–358
    homozygous missense, 357
    physiological phenotypes in patients with
      deficiency, 354t–355t
    R540Q mutant, 353, 356
Disposition, Se forms, 141
Distal sequence element (DSE), in *Trsp* gene,
    59–60
DMBA (7,12-dimethylbenz[a]anthracene),
    147
DNA methyltransferase (DNMT), 121
DNCB (1-chloro-2,4-dinitrobenzene), 176
DNMT (DNA methyltransferase), 121
Docetaxel, inhibitory effect of, 124
Dox (doxycycline), 43, 63
Doxorubicin
  in cancer cells, 124
  chemotherapeutic effect, 124
  resistant lung cancer cells, 123
Doxycycline (Dox), 43, 63
DPMSC (diphenylmethyl selenocyanate),
    148
Drosophila melanogaster, 32, 38, 63–64
Drug(s)
  chemotherapeutic, cytotoxic effects of
    selenite with
    mechanisms and evidence for
      potentiation, 123–124
  developments, 100–103, 102f
    Bolton-Hunter reagent, 101–102, 101f,
      102f
    TMZ, 101, 101f
DSE (distal sequence element), in *Trsp* gene,
    59–60
DSS (dextran sodium sulfate), 191, 196, 296
DT cells, 180
DU-145 prostate cancer cells, 98
Dystrophy, axial muscular, 360–361

E

E3 allele, 311
E4 allele, 311
EATC (Ehrlich ascites tumor cells), 87, 114
Ecosystem, metabolic
    integration of systems into, 284–285, 284f
EEFSec (Sec-specific elongation factor)
    for Sec incorporation, 32, 33, 38–39, 344, 346
Ehrlich, Paul, 95, 97f
Ehrlich ascites tumor cells (EATC), 87, 114
EIF2 (eukaryotic initiation factor 2), inhibition
    of, 118
EIF4a3 (eukaryotic initiation factor 4A), for Sec
    incorporation, 41
EIF4E, ser209 phosphorylation of, 118–119
Electrophilic response elements (EpRE), 281
Elemental Se, 25–26
Elucidation, function
    selenoproteins in cancer, health, and
        development
        mouse models involving tRNA[Ser] Sec gene
            (*Trsp*) and transgene (*Trsp^t*), 65–71,
            See also tRNA[Ser] Sec gene (*Trsp*) and
            transgene (*Trsp^t*), mouse models
Emergence, Se
    in nutrition and health, 3
Energy dissipation, iodothyronine deiodinases in,
    257–258
Energy metabolism, Se in
    epidemiological studies, 249–250
    selenoprotein synthesis factors role, 250–252
        selenocysteine lyase, 251–252
        tRNA[Ser]Sec gene, 250–251
    specific selenoproteins, 252–264, 253t–255t
        glutathione peroxidases, 259–261
        iodothyronine deiodinases, 255–259,
            256f, *See also* Iodothyronine
            deiodinases
        SelM, 253, 254t, 262
        SelN, 253, 255t, 262
        SelS, 253, 255t, 263
        SelT, 253, 255t, 263–264
        Sepp1, 253, 255t, 262–263
        TrxR1, 253, 255t, 264
Enzymes
    ATP sulfurylase, in APS formation, 23
    GPx, *See* Glutathione peroxidases (GPx)
    selenoenzymes, *See* Selenoenzymes
    selenoproteins function as redox-active, 117
Epidemiological studies
    forms of Se in, 151–152
    Se in energy metabolism, 249–250
Epigenetic signature, modulation by selenite,
    121–123, 122f
EpRE (electrophilic response elements), 281
ERF1 (eukaryotic release factor 1), 34

Erk phosphorylation, 362
Escherichia coli, 33, 61, 63, 64, 87, 89, 274
Etiological factors, of insulin resistance, 219
Etiology, Keshan disease, 6
Eukaryotes, proteins
    molecular regulation of SeCys into, *See*
        Molecular regulation, of SeCys
Eukaryotic initiation factor 2 (eIF2), inhibition
    of, 118
Eukaryotic initiation factor 4A (eIF4a3), for Sec
    incorporation, 41
Eukaryotic release factor 1 (eRF1), 34
Euplotes, 33
Excretion, urine
    as selenosugars, 21
Expression
    gene
        regulation, 32
        SEP15 and cancer, 205–206
    GPx2, dual role of factors regulating,
        190–193
        Nrf2, 190–191
        Wnt, 192–193
    maximal, in SePP, 5
    overexpression
        GPx1 in diabetes, 224
        GPx1 in T2DM, 225
        selenoenzyme in cancers, 175
    patterns, Gad67, 351
    selenoprotein, regulation, 43–47
        NMD, 44–46, 45f
        Se availability, 43–44
        subcellular localization, 46–47
    SEP15 gene, cancer and, 205–206
    suboptimal, of selenoenzymes, 9

F

Fasting plasma glucose (FPG), developing
    impaired, 221
FdhF (formate dehydrogenase) gene, *TGA* in, 61
Firmicutes, 32
Fishes
    selenoneine in, 24–25
    Se source, 24
Flohé, Leopold, Dr., 85, 91
Food Standards Agency, 8
Forkhead box A2 (FOXA2), 228
Forkhead box O1 (FoxO1a/FoxO1), 225
Formate dehydrogenase (*fdhF*) gene, *TGA* in, 61
Forms of Se, in cancer prevention, 137–161
    biomarker studies with healthy individuals,
        156–160, 158t–159t
    cellular processes, modification, 160
    in clinical trials, 152–156, 154t–155t
    epidemiological studies, 151–152
    future research, 160–161, 161f

in nature, 139–140
overview, 137–139
in preclinical studies, 140–150
    chemical structures, 146f
    mechanism of action, 149–150
    metabolism and disposition, 141
    in vitro studies, 141–147, 142t–144t, 146f
    in vivo studies, 147–149
Forrest E. Shaklee Corporation Research Center,
    87
FOXA2 (forkhead box A2), 228
FoxO1a/FoxO1 (forkhead box O1), 225
FPG (fasting plasma glucose), developing
    impaired, 221
Fredga, Arne, Dr., 91
Fridovich, Irwin, 86–87
5-FU, in colon cancer, 124
Fungi, nutritional source, 19

G

Gad67 expression patterns, 351
Ganther, Howard, 86, 90, 93, 94
Garlic, 140
Gbp-1 (guanylate binding protein-1), 209
Gcl (Glu-Cys ligase), 275
GCL (glutamate-cysteine ligase), 191
GDM (gestational diabetes mellitus), 219
Gene expression
    regulation, 32
    SEP15 and cancer, 205–206
Genentech, Inc., 102
Genes
    Cyb5r3, 251
    Dhcr24, 251
    Ldlr, 251
    NFE2L2, 191
    polymorphisms in *SEP15* gene
        colorectal cancers, 208
        prostate cancer, 207–208
    SECISBP2, *See SECISBP2*
    selenoproteins, genetic polymorphisms in,
        See Genetic polymorphisms
    TGA in
        fdhF, 61
        Gpx1, 60–61
    tRNA[Ser]Sec
        in energy metabolism, 250–251, 253t
    tRNA[Ser Sec] gene *Trsp* and transgene (*Trsp'*)
        elucidate functions of selenoproteins in
            cancer, health, and development, 65–71
    Trsp^cΔ mice with transgenes, 68, 71
    Trsp^Δ mice with transgenes, 68, 71
    Trsp gene
        hepatic disruption of, 251
        Sec-tRNA[Ser]Sec, 59–60
        in selenoprotein synthesis, 250–251

Trsp^tA34 mice, *Trsp^Δ* and *Trsp^cΔ* mice with
    transgenes, 68, 71
Trsp^tG37 mice, *Trsp^Δ* and *Trsp^cΔ* mice with
    transgenes, 68, 71
Trsp^t mice, *Trsp^Δ* and *Trsp^cΔ* mice with
    transgenes, 68, 71
Geneticin (G418), 63
Genetic interactions, for SNP significance,
    336
Genetic polymorphisms
    in *APOE* gene, 311
    in selenoprotein genes, 309–337
        overview, 309–311, 310f
        SNPs, *See* Single nucleotide
            polymorphisms (SNPs)
    SEP15, cancer and, 206–208, 207f
Genome Wide Association Studies (GWAS),
    312, 335
Genomics, nutritional, 310
Gestational diabetes mellitus (GDM), 219
GK/GK1 (glucokinase), 225
Glucokinase (GK/GK1), 225
Glucose metabolism, Se in, 218–235
    attempted use of compounds, 220–221
    beneficial effects, 221
    GPx1 and, 223–228
        paradoxical roles, 223–226
        regulatory mechanisms on insulin-
            regulated, 226–228, *See also*
            Regulatory mechanisms of GPx1
    others, 231–233
    overview, 218
    pathways and coordinated mechanisms for
        selenoproteins, 234f
    redox control of insulin secretion and
        signaling, 218–220
    Sepp1 and, 228–230
        elevation of circulating levels in T2DM
            patients, 230
        functions, 228–229
        as negative regulator of insulin signaling,
            229–230
    supranutritional intake as potential risk factor
        for T2DM, 221–223
        animal studies, 221–222, 222f
        cross-sectional studies, 222–223
        RCTs, 223
Glucose-stimulated insulin secretion (GSIS),
    219, 220
Glucose utilization, iodothyronine deiodinases
    in, 258–259
Glu-Cys ligase (Gcl), 275
GluMeSeCys (γ-glutamyl methylselenocysteine),
    biosynthesis, 19
    MeSeCys for, 24
Glutamate-cysteine ligase (GCL), 191
Glutamyl cysteine ligase, 157

γ-Glutamyl methylselenocysteine (GluMeSeCys),
  biosynthesis, 19
  MeSeCys for, 24
γ-Glutamylselenocystathionine, in
  monkey, 24
γ-Glutamyl-Se-methylselenocysteine, in nature,
  140
Glutaredoxin (Grx), 174
Glutathione (GSH)
  metabolism, H_2O_2 and selenite on, 180
  oxidation by isothiocyanates and
      isoselenocyanates, 94f
  Se reduction by, 22
  system
    in cellular redox homeostasis, 275
    Trx and, 276
  TrxR1$^{-/-}$ liver tumors, 177
Glutathione conjugated selenocoxib-1 form,
  148
Glutathione-disulfide (GSSG), 275
Glutathione peroxidase 1 (GPx1)
  diabetes and, 223–228
    overexpression of, 224
    paradoxical roles, 223–226
    regulatory mechanisms, 226–228, See
      also Regulatory mechanisms of GPx1
  in energy metabolism, 254t, 260
  SNPs in gene, disease risk and
    bladder cancer, 322t
    breast cancer risk, 319t–320t, 332–333
    cardiovascular disease, 323t
    CRC risk, 321t
    Kashin-Beck disease, 323t
    laryngeal cancer, 322t
    lung cancer, 321t–322t
    metabolic syndrome, 323t
    obesity, 323t
    Pro198Leu (rs1050450), disease risk and,
      319t–323t
    prostate cancer risk, 320t–321t, 333
    stroke, 323t
    TGA in, 60–61
  translated by Sec tRNA[Ser]Sec, 63
Glutathione peroxidase 2 (GPx2), 189–197
  expression, dual role of factors regulating,
    190–193
    Nrf2, 190–191
    Wnt, 192–193
  overview, 189–190
  SEP15 and, 210
  tumor stage–specific functions, 193–197
    inhibition of inflammation, 196–197
    modulation of proliferation and apoptosis,
      194–196, 195f
    redox regulation, 193–194
Glutathione peroxidase 3 (GPx3), in energy
  metabolism, 253, 254t, 260–261

Glutathione peroxidase 4 (GPx4)
  in energy metabolism, 253, 254t, 261
  SNPs in gene, disease risk and, 324t–331t
    breast cancer risk and, 324t
    CRC risk, 324t–325t, 334
    Kashin-Beck disease, 326t
    laryngeal cancer, 326t
    lung cancer, 326t
    prostate cancer, 325t
  translated by Sec tRNA[Ser]Sec, 63
Glutathione peroxidases (GPx)
  in energy metabolism, 259–261
  GPx1, See also Glutathione peroxidase 1
    (GPx1)
    diabetes and, 223–228
    in energy metabolism, 254t, 260
    SNPs in gene, disease risk and, 318–335
  GPx2, 189–197, See also Glutathione
    peroxidase 2 (GPx2)
    expression, dual role of factors regulating,
      190–193
    overview, 189–190
    SEP15 and, 210
    tumor stage–specific functions, 193–197
  GPx3, in energy metabolism, 253, 254t,
    260–261
  GPx4, See also Glutathione peroxidase 4
    (GPx4)
    in energy metabolism, 253, 254t, 261
    SNPs in gene, disease risk and, 324t–331t
    translated by Sec tRNA[Ser]Sec, 63
  maximal expression, 5
  from rat erythrocytes, 85
  redox-active enzyme, 4
Glutathione S-transferases (GSTs), 191, 279
Se-Glutathionylseleno-N-acetylgalactosamine
    (GSSeGalNAc), selenosugar, 23
Glycogen synthase kinase-3β (GSK3β),
    282–283
Goiter, 6
GPx, See Glutathione peroxidases (GPx)
GPx1, See Glutathione peroxidase 1 (GPx1)
GPx3 (glutathione peroxidase 3), in energy
    metabolism, 253, 254t, 260–261
Growth, in human SECISBP2 mutations, 359
Grx (glutaredoxin), 174
GSIS (glucose-stimulated insulin secretion), 219,
    220
GSK3β (glycogen synthase kinase-3β),
    282–283
GSSeGalNAc (Se-Glutathionylseleno-N-
    acetylgalactosamine), selenosugar, 23
GSSG (glutathione-disulfide), 275
GSTs (glutathione S-transferases), 191, 279
Guanylate binding protein-1 (Gbp-1), 209
GWAS (Genome Wide Association Studies),
    312, 335

H

HaCaT (human immortalized keratinocytes), 145
Hallmarks, of cancer
    Trx system and, 178–179
Hamilton, J. W., 86
Harman, Denham, Dr., 86–87
HCC (hepatocarcinoma cells), GPx2 in, 195
HDAC (histone deacetylase) inhibitor, 148,
    149–150
Health
    consequences, of low Se status, 6
    selenium in, 3
Hearing loss, 360
HeLa cells, selenodiglutathione in, 149
Heligmosomoides bakeri, 301–302
Helminth infections, Se in, 301–302
Hepatic Sepp1, 230
Hepatitis, 6
Hepatocarcinoma cells (HCC), GPx2 in, 195
Hepatocyte growth factor (Hgf), 176
Hepatocyte nuclear factor 4α (HNF4α), 230
Herceptin (Trastuzumab), 102, 102f
Hgf (hepatocyte growth factor), 176
HIF-1 (hypoxia-inducible factor-1), 210
    HIF-1α, modulation by selenite, 120
Histone deacetylase (HDAC) inhibitor, 148,
    149–150
HMG-CoA reductase (3-hydroxy-3-methyl-
    glutaryl-CoA reductase), 251
HNF4α (hepatocyte nuclear factor 4α), 230
Hoekstra, William, Dr., 85
HOMA (homeostasis model assessment) index,
    259
Homeostasis
    redox, GSH system in, 275
    TrxR1 maintaining normal cells in, 175
Homeostasis model assessment (HOMA) index,
    259
Housekeeping selenoproteins, 32
H-ras oncogene, 178
HSA (human serum albumin), cysteine residues
    of, 115
HT-29 cells, 195, 196
Human immortalized keratinocytes (HaCaT),
    145
Human *SECISBP2* mutations, disorders, 351–364
    axial muscular dystrophy, 360–361
    azoospermia with spermatogenic maturation
        arrest, 362
    biochemistry, 359
    cutaneous photosensitivity, 362–363
    developmental and neurological phenotype,
        360
    effect of treatment, 364
    growth and skeletal phenotype, 359
    hearing loss, 360

impaired T-cell proliferation and abnormal
    cytokine production, 363–364
metabolic phenotype, 361–362
SBP2 mutations, 351–359, 352t
    biallelic *SECISBP2* mutations, 358–359
    blood biochemistry, 353t
    heterozygous missense, 357–358
    homozygous missense, 357
    physiological phenotypes in patients with
        deficiency, 354t–355t
    R540Q mutant, 353, 356
Human serum albumin (HSA), cysteine residues
    of, 115
Human studies
    selenite in cancer, 124–129
        rationale for Se use, 124–125
        safety and maximal tolerable dose,
            determination, 129
        Se in combination with surgery,
            radiotherapy, and chemotherapy, 125,
            126t–127t, 128t
        Se use as chemotherapeutic drug, 125, 129
    selenoproteins in, 32
    Se requirement by, 4–5
Hydrogen peroxide (H$_2$O$_2$)
    degrading activity of GPx1, cellular
        protection through, 226
    on GSH metabolism, 180
    in redox regulation, 193
3-Hydroxy-3-methyl-glutaryl-CoA reductase
    (HMG-CoA reductase), 251
Hyperglycemia, 224, 225, 248
Hyperinsulinemia, 221–222, 225, 248
Hypothyroidism, 233
Hypoxia-inducible factor-1 (HIF-1), 210
    HIF-1α, modulation by selenite, 120

I

Ileocolitis
    development of, 194
    microbiota-related, 196
Illumina™ GoldenGate assay, 335
Impaired glucose tolerance (IGT), 221
Impaired T-cell proliferation, 363–364
Inducible nitric oxide synthase (iNOS), 298
Infections, helminth, 301–302
Infertility, male
    insufficient selenium and, 31
Inflammation
    inhibition of, GPx2, 196–197
    macrophages in
        anti-inflammation and, 296–298, 297f
        pathological, defined, 294
        resolution, 295–296
        response, overview, 294–295
Influenza, 6

Influenza strain, 6
Inhibition
    eIF2, 118
    of inflammation, GPx2, 196–197
Inorganic salts, of Se, 25–26
INOS (inducible nitric oxide synthase), 298
Institute of Medicine (IOM), 8
Insulin
    resistance, 225
        as metabolic syndrome hallmark,
            248–249
    secretion and signaling, redox control,
        218–220
    sensitivity, iodothyronine deiodinases in,
        258–259
    signaling, SEPP1 as negative regulator of,
        229–230
Insulin promoter factor 1 (Ipf1), 249
Insulin-regulated glucose metabolism
    regulatory mechanisms of GPx1 on, 226–228
        cellular protection through H_2O_2-
            degrading activity, 226
        outcome of overproduction and knockout
            alone/in combination with SOD1,
            226–228, 227f
Integration, of metabolic systems into "metabolic
    ecosystem," 284–285, 284f
Internal ribosome entry site (IRES), 42–43
International HapMap Project, 311
Intraocular lenses (IOLs), 99
In vitro studies
    chemotherapeutic agent, selenite, 112, 113t
    Se forms in preclinical studies, 141–147,
        142t–144t, 146f
In vivo studies
    chemotherapeutic agent, selenite, 113–115
    Se forms in preclinical studies, 147–149
        carcinogen-induced and xenograft
            models, 147–148
        transgenic models, 148–149
Iodine-deficiency diseases, 6
Iodothyronine deiodinase 2 (DIO2)
    Dio2 knockout mouse, 257–259
    at ER membrane, 233
    in glucose metabolism and diabetes, 231
Iodothyronine deiodinases, 255–259
    Dio2 knockout mouse, 257–259
    in energy dissipation via BAT, 257–258
    at ER membrane, 233
    extrathyroidal metabolism of thyroid
        hormones, 256f
    in insulin sensitivity and glucose utilization,
        258–259
    local thyroid hormone activation/inactivation
        on food intake, 258
    type 1 and 2, 256
IOLs (intraocular lenses), 99

IOM (Institute of Medicine), 8
Ipf1 (insulin promoter factor 1), 249
IRES (internal ribosome entry site), 42–43
Irinotecan, in colon cancer, 124
Isopentenylation reaction, 251
Isoselenocyanates (R-CNSe)
    GSH oxidation by, 94, 94f
    redox cycle with GSH, 101
Isothiocyanates (R-CNS), 94
    GSH oxidation by, 94f
    in malignant melanoma xenograft model, 148

J

Jukes, Thomas, 85
Jurkat cells, eIF-2α in, 118

K

Kashin-Beck disease, 6
    occurrence, 56
    SNPs in selenoprotein genes
        GPx4, 326t
        Gpx1 gene, 323t
        Sepp1, 328t
Kelch-like ECH-associated protein 1 (Keap1)
    stress response, 281–282
    thiol modification of, 190
Keshan disease
    etiology, 6
    incidence, 5
    occurrence, 56
    prevalence, 4–5
    prevention, 5
Kink-turn (K-turn), RNA-binding motifs, 35
Knockout mouse models
    Dio2, 257–259
    of GPx1 alone/in combination with SOD1,
        outcome, 226–228, 227f
    phenotypes, of specific selenoproteins,
        348t–349t
K-turn (Kink-turn), RNA-binding motifs, 35
Kwashiorkor, 6

L

L7Ae motifs
    binding, 35, 36
    type RNA interaction, 344
L-arginine (L-Arg), 298, 299
Laryngeal cancer, SNPs in selenoprotein genes
    Gpx1, 322t
    GPx4, 326t
    SEP15, 331t
LDL (low-density lipoprotein) receptors, 311
Ldlr (low-density lipoprotein receptor) gene, 251
Lecythis minor, 24

Le Province Medical, 95
Leukemia, selenocystine in, 110
Levander, Orville, 98
Lewis lung carcinoma (LLC1) cells, 209
Liliaceae family, 19
LINE-1 (long interspersed nucleotide elements), 121
Lipopolysaccharide (LPS), 294–295
Liver cancer
    Se supplementation for, 152
    TrxR1 deletion in promoting/sustaining, 176–177
LLC1 (Lewis lung carcinoma) cells, 209
Localization, subcellular
    in selenoprotein expression regulation, 46–47
Local thyroid hormone activation/inactivation, on food intake, 258
Long interspersed nucleotide elements (LINE-1), 121
Low-density lipoprotein receptor *(Ldlr)* gene, 251
Low-density lipoprotein (LDL) receptors, 311
LPS (lipopolysaccharide), 294–295
L-selenocysteine, 140
L-selenomethionine, 140
LS174T colon tumors in nude mice, 98
Lucigenin chemiluminescence (CL) assay, for superoxide detection, 90–93
Lung cancer
    doxorubicin resistant cells, 123
    incidence, 153
    LLC1 cells, 209
    SNPs in selenoprotein genes
        Gpx1, 321t–322t
        GPx4, 326t
        SEP15, 330t, 331t

M

Mackerel, selenoneine in, 140
Macrophage polarization, Se in, 293–302
    in asthma and helminth infections, 301–302
    in inflammation
        anti-inflammation and, 296–298, 297f
        resolution, 295–296
        response, overview, 294–295
    overview, 293–294
    phenotypes, 298–301, 299f
MAFA, 259
"Magic bullet" concept
    redox cycling Se for cancer, 95–98, 97f
    selenides, 98–100, 100f
Maio-Lin Hu, 90
Male infertility, insufficient selenium and, 31
Mammalian selenoproteins, function, 347–351
    deletion of *SECISBP2*, 350–351
    phenotypes in knockout mouse models, 348t–349t
Marasmus, 6

Marine reptiles, selenoneine in, 24–25
Marsden, William, 84
Martin, John L., 86
MarvinSketch, 145
Mathews, Steven, 99
Matrix metalloprotease 9 (MMP-9), 195
Matsueda, Gary, Dr., 95
Maximal tolerable dose, determination, 129
MB (methylene blue), reduction by selenite and selenocystine, 89–91, 93
McArdle 7777 cells, 41
McConnell, Kenneth, 86
McCord, Joe M., 86–87
Measles, 6
Mechanism(s)
    of action, of Se forms, 149–150
    TrxR1, 175–178
Mechanistic studies, SEP15
    cancer and, 208–210
MEFs (mouse embryonic fibroblasts), 178
Meso 6, 210
Met (methionine), 19
Metabolic activities coordinator, TrxR1, 273–285
    integration of systems into "metabolic ecosystem," 284–285, 284f
    other metabolic pathways, 277–279, 278f
    overview, 273–274
    potential interacting regulators, 280–284
        AMPK, 283
        Nrf2/Keap1 stress response, 281–282
        p53, 283–284
        PTP1B, 282–283
        Txnip, 280–281, 280f
    redox metabolism, 274–276
        GSH and Trx system, 276
        GSH system in cellular redox homeostasis, 275
        Trx system in cellular redox homeostasis, 275–276
        universal need for reductive systems, 274–275
Metabolic ecosystem, integration of systems into, 284–285, 284f
Metabolic phenotype, *SECISBP2* mutations, 361–362
Metabolic syndrome, selenoproteins and, 248–264
    hallmark, insulin resistance as, 248–249
    overview, 248
    selenium and, 249–252
        in energy metabolism, 249–264, *See also* Energy metabolism
    SNPs in genes, 323t
    specific selenoproteins and energy metabolism, 252–264
        glutathione peroxidases, 259–261
        iodothyronine deiodinases, 255–259, 256f

SelM, 253, 254t, 262
SelN, 253, 255t, 262
SelS, 253, 255t, 263
SelT, 253, 255t, 263–264
Seppl, 253, 255t, 262–263
TrxR1, 253, 255t, 264
Metabolism, 19–26
  energy, Se in, *See also* Energy metabolism
    epidemiological studies, 249–250
    selenoprotein synthesis factors role,
      250–252, *See also* Selenoprotein
      synthesis factors
    specific selenoproteins, 252–260,
      253t–255t
  glucose, Se in, 218–235, *See also* Glucose
    metabolism
  GSH, hydrogen peroxide and selenite
    on, 180
  nutritional sources, 21–26
    animal metabolites, 22–23
    biological sources, 24–25, 25f
    elemental and inorganic salts of Se, 25–26
    plant metabolites, 23–24
  overview, 19–21, 20f, 21f
  pathway
    in animals, 19, 20f, 21
    in plants, 19, 21, 21f
  PLP, 64
  redox, TrxR1 in, 274–276
    GSH and Trx system, 276
    GSH system in cellular redox
      homeostasis, 275
    Trx system in cellular redox homeostasis,
      275–276
    universal need for reductive systems,
      274–275
  SeCys, 9
  Se forms, 141
  TrxR1 in, 175
Methionine (Met), 19
Methylene blue (MB), reduction by selenite and
  selenocystine, 89–91, 93
Methylseleninic acid
  chemopreventive properties against prostate
    cancer, 148–149
  HDAC activity, inhibition, 150
  macrophage phenotype shunting, 300
Methylselenocysteine (MeSeCys)
  biosynthesis, 19
    of GluMeSeCys, 24
  fromation, 24
Se-methylselenogalactosamine (MeSeGalNH₂),
  selenosugar, 23
Methylselenol, 7, 91
Se-Methylseleno-*N*-acetyl-galactosamine
  (MeSeGalNAc), urinary metabolite,
  21, 23

Se-Methylseleno-*N*-acetyl-glucosamine
  (MeSeGlcNAc), selenosugar, 23
Mice lacking
  Gclc, 275
  GR, 275
Mills, Gordon, 85
Milner, John, 87
Misra, H. P., 94
Mitotic inhibitors, 123
MMP-9 (matrix metalloprotease 9), 195
MMSe (monomethylselenol), animal metabolite,
  22
Modified base synthesis, of Sec-tRNA^[Ser]Sec,
  57–59, 58f
Modulation
  of cancer-associated signaling pathways by
    selenite, 118–123
    AP-1 transcription factor, 121
    cell cycles, 119
    epigenetic signature, 121–123, 122f
    HIF-1α, 120
    NF-κB, 120–121
    p53, 119–120
    translational machinery, 118–119
  of proliferation and apoptosis, GPx2 and,
    194–196, 195f
Molecular chemoprevention, as guiding principle,
  139
Molecular regulation, of SeCys, 31–48
  overview, 31–33
  Sec insertion process, 33–43, *See also* Sec
    insertion process
    containing proteins, 31–32
    eIF4a3, 41
    essential factors, 34–39
    mechanism, recoding event, 33–34
    other factors, 42–43
    RPL30, 40–41
    SECISBP2L, 39–40
    SEPP1, 32
    SRE, 42
  selenoprotein expression, regulation, 43–47
    NMD, 44–46, 45f
    Se availability, 43–44
    subcellular localization, 46–47
MondoA, 281
Monomethylselenol (MMSe), animal metabolite,
  22
Monoselenides, in superoxide generation, 91
Mouse embryonic fibroblasts (MEFs), 178
Mouse models
  azo dye–induced liver carcinogenesis,
    113–115
  CDF mice, 114
  db/db mice
    hyperglycemia in, 224
    T2DM in, 220

involving tRNA[Ser] Sec gene (*Trsp*) and
transgene (*Trsp^t*) elucidate functions
of selenoproteins in cancer, health,
and development, 65–71
Trsp^Δ and *Trsp^cΔ* mice, 68, 69t–70t
Trsp^Δ and *Trsp^cΔ* mice with transgenes,
Trsp^t, *Trsp^tG37*, *Trsp^tA34*, 68, 71
Trsp^t, *Trsp^tG37*, and *Trsp^tA34* mice, 66,
67t–68t
knockout
Dio2, 257–259
of GPx1 alone/in combination with SOD1,
outcome, 226–228, 227f
phenotypes, of specific selenoproteins,
348t–349t
LS174T colon tumors in nude mice, 98
mouse colon carcinoma CT26 cells, 208–209
selenite, chemotherapeutic agent
in vivo studies, 113–115
MSA (methylseleninic acid), in macrophage
phenotype shunting, 300
Multifaceted and intriguing effects, of Se and
selenoproteins, 218–235
in glucose metabolism and diabetes,
218–235
attempted use of compounds, 220–221
GPx1 and, 223–228, *See also* Glutathione
peroxidase 1 (GPx1), diabetes and
others, 231–233
Seppl and, 228–230
supranutritional intake as potential risk
factor for T2DM, 221–223
overview, 218
redox control of insulin secretion and
signaling, 218–220
Muscular dystrophy, axial, 360–361
Mutations, in *SECISBP2*, 343–366
human, 351–364
axial muscular dystrophy, 360–361
azoospermia with spermatogenic
maturation arrest, 362
biochemistry, 359
cutaneous photosensitivity, 362–363
developmental and neurological
phenotype, 360
effect of treatment, 364
growth and skeletal phenotype, 359
hearing loss, 360
impaired T-cell proliferation and
abnormal cytokine production,
363–364
metabolic phenotype, 361–362
SBP2 mutations, 351–359, 352t, 353t,
354t–355t
mammalian selenoproteins, function,
347–351
deletion, 350–351

phenotypes in knockout mouse models,
348t–349t
overview, 343–344
SBP2, structure and function, 344–346, 345f
selenoprotein deficiencies and clinical
phenotypes, 366t
Mycoplasma, 33
Myxedematous cretinism, 6

N

N-acetylcysteine (NAC), administration, 279
N-acetyl-*p*-benzoquinone imine (NAPQI),
278–279
NADPH oxidase 1 (NOX1), for Wnt pathway,
193–194
NADPH quinone oxidoreductase 1 (NQO1), 191
NAPQI (N-acetyl-*p*-benzoquinone imine),
278–279
National Cancer Institute Developmental
Therapeutics Program, 205–206
National Health and Nutrition Examination
Survey (NHANES), 153, 250
Nature, Se forms in, 139–140
Negative Biopsy Trial (NBT), 153
NES (nuclear export signal) in SBP2, 46–47,
344, 346
Neurological phenotype, human *SECISBP2*
mutations, 360
NFE2L2 gene, 191
NHANES (National Health and Nutrition
Examination Survey), 153, 250
NHL (non-Hodgkin lymphoma), 125
NIH 3T3 cells, 179, 180
Nitric oxide, production, 298
No adverse effect level (NOAEL), 8
Non-Hodgkin lymphoma (NHL), 125
Nonmelanoma skin cancer, 152
Nonsense mediated decay (NMD)
in selenoprotein expression regulation,
44–46, 45f
selenoprotein mRNAs for, 41, 346
Nonsense/opal codon, defined, 57
Nonsteroidal anti-inflammatory drugs (NSAIDs),
295, 298
NOX1 (NADPH oxidase 1), for Wnt pathway,
193–194
Nox4, 226
ΔNp63γ, transcription factor, 192
NSAIDs (nonsteroidal anti-inflammatory drugs),
295, 298
Nuclear export signal (NES) in SBP2, 46–47,
344, 346
Nuclear factor–kappa B (NF-κB)
enhanced activation of, 298
modulation by selenite, 120–121
PPARγ and, 296

Nuclear factor (erythroid-derived 2) (Nrf2)
  expression of antioxidant proteins, 177
  GPx2 expression, 190–191
  stress response, 281–282
  Txnip by, 281
Nuclear localization signal (NLS), sequences in
    SBP2, 46, 344, 346
Nuclear magnetic resonance (NMR), 36
Nucleolin protein, 42
Nucleoredoxin (NRX), 193
Nutrition, Se emergence, 3
Nutritional genomics, 310
Nutritional Prevention of Cancer (NPC) trial
  baseline levels of Se in, 153
  carcinomas and cancer mortality and, 7–8
  chemopreventive effect of selenized yeast, 139
  with selenized yeast, 156
  T2DM, increased risk of, 223
Nutritional sources, Se, 21–26
  animal metabolites, 22–23
  biological sources, 24–25, 25f
  elemental and inorganic salts, 25–26
  plant metabolites, 23–24
  selenate, 25
  selenite, 25
  selenocyanate, 26
  selenosulfate, 26
Nuts, selenocystathionine in, 140

O

Obesity, 225, 248–249, 323t
Onions, 140
Oonopsis genus, 84
O-phosphohomoserine, SeCys with, 24
Opn1 (osteopontin), 150, 176
Organoselenium
  agents, 152
  selenomethionine and selenized yeast testing
      for, 156
Osteoarthropathy, Kaschin-Beck disease, 6
Osteopontin (Opn1), 150, 176
Outcome, of overproduction and knockout of
    GPx1, 226–228, 227f
Overproduction, of GPx1 alone/in combination
    with SOD1, 226–228, 227f
Oxaliplatin, in colon cancer, 124
Oxidation, GSH
  by isothiocyanates and isoselenocyanates, 94f
Oxidative stress biomarkers, in prostate cancer,
    157

P

P53, tumor-suppressor protein
  modulation by selenite, 119–120
  TrxR1-interacting regulators, 283–284

PACAP (pituitary adenylate cyclase-activating
    peptide), 264
PAMPs (pathogen-associated molecular
    patterns), 297
Pancreatic duodenal homeobox 1 (PDX1)
  insulin synthesis and function, 228
  promoter, 225
Paracelsus, 85
Paradoxical roles, of GPx1 in diabetes, 223–226
Pathak, Keshar, Dr., 91
Pathogen-associated molecular patterns
    (PAMPs), 297
Pathological inflammation, defined, 294
PBMCs (peripheral blood mononuclear cells),
    363
Pck1, 229
Pctp (phosphatidylcholine transfer protein), 251
Peripheral blood mononuclear cells (PBMCs), 363
Peroxiredoxin 1, 191
Peroxisomal proliferator-activated receptor-γ
    coactivator 1α (PGC1α), 225, 230
Peroxisome proliferator activator receptor
    (PPAR)-γ
  ligand-dependent activation of, 296
  in macrophage differentiation, 300
  PGE$_2$ metabolites, 295, 298
  PPRE, 300
Phase III clinical trials
  cost, 161
  expense and lack of success, 156
  initiation of future, 149
  organoselenium testing in, 157
  selenomethionine for, 149
Phase I trial, for safety assessment, 129
Phenotypes
  in human *SECISBP2* mutations
    clinical, selenoprotein deficiencies and,
        366t
    developmental and neurological, 360
    metabolic, 361–362
    skeletal, 359
  in knockout mouse models of specific
      selenoproteins, 348t–349t
  macrophage, 298–301, 299f
  physiological, in patients with SBP2
      deficiency, 354t–355t
1,4-phenylenebis (methylene) selenocyanate
    (*p*-XSC), 141, 300
Phosphatase and tensin homolog protein (PTEN),
    227, 282
Phosphatidylcholine transfer protein *(Pctp)*, 251
Phosphoenolpyruvate carboxykinase (PEPCK/
    PCK), 225, 229
Phosphorylation, Erk and Akt, 362
Phosphoseryl-tRNA kinase *(Pstk)* gene, 61
Photosensitivity, cutaneous
    SECISBP2 mutations, 362–363

Physiological factors, for SNP significance, 336
Piette, Larry, Dr., 88–89
Pituitary adenylate cyclase-activating peptide
    (PACAP), 264
Plants
    metabolic pathway, 19, 21, 21f
    metabolites, 23–24
Plant-specific selenoamino acids
    GluMeSeCys, *See* γ-Glutamyl
        methylselenocysteine (GluMeSeCys)
    MeSeCys, *See* Methylselenocysteine
        (MeSeCys)
Plasma selenium, 156–157
    concentrations, 157
    proteomic profiling, 156
Plasmodium, 59
PLP (pyridoxal phosphate) metabolism, 64
Polymethylmethacrylate (PMMA) polymer, 99
Polymorphisms, genetic
    in selenoprotein genes, *See* Genetic
        polymorphisms
    SelS 5227GG, 232
    SEP15, cancer and, 206–208, 207f
PPARγ response element (PPRE), 300
Preclinical studies, Se forms in, 140–150
    chemical structures, 146f
    mechanism of action, 149–150
    metabolism and disposition, 141
    in vitro studies, 141–147, 142t–144t, 146f
    in vivo studies, 147–149
        carcinogen-induced and xenograft
            models, 147–148
        transgenic models, 148–149
Pregnancy, increasing fasting glucose, 221
Premature termination codons (PTC), 44–46
Prevention
    cancers
        Se forms in, 137–161, *See also* Forms of
            Se
        TrxR1 in, 182f
    diabetes, attempted use of compounds,
        220–221
    Keshan disease, 5
Pro198Leu (rs1050450), in *GPX1* and disease
    risk, 319t–323t
Proliferation
    modulation, GPx2 and, 194–196, 195f
    T-cell, impaired, 363–364
Promoting cancer, TrxR1 in
    effects and mechanisms, 175–178
    deletion, for cancer protection, 176
    deletion in liver cancer, 176–177
    in other cancer, 178
    varies in different tissues providing
        avenues for cancer therapy, 178
Pro-oxidant functions, antioxidant and, selenite,
    111–112, 111f

Prostaglandin D_2 (PGD$_2$), 295
Prostate cancer
    antibody, 97–98
    biomarkers of oxidative stress in, 157
    cell lines, 193–194
    DU-145 cells, 98
    G2/M arrest in, 194
    GPx2 in, 190, 193–194, 195
    incidence, 153
    marker α-1-antitrypsin, 156
    polymorphisms in *SEP15* gene, 207–208
    protective effects of Se, 7
    PSA levels, use, 156
    selenized yeast supplementation in, 157
    selenocoxib-1 in, 148
    SeMet and vitamin E for, 153
    Se-methylselenocysteine, chemopreventive
        properties, 148, 149
    SNPs in selenoprotein genes and
        GPx4, 325t
        Gpx1 gene, 320t–321t, 333
        SEP15, 325t, 330t, 331t, 333
        Sepp1, 328t, 329t
Proteins, *See also specific* entries
    eukaryotes, molecular regulation of SeCys into,
        See Molecular regulation, of SeCys
    GluMeSeCys, *See* γ-Glutamyl
        methylselenocysteine (GluMeSeCys)
    MeSeCys, *See* Methylselenocysteine
        (MeSeCys)
    nucleolin, 42
    p53, tumor-suppressor protein
        modulation by selenite, 119–120
        TrxR1-interacting regulators, 283–284
    Pctp, 251
    protein-deficiency diseases, 6
    Pyl, 33
    Sec-containing proteins, 31–32
    Sec insertion into, 64, 65f
    Se-containing, 19
    SECp43, 42
    selenoproteins, *See* Selenoproteins
    SeMet residues, 22
    SePP, 4
    translation process, SeMet in, 19
    YB1, 42
Protein-tyrosine phosphatase 1B (PTP1B), 219,
    260, 282–283
Protein tyrosine phosphatases (PTPs), 279
Proteomic profiling, of plasma, 156–157
Proximal sequence element (PSE), in *Trsp* gene,
    59–60
Psammomys obesus, 231
PSE (proximal sequence element), in *Trsp* gene,
    59–60
Pstk (phosphoseryl-tRNA kinase) gene, 61
PTC (premature termination codons), 44–46

PTEN (phosphatase and tensin homolog protein),
    227, 282
PTP1B (protein-tyrosine phosphatase 1B), 219,
    260, 282–283
PTPs (protein tyrosine phosphatases), 279
P-XSC (1,4-phenylenebis (methylene)
    selenocyanate), 141, 300
Pyl (pyrrolysine), 33
Pyridoxal phosphate (PLP) metabolism, 64
Pyrrolysine (Pyl), 33

R

Rabbit reticulocyte lysate (RRL), 34
    translation system, 40
Radiotherapy, Se in combination, 125, 126t–127t,
    128t
Randomized controlled trials (RCTs), 7
    supranutritional Se intake as potential risk
        factor for T2DM, 223
Raphanus sativus, 24
Rationale, for Se use in cancer, 124–125
Raynaud's disease, 364
RBD (RNA binding domain), 36, 39–40, 344
Reactive oxygen species (ROS)
    formation of, 111–112
    generation, 117–118, 260
    in insulin signaling and secretion, 219–220
    nonstoichiometric production, 111, 111f
    redox balance, component of, 219
    on UCP2 production, 228
RECIST (Response Evaluation Criteria In Solid
    Tumors) criteria, 129
Recoding event, 33–34
Red blood cells (RBCs)
    AE1 in, 115–116
    selenide from, 22
    selenite uptake by, 115–116
Redox control, of insulin secretion and signaling,
    218–220
Redox cycling Se, 83–103
    for cancer, 95–98
    concept, 94
    defined, 90
    dental sealants containing, 100, 100f
    other and recent drug developments, 100–103
        Bolton-Hunter reagent, 101–102, 101f, 102f
        TMZ, 101, 101f
    overview, 83–85
    selenide, 93, 93f
Redox metabolism, TrxR1 in, 274–276
    cellular redox homeostasis
        GSH and Trx system, 276
        GSH system in, 275
        Trx system in, 275–276
    universal need for reductive systems, 274–275
Redox regulation, GPx2, 193–194

Reduction
    MB by selenite, 89–91, 93
    Se
        cancer risk, 7–8
        by GSH, 22
    selenate
        mechanism, 22
        to selenite, 23
Regulation
    gene expression, 32
    molecular, of SeCys, *See* Molecular
        regulation, of SeCys
    redox, GPx2, 193–194
    selenoprotein expression, 43–47
        NMD, 44–46, 45f
        Se availability, 43–44
        subcellular localization, 46–47
Regulatory mechanisms of GPx1
    on insulin-regulated glucose metabolism,
        226–228
        cellular protection through H_2O_2-
            degrading activity, 226
        outcome of overproduction and knockout,
            226–228, 227f
Reid, Ted, Dr., 99
Reptiles, marine
    selenoneine in, 24–25
    Se source, 24
Requirements, Se
    amount, 5
    by humans, 4–5
Resistance, insulin, 225
    as metabolic syndrome hallmark, 248–249
Resolution, inflammation, 295–296
Respiration, DMSe excreted by, 141
Response, inflammation, 294–295
Response Evaluation Criteria In Solid Tumors
    (RECIST) criteria, 129
Ribonucleotide reductase (RNR), 274
Ribosomal protein L30 (RPL30), for Sec
    incorporation, 40–41
Rickettsia parkeri, 38–39
Risk, cancer, 7–8
RNA binding domain (RBD), 36, 39–40, 344
RNA virus, Coxsackie B4, 6
RNR (ribonucleotide reductase), 274
Robinson, Edward G., 95, 97f
Rotruck, John, 85
RPL30 (ribosomal protein L30), for Sec
    incorporation, 40–41
R540Q mutant SBP2, 353, 356

S

SAA (serum amyloid A), 231
S-adenosylmethionine (SAM), dependent
    methylation, 141

Safety, 8
assessment, phase I trial for, 129
SAM (S-adenosylmethionine), dependent
methylation, 141
SARS (severe acute respiratory syndrome), 6
Schistosoma mansoni, 301
Schrauzer, Gerhard, 89, 91
Schwarz, Klaus, 85, 91
Science, 87
Se-75, 86
Se-adenosylselenomethionine (SeAM), in animal
sample, 22
Sec, *See* Selenocysteine (SeCys/Sec)
Sec codon redefinition element (SRE), for
incorporation, 42
Sec incorporation domain (SID), 36, 39–40, 344
Sec insertion process, 31–43
eIF4a3, 41
essential factors, 34–39
eEFSec, 32, 33, 38–39
SBP2, 32, 33, 36–37
SECIS elements, 34–36
Sec-tRNA, 32, 33–34, 39
high reactivity of, 32
mechanism, recoding event, 33–34
other factors, 42–43
overview, 31–33
RPL30, 40–41
SECISBP2L, 39–40
SEPP1, 32
SRE, 42
SECIS binding protein 2 (SBP2)
in glucose metabolism and diabetes, 233
localization of, 346
mutations, 351–359, 352t
biallelic *SECISBP2* mutations, 358–359
blood biochemistry, 353t
heterozygous missense, 357–358
homozygous missense, 357
physiological phenotypes in patients with
deficiency, 354t–355t
R540Q mutant, 353, 356
for Sec incorporation, 32, 33, 36–37
CRD, 47, 344
NES in, 46–47, 344, 346
NLS sequences in, 46, 344, 346
for selenoprotein synthesis, 343
structure and function, 344–346, 345f
SECIS-binding protein 2L (SECISBP2L), for Sec
incorporation, 39–40
SECIS-binding proteins, 42
SECISBP2
deletion of, 350–351
in mice, 36–37
reduced abundance of selenoprotein
mRNAs in, 47
mutations, 343–366

human, disorders, *See* Human *SECISBP2*
mutations
mammalian selenoproteins, function, *See*
Mammalian selenoproteins
overview, 343–344
SBP2, structure and function, 344–346,
345f
selenoprotein deficiencies and clinical
phenotypes, 366t
SECISearch, 356
Sec lyase/Scly (selenocysteine lyase), in energy
metabolism, 251–252
Secondary structure, of Sec-tRNA[Ser]Sec, 57–59,
58f
Se-containing protein, 19
SECp43 protein, 42
Sec redefinition element, 344
Secretion, insulin
redox control of, 218–220
Sec-specific elongation factor (eEFSec)
for Sec incorporation, 32, 33, 38–39, 344, 346
SecTRAPs (selenium-compromised thioredoxin
reductase-derived apoptotic proteins),
181
Sec-tRNA[Ser]Sec, *See* Selenocysteine tRNA (Sec-
tRNA[Ser]Sec)
SeCys, *See* Selenocysteine (SeCys/Sec)
SeCys insertion element (SECIS), 4
defined, 32, 310
eukaryotic elements, 64
with SBP2, 343
for Sec incorporation, 34–36
SECISBP2L, for Sec incorporation,
39–40
SeHcy (selenohomocysteine), 24
Seko, Yushi, Dr., 88–89
SELECT (Selenium and Vitamin E Cancer Trial),
7–8, 138–139, 223, 250
Selective 2′-hydroxyl acylation analyzed by
primer extension (SHAPE) analysis,
36
Selenate
biomarkers studies, 159t
nutritional source of Se, 22
inorganic form, 25
inorganic forms, 25
by plants from soil, 23
reduction mechanism, 22
reduction to selenite, 23
Selenenylsulfide, 275
Selenides
demethylation, 22–23
GSH to, 22
"magic bullet" concept, 98–100, 100f
from RBCs, 22
redox cycling, 93, 93f
Se donor in plants, 24

Selenite
    biomarkers studies, 158t, 159t
    for cancer, 95, 105–129
        antioxidant and pro-oxidant functions,
            111–112, 111f
        human studies, 124–129
        modulation of cancer-associated
            signaling pathways, 118–123, *See also*
            Modulation
        molecular basis of cytotoxicity, 115–124,
            See also Cytotoxicity, selenite
        overview, 110
        potent chemotherapeutic agent, 112–115,
            113t
        rationale for Se use, 124–125
        safety and maximal tolerable dose,
            determination, 129
        Se in combination with surgery,
            radiotherapy, and chemotherapy, 125,
            126t–127t, 128t
        Se use as chemotherapeutic drug, 125, 129
        transport in cells, 115–117
    in dietary supplemental yeast, 89
    on GSH metabolism, 180
    MB reduction by, 89–91, 93
    nutritional source of Se, 25
    selenate reduction to, 23
    toxicity in different cell lines of multiple
        origins, 112, 113t
    uptake by RBCs, 115–116
Selenium (Se)
    beneficiary, 9–10
    in cancer, *See* Cancers
    chemopreventive properties, 31
    in combination with surgery, radiotherapy,
        and chemotherapy, 125, 126t–127t,
        128t
    deficiency, 5–6
    discovery, 83–85
    emergence, in nutrition and health, 3
    forms in cancer prevention, 137–161, *See also*
        Forms of Se
    free forms, of TrxR1 in cancer, 180–181
    health consequences, 6
    low blood Se levels, 6
    in macrophage polarization, *See* Macrophage
        polarization
    metabolic syndrome and, 249–252, *See also*
        Metabolic syndrome
    metabolism, *See* Metabolism
    multifaceted and intriguing effects, *See*
        Multifaceted and intriguing effects
    necessity, 4
    other and recent drug developments, 100–103,
        101f, 102f
    redox cycling and toxicity, *See* Redox cycling;
        Toxicity

requirement
    amount, 5
    by humans, 4–5
    safety, 8
    SeCys, 4
    supranutritional intakes, 7
    use as chemotherapeutic drug, 125, 129
Selenium and Vitamin E Cancer Trial (SELECT),
    7–8, 138–139, 223, 250
Selenium-compromised thioredoxin reductase-
    derived apoptotic proteins
    (SecTRAPs), 181
Selenium-Tellurium Development Association,
    85
Selenized milk protein, 147
Selenized yeast
    biomarkers studies, 157, 158t–159t
    dosage, 157
    low Se in soil, 152
    NPC trial with, 156
    in selenomethionine form, 148
    supplementation in prostate cancer, 157
    testing for organoselenium, 156
Selenobetaine, 25
Selenocoxibs, 148
Selenocyanate, 26
Selenocystamine, superoxide generating by, 93,
    93f
Selenocystathionine
    in Astragalus species, 140
    formation, 24
    in nuts, 140
    from Se accumulator plants, 86
Selenocysteine (SeCys/Sec)
    administration, 129
    in animal foods, 140
    biosynthesis, 61–63, 62f
    discovery, 4
    GluMeSeCys, *See* γ-Glutamyl
        methylselenocysteine (GluMeSeCys)
    incorporation, 310
        sec-tRNA[Ser]Sec for, 32, 33–34, 39
    insertion into protein, 64, 65f
    for leukemia, 110
    L-selenocysteine, 140
    MB reduction by, 89–91, 93
    in meat and milk proteins, 22
    MeSeCys, *See* Methylselenocysteine
        (MeSeCys)
    metabolism, 9
    in Met biosynthetic pathway, 24
    molecular regulation, *See* Molecular
        regulation, of SeCys
    residue for selenoenzymes, 174
    residue in TrxR1 function, 180–181
    SECIS, *See* SECIS binding protein 2
        (SBP2)

selenoproteins
  sequences, 19
  for synthesis, 137–138
side effects, 129
tRNA, biosynthesis of, 24
UGA codon for, 60–61
Selenocysteine lyase (Sec lyase/Scly), in energy
    metabolism, 251–252, 253t
Selenocysteine tRNA (Sec-tRNA[Ser]Sec)
  for Sec incorporation, 32, 33–34, 39
  in selenoprotein biosynthesis, 55–72
    early history and modified bases, 57
    features, 57
    gene (Trsp) and its transcription, 59–60
    mouse models involving tRNA[Ser] Sec gene
        (Trsp) and transgene (Trsp'), 65–71,
        See also tRNA[Ser] Sec gene (Trsp) and
        transgene (Trsp')
    overview, 56–57
    Sec, insertion into protein, 64, 65f
    Sec biosynthesis, 61–63, 62f
    secondary structure and modified base
        synthesis, 57–59, 58f
    SPS1, SelD paralog, 63–64
    UGA, codon for Sec, 60–61
Selenodeiodinases, 6
Selenodiglutathione, 92
  in HeLa cells, 149
Selenoenzymes
  actions, 7
  activity of, 21
  antioxidant and redox regulator, 175
  in driving malignancy, 175
  overexpression in tumors and cancer cell, 175
  Sec residue for, 174
  selenosulfate in, 26
  suboptimal expression of, 9
Selenohomocysteine (SeHcy), 24
Selenohomolanthionine (SeHLan)
  biosynthesis, 19
  formation, 24
Selenomethionine (SeMet)
  biomarkers studies, 157, 158t, 159t
  in dietary supplemental yeast, 89
  dogs supplemented with, 148
  dosage, 157
  formation, 24
  in general protein sequences, 19
  HDAC activity and, 150
  L-selenomethionine, 140
  macrophage phenotype shunting, 300
  in MB reduction, 91
  in meat proteins, 22
  in nature, 140
  nonspecifically incorporated, 43
  protective effects, 153
  Se dominant food form, 9

selection for Phase III Clinical Trial, 149
  testing for organoselenium, 156
  toxicity of, 91–92
Selenoneine
  biosynthetic pathway, 25
  in bluefin tuna, marine reptiles and fishes, 24–25
  in human blood cell, 25
Selenophosphate
  Se donor in plants, 24
  synthesis, 61
Selenophosphate synthetase 1 (SPS1), SelD
    paralog, 63–64
Selenophosphate synthethase (SPS), 61, 252
15-kDa Selenoprotein (SEP15), 203–211
  cancer and, 205–210
    breast cancer risk, 324t, 333
    CRC, 330t, 331t, 334
    gene expression, 205–206
    genetic polymorphisms, 206–208, 207f
    laryngeal, 331t
    lung, 330t, 331t
    mechanistic studies, 208–210
    prostate, 325t, 330t, 331t, 333
  combined knockdown, of SEP15 and TrxR1,
      210–211
  GPx2 and, 210
  overview, 203–205
  SNPs in gene, disease risk and, 324t–331t
    CRC, 330t, 331t, 334
    laryngeal cancer, 331t
    lung cancer, 330t, 331t
    prostate cancer, 325t, 330t, 331t, 333
  TrxR1 and, 210
    combined knockdown, 210–211
Selenoprotein B (SelB)
  defined, 38
  in eEFSec function, 38
Selenoprotein D (SelD)
  in eEFSec function, 38
  paralog, SPS1, 63–64
Selenoprotein M (SelM), 203
  in energy metabolism, 253, 254t, 262
Selenoprotein N (SelN)
  biochemical and physiological role, 350
  in energy metabolism, 253, 255t, 262
Selenoprotein P (Sepp1)
  diabetes and, 228–230
    elevation of circulating levels in T2DM
        patients, 230
    functions, 228–229
    as negative regulator of insulin signaling,
        229–230
  in energy metabolism, 253, 255t, 262–263
  hepatic, 230
  prostate cancer risk, 325t, 333
  Sec codons, 32
  selenium transport protein, 115

SNPs in gene, disease risk and, 324t–331t
    breast cancer, 327t, 328t
    CRC, 327t, 329t, 334
    Kashin-Beck disease, 328t
    prostate cancer, 328t, 329t
Selenoproteins
    amount, 21
    biosynthesis, 9
        Sec-tRNA in, *See* Selenocysteine tRNA
            (Sec-tRNA[Ser]Sec])
    with *Clostridia*, 32
    deficiencies and clinical phenotypes,
        366t
    defined, 203, 310
    with *Deltaproteobacteria*, 32
    expression, regulation, 43–47
        NMD, 44–46, 45f
        Se availability, 43–44
        subcellular localization, 46–47
    function as redox-active enzymes, 117
    genetic polymorphisms in genes, 309–337
        SNPs, *See* Single nucleotide
            polymorphisms (SNPs)
    in glucose metabolism and diabetes, *See*
        Diabetes; Glucose metabolism
    housekeeping, 32
    in humans, 32
    mammalian, function, 347–351
        deletion of *SECISBP2*, 350–351
        phenotypes in knockout mouse models,
            348t–349t
    metabolic syndrome and, *See* Metabolic
        syndrome, selenoproteins and
    in mice, 32
    mouse models involving tRNA[Ser] Sec] gene
        (*Trsp*) and transgene (*Trsp^t*), 65–71
        Trsp^Δ and *Trsp^cΔ* mice, 68, 69t–70t
        Trsp^Δ and *Trsp^cΔ* mice with transgenes,
            Trsp^t, *Trsp^tG37*, *Trsp^tA34*, 68, 71
        Trsp^t, *Trsp^tG37*, and *Trsp^tA34* mice, 66,
            67t–68t
    multifaceted and intriguing effects, *See*
        Multifaceted and intriguing effects
    pathways and coordinated mechanisms for,
        234f
    recovery, 21
    Sec insertion process, *See* Sec insertion
        process
SeCys in, 4
SelB
    defined, 38
    in eEFSec function, 38
SelD
    in eEFSec function, 38
    paralog, SPS1, 63–64
SelM, 203
    in energy metabolism, 253, 254t, 262

SelN
    biochemical and physiological role, 350
    in energy metabolism, 253, 255t, 262
SelS, 231–232, 231f, 252
    in energy metabolism, 253, 255t, 263
SelT, 231
    in energy metabolism, 253, 255t, 263–264
SEP15, *See* SEP15
SEPP1, *See* Selenoprotein P (Seppl)
Se sources for synthesis, 21
specific, in energy metabolism, 252–264,
        253t–255t
    glutathione peroxidases, 259–261
    iodothyronine deiodinases, 255–259,
        256f, *See also* Iodothyronine
        deiodinases
synthesis, 19, 20f, 24–25
    in animals, 24
    DMSe and TMSe for, 25
    factors, in energy metabolism, 250–252,
        See also Selenoprotein synthesis
        factors
TrxR1, *See* Thioredoxin reductase 1 (TrxR1)
type 2 deiodinase (D2), 253, 253t
type 3 deiodinase (D3), 253, 254t
Selenoprotein S (SelS), 231–232, 231f, 252
    in energy metabolism, 253, 255t, 263
    SelS 5227GG polymorphism, 232
Selenoprotein synthesis factors, role
    energy metabolism and, 250–252
        selenocysteine lyase, 251–252
        tRNA[Ser]Sec] gene, 250–251
Selenoprotein T (SelT), 231
    in energy metabolism, 253, 255t, 263–264
Selenosis, chronic, 8
Selenosugars
    animal metabolite, 22
    determination, 21
    GSSeGalNAc, 23
    identification, 21
    MeSeGalNAc, 21, 23
    MeSeGalNH₂, 23
    MeSeGlcNAc, 23
    urine excretion as, 21
Selenosulfate, 26
Selenotrisulfides, formation of, 90
SelH, in eEFSec function, 38
SelK, in eEFSec function, 38
Se-methylselenocysteine
    in Astragalus species, 140
    biomarkers studies, 158t, 159t
    chemopreventive properties against prostate
        cancer, 148, 149
    HDAC activity and, 150
    in nature, 140
    Se accumulator plants, 86
    toxicity of, 91–92

Se-methylselenoneine, in human blood cell, 25
Sensitivity, insulin
    iodothyronine deiodinases in, 258–259
SEP15, *See* SEP15
SEPN1-myopathies, 360–361
Seppl, *See* Selenoprotein P (Seppl)
SePP (Se-transport protein selenoprotein P)
    GPxs, isoform of, 4
    maximal expression, 5
Serine (Ser) activation, by acetyl-CoA, 24
Ser209 phosphorylation of eIF4E, 118–119
Serum amyloid A (SAA), 231
Se-transport protein selenoprotein P (SePP)
    GPxs, isoform of, 4
    maximal expression, 5
Severe acute respiratory syndrome (SARS), 6
SHAPE (selective 2′-hydroxyl acylation analyzed
    by primer extension) analysis, 36
Sharma, Arun, Dr., 101
SID (Sec incorporation domain), 36, 39–40, 344
Sigma Chemical Co., 91
Signaling
    cancer-associated pathways, *See* Cancer-
        associated signaling pathways
    insulin
        redox control of, 218–220
        SEPP1 as negative regulator of, 229–230
    Wnt signaling pathway, 192–193
Signal transducer and activator of transcription
    (STAT6), 300, 301
Single nucleotide polymorphisms (SNPs), 311–337
    origin of, 311–312
    in selenoprotein genes, 312–315, 313t–314t
        biomarkers in healthy populations,
            316t–318t
        breast cancer and, 319t–320t, 324t, 327t,
            328t, 332–333
        CRC risk, 321t, 324t–325t, 327t, 329t,
            333–335
        diseases and, 315–335, 316t–331t
        GPX4, Seppl, and *SEP15* genes and
            disease risk, 324t–331t
        pathway approach, 335
        Pro198Leu (rs1050450) in *GPX1* and
            disease risk, 319t–323t
        prostate cancer and, 320t–321t, 325t, 328t,
            329t, 330t, 331t, 333
    significance of, factors influencing, 336–337
        biomarkers of Se status in association
            studies, 336–337
        genetic interactions, 336
        physiological factors, 336
Skeletal phenotype, in human *SECISBP2*
    mutations, 359
SLC7A11, glutamate antiporter, 117
SNPs, *See* Single nucleotide polymorphisms
    (SNPs)

Sodium selenite, for insulin synthesis, 220
"Sparing" effect, of vitamin E, 4
Spermatogenic maturation arrest, azoospermia
    with, 362
SPS (selenophosphate synthethase), 252
SPS1 (selenophosphate synthetase 1), SelD
    paralog, 63–64
SPS2 (selenophosphate synthetase), 61
SRE (Sec codon redefinition element), for
    incorporation, 42
Stanleya genus, 84
Stanleya pinnata, 85f
STAT6 (signal transducer and activator of
    transcription), 300, 301
Stem cell theory, 193
Steroids, 295
Streptozotocin, 224, 229
Stress response, Nrf2/Keap1, 281–282
Stroke, SNPs in selenoprotein genes and, 323t
Subcellular localization, in selenoprotein
    expression regulation, 46–47
Suberoylanilide hydroxamic acid, 148
Sulfiredoxin 1, 191
Sulfite, production, 23
Sulforaphane, 191
Superoxide dismutase (SOD)
    determination, 224
    outcome of overproduction and knockout of
        GPx1, 226–228, 227f
Superoxide generation, Se toxicity cause, 88–100
    determination, 89–95
        GSH oxidation by isothiocyanates and
            isoselenocyanates, 94f
        in lucigenin CL assay, 90–93
        in MB assay, 89–91, 93
        selenide redox cycling, 93, 93f
        by selenocystamine, 93, 93f
        redox cycling Se for cancer, 95–98, *See also*
            Redox cycling
    selenides, "magic bullet" concept, 98–100,
        100f
Supplementation
    Se for liver cancer, 152
    selenized yeast, in prostate cancer, 157
Supranutritional Se, intake as potential risk factor
    for T2DM, 221–223
    animal studies, 221–222, 222f
    cross-sectional studies, 222–223
    RCTs, 223
Surgery, Se in combination, 125, 126t–127t, 128t
Sustaining cancer, TrxR1 in
    deletion in liver cancer, 176–177
    varies in different tissues providing avenues
        for cancer therapy, 178
Systemic inflammatory response syndrome (SIRS)
    multiorgan dysfunction with, 294
    Se deficiency, 293

T

TAG-72 monoclonal antibody, 98
TagSNP, genotyping of, 334–335
Tappel, Al, 86, 90
Tauroursodeoxycholic acid (TUDCA), 259
Taxol, chemotherapeutic effect of, 124
Taxoplasma gondii, 59
TBP2 (Trx-binding protein-2), 280
T-cell proliferation, impaired, 363–364
Tellurium, 84
Telomere shortening, defective T-cell
        proliferation, 363–364
Temozolomide (TMZ), 101, 101f
Testosterone (T), 148
Tetrahymena, 59
Tetraselenocyclophane (TSC), 141
Texas Tech Health Sciences Center, 101
TGR (thioredoxin/glutathione reductase), 174
Thiobarbituric acid, 221
Thioredoxin/glutathione reductase (TGR), 174
Thioredoxin-interacting protein (Txnip),
        280–281, 280f
Thioredoxin reductase 1 (TrxR1), 173–183
    in cancer prevention and progression, 182f
    classes, 173
    coordinator in metabolic activities, *See*
            Metabolic activities coordinator
    in energy metabolism, 253, 255t, 264
    maintaining normal cells in homeostasis, 175
    in metabolism, 175
    mitochondrial form, 174
    overview, 173–174
    Sec in, 62–63
    Se in cancer and, 179–181
        free forms, role in cancer, 180–181
        Sec residue in, 180–181
    SEP15 and, 210
        combined knockdown, 210–211
    Trx system and hallmarks of cancer, 178–179
    tumor-promoting effects and mechanisms,
            175–178
        deletion, for cancer protection, 176
        deletion, in promoting/sustaining liver
            cancer, 176–177
        effect in other cancer, 178
        varies in different tissues providing
            avenues for cancer therapy, 178
Thromboxane (TXA₂), 295
Thyroid-stimulating hormone (TSH), *SECISBP2*
        mutations and, 359
Thyroxine (T4), 233
    conversion of, 6
    SECISBP2 mutations and, 359
TIGAR (TP53-induced glycolysis and apoptosis
        regulator), 284

TLRs (toll-like receptors), 297
TMSe (trimethylselenonium ion), animal
        metabolite, 22
TMZ (temozolomide), 101, 101f
TNF-α (tumor necrosis factor-α), relative
        sensitivities of, 179
Toll-like receptors (TLRs), 297
Topoisomerase inhibitors, 123
Toxicity, 83–103
    cause
        superoxide generation, 88–100, *See also*
            Superoxide generation
        unanswered, 86–88
    chemistry and, 83–85
    discovery, 83–85
    other and recent drug developments,
            100–103
        Bolton-Hunter reagent, 101–102, 101f, 102f
        TMZ, 101, 101f
    overview, 83–85
    selenite in different cell lines of multiple
            origins, 112, 113t
    of SeMet, 91–92
    of Se-methylselenocysteine, 91–92
TP53-induced glycolysis and apoptosis regulator
        (TIGAR), 284
Tran, Phat, 101
Transcription, of *Trsp* gene, 59–60
Transcriptome, 277, 279
Transgenic mice encoding *Trsp*
    selenoproteins in cancer, health, and
            development, 65–71
        Trsp^t, Trsp^{tG37}, and *Trsp^{tA34}* mice, 66,
            67t–68t
Transgenic models, 148–149
Translational machinery, 118–119
Transport, selenite
    in cancer cells, 115–117
Trastuzumab (Herceptin), 102, 102f
Tribolium castaneum, 63–64
Triiodothyronine (T3), 233
    SECISBP2 mutations and, 359
Trimethylselenonium ion (TMSe), animal
        metabolite, 22
tRNA^{[Ser]Sec} gene, in energy metabolism,
        250–251, 250t
tRNA^{[Ser] Sec} gene (*Trsp*) and transgene (*Trsp^t*),
        mouse models
    elucidate functions of selenoproteins in
            cancer, health, and development,
            65–71
        Trsp^Δ and *Trsp^cΔ* mice, 68, 69t–70t
        Trsp^Δ and *Trsp^cΔ* mice with transgenes,
            Trsp^t, Trsp^{tG37}, Trsp^{tA34}, 68, 71
        Trsp^t, Trsp^{tG37}, and *Trsp^{tA34}* mice, 66, 67t–68t
Trp (tryptophan), 33, 63

Trsp^{cΔ} mice, 68, 69t–70t
  with transgenes, *Trsp^t*, *Trsp^{tG37}*, *Trsp^{tA34}*, 68, 71
Trsp^Δ mice, 68, 69t–70t
  with transgenes, *Trsp^t*, *Trsp^{tG37}*, *Trsp^{tA34}*, 68, 71
Trsp gene
  hepatic disruption of, 251
  Sec-tRNA$^{[Ser]Sec}$, 59–60
  in selenoprotein synthesis, 250–251
Trsp^{tA34} mice, 66, 67t–68t
  Trsp^Δ and *Trsp^{c}Δ* mice with transgenes, 68, 71
Trsp^{tG37} mice, 66, 67t–68t
  Trsp^Δ and *Trsp^{c}Δ* mice with transgenes, 68, 71
Trsp^t mice, 66, 67t–68t
  Trsp^Δ and *Trsp^{c}Δ* mice with transgenes, 68, 71
Trx-binding protein-2 (TBP2), 280
TrxR1, *See* Thioredoxin reductase 1 (TrxR1)
Trx system
  in cellular redox homeostasis, 275–276
  GSH and, 276
  hallmarks of cancer and, 178–179
Trypanosoma brucei, 60
Tryptophan (Trp), 33, 63
TSC (tetraselenocyclophane), 141
TSH (thyroid-stimulating hormone), *SECISBP2* mutations and, 359
TUDCA (tauroursodeoxycholic acid), 259
Tumor necrosis factor-α (TNF-α), relative sensitivities of, 179
Tumor-promoting effects, TrxR1, 175–178
  deletion
    for cancer protection, 176
    effect in other cancer, 178
    in promoting/sustaining liver cancer, 176–177
Tumors, selenoenzyme overexpression in, 175
Tumor stage–specific functions, GPx2, 193–197
  inhibition of inflammation, 196–197
  modulation of proliferation and apoptosis, 194–196, 195f
  redox regulation, 193–194
Tuna, selenoneine in, 24–25, 140
Tunicamycin, 204
Txnip (thioredoxin-interacting protein), 280–281, 280f
Type 2 deiodinase (D2), 253, 253t–254t
Type 3 deiodinase (D3), 253, 254t
Type 1 diabetes mellitus (T1DM)
  pathogenesis, ROS in, 219
  thiobarbituric acid in, 221
Type-2 diabetes mellitus (T2DM)
  in *db/db* mice, 220
  elevation of circulating SEPP1 levels in patients, 230

global GPx1 overexpression, 225
increasing risk of, 8, 218
metabolic disorders, development, 219
pathogenesis, ROS in, 219
with sodium selenite, 221
supranutritional Se intake as potential risk factor for, 221–223
  animal studies, 221–222, 222f
  cross-sectional studies, 222–223
  RCTs, 223
thiobarbituric acid in, 221

U

UCP-1 (uncoupling protein-1), 257
UCP-2 (uncoupling protein-2), 225
  production, ROS on, 228
UDP-glucose:glycoprotein glucosyltransferase (UGGT), 204
UGA coding
  in *E. coli*, 33
  in eubacteria and archaebacteria, 33
  in *Mycoplasma*, 33
  protein synthesis stop codon, 57
  recoding
    AMGs and, 43
    to Sec, 32, 33
    SECIS and SEPP1 and, 32
  in RPL30, 40, 41
  for Sec, 60–61
    incorporation, 34
  in Selenoprotein N, 42
  seryl-tRNA decoded, 57
  for Trp in mitochondria, 37
UGGT (UDP-glucose:glycoprotein glucosyltransferase), 204
UK PRECISE (United Kingdom Prevention of Cancer by Intervention of Selenium) study, 250
Uncoupling protein-1 (UCP-1), 257
Uncoupling protein-2 (UCP-2), 225
  production, ROS on, 228
United Kingdom Prevention of Cancer by Intervention of Selenium (UK PRECISE) study, 250
Untranslated region (UTR), of selenoprotein, 32, 33
Urinary metabolite
  GSSeGalNAc, 23
  MeSeGalNAc, 21, 23
  MeSeGalNH$_2$, 23
  MeSeGlcNAc, 23
Urine, excretion as selenosugars, 21
UTR (untranslated region), of selenoprotein, 32, 33

V

Vitamin D3 upregulated protein (Vdup1), 280
Vitamin E
    deficiency, Se for, 3
    protective effects, for prostate cancer, 153
    "sparing" effect, 4
Von Wassermann, August Paul, 95

W

Wendel, Albrecht, Dr., 88
Wnt signaling pathway, 192–193
World Health Organization (WHO), 8
Wright, Thomas, 84

X

XCT, glutamate antiporter, 117
Xenograft models, 147–148
Xenopus oocytes
    modified bases in tRNA[Ser]Sec in, 58
    Trsp gene transcription in, 59–60
Xylorrhiza genus, 84

Y

YB1 protein, 42
Yeast, nutritional source, 19, 21